International Association of Geodesy Symposia

Michael G. Sideris, Series Editor

For further volumes:
http://www.springer.com/series/1345

International Association of Geodesy Symposia

Michael G. Sideris, Series Editor

Symposium 101: Global and Regional Geodynamics
Symposium 102: Global Positioning System: An Overview
Symposium 103: Gravity, Gradiometry, and Gravimetry
Symposium 104: Sea SurfaceTopography and the Geoid
Symposium 105: Earth Rotation and Coordinate Reference Frames
Symposium 106: Determination of the Geoid: Present and Future
Symposium 107: Kinematic Systems in Geodesy, Surveying, and Remote Sensing
Symposium 108: Application of Geodesy to Engineering
Symposium 109: Permanent Satellite Tracking Networks for Geodesy and Geodynamics
Symposium 110: From Mars to Greenland: Charting Gravity with Space and Airborne Instruments
Symposium 111: Recent Geodetic and Gravimetric Research in Latin America
Symposium 112: Geodesy and Physics of the Earth: Geodetic Contributions to Geodynamics
Symposium 113: Gravity and Geoid
Symposium 114: Geodetic Theory Today
Symposium 115: GPS Trends in Precise Terrestrial, Airborne, and Spaceborne Applications
Symposium 116: Global Gravity Field and Its Temporal Variations
Symposium 117: Gravity, Geoid and Marine Geodesy
Symposium 118: Advances in Positioning and Reference Frames
Symposium 119: Geodesy on the Move
Symposium 120: Towards an Integrated Global Geodetic Observation System (IGGOS)
Symposium 121: Geodesy Beyond 2000: The Challenges of the First Decade
Symposium 122: IV Hotine-Marussi Symposium on Mathematical Geodesy
Symposium 123: Gravity, Geoid and Geodynamics 2000
Symposium 124: Vertical Reference Systems
Symposium 125: Vistas for Geodesy in the New Millennium
Symposium 126: Satellite Altimetry for Geodesy, Geophysics and Oceanography
Symposium 127: V Hotine Marussi Symposium on Mathematical Geodesy
Symposium 128: A Window on the Future of Geodesy
Symposium 129: Gravity, Geoid and Space Missions
Symposium 130: Dynamic Planet - Monitoring and Understanding...
Symposium 131: Geodetic Deformation Monitoring: From Geophysical to Engineering Roles
Symposium 132: VI Hotine-Marussi Symposium on Theoretical and Computational Geodesy
Symposium 133: Observing our Changing Earth

VII Hotine-Marussi Symposium on Mathematical Geodesy

Proceedings of the Symposium in Rome, 6-10 June, 2009

Edited by

Nico Sneeuw
Pavel Novák
Mattia Crespi
Fernando Sansò

Springer

Editors

Nico Sneeuw
Institute of Geodesy
Geschwister-Scholl-Str. 24D
70174 Stuttgart
Germany
sneeuw@gis.uni-stuttgart.de

Mattia Crespi
Università di Roma "La Sapienza"
via Eudossiana 18
00184 Rome
Italy
mattia.crespi@uniroma1.it

Pavel Novák
Research Institute of Geodesy
Ondrejov
Czech Republic
pavel.novak@pecny.cz

Fernando Sansò
Politecnico di Milano - Polo di Como
via Valleggio 11
22100 Como
Italy
fernando.sanso@polimi.it

ISBN 978-3-642-22077-7 e-ISBN 978-3-642-22078-4
DOI 10.1007/978-3-642-22078-4
Springer Heidelberg Dordrecht London New York

Library of Congress Control Number: 2011942604

Cover design: WMXDesign GmbH

Printed on acid-free paper

Springer is part of Springer Science+Business Media (www.springer.com)

Preface

This volume contains the proceedings of the VII Hotine-Marussi Symposium on Mathematical Geodesy, which was held from 6 to 10 July 2009. The symposium took place at the Faculty of Engineering of the Sapienza University of Rome, Italy, in the ancient *chiostro* of the Basilica of S. Pietro in Vincoli, famously known for its statue of Moses by Michelangelo.

The traditional name *mathematical geodesy* for the series of Hotine-Marussi Symposia may not fully do justice to the symposium's broad scope of theoretical geodesy in general. However, the name for the series has been used since 1965, i.e., the days of Antonio Marussi, which is a good reason to adhere to it. The venue of the Hotine-Marussi Symposia has traditionally been in Italy. The choice for Rome, if a reason is needed at all, was partially made because 2009 was the International Year of Astronomy. Two important astronomical events were commemorated: the publication of Kepler's *Astronomia Nova* in 1609, in which he published his first two laws of planetary motion, as well as the very first astronomical use of a telescope by Galileo and his discovery of Jupiter's moons. Besides one of the founding fathers of geodesy, the unit of Gal being named after him, he was one of the cofounders and an early member of the Accademia Nazionale dei Lincei in Rome. It was a pleasure, therefore, that a special session was organized by Fernando Sansò at the Villa Farnesina, located at the Academy. The special session was dedicated to the memory of Antonio Marussi (1908–1984), who was the driving force behind the series of Hotine (later Hotine-Marussi) Symposia.

Since 2006 the series is under the responsibility of the InterCommission Committee on Theory (ICCT), a cross-commission entity within the International Association of Geodesy (IAG). The overall goal of the Hotine-Marussi Symposia has always been the advancement of theoretical geodesy. This goal is aligned with the objectives of the ICCT, which has the developments in geodetic modeling and data processing in the light of recent advances of geodetic observing systems as well as the exchange between geodesy and neighboring Earth sciences as its central themes. Indeed, the current proceedings are testimony to the width and vibrancy of theoretical geodesy.

The symposium attracted 132 participants who contributed 75 papers (51 oral and 24 poster), organized in eight regular sessions plus the session at the Accademia Nazionale dei Lincei To a large extent, the sessions' topics were modeled on the study group structure of the ICCT. The chairs of the ICCT study groups, who constituted the Symposium's Scientific Committee, were at the same time responsible for organizing the sessions:

1. *Geodetic sensor systems and sensor networks*
   S. Verhagen
2. *Estimation and filtering theory, inverse problems*
   H. Kutterer, J. Kusche
3. *Time series analysis and prediction of multi-dimensional signals in geodesy*
   W. Kosek, M. Schmidt
4. *Geodetic boundary value problems and cm-geoid computational methods*
   Y.M. Wang, P. Novák
5. *Satellite gravity theory*
   T. Mayer-Gürr, N. Sneeuw
6. *Earth oriented space techniques and their benefit for Earth system studies*
   F. Seitz, R. Gross
7. *Theory, implementation and quality assessment of geodetic reference frames*
   Dermanis, Z. Altamimi
8. *Temporal variations of deformation and gravity*
   G. Spada, M. Crespi, D. Wolf

We want to express our gratitude to all those who have contributed to the success of the VII Hotine-Marussi Symposium. The aforementioned study group chairs (Scientific Committee) put much effort in organizing attractive sessions and convening them. They also organized the peer review process. We equally owe thanks to all reviewers. Although much of the review process itself remains anonymous, the complete list of the reviewers is printed in this volume as a token of our appreciation of their dedication.

Financial and promotional support was given by a number of agencies and institutions. Special thanks go to Federazione delle Associazioni Scientifiche per le Informazioni Territoriali e Ambientali (ASITA), Agenzia Spaziale Italiana (ASI), the European Space Agency (ESA), and the Faculty of Engineering of the Sapienza University of Rome.

But most of all we like to thank Mattia Crespi and his team (Gabriele Colosimo, Augusto Mazzoni, Francesca Fratarcangeli, and Francesca Pieralice) who hosted the symposium. It is well known that the quality of a Local Organizing Committee (LOC) is decisive to a successful scientific meeting. Beyond responsibility for website, registration, technical support, and all kinds of other arrangements, the LOC organized a great social event to the St. Nilus' Abbey, the archeological area of Monte Tuscolo and the Villa Grazioli in Frascati. Through their able organization and improvisation skills, Mattia Crespi and his team have done more than their share in bringing the VII Hotine-Marussi Symposium to success.

Stuttgart

Nico Sneeuw
Pavel Novák
Mattia Crespi
Fernando Sansò

Fifty Years of Hotine-Marussi Symposia

In 1959, Antonio Marussi, in cooperation with the Italian Geodetic Commission, started a series of symposia in Venice. The first three of these covered the entire theoretical definition of 3D Geodesy, as delineated in discussions with renowned contemporary scientists:

- 1959, Venice, 16–18 July, 1st Symposium on Three Dimensional Geodesy, published in Bollettino di Geodesia e Scienze Affini, XVIII, N° 3, 1959
- 1962, Cortina d'Ampezzo, 29 May-1 June, 2nd Symposium on Three Dimensional Geodesy, published in Bollettino di Geodesia e Scienze Affini, XXI, N° 3,1962
- 1965, Turin, 21–22 April, 3rd Symposium on Mathematical Geodesy, published by Commissione Geodetica Italiana, 1966

From the very beginning, Martin Hotine provided essential inspiration to these symposia. After his death in 1968, the following symposia bear his name:

- 1969, Trieste, 28–30 May, 1st Hotine Symposium (4th Symposium on Mathematical Geodesy), published by Commissione Geodetica Italiana, 1970
- 1972, Florence, 25–26 October, 2nd Hotine Symposium (5th Symposium on Mathematical Geodesy), published by Commissione Geodetica Italiana, 1973
- 1975, Siena, 2–5 April, 3rd Hotine Symposium (6th Symposium on Mathematical Geodesy), published by Commissione Geodetica Italiana, 1975
- 1978, Assisi, 8–10 June, 4th Hotine Symposium (7th Symposium on Mathematical Geodesy), published by Commissione Geodetica Italiana, 1978
- 1981, Como, 7–9 September, 5th Hotine Symposium (8th Symposium on Mathematical Geodesy), published by Commissione Geodetica Italiana, 1981

After Marussi's death in 1984, the symposia were finally named the Hotine-Marussi Symposia:

- 1985, Rome, 3–6 June, I Hotine-Marussi Symposium (Mathematical Geodesy)
- 1989, Pisa, June, II Hotine-Marussi Symposium (Mathematical Geodesy)
- 1994, L'Aquila, 29 May-3 June, III Hotine-Marussi Symposium (Mathematical Geodesy, Geodetic Theory Today), published by Springer, IAG 114
- 1998, Trento, 14–17 September, IV Hotine-Marussi Symposium (Mathematical Geodesy), published by Springer, IAG 122
- 2003, Matera, 17–21 June, V Hotine-Marussi Symposium (Mathematical Geodesy), published by Springer, IAG 127
- 2006, Wuhan, 29 May-2 June, VI Hotine-Marussi Symposium (Theoretical and Computational Geodesy, 1st time under ICCT), published by Springer, IAG 132
- 2009, Rome, 6–10 June, VII Hotine-Marussi Symposium (Mathematical Geodesy), published by Springer, IAG

List of Reviewers

Hamza	Alkhatib	*Małgorzata*	Letachowicz
Detlef	Angermann	*Ambrogio*	Manzino
Simone	Atzori	*Daniele*	Melini
Paolo	Baldi	*Franz*	Meyer
Valentina	Barletta	*Federica*	Migliaccio
Riccardo	Barzaghi	*Juergen*	Mueller
Oliver	Baur	*Frank*	Neitzel
Battista	Benciolini	*Ingo*	Neumann
Ludovico	Biagi	*Tomasz*	Niedzielski
Richard	Biancale	*Pavel*	Novák
Christian	Bizouard	*Roland*	Pail
Klaus	Börger	*Isabelle*	Panet
Claudio	Brunini	*Waldemar*	Popiński
Carine	Bruyninx	*Roelof*	Rietbroek
Aleksander	Brzezinski	*Marco*	Roggero
Benjamin Fong	Chao	*Fausto*	Sacerdote
Mattia	Crespi	*Daniele*	Sampietro
Maria Clara	de Lacy	*Fernando*	Sansò
Athanasios	Dermanis	*Michael*	Schmidt
Roberto	Devoti	*Steffen*	Schön
Indridi	Einarsson	*Wolf-Dieter*	Schuh
Will	Featherstone	*Florian*	Seitz
Paolo	Gasperini	*Enrico*	Serpelloni
Stefan	Gernhardt	*Frederik J.*	Simons
Erik	Grafarend	*Nico*	Sneeuw
Petr	Holota	*Giorgio*	Spada
Jianliang	Huang	*Paolo*	Stocchi
Urs	Hugentobler	*Salvatore*	Stramondo
Chris	Jekeli	*Robert*	Tenzer
Maciej	Kalarus	*Peter J.G.*	Teunissen
Boris	Kargoll	*Jose*	van den IJssel
Roland	Klees	*Sandra*	Verhagen
Wiesaw	Kosek	*Bert*	Vermeersen
Jürgen	Kusche	*Jenny*	Virtanen
Hansjörg	Kutterer	*Matthias*	Weigelt

Contents

Part I

Session at the Accademia Nazionale dei Lincei

Welcome to the Participants to the VII° Hotine-Marussi Symposium

1

M. Caputo

Good morning. I am Michele Caputo. The president of our Accademia prof. G. Conso could not come to the meeting. He asked me to present his greetings, to welcome you on his behalf and wish a good visit of the Accademia.

Our Academia was founded in by the young Federico Cesi, 18 years old, in the year 1603. The name of the Accademia comes from the lynx, the elegant feline, which was supposed to have excellent eyes and see well at incommensurable distances. Galilei observed the planets from the highest portion of the garden outside this building. He had joined the Accademia in 1625.

Few of you may know of the Pizzetti–Somigliana theory, but all know of the International Gravity Formula. It was all born and developed within the walls of this building.

In fact following the path indicate by Pizzetti in a series of papers published between 1894 and 1913, Somigliana (1929) developed the general theory of the gravity field of a rotating ellipsoid of revolution. At the same time Silva (1928, 1930) estimated the values to adopt for the parameters appearing in the formula from the average values obtained using the observed gravity on the surface of the Earth.

M. Caputo (✉)
Department of Physics, University of Rome La Sapienza, Piazzale A. Moro 2, 00185, Roma, Italy

Department of Geology and Geophysics, Texas A&M University, College Station, 77843, Texas
e-mail: mic.caput@tiscali.it

Finally Cassinis (1930) presented the series expansion of the original closed form formula at the 1930 IUGG Assembly in Stockholm which adopted the formula to be used for the normal values of gravity on the surface of the international ellipsoid of revolution. This ellipsoid had been adopted by the International Association of Geodesy in the 1923 assembly. 57 years later the closed form formula of the Pizzetti–Somigliana theory was extended to space, for whatever it may be useful, introducing the then available satellite data (Caputo and Benavidez 1987).

It was almost all discussed within the walls of the Accademia dei Lincei and published in its proceedings Now all theoretical geodesists who are familiar with the gravity field of the Earth know that Somigliana, Pizzetti and Cassinis were members of the Accademia where they often met and discussed of theoretical Geodesy. One more notable member of the Accademia was Antonio Marussi who was one of the most complete professionals of geodesy I knew in my life; he knew the use of the data resulting from the observations made with the Stark Kammerer theodolite and how to make sophisticated maps, at the time when the Brunswiga Addiermachine desk mechanical computer was the most advanced instrument to make multiplications and divisions; Marussi had the expertise of making accurate measurements as well as that to use differential geometry to model what is called intrinsic geodesy. And finally he made the extraordinary pendulums. We are here to honour him, as well as his colleague Hotine.

Thank you for coming to Accademia dei Lincei. I wish a good day of work.

N. Sneeuw et al. (eds.), *VII Hotine-Marussi Symposium on Mathematical Geodesy*, International Association of Geodesy Symposia 137, DOI 10.1007/978-3-642-22078-4_1, © Springer-Verlag Berlin Heidelberg 2012

References

Caputo M, Benavidez A (1989) The normal gravity field in space and a simplified model. Manuscripta Geodaetica 14:2–6, also in Rend Accad Naz Lincei, Cl Sc Fis Mat Nat, 1–9, 1987

Cassinis G (1930) Sur l'adoption d'une formule internationale pour la pesanteur normale. Bulletin Géodesique 26: 40–49

Cassinis G (1933) Tables des valeurs de la pesanteur normale internationale. Bulletin Géodesique 32

Cassinis G, Dore P, Ballarin S (1937) Tavole fondamentali per la riduzione dei valori osservati della gravità. Commissione Geodetica Italiana, Milano, Nuova Serie, 13

Marussi A (1960) I primi risultati ottenuti nella stazione per lo studio delle maree della verticale della Grotta *Gigante*. Bollettino di Geodesia e Scienze Affini 19, 4, 645–667

Pizzetti P (1913) Principi della teoria meccanica della figura dei pianeti. Spoerri Printer, Pisa

Pizzetti P (1984) Sulla espressione della gravità alla superficie del geoide, supposto ellissoidico. Atti R Accademia dei Lincei, Classe Sc Fis Mat Nat 3:166–172

Silva G (1928) Sulla definizione della gravità normale. R.C. Accademia Lincei 7(5):12

Silva G (1930) Sulla formula della gravità normale. R.C. Accademia Lincei 11(6):7

Somigliana C (1928) Sulla definizione della gravità normale. R.C. Accademia Lincei 7(6):3

Somigliana C (1929) Teoria generale del campo gravitazionale dell'ellissoide di rotazione. Memorie Società Astronomica Italiana 6:541–599

The Marussi Legacy: The Anholonomity Problem, Geodetic Examples

2

E.W. Grafarend

Antonio Marussi *died 16th April 1984 in Trieste, nearly exactly 25 years ago. He is the founder of the* Geodetic Anholonomity Problem *or the problem of integrability of geodetic observational functionals. My talk will try to* open your eyes by simple examples.

Top 1: Gravitostatics

Geodetic heights, better height differences are not integrable. For instance, every geodesist knows "dH", the infinitesimal change of geodetic heights. In my courses in Physics I learned the notion $đH$. In terms of *Planck notation* $đH$ *is not integrable*. But the *Gauss potential* – C.F. Gauss introduced the notion of *potential* – *is integrable*. We all know the transformation

$$dW = -\Gamma đH,$$

where we use the input "geometric height differential" (anholonomic) versus the physical height difference in terms of output of the potential differential (holonomic). The factor of integrability is the modulus of gravity, also called an element of the *Frobenius matrix*. *A. Marussi* recognized first this key problem and generalized it into *three- and four-dimensional Geodesy*, into space-time geodesy. Notable, the potential W consists of two quantities: the gravitational potential U and the centrifugal potential V. In contrast the *Euler rotational force* and the *Coriolis rotational force* are not integrable.

Geodetic height systems referring to the *Gauss-Listing Geoid* are founded on "potential heights". To my strong belief, the anholonomity problem established *Geodesy as a Science!*

Integrability or anholonomity problems are treated nowadays by *Cartan calculus*, also called *exterior calculus* or the calculus of *differential forms*, a calculus introduced in the twentieth century by *Elie Cartan*, a famous French scientist. *F.R. Helmert* knew already about the subject, *Heinz Draheim* of Karlsruhe University wrote an early paper about *Cartan calculus* and surface geometry. I learned it in *Thermodynamics* from the *Carnot circle* or *Carnot loop* in my *Physics Courses*.

Top 2: Gravitodynamics

I only mention the papers by *E. Doukakis*, his *Ph.D. Thesis*, to include space-time concepts on anholonomity problems, namely integrability, both in the space and in the time domain. There is no time to discuss this in more detail.

Top 3: Space-Time Geodesy

A. Marussi is the real founder of space-time Geodesy: He influenced Hotine (1969) to write his famous textbook with more than 5,000 formulae. He influenced also J. Zund (1988-1994) to write many beautiful papers on *Differential Geodesy* and the leg calculus. In addition he published *A. Marussi's works* in a remarkable Springer Edition. *J. Zund's book* on Differential Geodesy is another masterpiece of depth and wide range. (*J. Zund*, Differential Geodesy, Springer Verlag, Berlin 1994). At this point, another mathematician has to be mentioned who also applied *Cartan calculus* to the holonomity problem: Nathaniel (Nick) Grossmann from UCLA. He wrote remarkable papers on the geodetic *anholonomity problem*. He is a trained mathematician on *Cartan or exterior calculus*. See our reference list at the end.

Top 4: Refraction and Diffraction

There are excellent papers in Physics on this subject written in exterior calculus. For instance, I recall a paper by *P. Defrise et al* from Belgium.

N. Sneeuw et al. (eds.), *VII Hotine-Marussi Symposium on Mathematical Geodesy*, International Association of Geodesy Symposia 137, DOI 10.1007/978-3-642-22078-4_2, © Springer-Verlag Berlin Heidelberg 2012

Top 5: Continuum Mechanics

Traditionally, plasticity problems and nonlinear stress-strain relations are treated by *Cartan calculus* and exterior differential forms.

Top 6: Deformation Analysis

There is a special geometric property within *Cartan calculus*. When we transform within Gauss surface geometry a Riemann metric to orthogonal axes, we arrive at a picture of a circle: the orthonormal axes produce a *Cartan reference system* which is *anholonomic*. For *deformation analysis*, it is possible to transform a *left metric* into a *right metric*, namely from a *left circle* into a *right ellipse* or vice versa. This is the extended Cartan system when transforming two Riemann manifolds.

Top 7: Map Projections

The *Tissot ellipse* is the proper tool when we transform a left Riemann metric to a right Riemann metric. It is the extended *Cartan reference system* from a circle (left) to an ellipse (right). Reference has to be made to *C. Boucher, A. Dermanis, E. Livieratos* and many others. For more details, we refer to our book "Map Projections" (Springer Verlag, 750 pages, Berlin-Heidelberg 2006).

Top 8: Rotational motion by Cartan calculus and Omega quantities

E. Cartan introduced his new concept by referring to the *Euler kinematical equation*. You have to introduce the transformation from rotational velocities "Omega" to *Euler angles*: $\omega = \mathbf{M}(d\alpha, d\beta, d\gamma)$. "Omega" is the rotational vector which is mapped to *Euler angles*. ω is *not integrable*, $(d\alpha, d\beta, d\gamma)$ are *integrable*.

Top 9: Relativity

Hehl (1996) referred to more than 100 authors to establish *Einstein-Cartan geometry* with *spin degrees-of-freedom*. One part of the connection symbols are anti-symmetric characterizing *Cartan torsion* related to my *M.Sc. Thesis* in *Theoretical Physics*. We refer also to the correspondence between *Elie Cartan and Albert Einstein*, published by Springer Verlag.

What has happened meanwhile?

First, Cartan geometry was generalized to *Clifford algebra* and *Clifford analysis* in order to account for *symmetric differential forms or symmetric matrices and antisymmetric differential forms or antisymmetric matrices*. Nowadays we summarize to *multilinear algebra* and *multilinear analysis*. There are special conferences every year devoted to *Clifford algebra* and *Clifford analysis*. As a reference see my review "Tensor Algebra, Linear Algebra, Multilinear Algebra" (344 References), Stuttgart 2004. The famous papers by *W.K. Clifford* were published in 1878 and 1882.

Second, Henry Cartan, son of Elie Cartan, also professor at the Sorbonne, established with 50 French mathematicians the topic of *Structure Mathematics*. In a collective series they wrote more than 20 books, first in French, then in many other languages under the pseudonym "Nicholas Bourbuki". Basically they found out that there are only three basic structures based on advanced set theory and being in interference with each other:

- Order structure
- Topological structure
- Algebraic structure

Now it is time for my examples.

Example 1. Misclosure within a local triangular network and a threedimensional Euclidean space

By Figs. 2.1–2.4 and Tables 2.1–2.6 we present a triangular network within a *threedimensional Euclidean space*. Our target is the computation of the misclosures caused by three local vertical/horizontal directions at the points $\{P_\alpha, P_\beta, P_\gamma\}$ which differ from the geometric vertical/horizontal directions. These

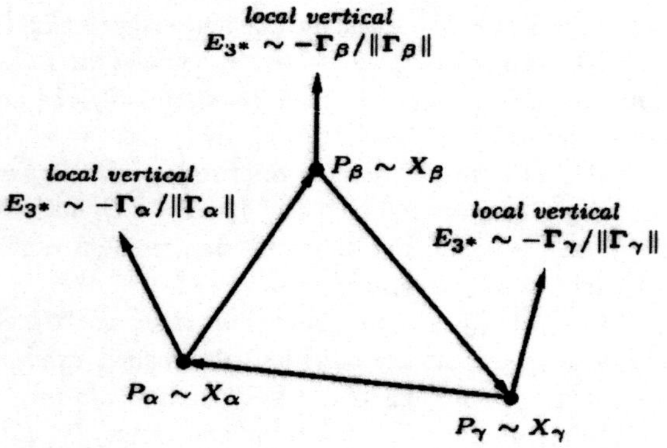

Fig. 2.1 Triangular network $\{P_\alpha, P_\beta, P_\gamma|\mathbf{O}\}$, placement vectors at the origin $\mathbf{O}$, local verticals $E_3(P_\alpha)$, $E_3(P_\beta)$, $E_3(P_\gamma)$, $\Gamma_\alpha, \Gamma_\beta, \Gamma_\gamma$ local gravity vectors

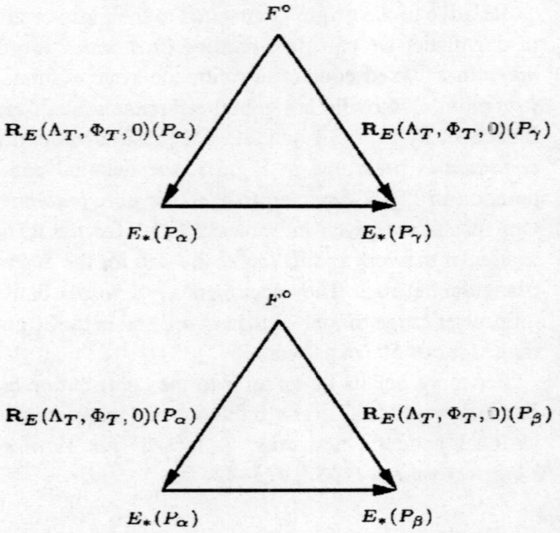

Fig. 2.2 Commutative diagrams: *moving horizon reference systems* E_* versus *fixed equatorial reference systems* F°

$$\overset{\alpha}{X}_{\alpha\beta} := \overset{\alpha}{X}_{\beta} - \overset{\alpha}{X}_{\alpha}, \ldots, \overset{\alpha}{Z}_{\gamma\alpha} := \overset{\alpha}{Z}_{\alpha} - \overset{\alpha}{Z}_{\gamma}$$

$$\overset{\alpha}{X}_{\alpha\beta} + \overset{\alpha}{X}_{\beta\gamma} + \overset{\alpha}{X}_{\gamma\alpha} = 0$$

$$\overset{\alpha}{Y}_{\alpha\beta} + \overset{\alpha}{Y}_{\beta\gamma} + \overset{\alpha}{Y}_{\gamma\alpha} = 0$$

$$\overset{\alpha}{Z}_{\alpha\beta} + \overset{\alpha}{Z}_{\beta\gamma} + \overset{\alpha}{Z}_{\gamma\alpha} = 0$$

Fig. 2.3 Holonomity condition in terms of relative coordinates in a fixed reference system, fixed to the reference point P_α

verticals/horizontals are *not parallel to each other* causing the *anholonomity problem* or the *misclosures*. Of course, we assume *parallelism in the Euclidean sense (Euclid's axiom number five)*.

Our two computations are based *first* on a holonomic reference system at the point P_α which is not operational and *second* on a realistic anholonomic reference system attached to the points $\{P_\alpha, P_\beta, P_\gamma\}$, separately. We use a local network of an extension of *25* m versus *500* m.

"anholonomity condition"

$$\overset{\alpha}{X}_{\alpha\beta} := \overset{\alpha}{X}_{\beta} - \overset{\alpha}{X}_{\alpha}, \ldots, \overset{\alpha}{Z}_{\gamma\alpha} := \overset{\alpha}{Z}_{\alpha} - \overset{\alpha}{Z}_{\gamma}$$

$$\overset{\alpha}{X}_{\alpha\beta} + \overset{\beta}{X}_{\beta\gamma} + \overset{\gamma}{X}_{\gamma\alpha} \neq 0 \quad \text{(misclosure)}$$

$$\overset{\alpha}{Y}_{\alpha\beta} + \overset{\beta}{Y}_{\beta\gamma} + \overset{\gamma}{Y}_{\gamma\alpha} \neq 0 \quad \text{(misclosure)}$$

$$\overset{\alpha}{Z}_{\alpha\beta} + \overset{\beta}{Z}_{\beta\gamma} + \overset{\gamma}{Z}_{\gamma\alpha} \neq 0 \quad \text{(misclosure)}$$

"representation of the base vectors in the horizon reference frame $\mathbf{E}^*$*"*

$$\mathbf{E}_1 \cdot X_{\alpha\beta} + \mathbf{E}_2 \cdot Y_{\alpha\beta} + \mathbf{E}_3 \cdot Z_{\alpha\beta}$$

"Direct and inverse transformation of Cartesian coordinates into spherical coordinates"

(horizontal coordinate $H_{\alpha\beta}$, vertical coordinate $V_{\alpha\beta}$, distance S, azimuth $A_{\alpha\beta}$, vertical angle $B_{\alpha\beta}$, horizontal orientiation unknown)

$$X_{\alpha\beta} = S_{\alpha\beta} \cos A_{\alpha\beta} \cos B_{\alpha\beta},$$

$$Y_{\alpha\beta} = S_{\alpha\beta} \sin A_{\alpha\beta} \cos B_{\alpha\beta},$$

$$Z_{\alpha\beta} = S_{\alpha\beta} \sin B_{\alpha\beta}$$

$$A_{\alpha\beta} := H_{\alpha\beta} + \Sigma_\alpha = \arctan\left(Y_{\alpha\beta}/X_{\alpha\beta}\right),$$

$$B_{\alpha\beta} := V_{\alpha\beta} = \arctan\left(Z_{\alpha\beta}/\sqrt{X_{\alpha\beta} + Y_{\alpha\beta}}\right)$$

$$S_{\alpha\beta} := \sqrt{X_{\alpha\beta}^2 + Y_{\alpha\beta}^2 + Z_{\alpha\beta}^2}$$

Point transformation
$\mathbf{E}_*^\alpha \to \mathbf{E}_*^\beta$ *and* $\mathbf{E}_*^\beta \to \mathbf{E}_*^\gamma$

"Euler angles"

$$\mathbf{E}^\alpha(P_\alpha) \to \mathbf{E}^\beta(P_\beta):$$

$$\mathbf{E}^\alpha(P_\alpha)\mathbf{R}_E(\Lambda_\alpha, \Phi_\alpha, 0)\mathbf{R}_E^T(\Lambda_\beta, \Phi_\beta, 0)$$

$$\Delta\Lambda := \Lambda_\beta - \Lambda_\alpha, \quad \Delta\Phi := \Phi_\beta - \Phi_\alpha$$

$$\mathbf{R}_E(\Lambda_\alpha, \Phi_\alpha, 0)\mathbf{R}_E^T(\Lambda_\beta, \Phi_\beta, 0) \doteq$$

$$\doteq \begin{bmatrix} 1 & -\Delta\Lambda \sin\Phi_\alpha & \Delta\Phi \\ \Delta\Lambda \sin\Phi_\alpha & 1 & \Delta\Lambda \cos\Phi_\alpha \\ \Delta\Phi & -\Delta\Lambda \cos\Phi_\alpha & 1 \end{bmatrix}$$

"antisymmetric matrix $\mathbf{A}$*"*

$$-\begin{bmatrix} \overset{\beta}{X}_{\beta\alpha} \\ \overset{\beta}{Y}_{\beta\alpha} \\ \overset{\beta}{Z}_{\beta\alpha} \end{bmatrix} = (\mathbf{I} + \mathbf{A}^T) \begin{bmatrix} \overset{\alpha}{X}_{\alpha\beta} \\ \overset{\alpha}{Y}_{\alpha\beta} \\ \overset{\alpha}{Z}_{\alpha\beta} \end{bmatrix}$$

analogue formulae for

$$X_{\gamma\alpha}, \ldots, Z_{\gamma\alpha}$$

analogue formulae for

$$A_{\beta\alpha},\ B_{\beta\alpha},\ S_{\beta\alpha},\ \ldots,\ A_{\gamma\alpha},\ B_{\gamma\alpha},\ S_{\gamma\alpha}$$

Fig. 2.4 Anholonomity in a moving frame at points $\{P_\alpha, P_\beta, P_\gamma\}$

Table 2.1 25 meter local network

$$\overset{\alpha}{X}_{\alpha\beta} = +30\ m,\ \overset{\alpha}{X}_{\beta\gamma} = +50\ m,\ \overset{\alpha}{X}_{\gamma\alpha} = -80\ m$$

$$\overset{\alpha}{Y}_{\alpha\beta} = +30\ m,\ \overset{\alpha}{Y}_{\beta\gamma} = -50\ m,\ \overset{\alpha}{Y}_{\gamma\alpha} = +20\ m$$

$$\overset{\alpha}{Z}_{\alpha\beta} = +5\ m,\ \overset{\alpha}{Z}_{\beta\gamma} = +15\ m,\ \overset{\alpha}{Z}_{\gamma\alpha} = -20\ m$$

$$\Lambda_{\alpha\beta} = 1'' \sim 4.85 \cdot 10^{-6}\ \text{RAD},$$

$$\Phi_{\alpha\beta} = -0.5'' \sim -2.42 \cdot 10^{-6}\ \text{RAD}$$

$$\Lambda_{\alpha\gamma} = -1'' \sim -4.85 \cdot 10^{-6}\ \text{RAD},$$

$$\Phi_{\alpha\gamma} = -2.5'' \sim -12.12 \cdot 10^{-6}\ \text{RAD}$$

$$\Phi_\alpha = 48.783^\circ$$

Relative to the origin **O** attached to the mass centre of our planet we calculate *relative Cartesian coordinates* in a "fixed equatorial reference system" transformed to a "moving horizontal reference system" as illustrated by Figs. 2.1 and 2.2. The basic *holonomity condition* is presented in Fig. 2.3, the detailed computation in Fig. 2.4 related to *realistic anholonomity*. Our results are given in Tables 2.1–2.3 for the *25* m triangular network and in *Tables* 2.4–2.6 for the *500* m triangular network: They document a *misclosure* in the millimeter range for our *25* m network and in the 30 cm range for our *500* m network.

For more details let us refer to the contribution by *E. Grafarend* (1987): The influence of local verticals in local geodetic networks, *Zeitschrift für Vermessungswesenv 112 (1987) 413–424.*

Table 2.2 25 meter local network, detailed computation

$$\begin{bmatrix} \overset{\beta}{X}_{\beta\gamma} \\ \overset{\beta}{Y}_{\beta\gamma} \\ \overset{\beta}{Z}_{\beta\gamma} \end{bmatrix} = \begin{bmatrix} 1 & +\Lambda_{\alpha\beta} \sin\Phi_\alpha & +\Phi_{\alpha\beta} \\ -\Lambda_{\alpha\beta} & 1 & -\Lambda_{\alpha\beta}\cos\Phi_\alpha \\ -\Phi_{\alpha\beta} & +\Lambda_{\alpha\beta}\cos\Phi_\alpha & 1 \end{bmatrix} \begin{bmatrix} \overset{\alpha}{X}_{\beta\gamma} \\ \overset{\alpha}{Y}_{\beta\gamma} \\ \overset{\alpha}{Z}_{\beta\gamma} \end{bmatrix} =$$

$$= \begin{bmatrix} 1 & +3.65 \cdot 10^{-6} & -2.42 \cdot 10^{-6} \\ -3.65 \cdot 10^{-6} & 1 & -3.19 \cdot 10^{-6} \\ +2.42 \cdot 10^{-6} & +3.19 \cdot 10^{-6} & 1 \end{bmatrix} \begin{bmatrix} +50\ m \\ -50\ m \\ +15\ m \end{bmatrix}$$

$$\begin{bmatrix} \overset{\gamma}{X}_{\gamma\alpha} \\ \overset{\gamma}{Y}_{\gamma\alpha} \\ \overset{\gamma}{Z}_{\gamma\alpha} \end{bmatrix} = \begin{bmatrix} 1 & +\Lambda_{\alpha\gamma} \sin\Phi_\alpha & +\Phi_{\alpha\gamma} \\ -\Lambda_{\alpha\gamma} & 1 & -\Lambda_{\alpha\gamma}\cos\Phi_\alpha \\ -\Phi_{\alpha\gamma} & +\Lambda_{\alpha\gamma}\cos\Phi_\alpha & 1 \end{bmatrix} \begin{bmatrix} \overset{\alpha}{X}_{\gamma\alpha} \\ \overset{\alpha}{Y}_{\gamma\alpha} \\ \overset{\alpha}{Z}_{\gamma\alpha} \end{bmatrix} =$$

$$= \begin{bmatrix} 1 & -3.65 \cdot 10^{-6} & -12.12 \cdot 10^{-6} \\ +3.65 \cdot 10^{-6} & 1 & +3.19 \cdot 10^{-6} \\ +12.12 \cdot 10^{-6} & -3.19 \cdot 10^{-6} & 1 \end{bmatrix} \begin{bmatrix} -80\ m \\ +20\ m \\ -20\ m \end{bmatrix}$$

Table 2.3 25 meter local network, misclosures

$$\overset{\beta}{X}_{\beta\gamma} = +50\ m - 0.22\ mm,\quad \overset{\gamma}{X}_{\gamma\alpha} = -80\ m + 0.17\ mm$$

$$\overset{\beta}{Y}_{\beta\gamma} = -50\ m - 0.23\ mm,\quad \overset{\gamma}{Y}_{\gamma\alpha} = +20\ m - 0.36\ mm$$

$$\overset{\beta}{Z}_{\beta\gamma} = +15\ m - 0.04\ mm,\quad \overset{\gamma}{Z}_{\gamma\alpha} = -20\ m - 1.03\ mm$$

$$\overset{\alpha}{X}_{\alpha\beta} + \overset{\beta}{X}_{\beta\gamma} + \overset{\gamma}{X}_{\gamma\alpha} \neq 0: \quad -0.22\ mm + 0.17\ mm = -0.05\ mm$$

$$\overset{\alpha}{Y}_{\alpha\beta} + \overset{\beta}{Y}_{\beta\gamma} + \overset{\gamma}{Y}_{\gamma\alpha} \neq 0: \quad -0.23\ mm - 0.35\ mm = -0.59\ mm$$

$$\overset{\alpha}{Z}_{\alpha\beta} + \overset{\beta}{Z}_{\beta\gamma} + \overset{\gamma}{Z}_{\gamma\alpha} \neq 0: \quad -0.04\ mm - 1.03\ mm = -1.07\ mm$$

Table 2.4 500 meter local network

$$\overset{\alpha}{X}_{\alpha\beta}=+500\,m,\ \overset{\alpha}{X}_{\beta\gamma}=+800\,m,\ \overset{\alpha}{X}_{\gamma\alpha}=-1300\,m$$
$$\overset{\alpha}{Y}_{\alpha\beta}=+500\,m,\ \overset{\alpha}{Y}_{\beta\gamma}=-800\,m,\ \overset{\alpha}{Y}_{\gamma\alpha}=+300\,m$$
$$\overset{\alpha}{Z}_{\alpha\beta}=+50\,m,\ \overset{\alpha}{Z}_{\beta\gamma}=+150\,m,\ \overset{\alpha}{Z}_{\gamma\alpha}=-200\,m$$

$$\Lambda_{\alpha\beta}=25''\sim 12.12\cdot 10^{-5}\ \text{RAD},$$
$$\Phi_{\alpha\beta}=-15''\sim -7.27\cdot 10^{-5}\ \text{RAD}$$
$$\Lambda_{\alpha\gamma}=-15''\sim -7.27\cdot 10^{-5}\ \text{RAD},$$
$$\Phi_{\alpha\gamma}=-45''\sim -2.18\cdot 10^{-4}\ \text{RAD}$$
$$\Phi_{\alpha}=48.783^\circ$$

Table 2.5 500 meter local network, detailed computation

$$\begin{bmatrix}\overset{\beta}{X}_{\beta\gamma}\\ \overset{\beta}{Y}_{\beta\gamma}\\ \overset{\beta}{Z}_{\beta\gamma}\end{bmatrix}=\begin{bmatrix}1 & +\Lambda_{\alpha\beta}\sin\Phi_\alpha & +\Phi_{\alpha\beta}\\ -\Lambda_{\alpha\beta} & 1 & -\Lambda_{\alpha\beta}\cos\Phi_\alpha\\ -\Phi_{\alpha\beta} & +\Lambda_{\alpha\beta}\cos\Phi\alpha & 1\end{bmatrix}\begin{bmatrix}\overset{\alpha}{X}_{\beta\gamma}\\ \overset{\alpha}{Y}_{\beta\gamma}\\ \overset{\alpha}{Z}_{\beta\gamma}\end{bmatrix}=$$
$$=\begin{bmatrix}1 & +9.12\cdot 10^{-5} & -7.27\cdot 10^{-5}\\ -9.12\cdot 10^{-5} & 1 & -7.99\cdot 10^{-5}\\ +7.27\cdot 10^{-5} & +7.99\cdot 10^{-5} & 1\end{bmatrix}\begin{bmatrix}+800\,m\\ -800\,m\\ +150\,m\end{bmatrix}$$
$$\begin{bmatrix}\overset{\gamma}{X}_{\gamma\alpha}\\ \overset{\gamma}{Y}_{\gamma\alpha}\\ \overset{\gamma}{Z}_{\gamma\alpha}\end{bmatrix}=\begin{bmatrix}1 & +\Lambda_{\alpha\gamma}\sin\Phi_\alpha & +\Phi_{\alpha\gamma}\\ -\Lambda_{\alpha\gamma} & 1 & -\Lambda_{\alpha\gamma}\cos\Phi_\alpha\\ -\Phi_{\alpha\gamma} & +\Lambda_{\alpha\gamma}\cos\Phi\alpha & 1\end{bmatrix}\begin{bmatrix}\overset{\alpha}{X}_{\gamma\alpha}\\ \overset{\alpha}{Y}_{\gamma\alpha}\\ \overset{\alpha}{Z}_{\gamma\alpha}\end{bmatrix}=$$
$$=\begin{bmatrix}1 & -5.47\cdot 10^{-5} & -2.18\cdot 10^{-4}\\ +5.47\cdot 10^{-5} & 1 & +4.79\cdot 10^{-5}\\ +2.18\cdot 10^{-4} & -4.79\cdot 10^{-5} & 1\end{bmatrix}\begin{bmatrix}-1300\,m\\ +300\,m\\ -200\,m\end{bmatrix}$$

Example 2. How to establish an orthonormal frame in Gauss surface geometry? Is the orthonormal frame anholonomic?

Table 2.7 Gauss surface geometry, Cartan surface geometry, orthonormal frame of reference, example of the sphere

$$\mathbf{x}(u,v)=r\cos u\cos v\mathbf{e}_1+r\sin u\cos v\mathbf{e}_2+r\sin v\mathbf{e}_3$$
$$\mathbf{g}_1=\frac{\partial\mathbf{x}}{\partial u}=r\cos v(-\mathbf{e}_1\sin u+\mathbf{e}_2\cos u)$$
$$\mathbf{g}_2=\frac{\partial\mathbf{x}}{\partial v}=-r\sin v\cos u\mathbf{e}_1-r\sin v\sin u\mathbf{e}_2+r\cos v\mathbf{e}_3$$
$$\mathbf{g}_3=\cos v(\mathbf{e}_1\cos u+\mathbf{e}_2\sin u)+\mathbf{e}_3\sin v$$
$$\|\mathbf{g}_1\|=r\cos v,\ \|\mathbf{g}_2\|=r,\ \|\mathbf{g}_3\|=1,\quad \langle\mathbf{g}_1|\mathbf{g}_2\rangle=0$$
$$\mathbf{c}_1:=\frac{\mathbf{g}_1}{\|\mathbf{g}_1\|},\quad \mathbf{c}_2:=\frac{\mathbf{g}_2}{\|\mathbf{g}_2\|},\quad \mathbf{c}_3:=\mathbf{g}_3$$

Here we concentrate to the question of *how to establish* an *orthonormal frame* $\{\mathbf{c}_1,\mathbf{c}_2,\mathbf{c}_3\}$, for instance for the *sphere* if we refer to *Gauss surface geometry*. Is the attached *orthonormal frame a coordinate base or not*? Is the orthonormal frame *anholonomic*?

Based on an *orthogonal reference frame* $\{\mathbf{g}_1,\mathbf{g}_2,\mathbf{g}_3\}$ with references on spherical longitude and spherical latitude called $\{u,v\}$ we compute an *orthonormal reference frame* $\{\mathbf{c}_1,\mathbf{c}_2,\mathbf{c}_3\}$, called *Cartan frame of reference* in Table 2.7. In Table 2.8 we introduce the *displacement* $\mathbf{dx}$ on the surface of the sphere, both in an *Gaussean frame of reference* and in the *Cartan frame of reference*. We ask the key question: Are the matrix components $\{\sigma^1,\sigma^2\}$ integrable? Table 2.8 is a very short introduction to "exterior calculus" or the *Cartan derivative*. The 3-index symbol is introduced and calculated for our example of the sphere. Naturally, the Cartan derivative is *not* integrable (Table 2.9)!

Table 2.6 500 meter local network, misclosures

$$\overset{\beta}{X}_{\beta\gamma}=+800\,m-83.9\,mm,\quad \overset{\gamma}{X}_{\gamma\alpha}=-1300\,m+27.2\,mm$$
$$\overset{\beta}{Y}_{\beta\gamma}=-800\,m-84.9\,mm,\quad \overset{\gamma}{Y}_{\gamma\alpha}=+300\,m-80.7\,mm$$
$$\overset{\beta}{Z}_{\beta\gamma}=+150\,m-5.8\,mm,\quad \overset{\gamma}{Z}_{\gamma\alpha}=-200\,m-297.8\,mm$$
$$\overset{\alpha}{X}_{\alpha\beta}+\overset{\beta}{X}_{\beta\gamma}+\overset{\gamma}{X}_{\gamma\alpha}\neq 0:\quad -83.9\,mm+27.2\,mm=-56.7\,mm$$
$$\overset{\alpha}{Y}_{\alpha\beta}+\overset{\beta}{Y}_{\beta\gamma}+\overset{\gamma}{Y}_{\gamma\alpha}\neq 0:\quad -84.9\,mm-80.7\,mm=-165.6\,mm$$
$$\overset{\alpha}{Z}_{\alpha\beta}+\overset{\beta}{Z}_{\beta\gamma}+\overset{\gamma}{Z}_{\gamma\alpha}\neq 0:\quad -5.8\,mm-297.8\,mm=-303.6\,mm$$

Table 2.8 Displacement vector of the surface of the sphere, Gaussean frame of reference versus Cartan frame of reference, integrability

"derivational equations of the first kind"

$$\mathrm{d}\mathbf{x} = \frac{\partial \mathbf{x}}{\partial u} du + \frac{\partial \mathbf{x}}{\partial v} dv =$$

$$= \mathbf{c}_1 r \cos v \, du + \mathbf{c}_2 r \, dv = \sigma^1 \mathbf{c}_1 + \sigma^2 \mathbf{c}_2$$

$$\begin{bmatrix} \sigma^1 \\ \sigma^2 \end{bmatrix} = \begin{bmatrix} r \cos v & 0 \\ 0 & r \end{bmatrix} \begin{bmatrix} du \\ dv \end{bmatrix} = \begin{bmatrix} a & b \\ c & d \end{bmatrix} \begin{bmatrix} du \\ dv \end{bmatrix}$$

$$\begin{bmatrix} du \\ dv \end{bmatrix} = \begin{bmatrix} a & b \\ c & d \end{bmatrix}^{-1} \begin{bmatrix} \sigma^1 \\ \sigma^2 \end{bmatrix} = \begin{bmatrix} (r \cos v)^{-1} & 0 \\ 0 & r^{-1} \end{bmatrix} \begin{bmatrix} \sigma^1 \\ \sigma^2 \end{bmatrix}$$

$$du = (r \cos v)^{-1} \sigma^1, \quad dv = r^{-1} \sigma^2 \longleftrightarrow$$

$$\sigma^1 = r \cos v \, du, \quad \sigma^2 = r \, dv$$

? integrability ?

$$\frac{\partial \sigma^1}{\partial v} - \frac{\partial \sigma^2}{\partial u} = -r \sin v \neq 0$$

$$\begin{bmatrix} \sigma^1 \\ \sigma^2 \end{bmatrix} =: \underline{\sigma}, \quad d\underline{u} := \begin{bmatrix} du \\ dv \end{bmatrix}$$

"if $\underline{\sigma} = A \, d\underline{u}$, then $d_C \sigma = dA \wedge d\underline{u}$"

Example 3. Discussion between *A. Marussi* and *C. Mineo* and the development of Differential Geodesy

Let us refer to the discussion of *A. Marussi* (1952): Intrinsic geodesy, The Ohio State Research Foundation, Project No. 485, Columbus/Ohio/USA 1952, *C. Mineo* (1955): Intrinsic geodesy and general properties of cartographic representations, Rend. Acc. Naz. Lincei, Cl. di Sc. Fis., Mat. e Nat., Serie 18, fasc. 6 and *A. Marussi* (1955): A reply to a note by *C. Mineo*, see C. Mineo *19*, fasc. 5 in order to document these discussions in the past to accept "Differential Geodesy" as a subject of science.

The subject of *Marussian Geodesy* was established in my paper *E. Grafarend* (1978): *Marussian Geodesy*, pages 209–247, Boll. di Geodesia e Scienze Affini, No. 23, April-Septembre 1978. Refer, in addition, to our contribution "Elie Cartan and Geodesy" by *F. Bocchio, E. Grafarend, N. Grossmann, J.G. Leclerc and A. Marussi* (1978): *Elie Cartan and Geodesy*. Boll. di Geodesia e Scienze Affini, No. 4, August-October 1978, presenting five papers given at *sixth symposium of mathematical geodesy* (third Hotine Symposium) held at *Siena/Italy*, April 2–5, 1975.

Example 4. Projective heights in geometry and gravity space, the work of *Antoni Marussi*

Satellite positioning in terms of Cartesian coordinates $(X, Y, Z) \in \mathbb{T}^2 \subset \mathbb{E}^3$ establishing a triplet of

Table 2.9 1-differential forms, exterior calculus, Cartan derivative

"1-differential form"

$$\sigma^1 = a \, du + b \, dv = a_\alpha du^\alpha$$

$$\sigma^2 = c \, du + d \, dv = b_\alpha du^\alpha$$

"exterior" or E. Cartan-derivative

$$d_C \sigma^\alpha = \sum_{\beta,\gamma=1}^{2} \frac{1}{2!} \Omega^\alpha_{\beta\gamma} \sigma^\beta \wedge \sigma^\gamma \sum_{\beta<\gamma} \Omega^\alpha_{\beta\gamma} \sigma^\beta \wedge \sigma^\gamma$$

! anti-symmetry !

$$du^\alpha \wedge du^\beta = -du^\beta \wedge du^\alpha$$

$$\sigma^\alpha = a^\alpha_\beta \, du^\beta \sim \underline{\sigma} = A \, d\underline{u}$$

three-index-symbol

$$d_C \sigma^\alpha = \sum_{\beta\gamma}^{\alpha} \sigma^\beta \wedge \sigma^\gamma$$

$$d\sigma^1_C = \frac{1}{2} \Omega^1_{12} \sigma^1 \wedge \sigma^2 + \frac{1}{2} \Omega^1_{21} \sigma^2 \wedge \sigma^1$$

$$= \frac{1}{2} (\Omega^1_{12} - \Omega^1_{21}) \sigma^1 \wedge \sigma^2 = \Omega^1_{12} \sigma^1 \wedge \sigma^2$$

$$d\sigma^2_C = \frac{1}{2} \Omega^2_{12} \sigma^1 \wedge \sigma^2 + \frac{1}{2} \Omega^2_{21} \sigma^2 \wedge \sigma^1$$

$$= \frac{1}{2} (\Omega^2_{12} - \Omega^2_{21}) \sigma^1 \wedge \sigma^2 = \Omega^2_{12} \sigma^1 \wedge \sigma^2$$

$$\boxed{\sigma^1 \wedge \sigma^2 = -\sigma^2 \wedge \sigma^1}$$

comparison of coefficients

$$\Omega^1_{12} = -\Omega^1_{21} = \tfrac{1}{r} \tan v$$

$$\Omega^2_{12} = -\Omega^2_{21} = 0$$

assumption: $d_C d\underline{u} = 0$

$$\text{proof: } d_C d\underline{u} = \left(\frac{\partial du}{\partial v} - \frac{\partial dv}{\partial v} \right) dv \wedge du = 0$$

$$d_C \underline{\sigma} = dA \wedge du = \begin{bmatrix} -r \sin v \, dv \wedge du \\ 0 \end{bmatrix}$$

$$d_C \sigma = \begin{bmatrix} -\tfrac{1}{r} \tan v \, \sigma^2 \wedge \sigma^1 \\ 0 \end{bmatrix} \neq 0$$

"summary"

$$d\sigma^\alpha_C = \sum_{\beta,\gamma=1}^{2} \Omega^\alpha_{\beta\gamma} \sigma^\beta \wedge \sigma^\gamma$$

$$= \sum_{\beta<\gamma} \Omega^\alpha_{\beta\gamma} \sigma^\beta \wedge \sigma^\gamma$$

Table 2.10 Projective heights in gravity space, geodesics

"stationary functional"

$$\delta \int_{s_1}^{s_2} ds = \delta \int_{\tau_1}^{\tau_2} \sqrt{g_{k\ell}(x^m) \frac{dx^k}{d\tau} \frac{dx^\ell}{d\tau}}\, d\tau = 0, \quad k, \ell, m \in \{1, 2, 3\}$$

subject to a conformally flat metric

$$g_{k\ell}(x^m) = \gamma^2(x^m)\delta_{k\ell}$$

"Lagrange equations"

$$\mathcal{L}\left(x^m, \frac{dx^m}{d\tau}\right) := \sqrt{\delta_{k_1\ell_1}\gamma^{k_1}\gamma^{\ell_1}} \sqrt{\delta_{k_2\ell_2}\frac{dx^{k_2}}{d\tau}\frac{dx^{\ell_2}}{d\tau}} =$$

$$= \sqrt{(\gamma^1)^2 + (\gamma^2)^2 + (\gamma^3)^2} \sqrt{\left(\frac{dx^1}{d\tau}\right)^2 + \left(\frac{dx^2}{d\tau}\right)^2 + \left(\frac{dx^3}{d\tau}\right)^2}$$

$$\delta \int_{\tau_1}^{\tau_2} \mathcal{L}\left(x^m, \frac{dx^m}{d\tau}\right) d\tau = 0 \Longleftrightarrow \frac{d}{d\tau}\left(\frac{\partial \mathcal{L}}{\partial\left(\frac{dx^k}{d\tau}\right)}\right) - \frac{\partial \mathcal{L}}{\partial x^k} = 0$$

$$\frac{\partial \mathcal{L}}{\partial\left(\frac{dx^k}{d\tau}\right)} = \frac{1}{2}\frac{\partial \mathcal{L}^2}{\partial\left(\frac{dx^k}{d\tau}\right)} = \frac{\sqrt{\delta_{k_1\ell_1}\gamma^{k_1}\gamma^{\ell_1}}}{\sqrt{\delta_{k_2\ell_2}\frac{dx^{k_2}}{d\tau}\frac{dx^{\ell_2}}{d\tau}}}\frac{dx^k}{d\tau} =$$

$$= \frac{\sqrt{(\gamma^1)^2 + (\gamma^2)^2 + (\gamma^3)^2}}{\sqrt{\left(\frac{dx^1}{d\tau}\right)^2 + \left(\frac{dx^2}{d\tau}\right)^2 + \left(\frac{dx^3}{d\tau}\right)^2}}\frac{dx^k}{d\tau}$$

$$\frac{\partial \mathcal{L}}{\partial x^k} = \frac{1}{2\mathcal{L}}\frac{\partial \mathcal{L}^2}{\partial x^k} = \frac{\sqrt{\delta_{k_2\ell_2}\frac{dx^{k_2}}{d\tau}\frac{dx^{\ell_2}}{d\tau}}}{\sqrt{\delta_{k_1\ell_1}\gamma^{k_1}\gamma^{\ell_1}}}\frac{1}{2}\partial_k(\delta_{k_3\ell_3}\gamma^{k_3}\gamma^{\ell_3}) =$$

$$= \frac{\sqrt{\left(\frac{dx^1}{d\tau}\right)^2 + \left(\frac{dx^2}{d\tau}\right)^2 + \left(\frac{dx^3}{d\tau}\right)^2}}{\sqrt{(\gamma^1)^2 + (\gamma^2)^2 + (\gamma^3)^2}}\frac{1}{2}\partial_k\gamma^2(x^m)$$

"transformation from τ to s"

(affine parameter)

$$\left.\begin{aligned} \frac{dx^k}{d\tau} &= \frac{dx^k}{ds}\frac{ds}{d\tau} = x^{\prime k}\frac{ds}{d\tau} \\ \frac{ds}{d\tau} &= \sqrt{g_{k\ell}(x^m)\frac{dx^k}{d\tau}\frac{dx^\ell}{d\tau}} \end{aligned}\right] \Rightarrow$$

$$\boxed{\begin{gathered} g_{k\ell}(x'')^\ell + [k\ell, m](x')^\ell(x')^m = 0 \\ \gamma^2(x'')^k + (\partial_\ell\gamma^2)(x')^k(x')^\ell - \tfrac{1}{2\gamma^2}\partial_k\gamma^2 = 0 \end{gathered}}$$

"Marussi gauge"

$$\delta_{k_1\ell_1}\gamma^{k_1}\gamma^{\ell_1} := \delta_{k_2\ell_2}\dot{x}^{k_2}\dot{x}^{\ell_2} \text{ or } (\gamma^1)^2 + (\gamma^2)^2 + (\gamma^3)^2 = (\dot{x}^1)^2 + (\dot{x}^2)^2 + (\dot{x}^3)^2$$

$$ds^2 = g_{k\ell}(x^m)dx^k dx^\ell = \gamma^2(x^m)\delta_{k\ell}dx^k dx^\ell = \gamma^2(x^m)((\dot{x}^1)^2 + (\dot{x}^2)^2 + (\dot{x}^3)^2)dt^2$$

$$\boxed{ds = \gamma^2(x^m)dt}$$

"transformation from s to t"

$$\boxed{\ddot{x}^k - \tfrac{1}{2}\partial_k\gamma^2(x^m) = 0}$$

"representation of the gradient of the factor of conformality in terms of gravity gradients"

$$\boxed{\begin{aligned} \tfrac{1}{2}\partial_k\gamma^2(x^m) &= \delta_{k_1\ell_1}\gamma^{k_1}\partial_k\gamma^{\ell_1} = \gamma^1\partial_k\gamma^1 + \gamma^2\partial_k\gamma^2 + \gamma^3\partial_k\gamma^3 = \\ &= \partial_1 w\partial_k\partial_1 w + \partial_2 w\partial_k\partial_2 w + \partial_3 w\partial_k\partial_3 w \end{aligned}}$$

Cartesian coordinates for *quantifying the position* of a topographic point requires a *complete redefinition* of geodetic projective heights in geometry and gravity space, namely with respect to a *deformable Earth body*. Such a redefinition has been presented *in two steps*:

(*i*) Projective heights are based upon projective lines which are
  (*i1*) *geodesics* (straight lines) in a Euclidean geometric space, *or*
  (*i2*) *geodesics* (plumblines/orthogonal trajectories with respect to a family of equipotential surfaces) in gravity space in a *conformally flat manifold*, the *Marussi manifold* with the modulus of gravity as the *factor of conformality*.

(*ii*) Projective heights are based upon a *minimal distance mapping* along those geodesics between a topographic point $(X, Y, Z) \in \mathbb{T}^2 \subset \mathbb{E}^3$ and a *reference surface*:
  (*ii1*) For projective heights in geometry space such as *standard reference surfaces* (two-dimensional Riemann manifolds) are the plane $\mathbb{P}^2$, the sphere $\mathbb{S}^2$ or the ellipsoid of revolution $\mathbb{E}^2_{a,b}$,
  (*ii2*) for projective heights in gravity space the standard reference surface is identifies by the reference equipotential surface, the *Geoid at some reference epoch* $t_0 \in \mathbb{R}$.

Here we review by Table 2.10 the variational calculus or the standard optimization routine to generate a *minimal distance mapping* between points on the topography and the reference surface, in particular the corresponding algorithm. We have referred to the *problem of holonomity of orthometric heights, normal orthometric heights* ("slightly anholonomic") for a "star-shaped gravity space" and of steric levelling heights ("pressure heights") in our contribution by *E. Grafarend, R. Syffus and R.J. You* (dedicated to the memory of *Antonio Marussi*) in "Allgemeine Vermessungsnachrichten (1995) 382–403".

Last, not least, I thank *Joseph Zund* for all previous discussions on *anholonomity*. We recommend to the reader to study his masterly written book *J. Zund* (1994): Foundations of Differential Geodesy, Springer Verlag, Berlin-Heidelberg-New York 1994 in which *Local Differential Geodesy* and *Global Geodesy in the Large* are elegantly described. We advice the reader also to study his *The work of Antonio Marussi*, Academia Nazionale dei Lincei, Atti dei Convegni Lincei, Report 91, Roma 1991, pages 9–20. Here the mathematical background as well as the geodetic background of A. Marussi based on interviews with *Mrs. Dolores Marussi de Finetti, Ian Reilly and his own research* are presented.

References

Bachmann D (2006) A geometric approach to differential forms. Birkhäuser, Boston

Bocchio F (1970) Su alcune applicazioni di interesse geodetico delle connessioni non simmetriche, Rendiconti della classe di Scienze Fisiche, matematiche e naturali, Accademia Nazionale dei Lincei (Roma), Ser. VIII, 48:343–351

Bocchio F (1974) From differential geodesy to differential geophysics. Geophys J Roy Astron Soc 39:1–10

Bocchio F (1975) The holonomity problem in geophysics. Bollettino di Geodesia e Scienze Affini, anno XXXIV, 453–459

Bocchio F (1978) Some of Marussi's contributions in the field of two-dimensional and three dimensional representation. Bollettino di Geodesia e Scienze Affini, anno XXXVII, 441–450

Bocchio F (1981) An inverse geodetic singularity problem. Geophys J Roy Astron Soc 67:181–187

Bocchio F (1982a) Geodetic singularities in the gravity field of a non-homogenious planet. Geophys J Roy Astron Soc 68:643–652

Bocchio F (1982b) Geodetic singularities, Reviews of Geophysics and Space Physics. In: Grafarend EW, Rapp RH (eds.) Advances in Geodesy, American Geophysical Union, vol 20. Washington, D.C., pp 399–409

Bode A, Grafarend EW (1982) The telluroid mapping based on a normal gravity potential including the centrifugal term. Bollettino di Geodesia e Scienze Affini, anno XLI, 22–56

Bourbaki N (1948) Éléments de mathematique, Livre II Algèbre, Chapitre 3, Hermann, Paris

Cartan E (1928) Leçons sur la géométrie des espaces de Riemann, Gauthier-Villars, Paris; enlarged 2nd edn. ibid (1946)

Cartan E (2001) Riemann geometry in an orthogonal frame, lectures delivered by Elie Cartan at the Sorbonne in 1926–27 (trans: Goldberg VV), foreword by S.S. Cheru, World Scientific, New Jersey

Cartan H (1967a) Calcul différentiel, Hermann, Paris = Differential calculus. Houghton-Mifflin, Boston

Cartan H (1967b) Formes différentiel, Hermann, Paris = Differential forms. Houghton-Mifflin, Boston (1970)

Choquet-Bruhat Y (1968) Géométrie différentielle et systèmes extérieurs. Dunod, Paris

Chovitz BH (1969) Hotines 'Mathematical geodesy'. In: Proceedings IV Symposium on Mathematical Geodesy, Trieste = 159–172 of Hotine (1991)

Chovitz BH (1972) Generalized three-dimensional conformal transformations. Bull Geod 104:159–163

Chovitz BH (1982) The influence of Hotine's Mathematical geodesy. Bollettino di Geodesia e Scienze Affini, anno XLI, 57–64

Darboux G (1915) Leçons sur la théorie générale des surfaces et applications géométriques du calcul infinitesimal, Deuxiè partie, 2nd edn. Gauthier-Villars, Paris

Defrise P (1975) Météorologie et géometrie différentielle, Publ. A 91, Institut Royal Météorologique de Belgique, Bruxelles

Defrise P (1978) Sur des applications de la géometrie différentielle en météorologie et en géodésie. Bollettino di Geodesia e Scienze Affini 37:185–196

Defrise P, Grafarend EW (1976a) Torsion and anholonomity of geodetic frames. Bollettino di Geodesia e Scienze Affini 35:153–160

Defrise P, Grafarend EW (1976b) Torsion and anholonomity of geodetic frames. Bollettino di Geodesia e Scienze Affini 35:81–92

Dietrich G (1936) Das ozeanische Nivellement und die Anwendung auf die Goldküste un die atlantische Küste der Vereinigten Staaten von Amerika. Z Geophys 12:287–298

Engels J, Grafarend EW (1992) A global representation of ellipsoidal heights - geoidal undulations or topographic heights - in terms of orthonormal functions, Part I: "amplitude-modified" spherical harmonic functions. Manuscripta Geodaetica 17:52–58

Engels J, Grafarend EW (1993) The gravitational field of topographic-isostatic masses and the hypothesis of mass condensation. Surv Geophys 140:495–524

Fischer I (1975) Does mean sea level slope up or down toward north? Bulletin Géodésique 49:17–26

Fischer I (1976) Does mean sea level slope up or down toward north? Bulletin Géodésique 50:79–80

Fischer I (1977) Mean sea level and the marine geoid - an analysis of concepts. Mar Geodes 1:37–58

Goenner H, Grafarend EW, You RJ (1994) Newton mechanics as geodesic flow on Maupertuis' manifold; the local isometric embedding into flat spaces. Manuscripta Geodaetica 19: 339–345

Grafarend E, Syffus R, You RJ (1995) Projective heights in geometry and gravity space. Allgemeine Vermessungsnachrichten, 382–403

Grafarend EW (1971) The object of anholonomity and a generalized Riemannian geometry for geodesy. Bollettino di Geofisica Teorica ed Applicata, 13:241–253

Grafarend EW (1973a) Gravity gradients and threedimensional net adjustment without ellipsoidal reference, The Ohio State University, Report No. 202, Columbus

Grafarend EW (1973b) Le theoreme de conservation de la courbure et la torsion or attempts at a unified theory of geodesy. Bulletin Géodésique, 109:237–260

Grafarend EW (1975) Threedimensional geodesy I - the holonomity problem. ZfV, 100:269–281

Grafarend EW (1976) Threedimensional geodesy III: Refraction. Bollettino di Geodesia e Scienze Affini 35: 153–160

Grafarend EW (1977) Geodäsie - Gaußsche oder Cartansche Flächengeometrie? Allgemeine Vermessungs-Nachrichten 4:139–150

Grafarend EW (1978) Marussian geodesy. Bollettino di Geodesia e Scienze Affini, anno XXXVII, 209–247

Grafarend EW (1980) The Bruns transformation and a dual setup of geodetic observational equations NOAA Technical Report, NOS 85 NGS 16, U.S. Department of Commerce, National Oceanic and Atmospheric Administration, National Ocean Survey, Rockville 1980

Grafarend EW (1987) Der Einfluss der Lotrichtung auf lokale geodätische Netze (The influence of the local vertical in local geodetic networks). Zeitschrift für Vermessungswesen 122:413–424

Grafarend EW (1994) What is a geoid? In: Vanicek P, Christou N (eds) Geoid and its geophysical interpretations. CRC Press, Boca Raton, pp 3–32

Grafarend EW (2004) Tensor algebra, linear algebra, multilinear algebra, Technical Report, 344 references, Department of Geodesy and Geoinformatics, Stuttgart University, Stuttgart

Grafarend EW, Engels J (1994) The convergent series expansion of the gravitational field of a starshaped body. Manuscripta Geodaetica 19:18–30

Grafarend EW, Lohse P (1991) The minimal distance mapping of the topographic surface onto the reference ellipsoid of revolution. Manuscripta Geodaetica 16:92–110

Grafarend EW, You RJ (1995) The Newton form of a geodesic in Maupertuis gauge on the sphere and the biaxial ellipsoid – Part I. Zeitschrift für Vermessungswesen 120:68–80

Grossman N (1974) Holonomic measurables in geodesy. J Geophys Res 79:689–694

Grossman N (1978a) Is the geoid a trapped surface? Bollettino di Geodesia e Scienze Affini anno XXXIV, 173–183

Grossman N (1978b) The pull-back operation in physical geodesy and the behaviour of plumblines. Manuscripta Geodaetica 3:55–105

Grossman N (1979) The nature of space near the Earth. Bollettino di Geodesia e Scienze Affini anno XXXV, 413–424

Grotemeyer KP (1969) Analytische geometrie. W. de Gruyter, Berlin

Hamilton WR (1828) Essay on the theory of systems of rays, Transactions of the Royal Irish Academy 15:69–174=1–106 paper I of Hamilton (1931)

Hamilton WR (1931) The mathematical papers of Sir William Rowan Hamilton: Volume I. In: Conway AW, Synge JL (eds.) Geometrical optics, Cambridge University Press, Cambridge (see Note 2: History of the theorem of malus, 463–464)

Hehl FW, von der Heide P, Kerlick GD, Nester JM (1976) General relativity with spin and torsion: foundations and prospects. Rev Mod Phys 48(3):393–416

Heitz S (1986) Grundlagen kinematischer und dynamischer Modelle der Geodäsie, 2. erw. Aufl. Mitt. a. d. Geod. Inst. der Rh. F. Wilhelms-Universität, No. 63, Bonn 1986

Herzberger M (1958) Modern geometrical optics. Interscience Publishers, New York

Holota P (1989) Boundary value problems and invariants of the gravitational tensor in satellite gradiometry. In: Sansò F, Rummel R (eds.) Theory of Satellite geodesy and gravity field determination. Lecture Notes in Earth Sciences, vol 25. Springer, Berlin, pp 447–458

Hotine M (1957a) Metrical properties of the earth's gravitational field, report to I.A.G. Toronto Assembly = 33–64 of Hotine (1991)

Hotine M (1957b) Geodesic coordinate systems, report to Venice Symosium = 65–89 of Hotine (1991)

Hotine M (1959) A primer on non-classical geodesy, report to Venice Symposium = 91–130 of Hotine (1991)

Hotine M (1965) Trends in mathematical geodesy. Bollettino di Geodesia e Scienze Affini, anno XXIV, 607–622 = 5–21 of Hotine (1991)
Hotine M (1966a) Geodetic applications of conformal transformations. Bulletin Géodésique 80:123–140
Hotine M (1966b) Triply orthogonal coordinate systems. Bulletin Géodésique 81:195–222
Hotine M (1966c) Note by the writer of the paper 'On Triply orthogonal coordinate systems'. Bulletin Géodésique 81:223–224
Hotine M (1967/1947) The orthomorphic projection of the spheroid, Empire Survey Review, 8, 62:300–311 (1946); 9, 63:25–35 (1947); 9, 64:52–70 (1947); 9, 65:112–123 (1947); 9, 66:157–166 (1947)
Hotine M (1969) Mathematical geodesy. U.S. Department of Commerce, Washington, D.C.
Hotine M (1991) Differential geodesy. (edited with commentary by J.D. Zund). Springer, Berlin
Kempin JH (1980) über den Aufbau großregionaler Gebrauchshöhennetze in den Niederungszonen äquatorialer Urwald-Flußsysteme unter besonderer Berücksichtigung des Amazonasgebiets durch hydrostatisches Nevellement, Habilitationsschrift, Universität Hannover, Hannover
Kummer EE (1860) Allgemeine Theorie der gradlinigen Strahlensysteme. Journal für die reine und angewandte Mathematik 57:189–230
Leclerc JG (1977) A computation of the geodetic object of anholonomity and the geodetic misclosures to a first order approximation. Bollettino di Geodesia e Scienze Affini 36:1–24
Leclerc JG (1979) Operational geodetic anholonomity. Bollettino di Geodesia e Scienze Affini 38:97–140
Levi-Cività T (1925) Lezioni di calcolo differenziale assoluto, Stock, Roma; = The absolute differential calculus: calculus of tensors; Persico E (ed.), (trans: Long M), Blackie & Son Limited, London (1926); reprinted by Dover Publications, New York (1989)
Livieratos E (1976) On the geodetic singularity problem. Manuscripta Geodaetica 1:269–292
Malus EL (1808) Mémoire sur l'optique. Journal de l'École polytechnique, 7 (14e Cahier):1–44
Malus EL (1811) Traité d'optique. Mémoires présentés a l'Institut des sciences par divers savants 2:214–302
Marussi A (1949) Fondements de géométrie différentielle absolue du champ potential terrestre. Bulletin Géodésique 14:411–439
Marussi A (1951a) Les principes de la géodésie intrinsèque. Bulletin Géodésique 19:68–76
Marussi A (1951b) Fondamenti di geodesia intrinseca. Publicazioni della Commissione Geodetica Italiana, Ser. III, 7:1–47 = 13–58 of Marussi (1985)
Marussi A (1951c) Su alcune proprietà integrali delle rappresentazioni conformi di superfici su superfici, Rendiconti della classe di Scienze fisiche, matematiche e naturali, Accademia Nazionale dei Lincei (Roma), Ser. VIII, 10:307–310 = 149–152 in Marussi (1985)
Marussi A (1953) Un'analogia fra le leggi della propagazione delle luce in mezzi rifrangenti continui isotropi, e le rappresentazioni conformi. Atti Fondazione Giorgio Ronchi dell'Instituto Nazionale de Ottica, anno VIII, 2–7 = 169–172 in Marussi (1985)
Marussi A (1959) Dalla geodesia classica alla geodesia in tre dimensioni. Bollettini di Geodesia e Scienze Affini, anno XVIII, 485–495 = 3–12 in Marussi (1985)
Marussi A (1979a) Natural reference systems and their reciprocals in geodesy, Publ. T.J. Kukkamäki 70th Birthday, Publ. Finnish Geodetic Institute, Nr. 89, Helsinki
Marussi A (1979b) The tidal field of a planet and the related intrinsic reference systems. Geophys J Roy Astron Soc 56:409–417 = 75–83 of Marussi (1985)
Marussi A (1985) Intrinsic geodesy (trans: Reilly WI). Springer, Berlin
Marussi A (1988) Intrinsic geodesy (a revised and edited version of his 1952 lectures by J.D. Zund), Report No. 390, Department of Geodetic Science and Surveying, The Ohio State University, Columbus
Milnor J (1965) Topology from the differentiable viwepoint, based on notes by D.W. Weaver, The University Press of Virginia, Charlottesville
Monge G (1809) Application de l'analyse a la géométrie, 4th edn., M.V. Bernard, Paris; 5th edn., Liouville J (ed.) with supplementary notes and memoir by Gauss Bachelier, Paris (1850)
Moritz H (1994) The Hamiltonian structure of refraction and the gravity field. Manuscripta Geodaetica 20:52–60
Nirenberg L (1953) The Weyl and Minkowski problems in differential geometry in the large. Comm Pure Appl Math 6:337–394
Nitsche JCC (1975) Vorlesungen über Minimalflächen. Springer, Berlin
Pizzetti P (1901) Un principio fondamentale nello studio delle studio delle superfici di livello terrestri. Rendiconti della Reale Accademia dei Lincei (Roma), 10:35–39
Pizzetti P (1906) Höhere Geodäsie, Encyklopädie der Mathematischen Wissenschaften Band VI, Erster Teil: Geodäsie und Geophysik, 125–239, B.G. Teubner, Leipzig
Pizzetti P (1913) Principii della teoria mecannica della figura dei pianeti. E. Spoerri, Pisa
Plückner J (1868/69) Neue Geometrie des Raumes, gegründet auf die Betrachtung der geraden Linie als Raumelement, Teil I & II, B.G. Teubner, Leipzig
Ricci G (1898) Lezioni sulla teoria delle superficie. F. Drucker, Verona-Padova
Ricci G (1956/57) Opere, two volumes, Edizioni Cremonese, Roma
Ricci G, Levi-Cività T (1901) Methodes de calcul differentiel absolu et leurs applications. Math Ann 54:125–201
Roberts S (1872) On the parallel surfaces of conicoids and conics. In: Proceedings of the London Mathematical Society, Ser. 1, 4:57–91
Roberts S (1873) On parallel surfaces. In: Proceedings of the London Mathematical Society, Ser. 1, 4:218–235
Rummel R (1985) Satellite gradiometry: a promising concept for 25 years, Die Arbeiten des Sonderforschungsbereiches 78 Satellitengeodäsie der Technischen Universität München 1984 und 1985, vorgelegt von M. Schneider, Bayrische Kommission für die internationale Erdmessung, N. 48: 123–139
Rummel R (1986) Satellite gradiometry. In: Sünkel H (ed.) Mathematical and numerical techniques in physical geodesy. Lecture notes in Earth sciences, vol 7. Springer, Berlin, pp 317–363

Rummel R, Van Gelderen M (1992) Spectral analysis of the full gravity tensor. Geophys J Int 111:159–169

Schouten JA (1951) Tensor analysis for physicists. Oxford at the Clarendon Press, Oxford, 2nd edn. ibid (1954)

Schouten JA (1954) Ricci-calculus: an introduction to tensor analysis and its geometrical allpications, 2nd edn., Springer, Berlin

Stäckel P (1891) über die Integration der Hamilton-Jacobischen Differentialgleichung mittelst Separation der Variablen, Habilitationsschrift, Halle

Stäckel P (1893) über die Bewegung eines Punktes in einer n-fachen Mannigfaltigkeit. Math Ann 42:537–563

Synge JL (1937) Geometrical optics. Cambridge University Press, Cambridge, 110 p

Thomson RE, Tabata S (1987) Steric height trends at ocean station Papa in the Northeast Pacific Ocean. Mar Geodes 11:103–113

Van Der Waerden BL (1966) Algebra, Erster Teil. 7th edn. of Moderne Algebra - I originally published in 1930; Springer, Berlin

Vanicek P (1982) Letter to the editor. Canadian Surveyor 36:122–123

Weyl H (1916) über die Bestimmung einer geschlossenen konvexen Fläche durch ihr Linienelement. Vierteljahresschrift der Naturforschenden Gesellschaft in Zürich 61:40–72 = 148–178 (mit Nachtrag by author in 1955) Selecta Hermann Weyl. Birkhäuser Verlag Basel (1956)

Weyl H (1918) Raum-Zeit-Materie: Vorlesungen über allgemeine Relativitätstheorie, five editions (1918–1924) Springer, Berlin; reprint ibid (1970)

Weyl H (1928) Gruppentheorie und Quantenmechanik, Verlag S. Hirzel, Leipzig = Theory of groups and quantum mechanics, (2nd edn. trans: Robertson HP), E.P. Dutton, New York (1932); reprint by Dover Publications Inc., New York (1950)

Wilkes JM, Zund JD (1982) Group-theoretical approach to the Schwarzschild solution. Am J Phys 50:25–27

Willmore TJ (1959) An introduction to differential geometry. Oxford at the Clavendon Press, London

Zund J (1994) Foundations of differential geodesy. Springer, Berlin

Zund JD (1988a) Tensorial methods in classical differential geometry - I: Basic principles. Tensor NS 47:74–82

Zund JD (1988b) Tensorial methods in classical differential geometry - I: Basic surface tensors. Tensor NS 47:83–92

Zund JD (1989) Differential geodesy of the Eötvös torsion balance. Manuscripta Geodaetica 14:13–18

Zund JD (1990a) An essay on the mathematical foundations of the Marussi-Hotine approach to geodesy. Bollettino di Geodesia e Scienze Affini, anno XLIX, 133–179

Zund JD (1990b) The assertion of Hotine on the integrability conditions in his general (ω, ϕ, N) coordinate system. Manuscripta Geodaetica 15:373–382

Zund JD (1990c) The Hotine problem in differential geodesy. Manuscripta Geodaetica 15:373–382

Zund JD (1992) The mathematical foundations of the Hotine-Marussi approach to geodesy. Bollettino di Geodesia e Scienze Affini, anno LI, 125–138

Zund JD (1993a) The Bruns curvature equations and tangential differential in differential geodesy. Bollettino di Geodesia e Scienze Affini, anno LII, 191–205

Zund JD (1993b) Hotine's (ω, ϕ, N) coordinate system, Phillips Laboratory Research Report, PL-TR-93-2174, 23 p

Zund JD, Moore W (1987) Hotine's conjecture in differential geodesy. Bulletin Géodésique 61:209–222

Zund JD, Wilkes JM (1988) The significance and generalization of two transformation formulas in Hotine's Mathematical geodesy. Bollettino di Geodesia e Scienze Affini, anno XLVII, 77–85

The Shielding of Gravitation: An Old Unresolved Puzzle

3

M. Caputo

Abstract

In the memory of Marussi and its memorable pendulums, made for the observations of Earth tides but used for the unexpected observations of the free modes of the Earth and the also unplanned attempt to observe the absorption of gravitation. Since we are concerned with geodesy, I will recall the result which, perhaps, is one of the most important observations obtained with the pendulums: that concerning the absorption of gravitation. I will give no new results.

1 Introduction

Nicolas Fatio de Duiller presented his theory for explaining gravitation at the Royal Society in 1690 (Launteren 2002), it had many opposites and later some support from Newton himself. Among the supporters was Le Sage (Edwards 2002a, b) who somewhat resumed the theory, discussed it with many scientists of its epoch such as Huigens and Leibniz and expanded and publicized it. It had also notable opponents as Eulero, Maxwell and Poincaré. It is presently known as Le Sage theory (Le Sage 1784).

Looking into the matter concerning this theory one sadly notes that, as in the past centuries when mathematical duels were held in churches, how bitter was the fight for priority in finding theories and mathematical discoveries. Especially when the theory was apparently new. It is notable that in the past the opponents of the new theories often expressed irrelevant personal opinions without scientific support also because of the limited knowledge of physics at that time. The opposition to Marconi is a good example of this type of arguing. Neither Marconi nor the opponents had any knowledge of the ionosphere essential for the propagation of radio-waves in many circumstances.

Le Sage (1784) theory states that a flow of particles called gravitons with isotropic distribution of directions permeates the universe. It hits the planets and, in the case of a system formed by 2 planets, they shade one another from the particles which in turn generate a push of each of the bodies towards the other. The field generated by this force is of the type of the inverse square law as shown by Le Sage himself (Le Sage 1784). In the case of 3 bodies, when they are aligned the central body is shaded on both sides in the direction of the other two and is less subject to gravitation as indicated in Fig. 3.1 in the case of a particular relative position. This is the shielding of gravity in Le Sage's (1784) theory.

The Le Sage (1784) theory arrives to the true basis of the problem: which is the physical mechanism of gravitation? Rather than Newton's "Hypothesis non fingo", that is "It all happens as if" of Newton or

M. Caputo (✉)
Phsycis department, University La Sapienza, Rome, Italy

Department of Geology and Geophysics, Texas A&M University, College Station, Texas

N. Sneeuw et al. (eds.), *VII Hotine-Marussi Symposium on Mathematical Geodesy*, International Association of Geodesy Symposia 137, DOI 10.1007/978-3-642-22078-4_3, © Springer-Verlag Berlin Heidelberg 2012

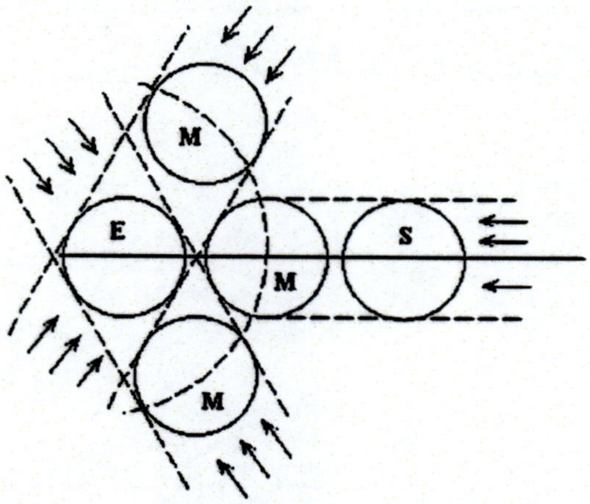

Fig. 3.1 Imaginative successive relative positions of 3 celestial bodies: Sun (S), Moon (M) and Earth (E) and Le Sage flux

the generic assumptions of others such as Majorana (1919, 1920).

In Majorana's theory the bodies emit an energy flux of some kind which produces gravitational effects on other bodies and produces self-shielding for which the gravitational mass is different from the inertial one. As in Le Sage's theory the flux it attenuated when passing through solid or fluid matter.

Majorana made many sophisticated experiment in Bologna and thought to have seen a shielding affect (Caputo 1962) which however was not confirmed in subsequent more sophisticated experiments. For a recent review of this matter see Duif (2004) or Edwards (2002a, b). Obviously all theories are valid with the limitations of our measuring instrument and philosophy.

According to the simple Majorana theory the screening of the moon would cause a decrease of gravitation at the site where the occultation of the Sun occurs. According to the Le Sage (1784) theory the Earth, at the time of occultation of the Sun, because of the position of the Sun behind the Moon, at the time of total eclipse would not feel a smaller effect of gravitation from the Moon and feel an acceleration towards the Moon.

According to Majorana the shielding of gravity would happen according to the formula

$$g = g_0 \int_0^h \exp(-\lambda\rho(l))dl$$

where ρ is the density of the shielding layer, λ the absorption coefficient and g and g_0 the resulting and the incoming gravity respectively.

This difference of perspective effects did not make much difference in the search of an anomaly in the data at the time of the 1960 solar eclipse. In fact the anomaly in the data was generally searched without making any assumption of the theoretical model of gravitation. In most cases the search concerned only an unspecified anomaly to be detected with correlations between the data and perspective model signals. The existence of a correlation and its sign would only be a constraint on the discussion of the perspective model of gravitation.

The aspects of the two theories trying to model gravitation and their shielding consequences attracted the interest of many scientists especially in geodesy astronomy and geophysics because in these fields was available the appropriate instrumentation for observing the shielding phenomenon and with appropriate approaches one would have a proof of the theory under scrutiny.

This presentation concentrates on the observations made with the horizontal pendulums in the Grotta Gigante in Trieste during the Solar eclipse of 1960; a brief discussion on the observations made in Florence with a La Coste gravimeter during the same eclipse will follow.

Other work done in the field of gravitation absorption made by other researchers, in spite of their t relevance, mostly in the field of Astronomy, should be considered; we are not pretending to write a review of the work done in this field. We wish to celebrate Marussi's pendulums. For an excellent review of the work done in the field of gravitation absorption see Edwards (2002a, b), for more references see Caputo (1962, 1977, 2006).

2 The Grotta Gigante in Trieste, the Great Pendulums and the Observations Made During the Solar Eclipse of 1960

1. In the theory of gravitation absorption it is suggested that a gravitational ray l of intensity g_0 be weakened after crossing a layer of material with density δ according to the law suggested by Bottlinger (1912)

$$g = g_0 \exp\left(-\lambda \int_l \delta dl\right) \tag{3.1}$$

where λ is the so called absorption coefficient. The experiment made so far allow to assume that $\lambda < 10\text{–}14\,\mathrm{g}^{-1}\,\mathrm{cm}^{-2}$; for layer thickness and density sufficiently small we may then write

$$g = g_0 \left(1 - \lambda \int_l \delta dl\right) \tag{3.2}$$

A straight forward method to verify the reality of the phenomenon or to estimate λ is to measure the weakening of a gravitational ray caused by the crossing of a layer of known density and thickness. The first physicists who investigated this phenomenon and estimate λ followed this path.

In 1897 Austin and Thwing set some screens of different density between the fixed and the mobile masses of a Cavendish balance; however the did not succeed to observe the absorption within a limit of the accuracy of the experiment which was 2% of the acting force.

In the first years of the following century, experiments with the same method were repeated without positive results by Kleiner (1905), by Cremieux (1906) and by Erisman (1908). Erisman (1908) reached the precision of 0.08% of the acting forces. Lager (1904) used a regular balance to weigh a spherical silver ball weighting 1.5 g, alternatively surrounded or not by a spherical lead layer, however he did not observe any weight variation larger than 0.01% of the acting forces.

Obviously the phenomenon interested also the astronomers; in 1911 some irregularities of the motion of the Moon, for which no causes could be found, were object of a prize emitted by the University of München. The prize was assigned in 1912 to Bottlinger (1912), who showed that the phenomenon of gravity absorption could explain the those irregularities. During Moon eclipses because of the absorption the Moon would be subject to an impulsive force due to the gravity absorption.; Bottlinger (1912) showed the these forces would cause a periodic variation of the mean longitude with a period of about 19 years; assuming that $\lambda = 3\,10\text{–}15$ the estimated variation would be in agreement with the observed ones. In the same year appeared also a paper by De Sitter who reached the same conclusions of Bottlinger.

Today it is believed that the findings of Bottlinger (1912) and De Sitter be invalidated by the poor knowledge of the time in the epoch of Bottlinger and DeSitter since then the knowledge of the time was related to the Earth rotation which suffers of periodic and a-periodic irregularities which may interfere with the supposed irregularities of the Moon motion.

In 1919 Majorana began a series of studies and laboratory experiments whose results are presented in a set of 18 notes appeared in the Proceedings of Accademia Nazionale dei Lincei from 1919 to 1922. First he made some theoretical studies where he suggested that the substances composing the Sun appear to us as masked in the gravitational effects by the exterior layers, that is due to the supposed effect of the gravitation absorption; in reality the mass of the Sun would appear to us as smaller than shows the classical theory. He showed that if we assume that the Sun is homogeneous.

With a density of $2\,\mathrm{g\,cm^{-3}}$ (respectively $20\,\mathrm{g\,cm^{-3}}$), the value sof the absorption coefficient λ assumes the value 1.11 10–11 (respectively 2.90 10–11).

Subsequently Majorana began a series of laboratory experiments performed with very refined techniques seeking to observe the variation of weight of a 1.3 kg led sphere when it was screened from the effect of Earth's gravitational field with other masses. As screen he used a 114 kg Mercury cylinder, then a 9.8 kg Led cube which surrounded completely the sphere.

From the results of these experiments Majorana was induced to state that the absorption effect existed. The experiments made with the Mercury cylinder led to conclude that to the value $\lambda = 7\,10\text{–}12$, while the experiments made with the Led cube gave $\lambda = 2\,10\text{–}12$.

Following the publication of the first results of Majorana's experiments, Russel (1921) showed that because of the gravity absorption the inertial mass of the planets could not be proportional to their gravitational mass and that consequently their motion should differ notably from what observed in reality. According to Russel this conditions the value of λ to be smaller than the values given by Majorana by a factor 10–4; he further suggested that the phenomenon observed by Majorana was not due to absorption but possibly to a relativistic effect.

Many years went by before the research on this matter would be resumed. In 1954 Brein (1954), using an idea of Tomaschek (1937), tried to observe the gravitation absorption during a Solar eclipse which occurred in Central Europe and the same was done by Tomaschek (1955) himself in the Shetland Islands. In that circumstance the Moon would have served as a screen relative to the Sun and relative apparent increase of gravity would have been observed. The experiment was made with a high sensitivity recording gravimeter, but few perturbations of difficult interpretation occurred and made results uncertain. However Brein (1957) from the results of the experiment inferred for λ the limit λ < 3 10–15 which is not in disagreement with the results of Bottlinger (1912).

2. The total Sun eclipse of February 15th 1961 was another circumstance to attempt a verification of the phenomenon. Experiments were made in Sofia and Kiev with Askania recording gravimeters by Venedikov (1961) and Dobrokhotov et al. (1961) and in Berchtesgaden with horizontal pendulums by Sigl and Eberhard (1961).

The results of these observations were presented at the IV Symposium on Earth Tides in Bruxelles in 1961; no evident effect of absorption was reported moreover no limits for the coefficient were given. The same type of observations in the circumstance of the 1961 Solar eclipse of 1961, with the suggestion of Marussi were made also with the great horizontal pendulums installed since 1958 in the Grotta Gigante near Trieste for the study of the tides of the Earth's crust (Marussi 1960a, b).

The circumstance was exceptional since the totality was in near proximity of the station since the minimum distance of the two bodies at totality was only 58″; moreover the height of the Sun at totality was 13° 30′ and the effect on the horizontal pendulums was very near the maximum one could hope. In order to observe the phenomenon the sensitivity of the pendulums was taken from 463 s to 657 s for the EW component and form 500 s to 580 s for the NS component. The longer period of oscillation implies in the recordings a ratio of 2.185 mm/msec and 1.702 mm/msec respectively. The speed of the recording photographic film was taken to 3.8 cm/h. Since the reading resolution is 0.1 mm, follows that the reading have an uncertainty of about 5 10–5 arcsec.

The recordings of Earth's tides during February 15th were favoured by excellent environmental and meteorological conditions: the barometric pressure, which could cause very small inclinations around the Dinaric axis, had no appreciable variations nor were recorded disturbances due to the flow of Karst waters.

The analogue recordings of Earth's tides during February 15th for both components EW and NS were digitised with readings every 12 min. The values were then fit to a set of sinusoids with various phased in order to eliminate the effect of Earth tides.

These time series were examined in a time interval of 12 h centred at totality of the eclipse and further filtered in order to eliminate the seiches of the Adriatic sea (Caloi 1938; Polli 1958, 1961).

The resulting residual time series were analysed with the χ^2 test which gave confidence levels of 78% and 85% for the two components. Since this result was not considered sufficiently significant the time series were filtered again to smoothing the data. The resulting curves CNS, CEW are shown in the Fig. 3.1. The deviations with respect to these curves have a level of randomness of 83% and 99% respectively, which we considered acceptable.

3. We compared the curves CNS, CEW with those which presumably would represent an absorption effects.

The variations on the horizontal components of the Lunisolar attraction due to gravity absorption in the case of Majprana's model during the successive phase of the eclipse have been computed for the particular case which we are considering with a process of graphics integration which ensure a precision of 2%; to this purpose we considered the Moon homogeneous with for density 3.34 while the density of the Sun we adopted the values, as function of the distance from its centre, given in the tables of Landolt and Börnstein (1952).

With g = 980.63 gal at the Grotta Gigante the components of the deflection of the vertical (expressed in milliseconds of arc) which represent North-South and the East-west components CNS, CEW are reproduces in the Fig. 3.2 for λ = 3 10–15 (curve c), λ = 10–15 (curve b), λ = 0.5 10–15 (curve a). The comparison of the theoretical curves with the experimental ones suggests the following considerations: no effect is seen in the NS component, while in the EW component the flattening of the oscillation towards East, which occurs at the time of the maximum of the eclipse,, could be due to the presence of the supposed gravitational absorption with a value of λ < 0.6 10–15.

Taking into account that this component should have a greater reliability because of the possible effect of absorption, which on this component should be 1.9 times larger than in the other, and also because

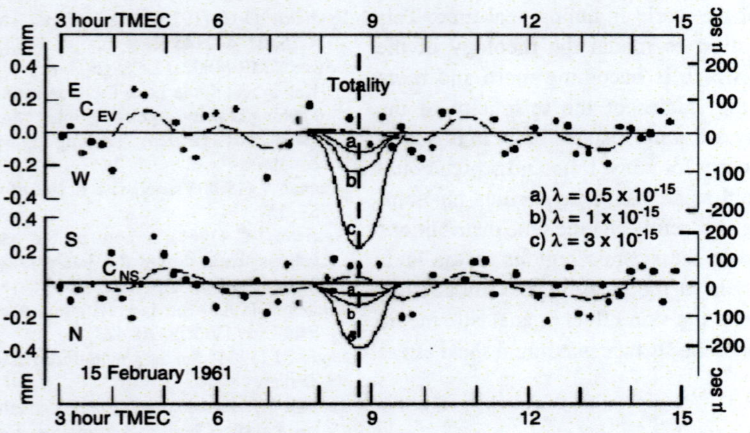

Fig. 3.2 Final filtered data of the CEW (top) and CNS components (bottom) of the Earth tide station in the Grotta Gigante during the 1960 solar eclipse

the confidence level with which have been eliminated the accidental departures is larger than for the other component, we will assume the limit for λ which results from it that is λ < 0.6 10–15. We note that this limit is 1/5 of that so far admitted and lower than the forecast of Russel (1921).

3 The Observations Made in Florence with a La Coste Gravimeter During the 1960 Solar Eclipse

Concerning the method in which the Moon is used as a screen it is of interest the experimental limit set for λ by Slichter et al. (1965) with observations, taken with a La Coste tidal gravimeters during the solar eclipse of February 15th 1961 is λ = 8.3 10–16.

The method using the Earth as a screen was introduced by Harrison (1963) who used. tidal gravimeters observation and found λ < 10–15. A better result is that of Unnikrishnan et al. (2002) with λ < 2 10–17 who used the same method and analyzed 11,000 min of data taken by Wang et al. (2000) with a gravimeter during the 1997 total eclipse in China.

4 The Work in Astronomy

Finally studying the fluctuations of the Moon motion Crawley et al. (1974) found λ < 6.3 10–15. However the most stringent limit on gravitation absorption, is that of Eckhardt (1920) who, in a brief note, reported that using the Laser ranging to the Moon data of Williams et al. (1976), gave what is now the best upper limit λ < 2 10–21.

Conclusion

Geodesy is gone long way since the glorious time of Pizzetti, Somigliana, Cassinis and Silva. It is now deeply involved in the survival of our planet, in essential social responsibilities and evolved towards developments unexpected a couple of decades ago, as it happens in many fields of science, and has shown its vitality in most countries. Geodesy has new very important tools and, as usual, the support of its contemporary mathematics which has developed new very efficient methods of analysis and modelling. Some experiments made with the classical instruments of geodesy could be repeated using the new more accurate instrumentation now available, also few of the results obtained in the studies of gravitation absorption could be reviewed and refined in view of the new mathematical tools presently in use.

Time has entered geodesy since long but we should take this more seriously since all moves in the Earth's system. We have a reference system and we may monitor all the movements almost in real time. The archive of these movements in the same reference system is available, as is the catalogue of stars in the Galaxy and outside it, as is the catalogues of earthquakes of the world in almost real time. This archive may allow to model the rheology of the Earth's crust, which is becoming more and more important in the studies of the seismicity of the Earth. Geodesy has some responsibility in this field which I tried carry for some time and which could possibly explain some rather mysterious phenomena such as the apparently chaotic time distributions of earthquakes in the seismic regions or the inefficient stress field at the base of peak mountains, apparently due to the skin effect in anelastic media which diminishes the surface maximum shear stress (Caputo 1995).

References

Bottlinger KF (1912) Die Gravitationstheorie und die Bewegung des Mondes, Innauguraldiss, Univ. München. C. Troemer, Freiburg Br., 57

Brein R (1954) Photographische Registrierung der Erdgezeiten mit einem Gravimeter, Deutsche Geolog. Comm., Reihe B, H. 34, Frankfurt a. M.

Brein R (1957) Beobachtungen zur Sonnenfinsternis 1954 in Sudnorwegen, Deutsche Geolog. Comm., Reihe B, H. 21, Frankfurt a. M.

Caloi P (1938) Sesse dell'Alto Adriatico con particolare riguardo al Golfo di Trieste, Memoria CCXLVII del R. Com. Talassografico It. Del C.N.R.

Caputo M (1962) Un nuovo limite superiore per il coefficiente di assorbimento della gravitazione. Accademia Nazionale dei Lincei. (A new upper limit for the coefficient of absorbtion in gravitation, *Proceedings of Accademia Nazionale dei Lincei*) 8, 32, 4, 509–515, 1962. Also in English: http://www.space-time.info/gravianom/literaturelist.html

Caputo M (1977) Experiments concerning gravitational shielding, Simposio Internazionale: Experimental Gravitation, Pavia 1976. Atti dei Convegni Lincei 34:193–211

Caputo M (1995) The skin effect of an elastic Earth. Revista Geofisica 43:145–162

Caputo M (2006) On new limits of the coefficient of gravitation shielding. Journal of Astrophysics and Astronomy 27: 439–441

Crawley R-J, Woodward J-F, Yougrau W (1974) Astronomische Nachrichten 295:203

Cremieux V (1906) Récherches sur la gravitation, Comptes Rendus Ac Sc 143:887–889

Dobrokhotov YuS, Pariisky NN, Lysenko VL (1961) Observation of tidal variation of gravity in Kiev during the solar eclipse of February 15 1961, IV symposium Intern. sur les Marées Terrestres, Comm. Observ. Royal de Belgique, Série Géophysique, 58, Bruxelles, 66

Duif PC (2004) A review of conventional explanations of anonalous observations during solar eclipses. http://arxiv.org/ftp/gr-qc/papers/0408/0408023.pdf

Eckhardt D-H (1920) Gravitational shielding. Phys Rev D 42(6):2144–2145

Edwards MR (2002a) Le Sage Theory of Gravity: the revival by Kelvin and Some Later Developments. In: Edwards MR (ed) Pushing Gravity. Apeiron Printer, Montreal, pp 65–78

Edwards MR (2002b) Pushing gravity. Apeiron Printer, Montreal

Erisman V (1908)Vierteljarisr. d. Naturforsh. Gesellsch. Zurich, 53, 157

Harrison J-C (1963) A note on the paper "Earth-tide observations made during the International Geophysical Year". J Geophys Res 68(5):1517–1519

Kleiner A (1905) Sur la gravitation. Archiv de Sc Phys et Nat Bibl Univ IV(20):420–423

Lager M (1904) Sur la gravitation. Dissertation, Université de Zurik

Landolt H, Börnstein R (1952) Zahlenwerte und Funktionen aus Physik, Chemie, Astronomie, Geophysik und Technik, III, Astronomie und Geophysik, Berlin-Gottingen- Heidelberg, 125

Launteren F (2002) Nicolas Fatio de Duiller on the Mechanoical Cause of Universal Gravitation. In: Edwards MR (ed) Pushing gravity. Apeiron Printer, Montreal, pp 41–60

Le Sage GL. Lucrèce Newtonien, Memoire de l'Academie Royale des Sciences et Belles Lettres de Berlin 1–28, 1784

Majorana Q (1919) Sulla gravitazione. Accad. Nazionale Lincei, Rendiconti Classe Scienze Fis. Mat. Naturali 2, Papers: I, 165–174; II, 221–223; III, 313–317; IV, 416–427; V, 480–489

Majorana Q (1957a) Ipotetiche conseguenze dell'assorbimento della gravitazione. Rend Accad Naz Lincei Cl Sc Fis Mat Nat 1:397–402

Majorana Q (1957b) Sull'ipotesi dell'assorbimento della gravitazione. Rend Accad Naz Lincei Cl Sc Fis Mat Nat 1:392–397

Marussi A (1960a) The University of Trieste station for the study of the tides of the vertical in the Grotta Gigante, III Intern. Sump. On Earth Tides, Ist. Topografia e Geodesia dell'Università di Trieste, Trieste, 40, 45

Marussi A (1960b) I primi risultati ottenuti nella stazione per lo studio delle maree della verticale della Grotta Gigante. Bollettino di Geodesia e Scienze Affini 19(4):645–667

Polli S (1958) Le sesse (seiches) dell'Adriatico. Annali di Geofisica XI(1):69–76

Polli S (1961) Sui periodi della oscillazioni dell'Adriatico, Atti XI Convegno Ass. Geofisica Italiana, Roma

Russel HN (1921) On Majorana's theory of gravitation. Astrophys J 54:334–340

Sigl R, Eberhard O (1961) Horizontalpendel Beobachtungen in Berchstesgaden während der Sonnenfinsterniss vom 15 February 1961, IV symposium Intern. sur les Marées Terrestres, Comm. Observ. Royal de Belgique, Série Géophysique, 58, Bruxelles 70

Slichter LB, Caputo M, Hager CL (1965) An Experiment concerning gravitational shielding. J. Geophys Res 70(6): 1541–1551

Tomaschek R (1937) Schwerkraftmessungen. Naturwissenschften 25:17–185, 7

Tomaschek R (1955) Schwerkraftmessungen. Naturwissenschften, 25:17–185, 7, Tidal gravity measurements in the

Shetlands; effect of the total eclipse of June 1954. Nature 175:937–942

Unnikrishnan CS, Mohapatra AK, Gillies GT (2002) Anomalous gravity data during the 1997 total solar eclipse do not support gravitational shielding. Phys Rev D 63:062002.

Venedikov A (1961) Premiers enregistrements des marées terrestres à Sofia, IV symposium Intern. sur les Marées Terrestres, Comm. Observ. Royal de Belgique, Série Géophysique, 58, Bruxelles 70

Wang Q-S, Yang XS, Wu CZ, Liu HC, Hua CC (2000) Precise measurement of gravity variation during a total solar eclipse. Phys Rev D 62:041101(R)

Williams JG, Dicke RH, Bender PL, Alley CO, Carter WE, Currie DG, Eckardt DH, Faller JE, Kaula WM, Mulholland JD, Plotkin HH, Poultney SK, Shelus PJ, Silverberg EG, Sinclair WS, Slade MA, Wilkinson DT (1976) New test of the equivalence principle from lunar laser ranging. Phys Rev Lett 36:551–556

Marussi and the First Formulation of Physical Geodesy as a Fixed-Boundary-Value Problem

4

F. Sansò and F. Sacerdote

Abstract

The famous work of A. Marussi on the geometry of the Earth's gravity field is known as intrinsic geodesy. This was aiming at describing all the relevant geodetic quantities in terms of the so called intrinsic coordinates (Λ, Φ, W) and of their reciprocal relation.

This has been done in a masterly way including all the interesting variables related to the curvatures of equipotential surfaces, which were in a sense the focus of this attention.

However, hidden in Marussi's equations, is the first formulation of the geodetic boundary-value problem in terms of a fixed-boundary problem. This requires a proper understanding of the nature of such equations, as explained in the paper.

Keywords

Intrinsic geodesy · Geodetic boundary-value problems · Gravity field geometry

1 Introduction

Antonio Marussi was mainly interested in the geometric structure of the gravity field. In his early pioneering work he introduced "intrinsic" coordinates (i.e. based uniquely on observable quantities, the astronomic longitude and latitude) on the level surfaces of the gravity potential; furthermore, using the fact that the gravity vector, i.e. the direction of the normal to level surfaces and the intensity of the variation of the gravity potential in space, is itself an observable, he set up an instrument to move from individual level surfaces into three-dimensional space, introducing as third coordinate the potential itself (Marussi 1951; Marussi 1985). Yet, no mention was made, at that time, of the dynamics of the gravity field, described by Poisson and Laplace equations. Only in a much later paper (Marussi 1975) it was introduced in the form of a first-order differential equation for the gravity modulus, together with a system of differential equations describing the geometry of the field, involving the gravity modulus too.

A first attempt to formulate a boundary-value problem for the gravity potential in the framework of intrinsic coordinates was made by Sansò (1981), who obtained an oblique-derivative problem for a suitably defined auxiliary potential.

F. Sansò (✉)
DIIAR, Politecnico di Milano, Polo Regionale di Como, via Valleggio 11, I-22100 Como, Italy
e-mail: fernando.sanso@polimi.it

F. Sacerdote
DICEA, Università di Firenze, Via di S.Marta 3, I-50139 Firenze, Italy
e-mail: fausto@dicea.unifi.it

N. Sneeuw et al. (eds.), *VII Hotine-Marussi Symposium on Mathematical Geodesy*, International Association of Geodesy Symposia 137, DOI 10.1007/978-3-642-22078-4_4, © Springer-Verlag Berlin Heidelberg 2012

In the present paper a new formulation of the boundary-value problem is given, with a Dirichlet boundary condition, directly deduced from Marussi's equations.

2 Equations for the Gravity Field in Intrinsic Coordinates

The starting point of Marussi's theory is the introduction of a general coordinate system $\{x^i\}$. Correspondingly a triad of tangent vectors along coordinate lines, $\underline{\nu}_i = \frac{\partial P}{\partial x^i}, i = 1, 2, 3$ is defined. Such vectors are generally neither mutually orthogonal nor normalized. Specifically Marussi introduced the coordinates Φ, Λ, W, where Φ, Λ are the astronomical latitude and longitude, W is the gravity potential. In order to investigate the spatial variation of the vectors $\underline{\nu}_i$, an orthonormal local triad $\mathbf{i}_k$, $k = 1, 2, 3$, where $\mathbf{i}_1$ points toward north, $\mathbf{i}_2$ toward east, $\mathbf{i}_3$ up, is defined, and the variation of the quantities

$$D = -\underline{\nu}_1 \cdot \mathbf{i}_1,\ D' = -\frac{1}{2}(\underline{\nu}_1 \cdot \mathbf{i}_2 \cos\Phi + \underline{\nu}_2 \cdot \mathbf{i}_1),$$
$$D'' = -\underline{\nu}_2 \cdot \mathbf{i}_2 \cos\Phi \tag{4.1}$$

is investigated. Note that these quantities are essentially related to the components of the coordinate vectors on the equipotential surfaces. Using the integrability conditions $\frac{\partial \underline{\nu}_i}{\partial x^j} = \frac{\partial \underline{\nu}_j}{\partial x^i}$, the equations

$$\frac{\partial D}{\partial W} - \left(1 + \frac{\partial^2}{\partial \Phi^2}\right)\frac{1}{g} = 0$$
$$\frac{\partial D'}{\partial W} - \left(\tan\Phi\frac{\partial}{\partial \lambda} + \frac{\partial^2}{\partial\Phi\partial\Lambda}\right)\frac{1}{g} = 0 \tag{4.2}$$
$$\frac{\partial D''}{\partial W} - \left(\cos^2\Phi - \sin\Phi\cos\Phi\frac{\partial}{\partial\Phi} + \frac{\partial^2}{\partial\Lambda^2}\right)\frac{1}{g} = 0$$

(g = gravity modulus)
(together with others that are not reported here, as they will not be used in the sequel) are obtained. These equations describe the geometric structure of the equipotential surfaces and their variations in terms of the reciprocal of the gravity modulus and its derivatives along the equipotential surfaces themselves.

The dynamic equation is derived from Poisson equation $\nabla^2 W = 2\omega^2 - 4\pi G\rho$ (where ρ is the mass density and ω is the angular velocity), using the intrinsic form of the Laplacian:

$$\nabla^2 W = \frac{1}{\sqrt{\det\{g^{ij}\}}}\frac{\partial}{\partial x^i}\left(\sqrt{\det\{g^{ij}\}}g^{ij}W|_j\right) \tag{4.3}$$

(g^{ij} = metric tensor)

Its expression is

$$-g\frac{\partial}{\partial W}\frac{1}{g} = \frac{2\omega^2 - 4\pi G\rho}{g^2} + \frac{H}{g}$$
$$-\frac{g}{\Delta}\left[D''\left(\frac{\partial}{\partial\Phi}\frac{1}{g}\right)^2\right.$$
$$-2D'\left(\frac{\partial}{\partial\Phi}\frac{1}{g}\right)\left(\frac{\partial}{\partial\Lambda}\frac{1}{g}\right)$$
$$\left.+D\left(\frac{\partial}{\partial\Lambda}\frac{1}{g}\right)^2\right] \tag{4.4}$$

where $\Delta = DD'' - D'^2$, $H = -\frac{D''+D\cos^2\Phi}{\Delta}$. It is remarkable that this equation too involves the geometric quantities D,D', D'' and the inverse of the gravity modulus.

3 Formulation and Linearization of the Boundary-value Problem

In order to formulate a boundary-value problem it is convenient to reduce this system of equations to one single equation in the unknown $v = \frac{1}{g}$. This result can be obtained in a simple way in a perturbative approach, carrying out a linearization with the spherical solution as starting point.

The dynamic equation is considered outside masses ($\rho = 0$) and without centrifugal term ($\omega = 0$).

From the expression of the spherical solution: $W_0 = \frac{\mu}{r}$; $g_0 = \frac{\mu}{r^2} = \frac{W^2}{\mu}$ one obtains

$$v_0 = \frac{\mu}{W^2} \tag{4.5}$$

Furthermore, it follows from (4.1) or (4.2) that

$$D_0 = -\frac{\mu}{W};\quad D_0' = 0;\quad D_0'' = -\frac{\mu\cos^2\Phi}{W}$$
$$\Rightarrow H_0 = \frac{2W}{\mu} \tag{4.6}$$

Now equation (4.4) is linearized with respect to the increments of the quantities

$$\begin{aligned} v &= \frac{\mu}{W^2} + \delta v \\ D &= -\frac{\mu}{W} + \delta D \\ D' &= \delta D' \\ D'' &= -\frac{\mu \cos^2 \Phi}{W} + \delta D'' \end{aligned} \tag{4.7}$$

From $W = \frac{\mu}{r} + 0(r^{-3})$, $g = \frac{\mu}{r^2} + 0(r^{-4})$ one obtains

$$\begin{aligned} g &= \frac{W^2}{\mu} + 0(W^4) = \frac{W^2}{\mu}(1 + 0(W^2)) \\ \Rightarrow v &= \frac{\mu}{W^2}(1 + 0(W^2)) = \frac{\mu}{W^2} + 0(1) \end{aligned} \tag{4.8}$$

Consequently δv turns out to be a bounded quantity.

Furthermore, the last term in the dynamic equation (4.4),

$$\frac{1}{\Delta}\left[D''\left(\frac{\partial}{\partial \Phi}\frac{1}{g}\right)^2 - 2D'\left(\frac{\partial}{\partial \Phi}\frac{1}{g}\right)\left(\frac{\partial}{\partial \Lambda}\frac{1}{g}\right) + D\left(\frac{\partial}{\partial \Lambda}\frac{1}{g}\right)^2\right]$$

is of second order.

Consequently, taking only first order terms, (4.4) is reduced to

$$\begin{aligned} &-\frac{\partial v}{\partial W} = Hv^2, \quad \text{i.e.} \\ &-\frac{\partial}{\partial W}(v_0 + \delta v) = H_0 v_0^2 + \delta H \cdot v_0^2 + 2H_0 v_0 \delta v \end{aligned} \tag{4.9}$$

where it can be easily seen that the expression for δH is

$$\delta H = \frac{W^2}{\mu^2 \cos^2 \Phi}(\delta D'' + \cos^2 \Phi \delta D) \tag{4.10}$$

Consequently

$$\begin{aligned} -\frac{\partial}{\partial W}\delta v &= \frac{W^2}{\mu^2 \cos^2 \Phi}(\delta D'' + \cos^2 \Phi \delta D)\frac{\mu^2}{W^4} \\ &+ 2\frac{2W}{\mu}\frac{\mu}{W^2}\delta v \end{aligned}$$

that finally leads to

$$W^2 \frac{\partial}{\partial W}\delta v + 4W\delta v = -\left(\frac{\delta D''}{\cos^2 \Phi} + \delta D\right) \tag{4.11}$$

Applying a further derivation with respect to W one obtains

$$\begin{aligned} &\frac{\partial}{\partial W}\left(W^2 \frac{\partial}{\partial W}\delta v\right) + 4W\frac{\partial}{\partial W}\delta v + 6\delta v - \tan \Phi \\ &\times \frac{\partial}{\partial \Phi}\delta v + \frac{\partial^2}{\partial \Phi^2}\delta v + \frac{1}{\cos^2 \Phi}\frac{\partial^2}{\partial \Lambda^2}\delta v = 0 \end{aligned} \tag{4.12}$$

This is a single equation in the unknown δv, which is assumed to be known on the boundary $W = W(\Phi, \Lambda)$ if the starting point is the vector Molodensky problem ($\mathbf{g}$ and W known on the unknown Earth's surface), so that this formulation leads to a Dirichlet problem.

4 Expansion into Spherical Harmonics in Cartesian-Marussi Coordinates

Equation (4.11) can be easily treated introducing the so-called Cartesian Marussi coordinates ξ_j:

$$\begin{aligned} \xi_1 &= W \cos \Phi \cos \Lambda \\ \xi_2 &= W \cos \Phi \sin \Lambda \\ \xi_3 &= W \sin \Phi \end{aligned} \tag{4.13}$$

indeed, it can be simply written as

$$W^2 \nabla_\xi^2 \delta v + 4W\frac{\partial}{\partial W}\delta v + 6\delta v = 0. \tag{4.14}$$

Assuming a solution of the form

$$\delta v = \sum_{n,m} \delta v_{nm} W^{\lambda_n} Y_{nm}(\Phi, \Lambda), \tag{4.15}$$

it leads to the characteristic equation

$$\lambda_n(\lambda_n + 1) - n(n+1) + 4\lambda_n + 6 = 0$$

whose solution is $\lambda_n = \left\langle \begin{matrix} -2 + n \\ -3 - n \end{matrix} \right.$.

Only the upper solution can be accepted, as δv must be bounded at the origin.

The first term, with $n = 0$, has the form $\frac{\delta v_0}{W^2}$, and represents a small correction to the coefficient of the spherical solution used as starting point (corresponding to a mass variation).

The second term, with $n = 1$, has the form $\sum_m \delta v_{1m} \frac{1}{W} Y_{1m}(\Phi, \Lambda)$, and corresponds to the variation of the spherical solution $v = \frac{r^2}{\mu}$ due to a displacement of the center of mass.

Indeed, in spherical approximation,

$$\delta \frac{1}{g} = -\frac{1}{g^2} \frac{\mathbf{g}}{g} \cdot \mathbf{W} \delta \mathbf{r} = \frac{r^4}{\mu^2} \frac{\mathbf{r}}{r} \cdot \left(-\frac{\mu}{r^3} (\mathbf{I} - 3\mathbf{P}_r) \right) \delta \mathbf{r}$$
$$= -2 \frac{r}{\mu} \frac{\mathbf{r}}{r} \cdot \delta \mathbf{r},$$

where $\mathbf{W}$ is the matrix of the second derivatives of W. The components of $(\mathbf{r}/r)$ are exactly proportional to the first order harmonics.

The first two terms can be assumed to vanish if the mass and the center of mass of the perturbed solution are coincident with those of the approximate solution.

5 Determination of the Boundary Surface

The position of the boundary points in the ordinary space can be recovered by means of the introduction of an auxiliary potential, the same already used by Sansò (1981): $\varphi = x_i \xi_i$.

Taking into account that

$$g_i = \frac{\partial W}{\partial x_i} = \frac{\partial W}{\partial \xi_k} \frac{\partial \xi_k}{\partial x_i} \Rightarrow \frac{\partial x_i}{\partial \xi_k} g_i = \frac{\partial W}{\partial \xi_k} = \frac{\xi_k}{W},$$

and that, consequently

$$g_i = g \frac{\xi_i}{W} \Rightarrow \frac{\partial x_i}{\partial \xi_k} \xi_i = \frac{\xi_k}{g},$$

it is possible to obtain

$$\frac{\partial \varphi}{\partial \xi_j} = x_j + \xi_i \frac{\partial x_i}{\partial \xi_j} = x_j + \frac{\xi_j}{g}$$

from which, finally

$$x_j = \frac{\partial \varphi}{\partial \xi_j} - \frac{\xi_j}{g} \tag{4.16}$$

In order to recover the auxiliary potential φ, one can start from (4.16), expressed in vector form, $\frac{\underline{\xi}}{g} = \nabla_\xi \varphi - \mathbf{x}$ and take its scalar product by $\underline{\xi}$, obtaining $\frac{W^2}{g} = W \frac{\partial \varphi}{\partial W} - \varphi$. Consequently

$$v = \frac{1}{g} = \frac{1}{W^2} \left(W \frac{\partial \varphi}{\partial W} - \varphi \right) = \frac{\partial}{\partial W} \left(\frac{\varphi}{W} \right) \tag{4.17}$$

Subtracting the spherical solution $v_0 = \frac{\mu}{W^2} \Rightarrow \varphi_0 = -\mu$ one obtains simply $\delta v = \frac{\partial}{\partial W} (\frac{\delta \varphi}{W})$.

The required result can be obtained by integration:

$$\delta \varphi = W \int_0^W \delta v(\tau, \sigma) d\tau (\sigma = (\Phi, \Lambda)) \tag{4.18}$$

6 Concluding Remarks

The present short note shows that some kind of formulation of the basic boundary-value problem of physical geodesy, with Dirichlet boundary condition, can be obtained using intrinsic coordinates, starting from the developments introduced by A. Marussi in order to investigate the geometric properties of the gravity field. Obviously this formulation is exceedingly simplified, as it uses as starting point for linearization the spherical solution. A further step might be carried out starting from an approximate solution with ellipsoidal symmetry, whose geometric properties were investigated by A. Marussi himself in his 1975 lectures. These developments throw in some sense a bridge between the studies on geometric properties, in which A. Marussi obtained his most relevant scientific results, according to the tradition of the Italian mathematical school in differential geometry, and the more recent investigations on physical geodesy, which have assumed a prominent role in last decades. In addition, one has to be aware that, in order to get sensible results, one has to exclude the rotational potential from W. As a matter of fact, this remark leads back to the argument that the actual gravity potential cannot be a one-to-one overall coordinate through the outer space if the centrifugal potential is left in it.

References

Marussi A (1949) Fondements de géométrie différentielle absolue du champ potentiel terrestre. Bull Géod 14:411–439

Marussi A (1975) Elements of tensor calculus, Part I and II (Geodetic applications). In: Brosowski B, Martensen E (eds) Methoden und Verfahren der Mathematischen Physik/Band 13. Bibliographisches Institut Mannheim

Marussi A (1985) Intrinsic geodesy. Springer-Verlag, Berlin

Sansò F (1981) The geodetic boundary-value problem and the coordinate choice problem. Bull Géod 55: 17–30

Part II

Geodetic Sensor Systems and Sensor networks

The Future of Single-Frequency Integer Ambiguity Resolution

5

Sandra Verhagen, Peter J.G. Teunissen, and Dennis Odijk

Abstract

The coming decade will bring a proliferation of Global Navigation Satellite Systems (GNSSs) that are likely to enable a much wider range of demanding applications compared to the current GPS-only situation. One such important area of application is single-frequency real-time kinematic (RTK) positioning. Presently, however, such systems lack real-time performance. In this contribution we analyze the ambiguity resolution performance of the single-frequency RTK model for different next generation GNSS configurations and positioning scenarios. For this purpose, a closed form expression of the single-frequency Ambiguity Dilution of Precision (ADOP) is derived. This form gives a clear insight into how and to what extent the various factors of the underlying model contribute to the overall performance. Analytical and simulation results will be presented for different measurement scenarios. The results indicate that low-cost, single-frequency Galileo+GPS RTK will become a serious competitor to its more expensive dual-frequency cousin.

Keywords

ADOP • Ambiguity resolution • Single-frequency RTK

1 Introduction

Global Navigation Satellite System (GNSS) ambiguity resolution (AR) is the process of resolving the unknown cycle ambiguities of the carrier phase data as integers. It is the key to high-precision GNSS parameter estimation. In order for AR to be successful, the probability of correct integer estimation needs to be sufficiently close to one. Whether or not this is the case depends on the strength of the underlying GNSS model and therefore on the number and type of signals observed, the number of satellites tracked, the relative receiver-satellite geometry, the length

S. Verhagen (✉)
Delft Institute of Earth Observation and Space Systems, Delft University of Technology, Delft, The Netherlands
e-mail: A.A.Verhagen@TUDelft.nl

P.J.G. Teunissen
Department of Spatial Sciences, Curtin University of Technology, Perth, Australia

Delft Institute of Earth Observation and Space Systems, Delft University of Technology, Delft, The Netherlands

D. Odijk
Department of Spatial Sciences, Curtin University of Technology, Perth, Australia

N. Sneeuw et al. (eds.), *VII Hotine-Marussi Symposium on Mathematical Geodesy*, International Association of Geodesy Symposia 137, DOI 10.1007/978-3-642-22078-4_5, © Springer-Verlag Berlin Heidelberg 2012

of the observational time window, the measurement precision, the dynamics of the positioning application and the need of having to include additional parameters like troposphere and/or ionosphere delays.

The coming decade will bring a proliferation of GNSSs (modernized GPS, Glonass, Galileo, Compass) that are likely to enable a much wider range of demanding applications compared to the current GPS-only situation due to the availability of many more satellites and signals. This contribution considers the application area of single-frequency real-time kinematic (RTK) positioning. Presently, low-cost single-frequency RTK systems lack real-time performance due to the weaknesses of the single-frequency GPS-only model, see e.g. Milbert (2005); Odijk et al. (2007); Takasu and Yasuda (2008). If low-cost single-frequency RTK would become feasible, a whole range of exciting applications awaits in e.g. the fast-evolving field of mobile Location Based Services, precision agriculture, surveying and mapping, e.g. Wirola et al. (2006); Denham et al. (2006); Saeki and Hori (2006); Millner et al. (2005).

In this contribution we analyze the ambiguity resolution performance of the single-frequency RTK model for different next generation GNSS configurations and for different positioning scenarios. For this purpose, first a closed form expression of the single-frequency Ambiguity Dilution of Precision (ADOP) is derived in Sect. 2. A performance analysis based on the ADOPs as well as empirical success rates is presented in Sect. 3. These results allow us to identify the circumstances that make successful single-frequency AR possible, as will be shown in the final Sect. 4.

2 Ambiguity Resolution

The key to rapid and high-precision GNSS positioning is the use of carrier-phase observations, which have mm-level precision while code observations only have a precision at the dm-level. In order to exploit the very precise carrier-phase measurements, first the unknown integer number of cycles of the observed carrier phase has to be resolved. The linearized double-difference GNSS model can be written as:

$$y = Bb + Aa + e, \quad b \in \mathbb{R}^{\nu}, \ a \in \mathbb{Z}^{n} \tag{5.1}$$

where y is the vector with double-differenced code and phase observables; b is the ν-vector with unknown real-valued parameters, such as the baseline increments, ionosphere and troposphere parameters; a is the n-vector with the unknown integer ambiguities; e is the noise vector. The matrices B and A link the unknown parameters to the observables. It is generally assumed that y follows the normal distribution, with zero-mean noise and the associated variance matrix Q_{yy} capturing the measurement precision.

Solving model (5.1) in a least-squares sense provides the so-called float solution, where the integer constraint on the carrier-phase ambiguities, i.e. $a \in \mathbb{Z}^n$, is not considered. This is done in a second step, the ambiguity resolution (AR) step, based on the float ambiguities $\hat{a}$ and associated variance matrix $Q_{\hat{a}\hat{a}}$. The integer least-squares (ILS) estimator is proven to be optimal in the sense that it maximizes the probability of correct integer estimation, Teunissen (1999). A well-known and efficient implementation of the ILS-principle is the LAMBDA method, Teunissen (1995). After resolving the integer ambiguities $\check{a}$, the final step is to adjust the float solution of b conditioned on the fixed integer solution. This provides the fixed baseline solution $\check{b}$.

Correct integer estimation is essential to guarantee that $\check{b}$ will have cm-level precision. Hence, the probability of correct integer estimation, called success rate, is a valuable measure to assess the positioning performance. Unfortunately, no analytical expression is available to compute the ILS success rate exactly. Several approximations were proposed in the past, see Verhagen (2005). In this contribution empirical success rates based on Monte Carlo simulations will be used.

In Teunissen (1997) the Ambiguity Dilution of Precision (ADOP) was introduced as an AR performance measure. It is defined as:

$$\text{ADOP} = \sqrt{|Q_{\hat{a}\hat{a}}|^{\frac{1}{n}}} \tag{5.2}$$

The ADOP measure has the unit of cycles, and it is invariant to the decorrelating Z-transformation of the LAMBDA method. It is equal to the geometric mean of the standard deviations of the ambiguities if these would be completely decorrelated. Hence, the ADOP approximates the average precision of the transformed

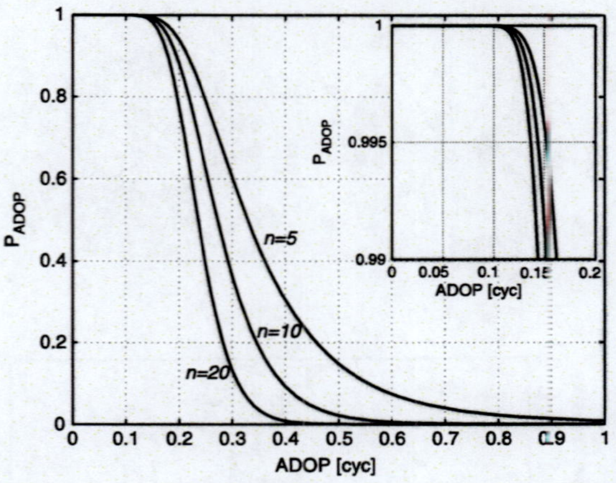

Fig. 5.1 P_{ADOP} as function of ADOP

ambiguities. The ADOP can also be used to get an approximation of the ILS success rate:

$$P(\check{a}=a) \approx P_{\mathrm{ADOP}} = \left(2\Phi(\frac{1}{2\mathrm{ADOP}})-1\right)^n \quad (5.3)$$

Figure 5.1 shows the relation between ADOP and P_{ADOP} for different values of n. From this figure it can be concluded that for successful ambiguity resolution the ADOP should be smaller than 0.15 cycles.

It is possible to derive closed-form expressions for ADOP. In Odijk and Teunissen (2008) this was done for a hierarchy of multi-frequency single-baseline GNSS models. The closed-form expressions give a clear insight into how and to what extent the various factors of the underlying GNSS model contribute to the overall AR performance, see Odijk and Teunissen (2007). The closed-form expression for the ADOP of the single-frequency model corresponding to a moving receiver covering a short time span (no change in satellite geometry) can be derived as (see table 8 in Odijk and Teunissen (2008), use $j=1$):

$$\mathrm{ADOP} = \left[\frac{\sigma_\phi}{\lambda}\right][2sf]^{\frac{1}{2}}\left[\frac{\sum_{s=1}^m w_s}{\prod_{s=1}^m w_s}\right]^{\frac{1}{2(m-1)}} \times \left[1+\eta\cdot\frac{\kappa}{1+\kappa}\right]^{\frac{1}{2}}\left[1+\eta\cdot\frac{(2\kappa+1)^2}{1+\kappa(1+\eta)}\right]^{\frac{\nu}{2(m-1)}} \quad (5.4)$$

with:

σ_ϕ undifferenced phase standard deviation [m]
σ_p undifferenced code standard deviation [m]
σ_ι undifferenced standard deviation of ionosphere observables [m]
λ carrier wavelength [m]
sf variance scale factor
m number of satellites
w_s elevation dependent weights, $s=1,\ldots,m$
η $=\frac{\sigma_p^2}{\sigma_\phi^2}$ and $\kappa=\frac{\sigma_\iota^2}{\sigma_p^2}$

The ionosphere-weighted model, see e.g. Odijk (2002), is used where a priori information on the ionosphere delays is used in the form of ionosphere observables with standard deviation σ_ι depending on the baseline length. If the baseline is sufficiently short, the double difference ionosphere observables will become zero, and σ_ι is set to zero.

In (5.4) sf is a scale factor, if $sf<1$ this can be either due to enhanced measurement precision, or due to an increased number of epochs k. In the first case it is assumed that the variance of code and phase observations is improved with the same factor sf. In the second case the scale factor would be equal to:

$$sf = \frac{1+\beta}{k-(k-2)\beta} \quad (5.5)$$

where $\beta (0 \leq \beta < 1)$ describes the correlation parameter of a first-order autoregressive time process. Hence, $\beta=0$ means that time correlation is absent and

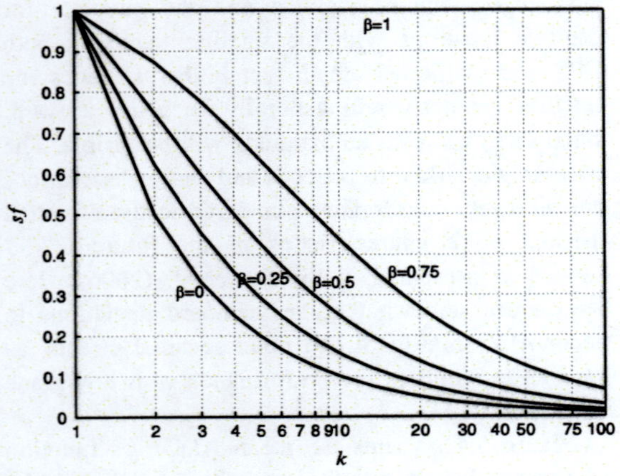

Fig. 5.2 Relation between the variance scale factor sf and number of epochs k for various time correlations β

$sf = \frac{1}{k}$, while $\beta = 1$ would mean that the observations are fully correlated between the epochs and $sf = 1$. Figure 5.2 shows the relation between the variance scale factor sf and the number of epochs k for various time correlations β.

3 Performance Analysis

An analysis of the ambiguity resolution performance is made based on the following assumptions:

$$\begin{aligned}
\sigma_\phi &= 2\,\text{mm}, \sigma_p = 20\,\text{cm} \\
\sigma_\iota &= 0, 4, 8\,\text{mm} \\
\lambda &= 25.48\,\text{cm (L5 frequency)} \\
w_s &= (1 + 10\exp(-e_s/10))^{-\frac{1}{2}} \\
\nu &= 3 \text{ (no troposphere parameters estimated)}
\end{aligned}$$

with e_s the elevation of satellite s in degrees. A mask angle of 10° is used.

The three values of σ_ι are assumed to correspond to baseline lengths of <5, 10 and 20 km, respectively.

The future Galileo constellation is considered, as well as the combined GPS+Galileo constelation, where for GPS the nominal constellation of 24 satellites is used. A time span equal to the repeat orbit period of Galileo, approximately 10 days, is considered. Two different geographical locations are considered, both at longitude 3°E and latitudes 45°N and 75°N, respectively. The mid-latitude location is selected because on average the least number of satellites are visible while at the higher latitude of 75°N the opposite is true. Figure 5.3 shows the number of visible satellites and the skyplots for the two locations with the satellite tracks of both GPS and Galileo. Note that at higher latitudes the satellite geometry will generally be better as well, since satellites from all azimuths will be visible. The standard deviations of the code and phase observations are relatively conservative compared to the expected thermal noise characteristics of the future GNSS signals as presented in Simsky et al. (2006). Here we choose somewhat higher standard deviations to account for multipath and other residual effects, as well as to simulate the performance with low-grade receivers.

Figure 5.4 presents the mean ADOP as function of the number of satellites m with $sf = 1$ (i.e. the mean for each m is calculated over all instances that m satellites are visible during the 10-day period). The

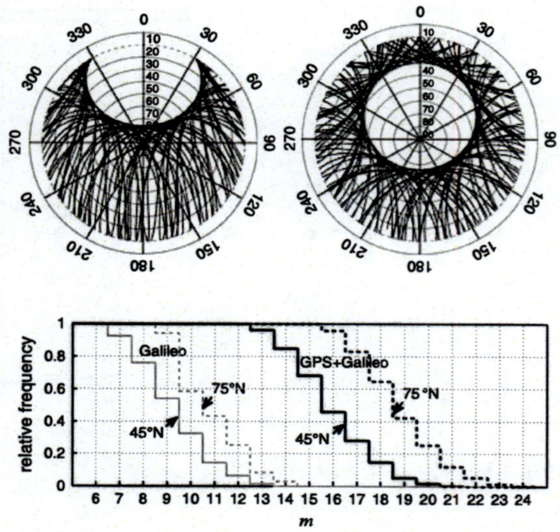

Fig. 5.3 *Top*: Skyplots (left for 45° latitude, right for 75° latitude) for one day with GPS and Galileo satellite tracks. *Bottom*: Relative frequencies that more than m satellites are visible for 10-day period

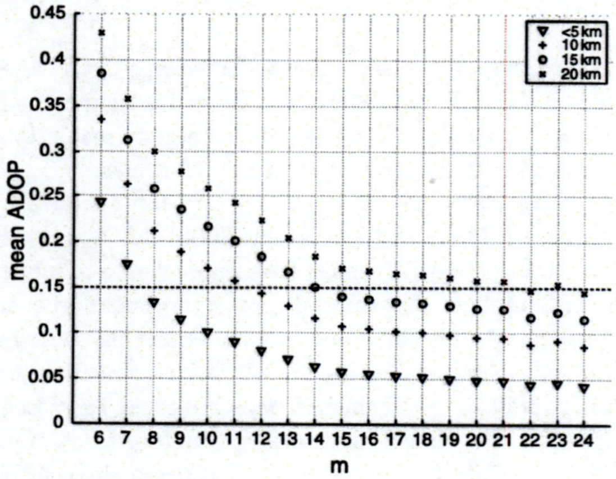

Fig. 5.4 Single-epoch, mean ADOPs [cycles] as function of number of satellites

average values of the two locations are shown, since it turned out that the impact of the satellite geometry on the ADOP – third term in (5.4) – is averaged out and thus the results are nearly identical for the two different locations.

From Fig. 5.1 it was concluded that an ADOP of 0.15 cycles was required for successful ambiguity resolution. Using this rule-of-thumb, it follows from Fig. 5.4 that 8 or more satellites are required with very

Table 5.1 Scale factor sf needed to obtain a success rate above 0.99 more than 99% of the time. The number between brackets is the corresponding number of epochs if $\beta = 0$

	Galileo		GPS+Galileo	
Baseline	45°N	75°N	45°N	75°N
<5 km	0.07 (*15*)	1 (*1*)	1 (*1*)	1 (*1*)
10 km	0.02 (*60*)	0.11 (*9*)	1 (*1*)	1 (*1*)
15 km	0.01 (*70*)	0.06 (*16*)	0.2 (*5*)	1 (*1*)

short baselines, more than 11 satellites with baselines of 10 km, and more than 14 satellites with baselines of 15 km. With longer baselines, single-epoch ambiguity resolution is generally not feasible. From Figs. 5.3 and 5.4 combined, it follows then that with very short baselines (<5 km) single-epoch, single-frequency RTK is possible with Galileo-only most of the time. However, for baselines up to 15 km this is only possible with GPS+Galileo.

Next, the AR performance is analyzed based on empirical success rates using Monte Carlo simulations, see e.g. Verhagen (2005). Table 5.1 presents the scale factor needed to obtain a success rate above 0.99 more than 99% of the time. The corresponding number of epochs if $\beta = 0$ is derived from Fig. 5.2, from which also follows that in the presence of time correlation more epochs are needed.

For baselines of 20 km and longer, single-frequency RTK is not feasible for large periods of time, and therefore the corresponding results are not shown in Table 5.1. Without time correlation and with 100 epochs of data, a success rate above 0.99 can be obtained during less than 75% of the time. However, for baselines shorter than 10 km instantaneous ambiguity resolution is possible with GPS+Galileo. At mid-latitudes the time to fix the ambiguities will often be longer with a baseline of 15 km, but is still rather short. With Galileo-only the time to fix depends very much on the satellite geometry and thus the location on Earth, but generally the time to fix will be more than 10 epochs with short baselines, and more than 50 epochs with baselines longer than 10 km.

4 Concluding Remarks

Single frequency RTK with the current GPS or future Galileo alone is only feasible with very short baselines (<5 km), and even then at some locations instantaneous ambiguity resolution will only be feasible for 65% of the time. At mid-latitudes more than 15 epochs of data are needed to guarantee a success rate above 0.99.

A dual-constellation GNSS will enhance the ambiguity resolution performance of single frequency RTK dramatically. Instantaneous success rates above 0.99 are obtained with baselines up to 15 km.

Acknowledgements The research of Sandra Verhagen is supported by the Dutch Technology Foundation STW, applied science division of NWO and the Technology Program of the Ministry of Economic Affairs.

Professor Teunissen is the recipient of an Australian Research Council Federation Fellowship (project number FF0883188). This support is greatly acknowledged.

References

Denham JL, Whitlock AK, Yule D (2006) Contemporary position and navigation needs of precision agriculture and victoria's GPSnet CORS network. In: Proceedings of IGNSS symposium, Holiday Inn Surfers Paradise, Australia

Milbert D (2005) Influence of pseudorange accuracy on phase ambiguity resolution in various GPS modernization scenarios. Navigation 52(1):29–38

Millner J, Hale M, Standen P, Talbot N (2004) The development and enhancement of a GNSS/GPS infrastructure to support location based service positioning systems in Victoria. In: Proceedings of international symposium on GNSS/GPS, Sydney, Australia

Odijk D (2002) Fast precise GPS positioning in the presence of ionospheric delays. Ph.D. thesis, Publications on Geodesy, 52, Netherlands Geodetic Commission, Delft

Odijk D, Teunissen PJG (2007) Sensititvity of ADOP to changes in the single-baseline GNSS model. Artificial Satellites 42(2):71–96

Odijk D, Teunissen PJG (2008) ADOP in closed form for a hierarchy of multi-frequency single-baseline GNSS models. J Geodes 82(8):473–492

Odijk D, Traugott J, Sachs G, Montenbruck O, Tiberius C (2007) Two approaches to precise kinematic GPS positioning with miniaturized L1 receivers. In: Proceedings of ION GNSS 2007, Fort Worth TX, pp 827–838

Saeki M, Hori M (2006) Positioning system using low-cost L1 GPS receivers. Comput Aided Civ Infrastruct Eng 21(4):258–267

Simsky A, Sleewaegen JM, Hollreiser M, Crisci M (2006) Performance assessment of Galileo ranging signals transmitted by GSTB-V2 satellites. In: Proceedings of ION GNSS 2006, Fort Worth TX

Takasu T, Yasuda A (2008) Evaluation of RTK-GPS performance with low-cost single-frequency GPS receivers. In: Proceedings of international symposium on GPS/GNSS 2008 Tokyo, Japan, pp 852–861

Teunissen PJG (1995) The least-squares ambiguity decorrelation adjustment: a method for fast GPS integer ambiguity estimation. J Geodes 70(1-2):65–82

Teunissen PJG (1997) A canonical theory for short GPS baselines. Part IV: Precision versus reliability. J Geodes 71: 513–525

Teunissen PJG (1999) An optimality property of the integer least-squares estimator. J Geodes 73(11):587–593

Verhagen S (2005) On the reliability of integer ambiguity resolution. Navigation 52(2):99–110

Wirola L, Alanen K, Kappi J, Syrjarinne J (2006) Bringing RTK to cellular terminals using a low-cost single-frequency AGPS receiver and inertial sensors. In: Proceedings IEEE/ION PLANS 2006, San Diego CA, pp 645–652

Integer Ambiguity Resolution with Nonlinear Geometrical Constraints

6

G. Giorgi, P.J.G. Teunissen, S. Verhagen, and P.J. Buist

Abstract

Integer ambiguity resolution is the key to obtain very accurate positioning solutions out of the GNSS observations. The Integer Least Squares (ILS) principle, a derivation of the least-squares principle applied to a linear system of equations in which some of the unknowns are subject to an integer constraint, was demonstrated to be optimal among the class of admissible integer estimators. In this contribution it is shown how to embed into the functional model a set of nonlinear geometrical constraints, which arise when considering a set of antennae mounted on a rigid platform. A method to solve for the new model is presented and tested: it is shown that the strengthened underlying model leads to an improved capacity of fixing the correct integer ambiguities.

Keywords

Constrained methods • GNSS • Integer ambiguity resolution

1 Introduction

The GNSS (Global Navigation Satellite System) observations are obtained tracking a number of satellites: both the code and carrier phase data are used to estimate the antennae positions. Because only the fractional part of the phase carrier observations can be measured, an ambiguity must be resolved for each incoming signal in order to fully exploit the capabilities of the GNSS positioning: by resolving the ambiguities one is able to achieve higher accuracies than using only the code data. The set of GNSS observations is usually cast into a (overdetermined) system of linearized equations, and the theory of Integer Least-Squares (ILS) (Teunissen 1993) is applied to solve for the linearized model in a least-squares sense, with a subset of the unknowns being integer-valued, namely the phase carrier ambiguities. An efficient implementation of the ILS was proposed in Teunissen (1994): the LAMBDA (Least-squares AMBiguity Decorrelation Adjustment) method is currently widely used for its high efficiency. For those applications where a subset of the real-valued unknowns is subject to geometrical constraints, one faces a substantial complication for the solution of the constrained ILS

G. Giorgi (✉) · S. Verhagen · P.J. Buist
Delft Institute of Earth Observation and Space Systems (DEOS), Delft University of Technology, 2600 GB Delft, The Netherlands
e-mail: gabriele.giorgi@tum.de

P.J.G. Teunissen
Delft Institute of Earth Observation and Space Systems (DEOS), Delft University of Technology, 2600 GB Delft, The Netherlands

Department of Spatial Sciences, Curtin University of Technology, Perth WA 6845, Australia

N. Sneeuw et al. (eds.), *VII Hotine-Marussi Symposium on Mathematical Geodesy*, International Association of Geodesy Symposia 137, DOI 10.1007/978-3-642-22078-4_6, © Springer-Verlag Berlin Heidelberg 2012

problem. A modification of the LAMBDA method was recently proposed in Teunissen (2006), Teunissen (2008), Teunissen (2010), Park and Teunissen (2003), Buist (2007), Park and Teunissen (2008), Giorgi et al. (2008) and Giorgi and Buist (2008) to solve for single-baseline constrained problems. We investigate in this contribution how to resolve for the integer ambiguities when a set of two or more antennae are mounted on the same rigid platform, with their relative positions known and constant. The problem was originally addressed in Teunissen (2007): the peculiar set of geometrical constraints posed on the baselines vectors is tackled by introducing a suitable parameterization of the baseline coordinates, and a modified cost function to be minimized in an ILS sense is introduced. It is shown here how to efficiently proceed for the search of the integer minimizer of the modified objective function, and a numerical evaluation of the capabilities of the constrained ILS is given: the single-frequency, single-epoch success rate is investigated.

2 Modeling of the GNSS Observables

Assuming two antennae tracking the same $n+1$ GNSS satellites, the set of single frequency, linearized double difference (DD) GNSS observations for the baseline at a given epoch is described via a Gauss-Markov model (Teunissen and Kleusberg 1998)

$$E(y) = Az + Gb \quad z \in \mathbb{Z}^n; b \in \mathbb{R}^p$$
$$D(y) = Q_y \qquad (6.1)$$

where $E(\cdot)$ is the expectation operator, y is the vector of code and carrier phase observables (order $2n$), z contains the n integer-valued ambiguities and b is the vector of remaining p real-valued unknowns. Here, we restrict ourselves to short baseline applications, assuming the three baseline coordinates as the only real-valued unknowns ($p = 3$). A and G are the design matrices which link the observables with the vectors of unknowns: A contains the carrier wavelengths, while G is the matrix of line-of-sight vectors.

$D(\cdot)$ is the dispersion operator: a Gaussian-distributed error is assumed on the vectors of observables, characterized by the variance-covariance (v-c) matrix Q_y.

We consider in this work a set of $m+1$ antennae tracking the same $n+1$ GNSS satellites: we cast the set of GNSS DD observations collected at the different m independent baselines into a unique frame, thus formulating a multivariate model (Teunissen 2007) as

$$E(Y) = AZ + GB \quad Z \in \mathbb{Z}^{n\times m}; B \in \mathbb{R}^{3\times m}$$
$$D(vec(Y)) = Q_Y \qquad (6.2)$$

where Y is the $2n$ by m matrix whose columns are the code and phase observations from each baseline, Z is the matrix containing the nm integer-valued ambiguities and B is the matrix of remaining $3m$ real-valued unknowns, i.e. the matrix whose columns are the coordinates of each baseline. The relative distances between the antennae are assumed to be short, so that the deviations between the different line-of-sight vectors as seen from each antenna can be disregarded and the same matrix of line-of-sight vectors G is used. The vec operator is here introduced in order to define the v-c matrix of the observables: it stacks the columns of the $2n$ by m matrix Y into a vector of order $2nm$. The dispersion of the vector $vec(Y)$ is characterized by the v-c matrix Q_Y.

We study in this contribution how to embed a set of nonlinear geometrical constraints posed on the $3m$ real-valued entries of B. We assume that the antennae are firmly mounted on the same rigid platform, and their relative distances are completely known. This results in two types of constraints to be considered: the baseline lengths and their relative orientation are known and constant. The hypothesis of constant length constrains the extremity of each baseline vector to lie on the surface of a sphere of radius equal to the baseline length; this reduces the number of independent baseline coordinates from $3m$ to $2m$. Due to the invariance of the antennae relative positions, the set of admissible baseline coordinates is described by a rigid rotation, and the real-valued unknowns to be determined are drastically reduced to three (two in the case of single-baseline) by virtue of the Euler's rotation theorem (Goldstein 1980). A suitable parameterization for the baseline coordinates is necessary to efficiently describe the characteristics of the baseline-constrained problem. To this purpose we introduce a frame of body axis $(u_1 u_2 u_3)$ defined by the antennae placement. The first body axis is aligned with the first baseline, the second body axis is perpendicular to the first, lying in the plane formed by the first two baselines, and the third body axis is directed so that $u_1u_2u_3$ form a right-handed orthogonal frame. The relation between the baseline coordinates expressed in the body frame $u_1u_2u_3$ (F) and a reference frame $x_1x_2x_3$ (B) under the hypothesis of rigid rotations is

$$B = R \cdot F \tag{6.3}$$

where the rotation matrix R, which describes the relative orientation of the two systems, defines a linear transformation $\mathbb{R}^{3\times m} \rightarrow \mathbb{R}^{3\times m}$. Due to the invariance of both the baselines lengths and their relative positions, the relation $B^T B = F^T F$ holds true; multiplying both the terms of (6.3) for B^T, we obtain $B^T B = F^T R^T R F$: hence the matrix R has to be orthogonal ($R^T R = I$). In order to avoid loss of generality when only two or three antennae are available, we define the rotation matrix as (Teunissen 2007)

$$m \geq 3: \quad RF = \begin{bmatrix} r_1, r_2, r_3 \end{bmatrix} \begin{bmatrix} f_{11} & f_{21} & f_{31} & \cdots & f_{m1} \\ 0 & f_{22} & f_{32} & \cdots & f_{m2} \\ 0 & 0 & f_{33} & \cdots & f_{m3} \end{bmatrix}$$

$$m = 2: \quad RF = \begin{bmatrix} r_1, r_2 \end{bmatrix} \begin{bmatrix} f_{11} & f_{21} \\ 0 & f_{22} \end{bmatrix}$$

$$m = 1: \quad RF = \begin{bmatrix} r_1 \end{bmatrix} \begin{bmatrix} f_{11} \end{bmatrix} \tag{6.4}$$

with r_i the i-th column of R and f_{ij} (scalar) the entries of F. We introduce for notational convenience the parameter q, to indicate the second dimension of R: $q = m$ for $m < 3$ and $q = 3$ for $m \geq 3$.

By the use of the rotation matrix, the problem of estimating the $3m$ baseline coordinates turns into the problem of estimating the $3q \leq 3m$ entries of an orthogonal matrix R, of which only three (two for a single baseline) are independent. The multivariate constrained model is then formulated as (Teunissen 2007):

$$E(Y) = AZ + GRF \quad Z \in \mathbb{Z}^{n\times m}; R \in \mathbb{O}^{3\times q}$$
$$D(vec(Y)) = Q_Y = P_m \otimes Q_y \tag{6.5}$$

where R describes the orientation of the body frame with respect to the frame wherein the GNSS measurements are obtained. The unknowns to be resolved are the nm integer-valued ambiguities and the three (or two in case of single-baseline) real-valued independent entries of R, which must belong to the class of 3 by q orthogonal matrices $\mathbb{O}^{3\times q}$. We assume that the different baseline observations are described by the same v-c matrix Q_y, and the dispersion of the matrix of observables Y is obtained via a Kronecker product between Q_y and the m by m matrix P_m, which defines the correlation between the baselines.

3 Constrained Integer Least-Squares

The Integer Least-Squares estimator for the solution of the system (6.1) was demonstrated to be optimal among the class of admissible integer estimators (Teunissen 1999). A closed-form solution of the ILS is not known: hence, a least-squares minimization implies an exhaustive search over a set of integer candidates. The LAMBDA method is a well-known and efficient implementation of the ILS, introduced in Teunissen (1993) and Teunissen (1995). The nonlinear constraints posed on the baseline coordinates strongly affects the resolution technique to be adopted, and a new formulation of the LAMBDA method is presented here. To express the model (6.5) in a vectorial form, we again make use of the vec operator:

$$E\,(vec(Y)) = \left[(I_m \otimes A)\ (F^T \otimes G)\right] \begin{pmatrix} vec(Z) \\ vec(R) \end{pmatrix}$$
$$Z \in \mathbb{Z}^{n\times m}; R \in \mathbb{O}^{3\times q}$$
$$D(vec(Y)) = P_m \otimes Q_y \tag{6.6}$$

We want to solve the system (6.6) in a least-squares sense, therefore minimizing the squared norm of the residuals with respect to the integer-valued matrix Z. The squared norm and its sum-of-squares decomposition reads (Teunissen 2007):

$$\left\| vec(Y) - (I_m \otimes A) vec(Z) - (F^T \otimes G) vec(R) \right\|^2_{P \otimes Q_y}$$
$$= \left\| vec(\hat{E}) \right\|^2_{P_m \otimes Q_y} + \left\| vec\left(Z - \hat{Z}\right) \right\|^2_{Q_{\hat{Z}}}$$
$$+ \left\| vec\left(\hat{R}(Z) - R\right) \right\|^2_{Q_{\hat{R}(Z)}} \tag{6.7}$$

where $\|\cdot\|^2_Q = (\cdot)^T Q^{-1} (\cdot)$ is the weighted squared norm and $\hat{Z}$ and $\hat{R}$ are the float solutions of the unknowns, i.e. the least-squares solution of (6.6) obtained without imposing any constraint on Z or R. $\hat{E}$ is the matrix of least-squares residuals, while $\hat{R}(Z)$ is the float estimator of R given the ambiguity matrix Z known. $Q_{\hat{Z}}$ is the v-c matrix of the float solution $vec(\hat{Z})$, while the v-c matrix $Q_{\hat{R}(Z)}$ defines the dispersion of $vec(\hat{R}(Z))$. Due to the constraints posed on Z and B, the last two terms of (6.7) cannot in general be made zero for any value of Z; thus the minimization problem must be taken with respect to both the integer matrix Z and the orthogonal matrix R:

$$\check{Z} = \arg\min_{Z \in \mathbb{Z}^{n \times m}} C(Z)$$

$$C(Z) = \left\| vec(Z - \hat{Z}) \right\|^2_{Q_{\hat{Z}}} + \left\| vec(\hat{R}(Z) - \check{R}(Z)) \right\|^2_{Q_{\hat{R}(Z)}} \tag{6.8}$$

with

$$vec(\check{R}(Z)) = \arg\min_{R \in \mathbb{O}^{3 \times q}} \left\| vec(\hat{R}(Z) - R) \right\|^2_{Q_{\hat{R}(Z)}} \tag{6.9}$$

The evaluation of the cost function $C(Z)$ involves the computation of two correlated terms: the first is the distance between Z and the float solution $\hat{Z}$, weighted by the v-c matrix $Q_{\hat{Z}}$, and the second is the distance between the conditional solution $\hat{R}(Z)$ and the minimizer of the constrained nonlinear least-squares problem (6.9).

The solution of the minimization problem (6.8) provides the fixed matrix of integer ambiguities $\check{Z}$ by taking advantage of the geometrical constraints expressed by the orthogonality of $\check{R}(Z)$. Solving the problem (6.9) for $Z = \check{Z}$ then gives the least squares estimation of the attitude of the body axis $\check{R}(\check{Z})$, i.e. the orientation of the set of m baselines with respect the frame of axes wherein the GNSS observation are taken. Since no analytical solution for the integer minimizer of (6.8) is known, a direct search method must be employed. The integer matrix which provides the smallest value for $C(Z)$ is exhaustively searched inside the set of integer candidates defined as

$$\Omega\left(\chi^2\right) = \{Z \in \mathbb{Z}^{n \times m} \mid C(Z) \leq \chi^2\} \tag{6.10}$$

where χ is a scalar chosen as to limit the search space $\Omega(\chi^2)$. The shape of set $\Omega(\chi^2)$ is driven by the matrices $Q_{\hat{Z}}$ and $Q_{\hat{R}(Z)}$ in (6.8): if $Q_{\hat{R}(Z)} \to 0$, the set would be ellipsoidal, as follows from the relation $\left\| vec(Z - \hat{Z}) \right\|^2_{Q_{\hat{Z}}} \leq \chi^2$. The tight relation between the two terms of (6.8) complicates the evaluation of the shape of the search space for $Q_{\hat{R}(Z)} \neq 0$.

We now focus on the three steps involved in the computation of the minimizer of (6.8): the derivation of the float solution, the search for the integer minimizer and the computation of the constrained nonlinear least-squares problem (6.9).

3.1 The Float Estimators

The float estimators $\hat{Z}$ and $\hat{R}$ are the least-squares solution of the system (6.6) when disregarding the integerness of the ambiguities and the orthogonality of R. These are obtained by solving the set of normal equations

$$N \begin{pmatrix} vec(\hat{Z}) \\ vec(\hat{R}) \end{pmatrix} = \begin{bmatrix} P_m^{-1} \otimes A^T Q_y^{-1} \\ F P_m^{-1} \otimes G^T Q_y^{-1} \end{bmatrix} vec(Y)$$

$$N = \begin{bmatrix} P_m^{-1} \otimes A^T Q_y^{-1} A & P_m^{-1} F^T \otimes A^T Q_y^{-1} G \\ F P_m^{-1} \otimes G^T Q_y^{-1} A & F P_m^{-1} F^T \otimes G^T Q_y^{-1} G \end{bmatrix} \tag{6.11}$$

The inversion of the normal matrix N provides the v-c matrices of the float solutions $vec(\hat{Z})$ and $vec(\hat{R})$:

$$\begin{bmatrix} Q_{\hat{Z}} & Q_{\hat{Z}\hat{R}} \\ Q_{\hat{R}\hat{Z}} & Q_{\hat{R}} \end{bmatrix} = N^{-1} \tag{6.12}$$

If we assume the matrix of ambiguities known, $\hat{R}(Z)$ and the associated v-c matrix are obtained as

$$vec(\hat{R}(Z)) = vec(\hat{R}) - Q_{\hat{R}\hat{Z}} Q_{\hat{Z}}^{-1} vec(\hat{Z} - Z)$$

$$Q_{\hat{R}(Z)} = Q_{\hat{R}} - Q_{\hat{R}\hat{Z}} Q_{\hat{Z}}^{-1} Q_{\hat{Z}\hat{R}} \tag{6.13}$$

Thus, the knowledge of the fixed matrix of ambiguities improves the precision of $\hat{R}(Z)$: the dispersion is reduced according to (6.13).

3.2 The Search for the Integer Ambiguities

As stated above, the minimization problem (6.8) can in principle be solved with an extensive search in the search space $\Omega\left(\chi^2\right)$: this is a non-trivial task if one aims to have an efficient and fast search. The choice for the scalar χ in (6.10) is critical: it must be large enough to guarantee the non-emptiness of $\Omega(\chi^2)$, but not too large to avoid onerous computational burdens due to the large number of integer candidates for which the solution of (6.9) must be evaluated. Setting the value of χ by picking up an integer matrix Z' and computing

$$\chi^2 = C(Z') \tag{6.14}$$

generally leads to unacceptable large values for χ, for which the computational burden is too heavy. This is due to the fact that the matrix $Q_{\hat{R}(Z)}$ is driven be the more precise phase measurements, and the second term of (6.8) largely amplifies the values of χ for any non-correct value of Z. An alternative approach to the extensive search in $\Omega(\chi^2)$ is to make use of approximating functions that are easier to evaluate than $C(Z)$, and a modification of the LAMBDA method is here proposed. In analogy with the bounding functions introduced for the single-baseline ($m = 1$) case in Teunissen (2006), we note that the expression (6.9) can be bounded via the smallest (λ_m) and largest (λ_M) eigenvalues of the matrix $Q^{-1}_{\hat{R}(Z)}$:

$$\begin{aligned} C_1(Z) &\le C(Z) \le C_2(Z) \\ C_1(Z) &= \left\|vec(Z-\hat{Z})\right\|^2_{Q_{\hat{Z}}} + \lambda_m \sum_{i=1}^{q} (\|\hat{r}_i(Z)\| - 1)^2 \\ C_2(Z) &= \left\|vec(Z-\hat{Z})\right\|^2_{Q_{\hat{Z}}} + \lambda_M \sum_{i=1}^{q} (\|\hat{r}_i(Z)\| + 1)^2 \end{aligned} \tag{6.15}$$

where $\hat{r}_i(Z)$ is the i-th column of $\hat{R}(Z)$ and the inequalities are derived from the rules of the scalar product between vectors. A clever strategy to quicken the search is to make use of these two bounds, and two efficient search strategies for the constrained ILS minimization have been developed (Buist 2007; Giorgi et al. 2008; Giorgi and Buist 2008): the methods were coined the *Expansion* approach and the *Search and Shrink* approach, respectively. The *Expansion* approach works by initially enumerating all the integer matrices contained in a small set of admissible candidates

$$\Omega_{exp}\left(\chi_0^2\right) = \{Z \in \mathbb{Z}^{n\times m} \mid C_1(Z) \le \chi_0^2\} \supseteq \Omega\left(\chi_0^2\right) \tag{6.16}$$

where the scalar χ_0 is initially chosen small enough and iteratively increased until, at step s, the set $\Omega_{exp}\left(\chi_s^2\right)$ turns out to be non-empty: as the evaluation of $C_1(Z)$ only involves the computation of two squared norms, the enumeration proceeds rather quickly. For each of the enumerated integer matrices in $\Omega_{exp}\left(\chi_s^2\right)$, the problem (6.9) is solved and the set $\Omega\left(\chi_s^2\right)$ is evaluated: if it is empty, the scalar χ_s is increased to $\chi_{s+1} > \chi_s$ and the enumeration in $\Omega_{exp}\left(\chi_{s+1}^2\right)$ repeated, otherwise the minimizer of $C(Z)$ is picked up.

A second strategy developed is a *Search and Shrink* approach: a second set is defined as

$$\Omega_{SaS}\left(\chi_0^2\right) = \{Z \in \mathbb{Z}^{n\times m} \mid C_2(Z) \le \chi_0^2\} \subseteq \Omega\left(\chi_0^2\right) \tag{6.17}$$

where χ_0 is chosen large enough to guarantee the non-emptiness of $\Omega_{SaS}(\chi_0^2)$. The search proceeds by iteratively shrinking the set, by means of searching for an integer matrix Z_{s+1} in $\Omega_{SaS}\left(\chi_s^2\right)$ which provides a smaller value for $\chi_{s+1}^2 = C_2(Z_{s+1}) < C_2(Z_s) = \chi_s^2$, until the minimizer of $C_2(Z)$ is found. The minimizer of $C(Z)$, which may differ from the one of $C_2(Z)$, is then extensively searched inside the shrunken set

$$\Omega\left(\overline{\chi}^2\right) = \{Z \in \mathbb{Z}^{n\times m} \mid C(Z) \le \overline{\chi}^2\} \supseteq \Omega_{SaS}\left(\overline{\chi}^2\right) \tag{6.18}$$

where $\overline{\chi}^2 = C_2(\overline{Z})$, being $\overline{Z}$ the minimizer of $C_2(Z)$. The two search strategies provide an efficient alternate way of performing the search for the integer minimizer of (6.8), overtaking both the issues of fixing the initial size of the search space and speeding up the search avoiding the computation of (6.9) a large number of times.

3.3 Solving the Nonlinear Least-Squares Problem

The evaluation of the function $C(Z)$ at a given point Z implies the solution of the nonlinear constrained least squares problem (6.9). Geometrically, it consists to find the closest point between a given data vector $vec(\hat{R}(Z))$ and a curved manifold of dimension $q+1$ embedded in the $3q$-dimensional space, where the metric is defined by the v-c matrix $Q_{\hat{R}(Z)}$. The manifold, which reflects the nonlinearity of the problem, is defined by the constraints equations $R^T R = I$. Making use of one of the representations that can be employed for the three-dimensional rotations needed to coalesce two orthogonal frames, such as the Gibbs vector, the Direct Cosine Matrix, the Quaternions or the Euler angles (Battin 1987), the vector $vec(\hat{R}(Z) - R)$ can be rewritten as a set of $3q$-nonlinear functions of a vector of independent unknowns γ, for which the orthogonal constraint on $R(\gamma)$ is implicitly fulfilled. The nonlinear least-squares problem can then be solved by an iterative technique such as the Gauss-Newton method.

4 Simulation Results

The proposed constrained ILS method was tested with simulated data: the simulation inputs are summarized in Table 6.1. Each of the 24 scenarios was processed with the unconstrained LAMBDA, disregarding the geometrical constraints, and the Constrained LAMBDA method, taking into account the orthogonality on R. The latter was applied on both a single baseline case and a two-baselines case: this to demonstrate the improvement when the number of geometrical constraints increases. Table 6.2 reports for the different methods the single-frequency, single-epoch success rate, which is defined as the ratio of correctly fixed matrix of ambiguities over the set of 10^5 samples simulated. The improvement in success rate was dramatic: especially for the weaker scenarios (lower number of satellite / higher noise levels) the difference between the methods was rather large, e.g. the weakest simulated dataset, with five available satellites and high noise values, showed an increment from a low 3% to 72% for the single baseline case, up to 99.6% for the two-baselines case. As expected, the strengthening of the underlying model due to the embedded geometrical constraints substantially affects the capacity of fixing the correct integer ambiguity matrix: only two baselines were indeed sufficient to obtain single-frequency, single-epoch success rates higher than 99% on all the data sets processed, obtaining a 100% success rate on 20 out of 24 data sets simulated.

Table 6.1 Simulation set up

Frequency	L1
Number of Satellite (PRNs) 5 / 6 / 7 / 8	Corresponding PDOP 4.19 / 2.14 / 1.92 / 1.81
Undifferenced code noise σ_p [*cm*]	30 - 15 - 5
Undifferenced phase noise σ_ϕ [*mm*]	3 - 1
Baselines f_i (x_1, x_2, x_3)	$f_1 = [1, 0, 0]$ m $f_2 = [-0.35, 1.97, 0]$ m
Samples simulated	10^5

Table 6.2 Simulation results: single-frequency, single-epoch success rates for the unconstrained and constrained LAMBDA methods. Success rates higher than 99.9% are stressed

σ_ϕ [*mm*]	3			1		
σ_p [*cm*]	30	15	5	30	15	5
N	*Single-baseline success rate, unconstrained LAMBDA* *Single-baseline success rate, Constrained LAMBDA* *Two-baselines success rate, Constrained LAMBDA*					
5	3.30	19.05	86.67	5.99	26.89	95.37
	72.43	88.86	99.63	96.54	**99.94**	**100**
	99.60	**99.94**	**100**	**100**	**100**	**100**
6	24.83	66.71	96.89	49.13	86.67	**99.99**
	95.75	99.18	99.90	**99.99**	**100**	**100**
	99.99	**100**	**100**	**100**	**100**	**100**
7	50.24	79.69	99.53	74.17	93.27	**100**
	99.34	**99.97**	**100**	**100**	**100**	**100**
	100	**100**	**100**	**100**	**100**	**100**
8	86.17	94.48	**99.99**	**99.97**	**99.99**	**100**
	99.80	**99.99**	**100**	**100**	**100**	**100**
	100	**100**	**100**	**100**	**100**	**100**

Conclusion

The problem of resolving the integer ambiguities which affect the GNSS carrier phase observations is the key to precise relative positioning. The LAMBDA method, which mechanizes the ILS principle, is used to efficiently and reliably fix the ambiguities. When the geometry of the antennae placement is known and constant, nonlinear constraints can be included in the theory, for the purpose of strengthening the model and improving the capacity of fixing the correct integer ambiguities. We proposed in this contribution a model for the GNSS observations which embeds the whole set of nonlinear geometrical constraints arising when considering frame of antennae of invariant relative positions. The cost function to be minimized in a ILS sense has been modified: in order to solve the minimization problem respecting both the integer and orthogonality constraints, a modification of the LAMBDA method is proposed and the integer matrix of ambiguities is searched via one of the two iterative search approaches depicted. Both the *Expansion* and the *Search and Shrink* algorithms can be applied to perform the search, resulting in a faster and more efficient approach than the extensive search. We tested the proposed method on different simulated data sets, investigating the influence of the number of available satellite and the noise levels on the code and phase observations: the difference when using the unconstrained LAMBDA and the Constrained LAMBDA is dramatic, with a large improvement in the capacity of resolving the correct integer matrix, especially for the scenarios characterized by lower number of available satellites/higher noise levels.

Acknowledgements Professor P.J.G. Teunissen is the recipient of an Australian Research Council Federation Fellowship (project number FF0883188): this support is greatly acknowledged.

The research of S. Verhagen is supported by the Dutch Technology Foundation STW, applied science division of NWO and the Technology Program of the Ministry of Economic Affairs.

References

Battin RH (1987) An introduction to the mathematics and methods of astrodynamics. AIAA Education Series, New York

Buist PJ (2007) The baseline constrained LAMBDA method for single epoch, single frequency attitude determination applications. In: Proceedings of ION GPS, 2007

Giorgi G, Buist PJ (2008) Single-epoch, single frequency, standalone full attitude determination: experimental results. In: 4th ESA workshop on satellite navigation user equipment technologies, NAVITEC, 2008

Giorgi G, Teunissen PJG, Buist PJ (2008) A search and shrink approach for the baseline constrained LAMBDA: experimental results. In: Proceedings of the international symposium on GPS/GNSS 2008, Tokyo, Japan, 2008

Goldstein H (1980) Classical mechanics. Addison-Wesley Pub. Co., Massachusetts

Park C, Teunissen PJG (2003) A new carrier phase ambiguity estimation for GNSS attitude determination systems. In: Proceedings of international GPS/GNSS symposium, Tokyo, 2003

Park C, Teunissen PJG (2008) A baseline constrained LAMBDA method for integer ambiguity resolution of GNSS attitude determination systems. J Contr Robot Syst (in Korean), 14(6):587–594

Teunissen PJG (1993) Least squares estimation of the integer GPS ambiguities. Invited lecture, Section IV theory and methodology, IAG General Meeting, Beijing. Also in: LGR series No.6, Delft Geodetic Computing Center, Delft University of Technology

Teunissen PJG (1994) A new method for fast carrier phase ambiguity estimation. In: Proceedings IEEE position location and navigation symposium, PLANS '94, pp 562–573

Teunissen PJG (1995) The least-squares ambiguity decorrelation adjustment: a method for fast GPS integer ambiguity estimation. J Geodes 70:65–82

Teunissen PJG, Kleusberg A (1998) GPS for geodesy. Springer, Berlin

Teunissen PJG (1999) An optimality property of the integer least-squares estimator. J Geodes 73(11):587–593

Teunissen PJG (2006) The LAMBDA method for the GNSS compass. Artificial Satellites. J Planet Geodes 41(3):88–103

Teunissen PJG (2007) A general multivariate formulation of the multi-antenna GNSS attitude determination problem. Artificial Satellites, 42(2):97–111

Teunissen PJG (2008) GNSS ambiguity resolution for attitude determination: theory and method. In: Proceedings of the international symposium on GPS/GNSS 2008, Tokyo, Japan, 11–14 November, 2008

Teunissen PJG (2010) Integer least squares theory for the GNSS compass. J Geodes, 84:433–447

Geodetic Sensor Systems and Sensor Networks: Positioning and Applications

7

Sandra Verhagen, Dorota Grejner-Brzezinska, Guenther Retscher, Marcelo Santos, Xiaoli Ding, Yang Gao, and Shuanggen Jin

Abstract

This contribution focuses on geodetic sensor systems and sensor networks for positioning and applications. The key problems in this area will be addressed together with an overview of applications. Global Navigation Satellite Systems (GNSS) and other geodetic techniques play a central role in many applications like engineering, mapping and remote sensing. These techniques include precise positioning, but also research into non-positioning applications like atmospheric sounding using continuously operating GNSS networks. An important research area is multi-sensor system theory and applications to airborne and land-based platforms, indoor and pedestrian navigation, as well as environmental monitoring. The primary sensors of interest are GNSS and inertial navigation systems. Furthermore, Interferometric Synthetic Aperture Radar (InSAR) is recognized as one of the most important state-of-the-art geodetic technologies used for generation of Digital Elevation Models and accurately measuring ground deformations.

Keywords

Current research issues • GNSS • InSAR

S. Verhagen (✉)
DEOS, Delft University of Technology, Delft, The Netherlands
e-mail: A.A.Verhagen@TUDelft.nl

D. Grejner-Brzezinska
Ohio State University, Columbus, OH, USA
e-mail: dorota@cfm.ohio-state.edu

G. Retscher
Vienna University of Technology, Vienna, Austria
e-mail: gretsch@pop.tuwien.ac.at

M. Santos
University of New Brunswick, New Brunswick, Canada
e-mail: msantos@unb.ca

X. Ding
Hong Kong Polytechnic University, Hong Kong
e-mail: lsxlding@inet.polyu.edu.hk

Y. Gao
University of Calgary, Calgary, Canada
e-mail: ygao@ucalgary.ca

S. Jin
CSR, University of Texas at Austin, Austin, USA
e-mail: sgjin@shao.ac.cn

1 Introduction

Global Navigation Satellite Systems (GNSS) play a central role in many applications like engineering, mapping and remote sensing. These techniques include precise positioning, as well as applications of reference frame densification and geodynamics, to address the

N. Sneeuw et al. (eds.), *VII Hotine-Marussi Symposium on Mathematical Geodesy*, International Association of Geodesy Symposia 137, DOI 10.1007/978-3-642-22078-4_7, © Springer-Verlag Berlin Heidelberg 2012

demands of precise, real-time positioning of moving platforms. Recognising the role of continuously operating GPS reference station network, research into non-positioning applications of such geodetic infrastructure is also pursued, such as atmospheric sounding. Thereby, other geodetic techniques should be considered as well.

An important research area is multi-sensor system theory and applications, with a special emphasis on integrated guidance, navigation, positioning and orientation of airborne and land-based platforms. The primary sensors of interest are GNSS and inertial navigation systems; however, the important role of other techniques used for indoor and pedestrian navigation, and environmental monitoring is also recognized.

Furthermore, Interferometric Synthetic Aperture Radar (InSAR) is recognized as one of the most important state-of-the-art geodetic technologies with applications like generation of Digital Elevation Models and accurately measuring ground deformations.

This contribution gives an overview of state-of-the-art technology and research issues for GNSS, multi-sensor systems and InSAR, respectively.

2 GNSS

2.1 High-Precision GNSS

Recent research and development activities in the field of high-precision GNSS have been in great extent driven for improved system performance with signals from multiple constellations and increased system cost-effectiveness and availability of high-precision GNSS. Some research subjects important to high precision GNSS applications are addressed in the following.

2.1.1 Augmentation with Multiple GNSS Signals

There are significantly increased efforts toward augmenting GPS-based systems with multiple GNSS signals. This comes with demands to further improve the positioning accuracy and reliability and increase continuous precise positioning availability in less desired observing environments such as urban canopy where significant signal blockages would make GPS-alone positioning very difficult (Cai and Gao 2009). Data processing technologies to support multiple GNSS signals from modernized GPS, GLONASS and Galileo systems are highly demanded. Benefits to system's robustness are particularly of interest to practical applications. This requires efforts to develop new signal combination strategies, modeling techniques and quality control measures (Feng and Rizos 2009; Fernandez-Plazaola et al. 2008).

2.1.2 Integration of PPP and Network-RTK

While Network RTK continues to receive increased adoption as more and more network infrastructures are being deployed and PPP is recognized as an attractive alternate to many high-precision applications, there are increased interests to integrate the two for combined advantages. Integration of PPP with Network RTK techniques may lead to improved position accuracy and reliability, operational flexibility and efficiency, particularly reduction in convergence time and network reference station density (Wubbena et al. 2005; Dixon 2006; Feng et al. 2007). State space corrections to support both Network RTK and PPP and their seamless integration should be investigated.

2.1.3 PPP for Single-Frequency Receivers

PPP was initially designed based on the use of dual-frequency GNSS receivers since dual-frequency observations are necessary in order to remove the effect of ionospheric refraction which is the biggest error source after the application of precise orbit and clock corrections. Increased research and development activities have been found in recent years towards single-frequency PPP. This is largely driven by the fact that the majority of GNSS applications are based on low-cost single-frequency receivers. Such efforts have already brought significant progress in methodology and product development of single-frequency precise point positioning based on precise correction data from the International GNSS Service (IGS) as well as Satellite-based Augmentation Systems (SBAS) (Chen and Gao 2008; Zhang and Lee 2008; Van Bree et al. 2009). Further, there is a great potential to significantly improve positioning accuracy with cheap GNSS chipsets. Technologies to process biased and noisy GNSS observations will be highly demanded and should be investigated.

2.1.4 Quality Control for High-Precision GNSS

Quality control is not new but becomes increasingly important for modern high-precision GNSS systems. This is particularly true for real-time systems such as Network RTK, PPP and other real-time systems

(Aponte et al. 2009). Quality measures should be developed to assess differential GNSS correction data and position solutions. Advanced techniques to ensure high reliability of on-the-fly ambiguity resolution are still a significant challenge for current high-precision RTK systems. This becomes even more critical when signals from multiple GNSS constellations are combined.

2.1.5 Availability of Precise Correction Data

Precise orbit, clock and further ionospheric correction data are essential for PPP and they will also contribute to RTK systems. Precise orbit and clock products have improved significantly in recent years and they are freely available over the Internet from organizations such as IGS. IGS real-time products are expected to be available in the near future. Further, ionospheric correction data is critical for single-frequency PPP and is highly demanded by the industry for product development. Increased availability of precise correction data will accelerate the development of real-time PPP products and reduce the time to market of new high-precision GNSS technologies.

2.2 Atmosphere Modelling

2.2.1 Ionosphere

The past years have seen an increasing effort in the collection of experimental data for monitoring of TEC and ionospheric scintillation studies. This effort has resulted in the deployment of dedicated networks of ground GNSS and scintillation receivers, at high and mid latitudes. There is also effort by means of satellite missions. For example, in situ measurements from GRACE K-Band ranging and CHAMP planar Langmuir probe (PLP) have been used for the validation of the International Reference Ionosphere (IRI) and occultation data used in combination with GNSS and satellite altimetry aiming at a combined global VTEC model (e.g. Todorova et al. 2008; Mayer and Jakowski 2009).

There has been effort put on enhancements in the spatial and temporal representation of TEC/VTEC, globally, regionally or locally. Another issue is that near- and real-time applications require the dissemination of predicted values of TEC. This brings to mind the SBAS, based on continental networks but regional or local systems may also support these applications.

Investigation into multi GNSS constellation and higher order (e.g., 3rd) determination TEC seem to be gaining momentum. Higher order ionospheric delay terms, which have been mostly disregarded in the dual-frequency world, can be taken into account in a multi-frequency reality, see e.g. (Hoque and Jakowski 2008; Hernández-Pajares et al. 2007).

2.2.2 Troposphere

The increasing use of Numerical Weather Models (NWM) has helped enhancing the prediction of neutral atmospheric models (Boehm et al. 2006). It has also become a source of neutral atmospheric delay that can be directly applied in GNSS processing, including PPP. If from one side NWMs contain a more realistic temporal representation of the delay than prediction models, from the other side the extraction of this information requires ray-tracing through the neutral atmosphere, a time consuming task if done properly. Fast and accurate algorithms are of fundamental necessity (Hobiger et al. 2008).

There has been an increasing emphasis of neutral-atmosphere delay monitoring by ground GNSS and satellite missions, with radio occultation consolidating itself as a solid technique (Wickert et al. 2009).

There is a continuing effort towards enhancements in the spatial and temporal representation of the neutral-atmosphere including its azimuthal asymmetry. Several models incorporating gradients, spherical harmonics, tomography, have been further tested including information from NWMs (Ghoddousi-Fard et al. 2009; Rohm and Bosy 2009).

2.3 GNSS Reflectometry

Reflected signals are normally a nuisance in case of precise positioning applications, since only the direct signals should be used for ranging. Recently, however, the GNSS reflected signals have given birth to new applications for various environmental remote sensing applications in atmosphere, ocean, land and cryosphere, e.g. (Jin and Komjathy 2010).

Surface multipath delay from the GNSS signal reflecting from the sea and land surface, could be used as a new tool in ocean, coastal, wetlands, Crater Lake, landslide, soil moisture, snow and ice remote sensing (e.g. Kamjathy et al., 2004). Together with information on the receiving antenna position and the medium, associating with the surface properties of the reflecting surface, the delay measurement can be used to determine such factors as wave height, wind speed, wind direction, and even sea ice conditions. Martin-Neira (1993) first proposed and described a bistatic ocean altimetry system utilizing the signal of GPS. Recently, a number of applications have been implemented using GPS signals reflected from the ocean surface, such as determining wave height, wind speed and wind direction of ocean surface, ocean eddy, and sea surface conditions.

Key topics of current research are:

- Extension of developments of current GPS reflected signal sensor techniques and their applications.
- Improvement of existing estimation algorithms and data processing for GPS reflected signals.
- Coordinated data collection campaigns and comparison with terrestrial and satellite remote-sensor observations.
- Investigation of multi-remote sensor integration and applications.

3 Multi-sensor Systems

3.1 Navigation and Mapping

Multi-sensor system theory and applications is an important research area as well. Here, we will put a special emphasis on integrated guidance, navigation, positioning and orientation of airborne and land-based platforms. The primary sensors of interest are GNSS and inertial navigation systems; however the important role of other techniques used for indoor and pedestrian navigation environmental monitoring is also recognized.

Key topics for further research in this field can be identified as:

- Technical advances in navigation sensors and algorithms, including autonomous vehicle navigation, based on:
  - GPS, pseudolites, INS, wheel sensors, ultrasonic and magnetic sensors
  - Cellular networks and their hybrid with GPS
- Technical advances in mapping sensors (CCD cameras, laser range finders, laser scanners, radar devices)
- Standardization of definitions and measurements of sensor related parameters
- Performance of stand alone and integrated navigation systems
- Non-linear estimation and information fusion methods
- Innovation in:
  - Algorithms, calibration, synchronization
  - Real-time processing and geo-referencing
  - Automated information extraction

3.2 Geotechnical and Structural Engineering

Nowadays extended multi-sensor deformation measurement systems consisting of terrestrial geodetic and geotechnical measurement as well as hydrological and meteorological instrumentation completed by the InSAR technique are mainly employed for multi-scale monitoring of landslide prone areas. Thereby InSAR is used for large-scale detection of landslide prone areas as well as for deformation measurements of the investigated landslide area. Such a complete measurement system is very suitable for the investigation of the kinematic behaviour of landslides and together with other (e.g. hydrological, meteorological, etc.) parameters for the study of the dynamics of landslides. The observation data is usually collected in GIS (see e.g. Lakakis et al. 2009; Mentes 2008) and used to develop Spatial Decision Support Systems (SDSS) and Early warning systems.

In the last years, Artificial Intelligence (AI) has become an essential technique for solving complex problems in Engineering Geodesy. AI is an extremely broad field – the topics range from the understanding of the nature of intelligence to the understanding of knowledge representation and deduction processes, eventually resulting in the construction of computer programs which act intelligently. Especially the latter topic plays a central role in applications (Reiterer and Egly 2008). Current applications using AI methodologies in engineering geodesy are: geodetic data analysis, deformation analysis, navigation, deformation network adjustment, and optimization of complex measurement procedures.

4 InSAR

Synthetic Aperture Radar (SAR) and Light Detection And Ranging (LiDAR) systems are very useful for geodetic applications, such as monitoring local area ground surface deformations due to volcanic and seismic activities, and ground subsidence associated with city development, mining activities, ground liquid withdrawal, and land reclamation.

InSAR is a very active field of research in the geo-detic research communities. The current research issues include the development of more effective methods/algorithms for InSAR solutions, the quality control and assurance of InSAR measurements, the study and mitigation of biases in InSAR measurements such as the atmospheric effects, integration of InSAR and other geodetic technologies such as GPS, and new and innovative applications of the technology in geodetic studies.

Acknowledgements This contribution is the result of a collaborative effort of IAG Commission 4, see http://enterprise.lr.tudelft.nl/iag/iag\comm4.htm.

The research of Sandra Verhagen is supported by the Dutch Technology Foundation STW, applied science division of NWO and the Technology Program of the Ministry of Economic Affairs.

References

Aponte J, Meng X, Dodson AH, Moore T, Hill C, Burbidge M (2009) Quality assessment for a network RTK GPS service in the UK. J Appl Geodes 3:25–34

Boehm J, Niell A, Tregoning P, Schuh H (2006) Global mapping function (GMF): a new empirical mapping func-tion based on numerical weather model data. Geophys Res Lett 33, L07304, doi:10.1029/ 2005GL025546

Chen K, Gao Y (2008) Ionospheric Effect Mitigation for Real-Time Single-Frequency Precise Point Positioning, Journal of Navigation, Institute of Navigation, 55(3)

Cai C, Gao Y (2009) A combined GPS/GLONASS navigation algorithm with limited satellite visibility, Journal of Navigation, Royal Institute of Navigation, 62(4)

Dixon K (2006) StarFireTM: A global SBAS for sub-decimetre precise point positioning. In: Proceedings of ION GNSS 2006, Fort Worth, Texas, pp 2286–2296

Feng YM, Rizos C (2009) Network-based geometry-free three carrier ambiguity resolution and phase bias calibration. GPS Solutions 13:43–56

Feng YM, Rizos C, Higgins M (2007) Multiple carrier ambiguity resolution and performance benefits for RTK and PPP positioning services in regional areas. In: Proceedings of ION GNSS 2007, 25–28 Sept, Fort Worth, TX, pp 668–677

Fernandez-Plazaola U, Martin-Guerrero TM, Entrambasaguas JT (2008) A new method for three-carrier GNSS ambiguity resolution. J Geodes 82:269–278

Ghoddousi-Fard R, Dare P, Langley RB (2009) Tropospheric delay gradients from numerical weather prediction models: effects on GPS estimated parameters. GPS Solutions 13: 281–291

Hernández-Pajares M, Juan JM, Sanz J, Orús R (2007) Second-order term in GPS: implementation and impact on geodetic estimates. J Geophys Res 112, B08417 (a correction appeared in Vol. 113, B06407)

Hobiger T , Ichikawa R, Kondo T, Koyama Y (2008) Fast and accurate ray-tracing algorithms for real-time space geodetic applications using numerical weather models. J Geophys Res 113(D203027):1–14

Hoque MM, Jakowski N (2008) Estimate of higher or-der ionospheric errors in GNSS positioning. Radio Science 43, RS5008, doi: 10.1029/2007RS003817

Jin SG, Komjathy A (2010) GNSS Reflectometry and Remote Sensing: New roles and Progresses, Adv Space Res 44

Lakakis K, Charalampakis M, Savaidis P (2009) A landslide definition by an integrated monitoring system. In: Fifth International Conference on Construction in the 21st Century (CITC-V), Collaboration and Integration in Engineering, Management and Technology. May 20–22, Istanbul Turkey. pp 1–8

Martin-Neira M (1993) A Passive Reflectometry and Interferometry System (PARIS): Application to Ocean Altimetry. ESA Journal 17(4):331–355

Mayer C, Jakowski N (2009) Enhanced E-layer ionization in the auroral zones observed by radio occultation measurements onboard CHAMP and Formosat-3/COSMIC. Ann Geophys 27:1207–1212

Mentes G (2008) Investigation of different possible agencies causing landslides on the High Loess Bank of the River Danube at Dunafldvr, Hungary. In: Proceedings of the Measuring the Changes, 13th FIG International Symposium on Deformation Measurements and Analysis, 4th IAG Symposium on Geodesy for Geotechnical and Structural Engineering, LNEC, Lisbon, Portugal, CD, May 12–15, pp 1–10

Reiterer A, Egly U (eds) (2008) Application of artificial intelligence in engineering geodesy. In: Proceedings of the First Worksop on AIEG 2008, 116 p, http://info.tuwien.ac.at/ingeo/Downloads/AIEG2008_Pro-ceedings.pdf

Rohm W, Bosy J (2009) Local tomography troposphere model over mountains areas. Atmos Res 93:777–785

Todorova S, Hobiger T, Schuh H (2008) Using the Global Navigation Satellite System and satellite altimetry for combined Global Ionosphere Maps. Adv Space Res 42:727–736, doi:10.1016/j.asr.2007.08.024

Van Bree RJP, Tiberius CCJM, Hauschild A (2009) Real time satellite clock corrections in precise point positioning. In: Proceedings of ION GNSS 2008, Savannah, Georgia, USA, September 22–24, 2009

Wickert J, Schmidt T, Michalak G, Heise S, Arras C, Beyerle G, Falck C, König R, Pingel D, Rothacher M (2009) GPS Radio Occultation with CHAMP, GRACE-A, SAC-C, TerraSAR-X, and FORMOSAT-3/COSMIC: Brief Review of Results from GFZ. In: Steiner AK, Pirscher B, Foelsche U, Kirchengast G (eds) New Horizons in Occultation Research. Springer, Berlin, pp 3–15, doi 10.1007/978-3-642-00321-9

Wubbena G, Schmitz M, Bagg A (2005) PPP-RTK: Precise Point Positioning using state-space representation in RTK networks. In: Proceedings of ION GNSS 2005, 13–16 September, Long Beach, California, pp 2584–2594

Zhang Y, Lee S (2008) Nexteq RT30 L1 GPS receiver for autonomous precise positioning. In: Proceedings of ION GNSS 2008, Savannah, Georgia, USA, September 16–19, 2008

Contribution of a Kinematic Station to the GNSS Network Solution for Real Time

8

T. Cosso and D. Sguerso

Abstract

GNSS networks for real time compute differential corrections using undifferenced equations to model observations biases. Actually there are a lot of NRTK service on the mainland, but it is very difficult to have a similar service offshore. The main goal of the present work is to analyse the possibility to insert in the network design that it could be also installed on a kinematic support. In the present work the feasibility of a network solution for real time with one kinematic station is analysed. For this aim it was investigated what kind of contribution the correlations, could provide in the estimation of the GNSS observations biases. Some simulations have been carried out and finally an experimental campaign has been performed and analysed by an innovative ad-hoc developed software.

Keywords

GNSS • NRTK • Undifferenced equations • Correlations • Kinematic solutions

1 Introduction

In the last years the concept of Network Real Time Kinematic (NRTK) positioning services grew up from local to national scale. A rover receiver can obtain a good position in real time, with phase or code observations, by receiving differential correction estimated from a network of permanent stations.

This kind of solution can be obtained on the mainland, but if we are interested in RTK positioning offshore it is very difficult to reach good precision due to the very long distances from the GNSS network.

T. Cosso (✉) · D. Sguerso
DICAT, Dip. di Ingegneria delle Costruzioni, dell'Ambiente e del territorio, Via Montallegro 1, 16145 Genova, Italy
e-mail: tiziano.cosso@unige.it; domenico.sguerso@unige.it

In the present work we want verify if it is possible to have a permanent station off-shore; because of the absence of static and stable supports for the stations, we suppose to create a permanent station installed on a kinematic structure with a multi-antenna system. Such system is that it would be useful to evaluate the attitude of the kinematic support as an important information to model its movements. As a consequence we have to verify the possibilities to use a kinematic multi-station as a vertex of the network to estimate its position, attitude and atmospheric parameters; in the meanwhile we have to verify if such multi-station has a positive or negative influence in the estimation of parameters related to the other stations. In the NRTK solution the main parametrized biases are estimated by an undifferenced approach: the system to solve is composed by one undifferenced equation for each observation. Thus, the estimated biases in correspondence of multi-station

N. Sneeuw et al. (eds.), *VII Hotine-Marussi Symposium on Mathematical Geodesy*, International Association of Geodesy Symposia 137, DOI 10.1007/978-3-642-22078-4_8, © Springer-Verlag Berlin Heidelberg 2012

allows to generate differential corrections offshore Some network simulations were performed, to evaluate the contribution of the correlations introduced by the multi-antenna system in the network solution with undifferenced equations. The obtained results have been verified by an ad-hoc developed software applied to real data.

The present work has to be considered as a feasibility analysis to extend the Network Real Time Service offshore. For this reason some simplificative hypothesis have been done, so to use combined iono-free code observation.

2 Network Simulations

Some network simulations were performed to analyse the of contribution that the kinematic multi-station brings in NRTK solution, through the contribution of the correlations introduced by common parameters, such as satellite clocks. Thus, we have to compare the estimations commons parameters in different network configurations.

Four simple configurations were analysed:

A-configuration: three permanent stations with mutual distances about 30 km.

B-configuration: A configuration plus an additional static permanent station.

C-configuration: three stations of A configuration plus a kinematic multi-station, composed by three jointed antennas-receivers systems, instead of the fourth single static station of B configuration.

C/bis-configuration: like C configuration, but with the kinematic multi-station composed by three joint antennas and only one receiver (hence one clock receiver).

In each configuration the parameters are estimated for single epochs; hence the movement of the multi-station is reproduces as sequence of instantaneous positions.

2.1 Hypothesis

As mentioned the present work deals with a preliminary analysis, so it will be presented here just a solution based upon code undifferenced observations, combined as a IONO-FREE to avoid at this stage, problems connected with ionosphere estimation. We know that these aren't usual condition for a network solution for real time, but we consider that we could obtain important directions about the behaviour of the network solution in this particular condition. In other words we assume to understand, also with these simplifications, the contribution of the kinematic multi-station to the entire network.

The observation equation for code measurements may be written as follows:

$$P_{(i)}^{j} = \sqrt{((x_i-x^j)^2+(y_i-y^j)^2+(z_i-z^j)^2)} + ct_i - ct^j + E_i^j + \frac{f_1^2}{f_2^2} I_i^j + T_i^j + cT_{GD} + M_i^j + \varepsilon_i^j \quad (8.1)$$

where, for each receiver i and satellite j:

E_i^j ephemerides error I_i^j ionosphere
T_i^j troposphere T_{GD} group delay
M_i^j multipath effect ε_i^j other errors

To simplify the simulation, the following hypothesis are imposed:

- We use ultra-rapid ephemerides, thus we assume that E_i^j could be neglected, because it is much lower than intrinsic precision of code observation.
- Ionosphere effects are neglected considering iono-free code observables.
- The multipath M_i^j effects, the group delay cT_{GD}, the effects related to the electronic behaviour of antennas and receivers and other noises sources ε_i^j are neglected.

Troposphere effects T_i^j may be modelled using a mapping function m_i^j depending on the elevation of each satellite, that multiplies a tropospheric zenith delay TZD_i depending on the receiver.

The equation so simplified is here reported:

$$P_{IF}^{j}(i) = \sqrt{(x^s - x_r)^2 + (y^s - y_r)^2 + (z^s - z_r)^2} + ct_i - ct^j + m_i^j\, TZD_i \quad (8.2)$$

The degrees of freedom of the spatial network are fixed by the precise ephemerides, but (8.2) is invariant for a temporal translation of the time scales of satellites and receivers, thus the two contributions t_i and t^j could not be estimated separately; for this reason the clock of one receiver or one satellite has to be fixed.

In the present work, the coordinates of one station were considered known to get a well-conditioned solution; then the clock and the troposphere parameters of the same receiver were fixed.

Table 8.1 Comparisons of the network simulations between A, B and C configuration

A	B	C	C/bis
Square root of the max value in the main diagonal			
6.30	3.80	3.60	4.00
Determinant			
52.5×10^{-5}	2.7×10^{-5}	0.8×10^{-5}	0.9×10^{-5}
Square root of the max eigenvalue			
10.80	6.00	5.50	6.20
Square root of the min and max eigenvalue ratio			
0.01	0.018	0.019	0.017

2.2 N^{-1} Comparisons

To evaluate the different behaviour of the four configurations, rows and columns concerning just the common parameters about to the three static permanent stations, were extracted from the N^{-1}, inverse of the normal matrix. The comparisons between the extracted matrices of the A, B, C and C/bis configurations were performed using these different criteria:

- Square root of the maximum value in the main diagonal
- Determinant
- Square root of the maximum eigenvalues
- Square root of the minimum and maximum eigenvalue ratio

The results are reported in Table 8.1.

Notice that the B configuration has all the values better than the A one; in particular the determinant decreases of one order and either the square root of the maximum value in the main diagonal and the square root of the maximum eigenvalue are halved.

The difference between A and B configurations is due to the covariances generated by the parameters related to the satellites; in particular, being ephemerides considered as known, the satellite clocks brings an important contribution in the solution. In fact, if the system is solved by fixing all the parameters related to the satellites, the covariances in the normal matrix would equal to 0 and so, no difference exists in the estimation of the parameters between A and B configurations.

Comparing B and C configurations, the increased correlations due to the multi-station make the "C" determinant one order lower than the "B" one, even if the square roots of the maximum eigenvalues are quite similar. Hence, the hyper-volume of the error hyper-ellipsoid is significantly reduced in the C configuration. To analyse more carefully the extracted N^{-1}, the main diagonal of the B and C configuration has been compared. The differences of the values in the main diagonal are included between 0 and 1.2; in general it can be asserted that:

$$[N_B^{-1}]_{ii} - [N_C^{-1}]_{ii} > 0$$

and so the C configuration, with variances lower than the B one, could be considered the better solution.

It is important to notice that, in all the three configurations, the square root of the higher values in the main diagonal varies from 4 to 6, and correspond to the height and the clock of the receivers, although DOP indexes are good (PDOP=2,0; VDOP=1,6; TDOP=1,1). Instead, the square root of the other values in main diagonal generally varies from 1 to 2.

The square root of the minimum and maximum eigenvalue ratio are quite similar and very far from unit.

A last test denominated C/bis has been carried out considering a unique clock for the multi-station, assuming to use three antennas with only one receiver. Comparing the "C" and "C/bis" values reported in the table, it is interesting to notice that, although the last solution has to estimate two unknown parameters less, with the same number of equations, the "C" solution gives better results. In fact, the strong correlation between the receiver's clocks of the multi-station, due to the pseudo-observations equations that introduce geometrical bond between the antennas, brings an important contribution. In this case, it seems that a greater number of unknowns is more useful than a higher redundancy, if such parameters create strong correlations.

3 New Software for Network Solution with a Kinematic Multi-station

Verified the positive contribution of the kinematic multi-station, an experimental campaign was set up to reproduce the behaviour of that multi-station. The test data-set was composed by observations coming from five Permanent Stations of the Polytechnic of Turin and a multi-station realized in Genoa in cooperation with Department of Naval Engineering of the University of Genoa; Fig. 8.1 shows the multi-station composed by three antennas collocated on a particular dynamic steel

Fig. 8.1 Structure of the kinematic multi-antenna used in the experimental campaign

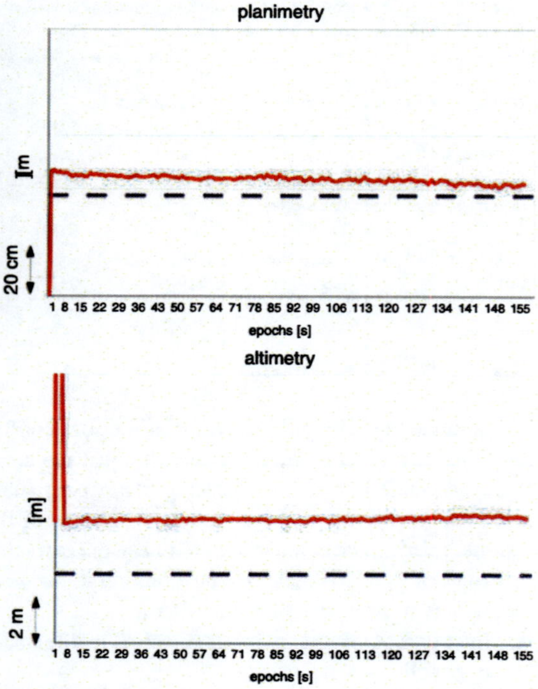

Fig. 8.2 Planimetry and altimetry of one station belonging to the network, calculated with the prototypal software implemented

structure. The whole system acquires GNSS data every 1 s. The reciprocal distances between the Permanent Station is of the order of 40 km.

At present, commercial software used in the management of the GNSS networks to produce real time services, does not allow to use dynamic receivers as permanent stations; so, to reach our aim, a software able to generate a network solution using the observations of also the multi-station, was written in MATLAB language.

Actually this prototypal software reads in input L1 and L2 code observations and ultra rapid ephemerides, preprocessed with Bernese software, involved in the elaborations sequentially, like if they arrive second by second in a real time mode. A Kalman Filter solution applied to the iono-free combination is implemented, to estimates epoch by epoch biases of each permanent stations, like in a real time kinematic procedure. The so obtained biases may be used as input informations to create differential corrections.

The same hypothesis of the simulation were considered, so the parameters estimated are coordinates, satellite and receiver clocks and tropospheric delay (total zenith delay).

The goodness of the solutions is evaluated by comparing the Permanent Stations coordinates in monograph with the ones obtained epochs by epochs with the prototypal software, either in planimetry and in altimetry.

In the Fig. 8.2 are represented planimetry and altimetry of one static permanent station included in the network solution; the continuous and the dash line represents the values calculated using the prototypal software and the reference values respectively.

The behaviour of the other stations is quite similar, so the previous graph could be considered representative of the entire network. Thee planimetric shift is about 20 cm, while altimetric one is about 2 m; this values can be considered satisfying considering that just code iono-free observations are used. Moreover, note the very short instability period, correspondent to the first three or four epochs.

4 Conclusion and Future Developments

In the present work has been investigated the opportunity to extend the NRTK service offshore; through a kinematic multi-station, verifying its contribution to the estimation of parameter related to the whole network. The network simulation of different configurations, using undifferenced equations for iono-free code observations, showed that a kinematic multi-station brings a positive contribution in the estimation of the entire network's parameters also if its attitude has to be estimated. It was underlined, although that it was a preliminary analysis with a lot of simplifying hypothesis, but we have point out the importance to introduce strongly correlated unknown parameters, also if it decreases the redundancy.

Thus an experimental campaign has been planned and carried out to create a dataset with real observations. Because commercial software cannot elaborate such dataset, a prototypal software able to create a GPS network solution for real time applications using as input data also a kinematic multi-station, was written in MATLAB language.

Good results have been provided; the coordinates of a permanent station have been estimated with a precision of about 20 cm in planimetry and 2 m in altimetry, if a multi-station is included in the network. Note that such results are satisfying because iono-free code observations are used. Moreover, the precision could be probably improved modelling better some effects neglected in the present version.

Hence, the test campaign results confirmed that the multi-station gives an important contribution to the GNSS network, maybe permitting longer distances between the permanent stations.

In the near future, some uploading and improvement of the software are expected. First of all, it will be necessary to implement the prototypal software in a compilable informatics language, and to complete it with the preprocessing data treatments.

Then, it will be uploaded so to generate a network solution with carriers measurements, where obviously the main difficulty will be the ambiguities resolution. Other improvements are needed, like to consider in the equations the PCV (Phase Centre Variation) parameters, the orientation of satellites, group delay, the ionospheric models and different kind of mapping functions of the tropospheric zenith delay. Also it could be useful to insert spatial models of the biases, evaluated for instance with other larger networks, so to make the solutions stronger; moreover, interpolations models will be implemented to obtain differential corrections from the punctual biases estimations.

Finally, at this moment the software simulates a real time solution by reading raw data in a post processing phase; hence, all the procedures to read in real time mode the input data to generate a "true" RTK network solution, has to be implemented.

Acknowledgements We want to thank Prof. A.M. Manzino, Polytechnic of Turin – DITAG, for his assistance by providing data coming from Polytechnic GNSS Permanent Station and for his precious scientific support, Prof G. Carrera of the University of Genoa DINAV and M. Goso e G. Grosso for the collaboration in the tests realization.

References

Borre K, Strang G (1997) Linear algebra, geodesy and GPS

Cina A, Manzino AM, Piras M, Roggero M (2004) Reti GPS per il tempo reale, le informazioni contenute nel segnale differenziale. Bollettino SIFET 4

Heroux P, Kouba J (2001) GPS precise point positioning using IGS orbit products. Phys Chem Earth A 26:573–578

Manzino AM (2003) Introduzione all'uso del filtro di Kalman in Geodesia. Bollettino SIFET 2:63–101

Niell AE (1996) Global mapping functions for the atmospheric delay at radio wavelengths. J Geophys Res 101(B2): 3227–3246

Sansò F (2006) Navigazione geodetica e rilevamento cinematico. Polipress, Milano

Teunissen PJG, Kleusberg A (1998) GPS for Geodesy. 2nd edn. Springer, Heidelberg

Wübbena G, Willgalis S (2001) State space approach for preciseee real time positioning in GPS reference networks. In: Proceedings of international symposium on kinematic systems in geodesy, geomatics and navigation, Banff, Canada, 5–8 June

Part III

Estimation and Filtering Theory, Inverse Problems

Approximation of Terrain Heights by Means of Multi-resolution Bilinear Splines

9

M.A. Brovelli and G. Zamboni

Abstract

The paper deals with the approximation of fields sampled at points irregularly distributed in a plane. Different algorithms both deterministic and stochastic have been developed in the last years and, among them, those based on spline functions. The interest of the authors with respect to this method is due to its suitability in reproducing fields as the ground surface topography to obtain the corresponding digital elevation models. The observations are modeled by the sum of a combination of spline functions and a white noise; a least squares adjustment approach is then used to estimate the unknown coefficients. One of the greatest drawbacks is the request of having almost regularly sampled data to avoid severe rank deficiency problems. In cases showing strong inhomogeneity in spatial distribution the coarsest resolution must be used, missing details where denser data are sampled to better describe the higher variability. To overcome this limit multi-resolution splines, without any particular orthogonality constraint, can be introduced. In the paper a multi-resolution least squares interpolator and a significant applicative example are presented. The main advantage of the proposed method consists in its ability to synthesize, within a certain accuracy, the behavior and shape of the field by means of a smaller number of coefficients compared with the count of starting observations.

Keywords

Field • Digital terrain model • Least squares interpolation • Multi-resolution

1 Introduction

Observing the world and trying to model the related phenomena, usually a classification between objects and fields is made (O'Sullivan and Unwin 2003). Objects are characterized by almost exact boundaries (e.g. buildings); fields on the opposite describe widespread phenomena continuously varying across the space. The latter definition can be simply translated formally saying that if $\underline{t}$ is a generic data location in a d-dimensional Euclidean space and $h(\underline{t}_0)$ at $\underline{t}_0$ spatial location is a random quantity, the field is described by some function $h(\underline{t})$, $\underline{t} \in R^d$.

The definition is quite general and it can be applied to a huge variety of phenomena, from the topographic

M.A. Brovelli (✉) · G. Zamboni
Politecnico di Milano, DIIAR, Como Campus via Valleggio 11, 22100 Como, Italy
e-mail: maria.brovelli@polimi.it

N. Sneeuw et al. (eds.), *VII Hotine-Marussi Symposium on Mathematical Geodesy*, International Association of Geodesy Symposia 137, DOI 10.1007/978-3-642-22078-4_9, © Springer-Verlag Berlin Heidelberg 2012

surface of the Earth (field of orthometric heights) to environmental variables.

Even if we suppose that fields are continuous in space, when using point-wise measurements we are obliged to sample them at a discrete point set. From these scattered observations we want finally to re-create a continuous piece of information, i.e. the continuous behavior of the field itself.

The problem is well known and can be dealt with applying exact or approximate interpolation. With the exact methods the interpolating surface passes exactly through all points whose values are known. But if data are affected by errors, it could be better to use, for instance, methods based on a least squares approach, which lead to filters and therefore reduce the effects of errors on the resulting surface. The latter case is named approximate interpolation or approximation.

A great number of both interpolation and approximation methods have been developed from the easiest and poorest "nearest neighboring" to the powerful but more complicate kriging. In the family of deterministic methods, the interpolation by means of spline functions (Moritz and Suenkel 1978; Unser 1999) has met with relatively good success probably due to its ease of use (few parameters to set and few conditions to be satisfied by the data) and at the same time its flexibility.

The paper concentrates on multi-resolution splines function approximation, which is particularly suitable for inhomogeneous spatially distributed data. Scattered exact or approximate data interpolation is a task not yet completely solved in an efficient way. Specifically our aim was to find a method allowing us to create the field, within a certain level of accuracy, starting from a small number of stored coefficients instead of restarting every time from the original observations. Such a solution could be of interest every time we need to store or transfer (think for instance the recently conceived web features and coverage services as proposed by the Open Geospatial Consortium: OGC 2005 and OGC 2008) information about fields consisting of very large amounts of observations. The algorithm was at the beginning studied and used by the authors for vector map warping based on automatically detected homologous pairs (Brovelli and Zamboni 2004). Here we concentrate on applications to digital terrain modelling.

Of course there is a wealth of methods serving the same purpose where the need to follow as exactly as possible observations is balanced against the wish to reduce noise influence. Typically methods stressing the closeness to data are less compressing with respect to methods putting more weight on the model (Hastie et al. 2001).

As an example we have compared our least squares approach with an approach proposed by Lee et al. (1997) where a large lattice of splines is built with a number of knots significantly higher than the number of data in order to obtain an accuracy better than a certain threshold. The mentioned method combines only local data and, as compared to ours, is much faster, though much less compressing (as an example in case of the TR1 dataset in paragraph 5 the same accuracy is reached with more than 250,000 splines instead of the 5,616 used in our approach).

The paper is organized as follows. Section 2 provides a short overview of interpolation by means of spline functions; Sect. 3 presents the main drawback related to such an interpolator, i.e. the problem of avoiding local rank deficiency in case of inhomogeneous spatial distribution. Section 4 illustrates the multi-resolution approach suitable for dealing with the previously mentioned problem. In Sect. 5 we present examples of application related to height field. Finally, conclusions and some remarks are shown.

2 Interpolation by Means of Spline Functions

The spline function interpolator here presented is a deterministic method based on a least squares approach. It is a global method, i.e. each observation contributes to the whole interpolating surface, but at the same time it shows a relatively short range of diffusion of the local information.

We suppose that the h field has been sampled at n locations $\underline{t}_1, \underline{t}_2, \ldots, \underline{t}_n$ and we model these observations $h_o(\underline{t})$ by means of a suitable combination of spline functions (deterministic model) and residuals ν_i seen as noises (stochastic model).

In a two-dimensional space each observation is described by:

$$h_0(\underline{t}_i) = \sum_{l=0}^{N_1-1} \sum_{k=0}^{N_2-1} \lambda_{lk} \varphi_\Delta(\underline{t}_i - \Delta \underline{\tau}_{lk}) + \nu_i \qquad (9.1)$$

where:

- N_1 and N_2, which represent the total number of knots of the splines in x and y directions, depend on the observation domain and on the chosen resolution; global rank deficiency is avoided by assuming more observations than spline coefficients;

- Δ is the grid spacing;
- $\underline{t}_{lk} = [l\ k]^T$ = knot indexes (l, k) of the grid;
- λ_{lk} is the coefficient of the spline at the knot $\underline{t}_{lk}$;
- $\varphi_\Delta(\underline{t})$ is the two dimensional spline function.

3 Inhomogeneous Spatial Distribution and Local Rank Deficiency

A simple interpolation with a regular lattice of splines is prone to local rank deficiency when the spatial distribution of the data is not homogeneous.

In Fig. 9.1a a one-dimensional sample of 30 observations is shown; using high resolution, the leftmost splines can not be determined because their coefficients never appear in the observation equations.

The trivial way to avoid local rank deficiency is to decrease the spline resolution but this decreases the interpolation details, specifically where the original field $h(t)$, showing higher variability, was higher sampled: the resulting coarse approximation curve is shown in Fig. 9.1b (continuous line).

An alternative is to add a further condition in the target function of the least-squares problem to express the "regularity" (for instance the continuity in the first derivative of the surface) of the estimated model (Brovelli et al. 2001). But in this solution, known as spline approximation with Tychonoff regularization, the numbers of unknowns (and therefore the dimension of the matrix to be inverted in the least squares approach) is not efficiently calibrated with the local density of observations. The local rank deficiency is avoided but the procedure, due to its blindness, leads to an increase in the computation time of the estimate just for adding few details. A more suitable solution has to consider the locations where more observations are available.

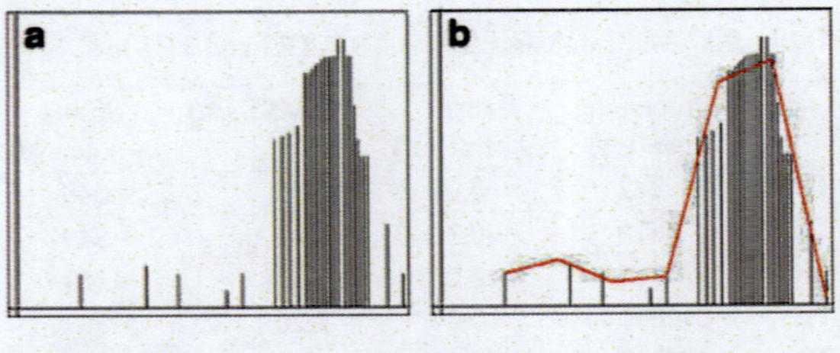

Fig. 9.1 Examples of one-dimensional spline approximation: data (a) and approximating curve (b)

4 Multi-resolution Approach

The main idea is to combine splines with different widths in order to guarantee in every region of the field the resolution adequate to data density, exploiting all available information implicitly stored in the sample.

Different "levels" of splines, corresponding to different halving steps, are considered. A new level corresponds with halving the width of the support of the previous level spline. Taking into account the global field domain $[t_{\min}, t_{\max}]$, the levels and corresponding ordered subintervals are shown in Table 9.1:

Each observation can be described as a linear combination of spline functions of decreasing (halving) Δ width:

$$h(t_i) = \sum_{h=0}^{M-1} \sum_{k=0}^{N_h-1} \lambda_{hk} \cdot \varphi\left(\frac{2^h(t_i - t_{\min})}{\Delta} - k\right) + \nu_i \tag{9.2}$$

where:

- M is the number of levels;
- N_h is the number of splines at level h ($N_h = 2^{h+1} + 1$);
- λ_{hk} is the spline coefficient at h level;
- φ is the one-dimensional spline function;
- $\Delta = (t_{\max} - t_{\min})/2$.

To appreciate the advantage of this approach the multi-resolution spline interpolator is shown in Fig. 9.2. Constraints must be introduced on λ_{hk} coefficients in order to avoid local rank deficiency. A general solution of this problem is till now under study and then, to be cautious, for the moment we have decided to adopt the following criterion: a generic kth spline function at h level

$$\varphi_{\Delta_h}(t - k_i \Delta_h - t_{\min}) \text{ where } \Delta_h = \frac{\Delta}{2^h} \tag{9.3}$$

Table 9.1 Levels of splines and corresponding knots

Level	Knots of the splines
1	$t_{\min}, t_{\min} + \Delta, t_{\max}$
2	$t_{\min}, t_{\min} + \Delta/2, t_{\min} + \Delta, t_{\min} + 3\Delta/2, t_{\max}$
3	$t_{\min}, t_{\min} + \Delta/4, t_{\min} + \Delta/2, t_{\min} + 3\Delta/4, t_{\min} + \Delta,$ $t_{\min} + 5\Delta/4, t_{\min} + 3\Delta/2, t_{\min} + 7\Delta/4, t_{\max}$
4	...

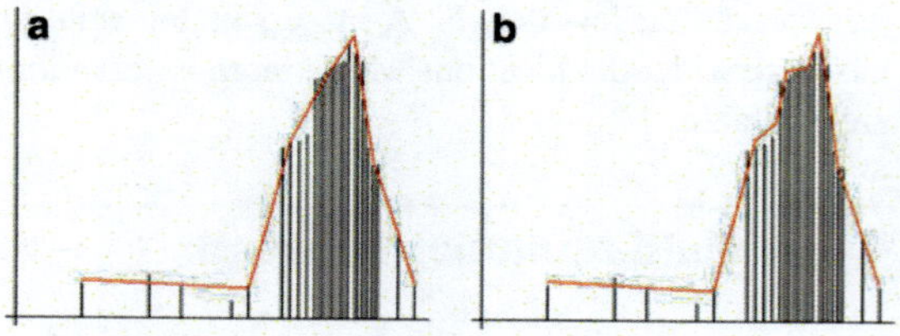

Fig. 9.2 Results of multi-resolution spline approximation with 4 (**a**) and 5 (**b**) levels

is active (i.e. $\lambda_{hk} \neq 0$) if:

- we have at least $f(f > 0)$ observations for each Δ_h half-support of the spline;
- it does not exist a spline at lower level having the same application point.

The bi-dimensional formulation can be directly obtained generalizing the mono-dimensional case.

We suppose that $h(t) = h(t_1,t_2)$ can be modeled as:

$$h(\underline{t}) = \sum_{h=0}^{M-1} \left[\sum_{l=0}^{N_{1h}-1} \sum_{k=0}^{N_{2h}-1} \lambda_{hlk} \varphi_{\underline{\Delta}_h} \left(\underline{t} - \underline{\Delta}_h \underline{\tau}_{lk} - \underline{t}_{\min} \right) \right] \tag{9.4}$$

where:

- $\underline{\Delta}_h = \begin{bmatrix} \Delta_{1h} & 0 \\ 0 & \Delta_{2h} \end{bmatrix}$ $\quad \Delta_{1h} = x$ grid resolution; $\Delta_{2h} = y$ grid resolution;
- $\varphi_{\underline{\Delta}_h}(\underline{t}) = \varphi_{\Delta 1h}(t_1) \cdot \varphi_{\Delta_{2h}}(t_2)$
- M = number of different resolutions used in the model;
- $\underline{\tau}_{lk} = [l\ k]^T$ = knot indexes (l,k) of the grid;
- λ_{hlk} = coefficient of the h resolution spline at the grid knot $\underline{\tau}_{lk}$;
- N_{1h} = number of x grid knots at the h resolution;
- N_{2h} = number of y grid knots at the h resolution.

To avoid local rank deficiency, we generalize the same criterion seen in 1D: at least f observations for each quarter of spline support are needed.

5 Some Tests

To evaluate the performance of the multi-resolution interpolator, we sample data from a LiDAR (Light Detection And Ranging) digital terrain model (DTM); it is a promontory overlooking the lake of Como in Northern Italy. The horizontal spacing of the grid is 2 m × 2 m and the fundamental vertical accuracy (Rood 2004) is of about 20 cm.

A TIN (Triangulated Irregular Network) is extracted from the grid in such a way that the maximum allowable difference in height between the grid and the TIN surface is less than a certain fixed tolerance.

By fixing the tolerance equal to 5 m, 2 m and 1 m, we create respectively the training datasets TR5, TR2 and TR1 containing scattered data. By fixing the tolerance equal to 20 cm (and removing TR5, TR2 and TR1), we create the test dataset TE (used for cross-validation). The original dataset is shown in Fig. 9.3. In Table 9.2 the statistics of the datasets are reported.

The application of the multi-resolution approximate interpolation leads to results summarised in Table 9.3. Moreover, as an example, in Fig. 9.4 the multi-resolution grid for TR5 is represented.

The analysis of results leads to the conclusion that in case of the first test (TR5, i.e., points extracted with a threshold of 5 m) we can reproduce completely the

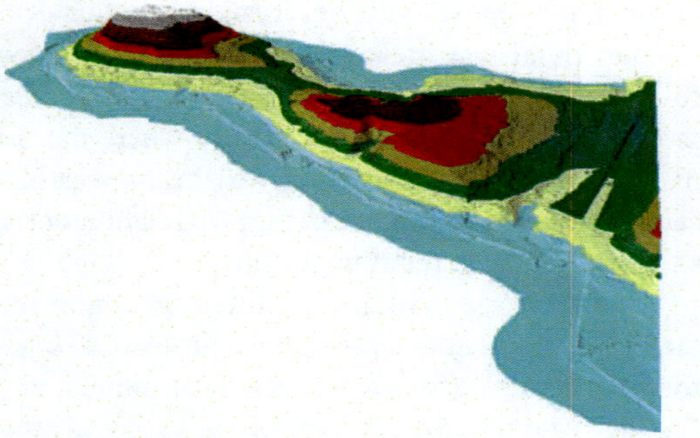

Fig. 9.3 The original dataset (2 m × 2 m DTM)

Table 9.2 Statistics of the DTM, training sets and testing set

	DTM	TR5	TR2	TR1	TE
Count	422610	3274	9256	21656	81869
Min (m)	197,44	197,44	197,44	197,44	197,47
Max (m)	332,27	332,27	332,27	332,27	332,23
Mean (m)	225,27	214,33	225,75	230,81	235,59
St. Dev. (m)	27,80	28,58	30,85	30,36	27,83

Table 9.3 Main statistics for the TR5, TR2 and TR1 test

	Processing time (ms)	Mean error (m)	RMSE (m)	Spline count (#)
TR5	782	0,00	3,39	349
TE	6875	−0,38	4,85	349
TR2	20265	0,00	1,82	1767
TE	44860	0,04	2,20	1767
TR1	314063	0,00	1,07	5616
TE	166109	0,01	1,41	5616

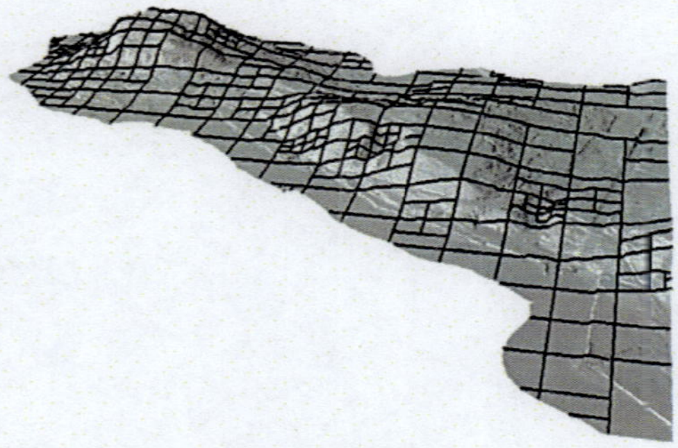

Fig. 9.4 Multi-resolution grid obtained for TR5

field within the same accuracy simply by using 349 spline functions. The time required both for computation of the unknowns and approximation on the large dataset (TE) is globally less than 7 s. In other cases we cannot stay within the thresholds, but in the former we are not so far from them. Obviously the number of spline functions increases and consequently also the time required for computations. The worst case with respect to the processing time is, as expected, the one corresponding to TR1. Another weak point is the value we got for the RMSE of the testing dataset (TE): it is too high and can not be accepted. Therefore we investigated the reasons for such unacceptable result and we individuated two possibilities; one is related to the morphology of the area: it is a varying and complex area, where the coast close to a vertical cliff in the north-easternmost part and a little plateau surrounded by steep slopes in the middle of the promontory give the largest errors. The exclusion of those points (0.6% of the total) leads to an RMSE equal to 1.19 m. The other reason deals with the sampling method which is not optimized with respect to our interpolator: data decimation, as mentioned, was simply done using a TIN algorithm.

Conclusion

A multi-level spline based method to efficiently approximate fields is proposed. The method is applied on terrain heights to derive the corresponding digital terrain model. The problems at the moment unsolved are:

- The theoretical estimation of local rank deficiency must be studied: we used $f = 1$ or $f = 2$ but it could be too conservative;
- The sampling method for the training sets must be better selected: data decimation was done using a TIN algorithm but an optimized sampling (with respect to our method) must be introduced.

Finally, to evaluate the performances of the interpolator, more comparisons with other methods have to be investigated.

References

Brovelli MA, Reguzzoni M, Sansò F, Venuti G (2001) Modelli matematici del terreno per mezzo di interpolatori a spline, Bollettino SIFET, supplemento speciale al n.2 - 2001, pp 55–80

Brovelli MA, Zamboni G (2004) Adaptive transformation of cartographic bases by means of multiresolution spline interpolation. Int Arch Photogram Rem Sens Spatial Inform Sci XXXV(part B2):206–211

Hastie T, Tibshirani R, Friedman J (2001) The elements of statistical learning data mining, inference, and prediction. Springer, New York

Lee S, Wolberg G, Shin SY (1997) Scattered data interpolation with multilevel B-splines. IEEE Trans Vis Comput Graph 3(3):228–244

Moritz H, Suenkel H (1978) Approximation methods in geodesy, Wichmann

OGC (2005) Web Feature Service (WFS) Implementation Specification Version 1.1.0. OGC 04–094. http://www.opengeospatial.org/standards/wfs.Accessed 15 Jan 2010

OGC (2008) Web Coverage Service (WCS) Implementation Standard Version 1.1.2. OGC 07–067r5. http://www.opengeospatial.org/standards/wcs. Accessed 15 Jan 2010

O'Sullivan D, Unwin D (2003) Geographic information analysis, Wiley, New York

Rood M (2004) ASPRS Guidelines Vertical Accuracy Reporting for Lidar Data. ASPRS (American Society for Photogrammetry and Remote Sensing) Lidar Committee (PAD)

Unser M (1999) Splines. A perfect fit for signal and image processing. IEEE Signal Process Mag 16(6):22–38

Flexible Dataset Combination and Modelling by Domain Decomposition Approaches 10

I. Panet, Y. Kuroishi, and M. Holschneider

Abstract

For geodetic and geophysical purposes, such as geoid determination or the study of the Earth's structure, heterogeneous gravity datasets of various origins need to be combined over an area of interest, in order to derive a local gravity model at the highest possible resolution. The quality of the obtained gravity model strongly depends on the use of appropriate noise models for the different datasets in the combination process. In addition to random errors, those datasets are indeed often affected by systematic biases and correlated errors.

Here we show how wavelets can be used to realize such combination in a flexible and economic way, and how the use of domain decomposition approaches allows to recalibrate the noise models in different wavebands and for different areas. We represent the gravity potential as a linear combination of Poisson multipole wavelets (Holschneider et al. 2003). We compute the wavelet model of the gravity field by regularized least-squares adjustment of the datasets. To solve the normal system, we apply the Schwarz iterative algorithms, based on a domain decomposition of the models space. Hierarchical scale subdomains are defined as subsets of wavelets at different scales, and for each scale, block subdomains are defined based on spatial splittings of the area. In the computation process, the data weights can be refined for each subdomain, allowing to take into account the effect of correlated noises in a simple way. Similarly, the weight of the regularization can be recalibrated for each subdomain, introducing non-stationarity in the a priori assumption of smoothness of the gravity field.

I. Panet (✉)
Institut Géographique National, Laboratoire LAREG – Marne-la-Vallée, & Institut de Physique du Globe de Paris, France
e-mail: isabelle.panet@ensg.eu

Y. Kuroishi
Geographical Survey Institute, Space Geodesy Research Division, Tsukuba, Japan

M. Holschneider
University of Potsdam, Department of Applied Mathematics, Potsdam, Germany

N. Sneeuw et al. (eds.), *VII Hotine-Marussi Symposium on Mathematical Geodesy*, International Association of Geodesy Symposia 137, DOI 10.1007/978-3-642-22078-4_10,

We show and discuss examples of application of this method for regional gravity field modelling over a test area in Japan.

Keywords

Regional gravity modeling • Wavelets • Domain decomposition methods

1 Introduction

The knowledge of the geoid is essential for various geodetic and geophysical applications. For instance, it allows the conversion between GPS-derived and levelled heights. It is also the reference surface for ocean dynamics. The geoid can be computed from an accurate gravity model merging all gravity datasets available over the studied area. With the satellite gravity missions GRACE and GOCE, our knowledge of the long and medium wavelengths of the gravity field is or will be greatly improved (Tapley et al. 2004; Drinkwater et al. 2007). The gravity models derived from those missions need to be locally refined using high resolution surface gravity datasets, to obtain the local high resolution models that will be used for geoid modeling. Such refinements also allow to underline possible biases of the surface gravimetry and to improve the local gravity models, provided that a proper combination with the satellite models is carried out, with an appropriate relative weighting of the datasets. Featherstone et al. (1998) provide an overview of methods developed to realize such combination, using the Stokes integration. Different weighting schemes have been proposed by various authors, see for instance Kern et al. (2003). Local functional representations of the gravity field can also be used (see Tenzer and Klees 2008, for an overview). They can be related to least-squares collocation in reproducing kernel spaces (Sansò and Tscherning 2003).

Here we show that wavelet representations of the gravity field can be very useful for that purpose. Because of their localization properties, the wavelets indeed allow a flexible combination of various datasets. We first explain how to compute a local wavelet model of the gravity field combining different datasets by an iterative domain decomposition approach. Then, we provide an example of application over Japan, an area where significant variations of the gravity field occur in a wide range of spatial scales.

2 Discrete Wavelet Frames

The gravity potential is modeled as a linear combination of wavelets. Wavelets are functions well localized both in space and frequency, which makes them interesting to combine data with different spatial and spectral characteristics. To model a geopotential, harmonic wavelets are well-suited (Freeden et al. 1998, Schmidt et al. 2005). We chose to use axisymmetric Poisson multipole wavelets, introduced by Holschneider et al. (2003). Because they can be identified with equivalent non-central multipolar sources at various depths, they are well-suited to model the gravity potential at a regional scale. A wavelet is described by its scale parameter (defining its width), its position parameter (defining its center in space), and its order (defining the multipoles, as explained in Holschneider et al. 2003). Here we use order three Poisson wavelets, which provide a good compromise between spatial and spectral localization.

A wavelet family is built by an appropriate discretization of the scale and position parameters, as explained in Chambodut et al. (2005), Panet et al. (2004, 2006). First, a sequence of scales is chosen in order to ensure a regular coverage of the spectrum. This leads to a dyadic sequence of scales. Then, for each scale, a set of positions on the mean Earth sphere is chosen, in order to ensure a regular coverage of the sphere. The number of positions increases as the scale decreases, because the dimension of harmonics spaces to be generated by the wavelets increases. The wavelets are thus located at the vertices of spherical meshes that are denser and denser as the scale decreases.

The wavelet family thus obtained forms a frame (Holschneider et al. 2003). It provides a complete and stable representation of the modeled field, that may also be redundant. The redundancy is evaluated by comparing the number of wavelets, approximated with band-limited functions, with the dimension of harmonic spaces to be generated (Holschneider

Table 10.1 Description of the wavelet frame used in gravity modelling over Japan.

Scale (km)	Number of wavelets	Area covered
300	380	25/49°N, 129/153°E
150	1406	25/49°N, 129/153°E
75	2,401	29/45°N, 133/149°E
38	9,604	29/45°N, 133/149°E
20	38,220	29/45°N, 133/149°E

et al. 2003). The wavelet family used in this study (see Table 10.1) is over-complete with a redundancy estimated to 1.4 at 10 km resolution.

Here, we build a wavelet family suitable for local gravity field modeling by refinement of a global geopotential model derived from GRACE data with a surface gravity dataset. We need to combine two datasets: the high resolution surface gravity one, and a dataset created at the ground level from the geopotential model up to degree 120, extending two degrees outside the surface data. We then select the wavelets as follows. First, the largest wavelet scale is limited by the size of the area covered with data. Scales larger than half of the width of the area indeed cannot be reliably constrained by local datasets. Second, wavelet positions, for each scale, are selected in the area covered by data. Potential data are modeled by large scale wavelets, and smaller scales are added to model the surface data. This leads to the wavelet set detailed in Table 10.1. Note that, although the central frequency of the smallest scale wavelets is 20 km, the spectrum is well covered down to 10 to 15 km resolution.

3 Domain Decomposition Methods

The coefficients of the wavelet representation of the gravity potential are computed by least-squares fit of the datasets. Each data type can be related to the potential by a functional relation, leading to the observation equations for each dataset i, with $i = 1, \ldots, I$. We obtain the following model:

$$b_i = A_i \cdot x + \varepsilon_i$$

Here, b_i is the measurement vector, A_i the design matrix relating the observations to the wavelet coefficients of the geopotential, and x the coefficients to be determined. The vector ε_i contains the data errors (comprising white noise and correlated errors), with covariance matrix W_i^{-1}. This matrix is not considered perfectly known a priori, and we will parameterize it with variance factors estimated in the computational process (see below). We then derive the normal system for each dataset: $N_i \cdot x = f_i$, where $N_i = A_i^t \cdot W_i \cdot A_i$ is the normal matrix, and $f_i = A_i^t \cdot W_i \cdot b_i$ is the associated right hand side. Summing the normals for all datasets, and adding a regularization term λK leads to the system to solve:

$$(N + \lambda K) \cdot x = f \qquad (10.1)$$

with $N = \sum_i N_i$ and $f = \sum_i f_i$. The regularization may be needed if the data distribution leads to an ill-posed problem, and also to stabilize the inversion if the wavelet family is too redundant.

To solve this problem and introduce flexibility, we apply iterative domain decomposition methods (see for instance Chan and Mathew (1994) and Xu (1992)). Here we briefly recall the principle of such approaches. The least-squares computation of a wavelet model can be viewed as a projection of the data vectors on the space $H = L^2(\sum)$ spanned by the wavelets, where $\sum$ stands for the Earth mean sphere. In the domain decomposition approaches, also named Schwarz algorithms, we split H into smaller subspaces named subdomains $\{H_k, k = 1, \ldots, \mathrm{p}\}$, that may be overlapping or not, so that we have $H = \Sigma_{k=1}^{p} H_k$. In order for the computation to converge fastly, it is interesting to choose not too correlated subdomains, and we naturally define subdomains spanned by the wavelets at a given scale (hereafter referred to as: *scale subdomains*). If the scale subdomains still comprise too many wavelets, which is the case at the smaller scales, we split them into smaller subdomains spanned by subsets of wavelets at the given scale. These are referred to as: *blocks subdomains*. They correspond to a spatial splitting of the area into blocks. To each scale level corresponds a block splitting, with only one block for the larger scales and an increasing number of blocks as the scale decreases. Here we used a simple definition of the blocks, limited by meridians and parallels, but one may consider general shapes, for instance following the physical characteristics of the area. We defined overlapping blocks subdomains, with the size of the overlap area depending on the scale level, in order to speed up the convergency of the computations. On the other hand, our scale subdomains are non-overlapping.

Finally, to each subdomain corresponds a subset of the total wavelet coefficient vector x that is to be computed.

Once the subdomains have been defined, the Schwarz algorithms consist in the following steps: (1) project the data vector and the normal systems on each subdomain, (2) compute the local wavelet coefficients by least-squares fit of the datasets for each wavelet subdomain, (3) gather these subsets of coefficients and update the global solution vector x, dropping the coefficients of wavelets located in the overlap areas and reweighting the coefficients, (4) update the right-hand side and iterate the computation. The coefficient weights are defined as the inverse of the number of overlapping blocks to which they belong. The Schwarz algorithms exist in two versions: the sequential one, where the subdomain solutions are computed sequentially, and the parallel one, where they are computed at the same time. In the case of multi-resolution representations based on wavelets, it is interesting to apply a hybrid algorithm, combining sequential Schwarz iterations on the scales subdomains with parallel iterations on the blocks. To design the iteration path over the scales, we followed the iteration sequences of multi-level iterative methods called multigrids. Multigrid methods (Wesseling 1991; Kusche 2001) are based on the resolution of successive projections of the normal system on coarse or fine grids, applying multi-level Schwarz iterations between subdomains corresponding to the grids. They are similar to a multi-scale resolution using wavelets, the wavelet coefficients at a given scale defining the details to add to a coarser grid approximation in order to obtain the finer grid approximation of the signal. We thus applied standard grid iterations schemes (from coarser to finer grids and vice versa) to design the wavelets scales iteration schemes (from larger to finer scales and vice versa). Figure 10.1 summarizes the approach.

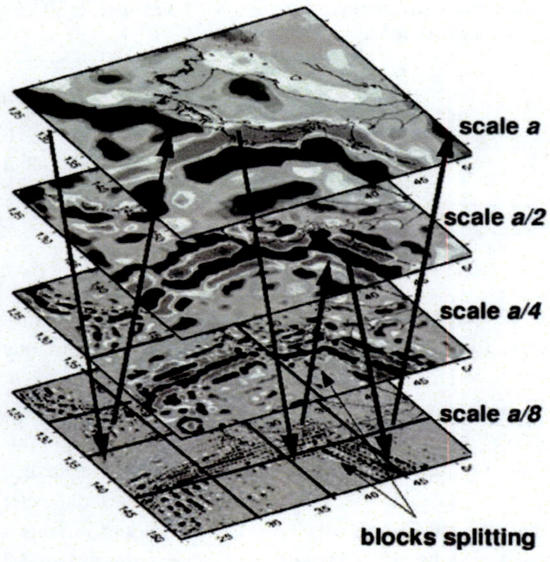

Fig. 10.1 The Schwarz iterative algorithm. Each layer correspond to a scale subdomain, and the solid lines define the spatial blocks. The gray arrows show the iterations over the scales.

In such iterative approach, it is possible to reweight the datasets and the regularization subdomain per subdomain. Following ideas by Ditmar et al. (2007) developed in the case of a Fourier analysis of data errors, we model the datasets systematic errors as a linear combination of wavelets, and add a white noise component. To model the systematic errors, we assume here that there exists a discrete orthonormal wavelet basis B sampled at the data points (it may be different from the Poisson wavelets frame). This requires a regular enough data sampling. Then, the covariance matrices of the errors W_i^{-1} may be written as:

$$W_i^{-1} = F_i^t \cdot D_i^{-1} \cdot F_i, \qquad (10.2)$$

where F_i and D_i are square matrices of size equal to the number of data in the dataset i. F_i is an orthogonal matrix containing the basis B wavelets sampled at the data points, and we have: $F_i^t \cdot F_i = I$. The weight matrix W_i thus verifies:

$$W_i = F_i^t \cdot D_i \cdot F_i. \qquad (10.3)$$

If the datasets errors can be considered locally stationary (without any abrupt variations) over the subdomains, then the projections of D_i over these subdomains can be approximated with a white noise of constant subdomain-dependent variance σ_k, leading to a block-diagonal structure of D_i. Inserting (10.3) into (10.1), and assuming a good enough decorrelation between the Poisson wavelets and the discrete wavelets of basis B for different scales and blocks, leads to a rescaling of the subdomains normals by a factor σ_k. In other words, the subdomains normals highlight different components of matrix W_i, and the scaling factors σ_k are roughly estimated using variance components analysis (Koch 1986; Kusche 2003) of a discrete wavelet transform of the residuals. The regularization may be reweighted in this way too. However, for the convergence of the iterations, a low condition number of the normal system is needed. This may require to increase the regularization weight. Thus, we chose to follow an iterated regularization approach (Engl, 1987), where an initially strong regularization is progressively removed by iterating, the number of iterations finally controlling the amount of regularization.

4 Application Over Japan

We validated the method on synthetic tests considering white and colored noise models, and then apply it to gravity field modeling over Japan, refining a GRACE-derived global geopotential model (EIGEN-GL04S by Biancale et al. 2005) with a local gravity model by Kuroishi and Keller (2005). We generated 5448 potential values at the Earth's surface from the EIGEN-GL04S model up to degree and order 120. The cumulative error is estimated to $0.8\,m^2/s^2$ in rms. The local gravity model is a 3 by 3 min Fayes anomaly grid at the Earth's surface (103,041 data), merging altimetry-derived, marine and land gravity anomalies (Fig. 10.2). The altimetry-derived gravity anomalies are the KMS2002 ones (Andersen and Knudsen 1998). In order to avoid aliasing from the highest frequencies of the gravity data, we removed the highest frequencies from the local model by applying a 10 km resolution moving average filter, corresponding to the wavelet model resolution. From both datasets, we removed the lower frequencies modeled by the lower degree components of the EIGEN-GL04S model, and the residuals are modeled using wavelets. This allows us to construct a hybrid spherical harmonics/wavelets model, refining locally the global EIGEN-GL04S model using wavelets. For the parametrization of the computation, we use 5 scales subdomains. For the scales 300 km and 150 km, there is only one block. For the scales 75 km, 38 km and 20 km, we split the area into 4, 16 and 36 blocks, respectively. We apply a few iteration cycles over the scale subdomains, and a few hundreds iterations over the blocks. We do not iterate our estimations of the datasets reweightings using variance components estimates, but carry out only one weight estimation at the end of the computation of the wavelet model. Indeed, as the potential data are perfectly harmonic, iterated variance components estimates tend to lead to a perfect fit of these data.

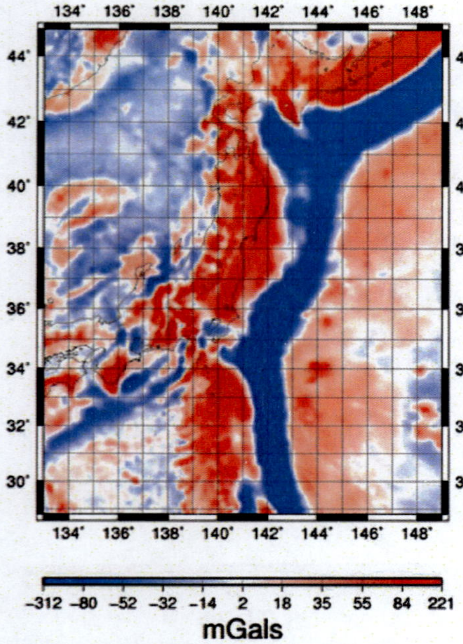

Fig. 10.2 Surface gravity model by Kuroishi and Keller (2005).

The results of a first computation, tightly constrained to the potential data for the large scale wavelets, and with a progressive increase of the weight of the surface data as the scale decreases, highlighted discrepancies between the two datasets, that we attributed to large scale systematic errors in the surface gravity model. Applying a low-pass filter to the residuals to the gravity anomaly data, we defined a corrector model and subtracted it from the surface gravity data. Applying the wavelet method on the corrected datasets allows to progressively improve the resulting wavelet model, and refine our corrector model. The final corrector thus obtained is represented on Fig. 10.3. It is consistent with results from Kuroishi (2009), underlining similar biases in the surface gravity model from a comparison with the GGM02C/EGM96 geopotential model. The residuals of the final wavelet model to the potential and gravity anomaly data are represented on Fig. 10.4, and the final wavelet model on Fig. 10.5. The RMSs of residuals

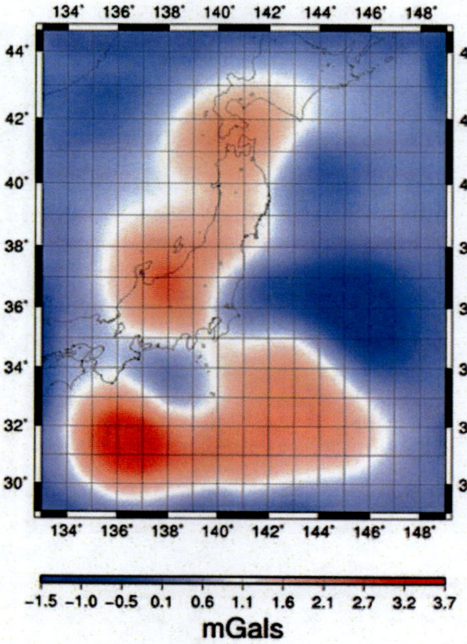

Fig. 10.3 The final corrector model to the surface gravity anomaly model, derived by low-pass filtering of the residuals of the wavelet model to the gravity anomaly data.

are $0.80\mathrm{m}^2/\mathrm{s}^2$ for the potential data, and 0.50 mGals at 15 km resolution for the corrected anomaly data. This is consistent with our a priori knowledge on the data quality. We also note that these residuals do not show any significant bias. The resolution of the wavelet model may be slightly coarser than that of the surface gravity model, which is why we observe very small scale patterns in the gravity anomaly residuals map. Small edge effects may also be present.

Conclusion

We developed an iterative method for regional gravity field modeling by combination of different datasets. It is based on a multi-resolution representation of the gravity potential using Poisson multipole wavelets. We define scale and blocks subdomains, and carry out the computation of the wavelet model subdomain per subdomain. This allows to introduce a flexible reweighting of the datasets in different wavebands and in different areas. Applying this approach to the example of gravity field modeling over Japan, a challenging area with important gravity undulations, allows

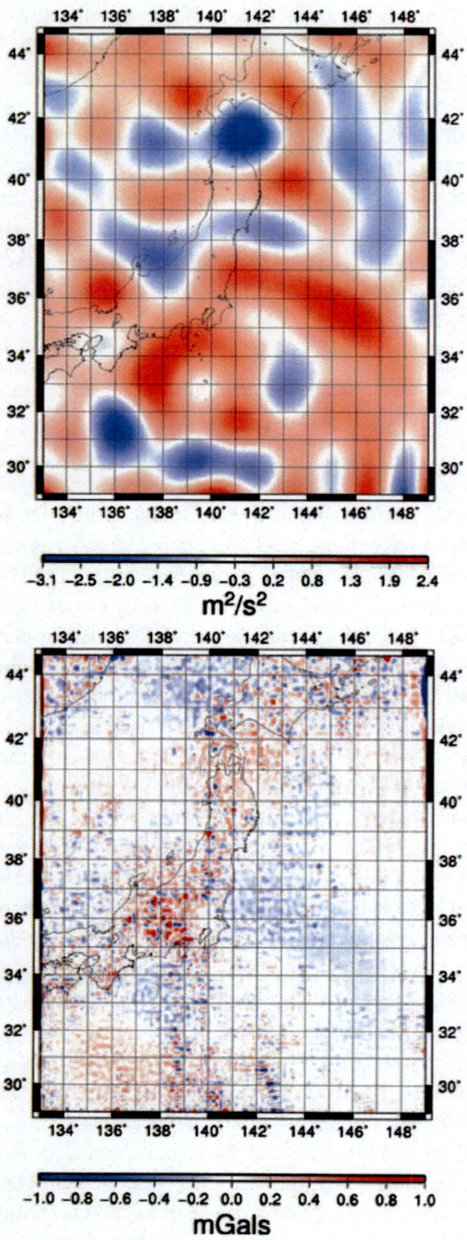

Fig. 10.4 Geographic distribution of residuals in the final combination. *Top panel*: potential residuals to degree 120. *Bottom panel*: residuals of corrected gravity anomalies at 15 km resolution.

to derive a hybrid spherical harmonics/wavelet model at about 15 km resolution, refining a global geopotential model with a local high resolution gravity model. Finally, the method can be used

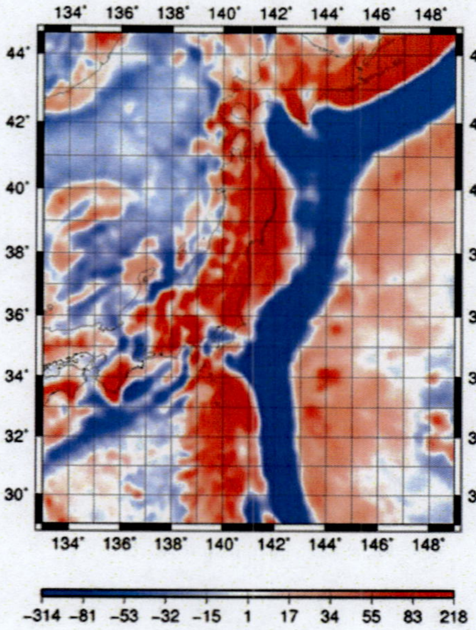

Fig. 10.5 Surface Fayes gravity anomalies computed from the final wavelet model obtained in the present study.

to regional modeling of the forthcoming GOCE level 2 gradient data, in combination with surface gravimetry.

References

Andersen OB, Knudsen P (1998) Global marine gravity field from the ERS-1 and Geosat geodetic mission altimetry. J Geophys Res 103:8129–8137

Biancale R, Lemoine J-M, Balmino G, Loyer S, Bruisma S, Perosanz F, Marty J-C, Gegout P (2005) Three years of decadal geoid variations from GRACE and LAGEOS data, CD-Rom, CNES/GRGS product

Chambodut A, Panet I, Mandea M, Diament M, Jamet O, Holschneider M (2005) Wavelet frames: an alternative to the spherical harmonics representation of potential fields. Geophys J Int 168:875–899

Chan T, Mathew T (1994) Domain decomposition algorithms. Acta Numerica 61–143

Ditmar P, Klees R, Liu X (2007) Frequency-dependent data weighting in global gravity field modeling from satellite data contaminated by non-stationary noise. J Geodes 81:81–96

Drinkwater M, Haagmans R, Muzzi D, Popescu A, Floberghagen R, Kern M, Fehringer M (2007) The GOCE gravity mission: ESA's first core explorer. Proceedings of the 3rd GOCE User Workshop, 6–8 November 2006, Frascati, Italy, pp 1–7, ESA SP-627

Engl HW (1987) On the choice of the regularization parameter for iterated Tikhonov regularization of ill-posed problems. J Approx Theory 49:55–63

Featherstone W, Evans JD, Olliver JG (1998) A Meissl-modified Vaníček and Kleusberg kernel to reduce the truncation error in gravimetric geoid computations. J Geodes 72(3): 154–160

Freeden W, Gervens T, Schreiner M (1998) Constructive approximation on the sphere (with applications to geomathematics), Oxford, Clarendon, Oxford

Holschneider M, Chambodut A, Mandea M (2003) From global to regional analysis of the magnetic field on the sphere using wavelet frames. Phys Earth Planet Inter 135: 107–124

Kern M, Schwarz P, Sneeuw N (2003) A study on the combination of satellite, airborne, and terrestrial gravity data. J Geodes 77(3–4):217–225

Koch K-R (1986) Maximum likelihood estimate of variance components. Bull Geodes 60:329–338

Kuroishi Y (2009) Improved geoid model determination for Japan from GRACE and a regional gravity field model. Earth Planets Space 61(7):807–813

Kuroishi Y, Keller W (2005) Wavelet improvement of gravity field-geoid modeling for Japan. J Geophys Res 110:B03402. doi:10.1029/2004JB003371

Kusche J (2001) Implementation of multigrid solvers for satellite gravity anomaly recovery. J Geodes 74:773–782

Kusche J (2003) A Monte-Carlo technique for weight estimation in satellite geodesy. J Geodes 76(11–12):641–652

Panet I, Jamet O, Diament M, Chambodut A (2004) Modelling the Earth's gravity field using wavelet frames. In: Proceedings of the Geoid, Gravity and Space Missions 2004 IAG Symposium, Porto

Panet I, Chambodut A, Diament M, Holschneider M, Jamet O (2006) New insights on intraplate volcanism in French Polynesia from wavelet analysis of GRACE, CHAMP and sea-surface data. J Geophys Res 111:B09403. doi:10.1029/2005JB004141

Sansò F, Tscherning CC (2003) Fast spherical collocation: theory and examples. J Geodes, 77, 101-112

Schmidt M, Kusche J, Shum CK, Han S-C, Fabert O, van Loon J (2005) Multiresolution representation of regional gravity data. In: Jekeli, Bastos, Fernandes (eds) Gravity, geoid and space missions. IAG Symposia 129, Springer, New York

Tapley B, Bettadpur S, Watkins M, Reigber C (2004) The gravity recovery and climate experiment: Mission overview and early results. Geophys Res Lett 31:L09607. doi:10.1029/2004GL019920

Tenzer R, Klees R (2008) The choice of the spherical radial basis functions in local gravity 709 field modeling. Studia Geophysica Geodetica 52(3):287–304

Wesseling P (ed) (1991) An introduction to multigrid methods. Wiley, New York, 294 p

Xu J (1992) Iterative methods by space decomposition and subspace correction. SIAM Rev 34(4):581–613

Optimal Hypothesis Testing in Case of Regulatory Thresholds

11

I. Neumann and H. Kutterer

Abstract

In this study hypothesis testing is treated, when neither the probability density function (pdf) of the test statistic under the null hypothesis nor the pdf of the test statistic under the alternative hypothesis are known. First, the classical procedure in case of random variability is reviewed. Then, the testing procedure is extended to the case when the uncertainty of the measurements comprises both random and systematic errors. Both types of uncertainty are treated in a comprehensive way using fuzzy-random variables (FRVs) which represent a combination of probability and fuzzy theory. The classical case of random errors (absence of systematic errors) is a special case of FRVs. The underlying theory of the procedure is outlined in particular. The approach allows the consideration of fuzzy regions of acceptance and rejection. The final (optimal) test decision is based on the utility theory which selects the test decision with the largest expected utility as the most beneficial one. An example illustrates the theoretical concept.

Keywords

Hypothesis testing • Decision making • Utility theory • Imprecise data • Fuzzy data analysis • Regulatory thresholds

1 Motivation

In hypothesis testing three important cases are of interest: In case (1) both the probability density functions (pdf) of the test statistics under the null and alternative hypothesis are known. A second case (2) is when the null hypothesis is much more probable and therefore the pdf of the alternative hypothesis is not (exactly) known. In case (3) neither the pdf of the test statistic under the null hypothesis nor the pdf of the test statistic under the alternative hypothesis are known. Whereas in case (1) no regions of acceptance or rejection are needed for the test decision (Luce and Raiffa 1989), in the cases (2) and (3) acceptance and rejection regions have to be defined. In case (3) there is no statistically founded method to define the region of acceptance. However, it can be constructed, e.g., based on expert knowledge what leads to regulatory thresholds. The test decision in case (2) and (3) is

I. Neumann (✉)
Institute of Geodesy, Geodetic Lab, University FAF Munich, Werner-Heisenberg-Weg 39, 85577 Neubiberg, Germany
e-mail: ingonewman@web.de

H. Kutterer
Geodetic Institute, Leibniz Universität Hannover, Nienburger Straße 1, 30167 Hannover, Germany

N. Sneeuw et al. (eds.), *VII Hotine-Marussi Symposium on Mathematical Geodesy*, International Association of Geodesy Symposia 137, DOI 10.1007/978-3-642-22078-4_11, © Springer-Verlag Berlin Heidelberg 2012

obtained by comparing a measured value with the region of acceptance and rejection.

In reality, the total uncertainty budget usually comprises at least two types of uncertainty: random variability which reflects uncontrollable effects during observation and data processing, and imprecision which is, e.g., due to remaining systematic errors between data and model. When uncertainty is present in the measurements, the test decision is not (always) clear. The general theory of decision making provides the required procedures for an optimal test decision. The most beneficial one (with the largest expected utility) is selected.

In this study, case (3) of hypothesis testing is of main interest. The classical procedure in case of random variability will be extended to imprecise data.

2 Uncertainty Modeling

In this paper two different aspects of uncertainty modeling are of interest. The first one concerns linguistic uncertainty and the second one measurement uncertainty.

Linguistic uncertainty can be modeled with the aid of fuzzy theory (Zadeh 1965). Here, LR-fuzzy intervals according to Dubois and Prade (1980) are used. An LR-fuzzy interval is a special case of a one-dimensional *fuzzy set* $\tilde{A}$ which is described by a *membership function* $m_{\tilde{A}}(x)$:

$$\tilde{A} := \{(x, m_{\tilde{A}}(x)) | x \in \mathbb{R}\} \text{ with } m_{\tilde{A}} : \mathbb{R} \to [0, 1]. \tag{11.1}$$

The *core* of a fuzzy set is the classical set of elements of $\tilde{A}$ with membership degree equal to 1.

An *LR-fuzzy interval* is then defined as a fuzzy set over $\mathbb{R}$ with a non-empty core. Its membership function is constructed by monotonously decreasing (left and right) *reference functions* L and R. For L and R the range of values is [0, 1]. For a graphical sketch see Fig. 11.1. LR-fuzzy intervals can be represented by $\tilde{X} = (x_m, r, c_l, c_r)_{LR}$. The *midpoint* is denoted by x_m. The *radius* of the interval representing the core is r. Together with the deterministic *spreads* c_l and c_r it serves as a measure of linguistic uncertainty. Linguistic uncertainty plays a key role in the definition of the regions of acceptance and rejection, see Sect. 3. Strategies to construct fuzzy numbers or fuzzy intervals based on expert knowledge are given in Nguyen and Kreinovich (1996) and Neumann (2009).

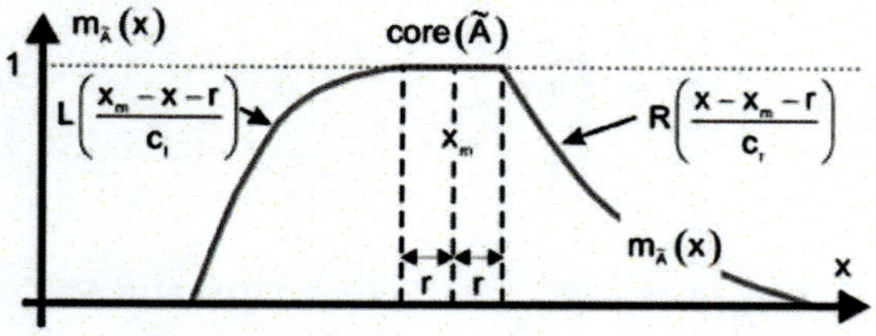

Fig. 11.1 LR-fuzzy interval with different reference functions

Based on $m_{\tilde{A}}(x)$ set-theoretical operations can be consistently extended to fuzzy sets: the intersection can be defined as $m_{\tilde{A}\cap\tilde{B}} = \min(m_{\tilde{A}}, m_{\tilde{B}})$ and the complement as $m_{\tilde{A}^c} = 1 - m_{\tilde{A}}$.

The combination of random variability and imprecision in measurement uncertainty is based on the theory of fuzzy-random variables (FRV), see, e.g., Kwakernaak (1978).

FRVs serve as basic quantities; they are an extension of the classical probability theory. For this reason, all statistical methods have to be extended to imprecise data and all statistical quantities are imprecise by definition.

Random variability is introduced through the midpoint of an LR-fuzzy interval which is modeled as a random variable and hence treated by methods of stochastics. In order to model both types of uncertainty in a comprehensive way, random variability is superposed by imprecision which is due to non-stochastic errors of the measurements and the physical model with respect to reality.

In general, this yields an *LR-fuzzy-random interval* $\underline{\tilde{X}} = (\underline{X}_m, X_r, c_l, c_r)_{LR}$ with a stochastic midpoint $\underline{X}_m$; the underline indicates a random variable. Actually, $\underline{\tilde{X}}$ is a special case of a fuzzy-random variable (Möller and Beer 2004). In contrast to the general case only the expectation value is considered as superposed by fuzziness but not the variance. Without imprecision the pure stochastic case is obtained ($X_r = c_l = c_r = 0$), see Dubois and Prade (1980) for examples. In case of normal distributed values for the random part, the standard deviation σ_x is the carrier of the stochastic uncertainty, and the *radius* X_r and *spreads* c_l and c_r are the carrier of imprecision.

A geometric interpretation of a FRV with $c_l = c_r = 0$ is given in Fig. 11.2. The lower and upper bound of the core define the variation range of the random midpoint.

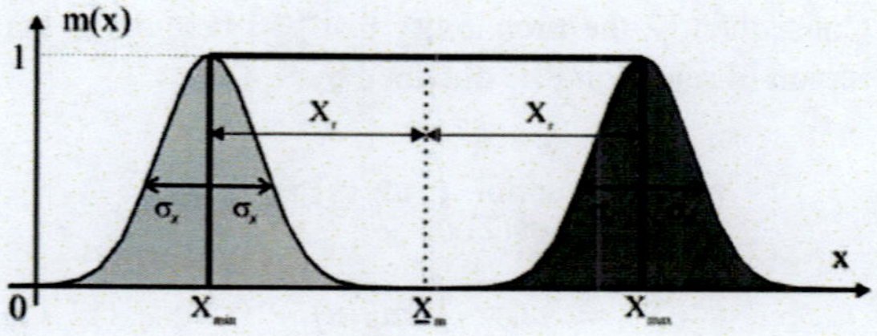

Fig. 11.2 FRV for $c_l = c_r = 0$ with the variation range of the random uncertainty component inside the core element

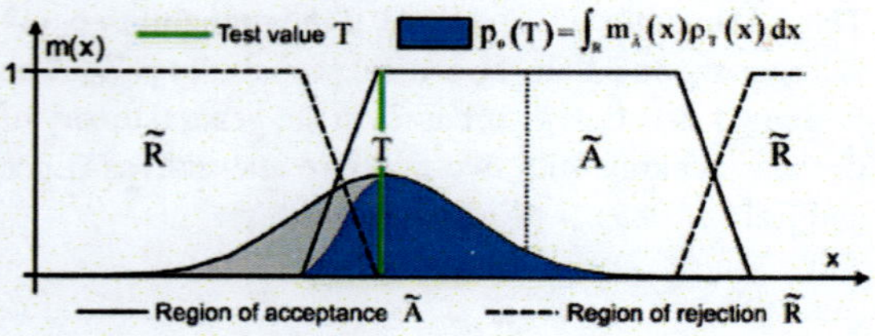

Fig. 11.3 Computation of the probability $p_0(T)$ that the test value T belongs to the imprecise region of $\tilde{A}$.

3 The Pure Stochastic Case

In the pure stochastic case of hypothesis testing in case of regulatory thresholds, the uncertainty of the test statistic T is described by a pdf: $T \sim \rho_T(x)$.

In the pure stochastic case, the regions of acceptance A and rejection R are defined through the intervals $[A] = [A^l, A^u]$ and $[R] = [R^l, R^u]$, respectively. The probabilities $p_0(T)$ and $p_1(T)$ that the test value T belongs to A or R can be computed by:

$$p_0(T) = \int_A \rho_T(x)dx \quad \text{and} \quad p_1(T) = 1 - \int_A \rho_T(x)dx. \tag{11.2}$$

Due to the linguistic imprecision or fuzziness of the formulated hypotheses such as "The allowed length of a machine axis is approximately..." the regions of acceptance and rejection may be fuzzy. This leads to the definition of regions of transition between strict acceptance and rejection of a given hypothesis.

The fuzzy region of acceptance $\tilde{A}$ is then defined with an LR-fuzzy interval by:

$$\tilde{A} := \left\{(x, m_{\tilde{A}}(x)) \,|x \in \mathbb{R}\right\}, \tag{11.3}$$

The imprecise region of rejection $\tilde{R}$ is the fuzzy-theoretical complement of $\tilde{A}$:

$$m_{\tilde{R}}(x) = 1 - m_{\tilde{A}}(x) \quad \forall x \in \mathbb{R}. \tag{11.4}$$

The membership function $m_{\tilde{A}}(x)$ is introduced in equation (11.2) to obtain the probability $p_0(T)$ that the test value T belongs to $\tilde{A}$, e.g., Klir (2006):

$$p_0(T) = \int_{\mathbb{R}} m_{\tilde{A}}(x)\rho_T(x)dx. \tag{11.5}$$

The probability $p_1(T)$ that T belongs to $\tilde{R}$ is:

$$p_1(T) = 1 - p_0(T) = \int_{\mathbb{R}} m_{\tilde{R}}(x)\rho_T(x)dx. \tag{11.6}$$

A graphical representation of the computed probability in (11.5) is given in Fig. 11.3.

In hypothesis testing four situations are possible in the final test decision. The main idea behind utility theory is to judge each possible decision with a utility value:

- $U_{0,0}$: utility for a correct choice of the null hypothesis.
- $U_{1,0}$: utility for an incorrect choice of the alternative hypothesis (type I error).
- $U_{1,1}$: utility for a correct choice of the alternative hypothesis.
- $U_{0,1}$: utility for an incorrect choice of the null hypothesis (type II error).

In the next step, the expected total utility K_0 and K_1 for the null and alternative hypothesis is computed:

$$\begin{aligned} K_0 &= p_0(T)U_{00} + p_1(T)U_{01} \\ &= p_0(T)(U_{00} - U_{01}) + U_{01} \\ K_1 &= p_0(T)U_{10} + p_1(T)U_{11} \\ &= p_0(T)(U_{10} - U_{11}) + U_{11} \end{aligned} \tag{11.7}$$

Finally, the test decision is based on the selection of the most beneficial hypothesis:

$$p_0(T)U_{00} + p_1(T)U_{01} \geq p_0(T)U_{10} + p_1(T)U_{11}. \tag{11.8}$$

This equation can be rearranged:

$$p_0(T) \geq p_{0,crit} = \frac{U_{11} - U_{01}}{U_{00} - U_{01} - U_{10} + U_{11}}. \tag{11.9}$$

The null hypothesis is selected, if the probability $p_0(T)$ is larger or equal than the critical probability $p_{0,crit}$. The presented test strategy is based on the general theory of decision making with two possible alternatives (Luce and Raiffa 1989).

4 The Extension to Imprecise Data

It is obvious that in the scenario described in Sect. 3 there is a strict (and unique) test decision. Nevertheless, in practical test situations imprecision of the data often superposes (and hence mitigates) this procedure (imprecise case). For this reason, the test decision is extended to the fuzzy-random-variables (FRVs) presented in Sect. 2.

The described procedure allows the treatment of multidimensional test statistics. The extension principle (Zadeh 1965) is used and the multidimensional case is mapped to a one-dimensional test statistic (Kutterer and Neumann 2007).

Due to the lack of space in this paper, *imprecision* is treated in terms of intervals. Hence, according to Fig. 11.2, the test statistic is a fuzzy-random variable $[\underline{\tilde{T}}]$, with $c_l = c_r = 0$ and $\underline{T}_m \sim \rho_{\underline{T}_m}(x)$ in the imprecise case. The situation with imprecise regions of acceptance and rejection is illustrated in Fig. 11.4.

When imprecision is considered in addition to the stochastic uncertainty component and when applying the calculation rules of FRVs, then the probability $[p_0(\underline{\tilde{T}})] = [p_0^l(\underline{\tilde{T}}), p_0^u(\underline{\tilde{T}})]$ that the test value $[\underline{\tilde{T}}]$ belongs to the region of acceptance $\tilde{A}$ is, see (Neumann 2009):

$$p_0^l(\underline{\tilde{T}}) = \min_{T_m \in [\underline{\tilde{T}}]} \int_{\mathbb{R}} m_{\tilde{A}}(x)\rho_{\underline{T}_m}(x)dx,$$

$$p_0^u(\underline{\tilde{T}}) = \max_{T_m \in [\underline{\tilde{T}}]} \int_{\mathbb{R}} m_{\tilde{A}}(x)\rho_{\underline{T}_m}(x)dx. \quad (11.10)$$

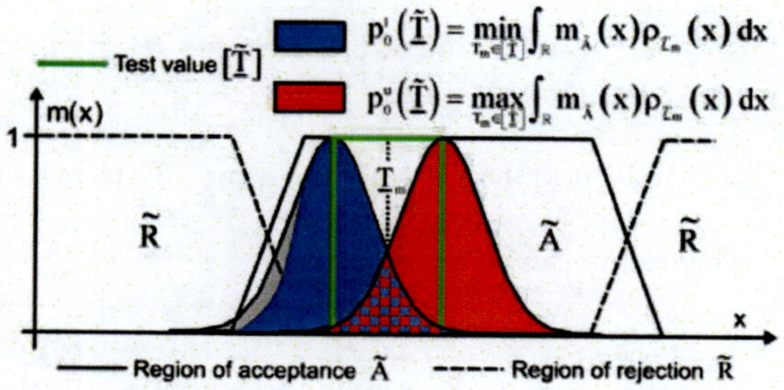

Fig. 11.4 The computation of the range of values for the probability $[p_0(\underline{\tilde{T}})]$ that the test value $[\underline{\tilde{T}}]$ belongs to $\tilde{A}$

Consequently, the probability that $[\underline{\tilde{T}}]$ belongs to the region of rejection $\tilde{R}$ is obtained by:

$$p_1^l(\underline{\tilde{T}}) = \min_{T_m \in [\underline{\tilde{T}}]} \int_{\mathbb{R}} m_{\tilde{R}}(x)\rho_{\underline{T}_m}(x)dx,$$

$$p_1^u(\underline{\tilde{T}}) = \max_{T_m \in [\underline{\tilde{T}}]} \int_{\mathbb{R}} m_{\tilde{R}}(x)\rho_{\underline{T}_m}(x)dx. \quad (11.11)$$

The expected total utility $[K_{0,\underline{\tilde{T}}}] = [K_{0,\underline{\tilde{T}}}^l, K_{0,\underline{\tilde{T}}}^u]$ for the null hypothesis is computed by (see (11.7)):

$$K_{0,\underline{\tilde{T}}}^l = \min_{p_0 \in [p_0]} (p_0(\underline{\tilde{T}})U_{00} + [1 - p_0(\underline{\tilde{T}})]U_{01}),$$

$$K_{0,\underline{\tilde{T}}}^u = \max_{p_0 \in [p_0]} (p_0(\underline{\tilde{T}})U_{00} + [1 - p_0(\underline{\tilde{T}})]U_{01}). \quad (11.12)$$

In case of unimodal distributions we obtain:

$$K_{0,\underline{\tilde{T}}}^l = p_0^l(\underline{\tilde{T}})(U_{00} - U_{01}) + U_{01},$$

$$K_{0,\underline{\tilde{T}}}^u = p_0^u(\underline{\tilde{T}})(U_{00} - U_{01}) + U_{01}. \quad (11.13)$$

Applying the same idea to obtain the expected total utility $[K_{1,\underline{\tilde{T}}}] = [K_{1,\underline{\tilde{T}}}^l, K_{1,\underline{\tilde{T}}}^u]$ for the alternative hypothesis leads for unimodal distributions to:

$$K_{1,\underline{\tilde{T}}}^l = p_0^u(\underline{\tilde{T}})(U_{10} - U_{11}) + U_{11},$$

$$K_{1,\underline{\tilde{T}}}^u = p_0^l(\underline{\tilde{T}})(U_{10} - U_{11}) + U_{11}. \quad (11.14)$$

The final test decision requires the comparison of two interval valued utilities. In 1951, Hurwicz developed a method, which allows the mapping of an interval into a single equivalent value. The so-called Hurwicz criterion (Hurwicz 1951) is a linear combination of the optimistic and pessimistic outcome of a decision:

$$\kappa K_{0,\underline{\tilde{T}}}^u + (1-\kappa)K_{0,\underline{\tilde{T}}}^l \geq \kappa K_{1,\underline{\tilde{T}}}^u + (1-\kappa)K_{1,\underline{\tilde{T}}}^l, \quad (11.15)$$

with $\kappa \in [0, 1]$. The parameter κ represents the user preferences with $\kappa = 1$ as most optimistic outcome and $\kappa = 0$ as most pessimistic outcome. Reformulating (11.15) leads to the unique decision criterion:

$$\kappa \begin{Bmatrix} \leq \\ > \end{Bmatrix} \phi_{crit} \in \mathbb{R} \Rightarrow \begin{Bmatrix} [\underline{\tilde{T}}] \text{ belongs to } \tilde{A} \\ [\underline{\tilde{T}}] \text{ belongs to } \tilde{R} \end{Bmatrix} \quad \text{with}$$

$$\phi_{crit} = \frac{p_0^u(\underline{\tilde{T}})(U_{10} - U_{11}) - p_0^l(\underline{\tilde{T}})(U_{00} - U_{01}) + U_{11} - U_{01}}{(p_0^u(\underline{\tilde{T}}) - p_0^l(\underline{\tilde{T}}))(U_{00} - U_{01} + U_{10} - U_{11})}. \quad (11.16)$$

The selection of κ depends on a particular application. In order to keep as many data as possible for further processing a conservative strategy is typically preferred in outlier testing; then $\kappa = 1$ is a proper choice. If, however, the result of the test is safety relevant, κ has to be small; best it is equal to 0.

5 Application Example

In practice, there are various possibilities for the application of the presented strategy. In the following, the survey of the length of a machine axis is of particular interest. The length is observed with two different instruments (*Leica* TCA 1101 and TCA 2003) The uncertainties of the length of the machine axis are computed based on the measurement uncertainties of the instruments. The results of an imprecise analysis according to (Neumann 2009) are given in Table 11.1. The random uncertainty component is Gaussian.

The region of acceptance is defined through linear reference functions L and R. The transition region starts at 9,997 mm and reaches the membership degree of 1 at 9,998 mm. At the value of 10,002 mm the membership degree decreases linearly to zero until 10,003 mm. The midpoint $\underline{T}_m$ of the test value is 9,999.2 mm. The radius of the test value depends on the particular instrument and is depicted in the right column of Table 11.1. The spread parameters c_l and c_r are zero. The midpoint of the test value lies inside the core element of $\tilde{A}$. Due to the lower uncertainty of the TCA 2003, the probability $[p_0(\tilde{\underline{T}})]$ that the test value $[\tilde{\underline{T}}]$ belongs to $\tilde{A}$ is larger than for the TCA 1101 [see (10) and (11)] (Table 11.2).

Table 11.1 Uncertainties of the length of the machine axis

Distance	Standard deviation (mm)	Interval radius T_r (mm)
TCA 1101	2.1	1.1
TCA 2003	0.9	0.8

Table 11.2 The probabilities $[p_0(\tilde{\underline{T}})]$ and $[p_1(\tilde{\underline{T}})]$

Probabilities	TCA 2003	TCA 1101
$p_0^l(\tilde{\underline{T}})$	0.8350	0.5956
$p_0^u(\tilde{\underline{T}})$	0.9905	0.7577
$p_1^l(\tilde{\underline{T}})$	0.0095	0.2423
$p_1^u(\tilde{\underline{T}})$	0.1650	0.4044

Table 11.3 Expected utilities for the two hypotheses

Utilities	TCA 2003	TCA 1101
$K^l_{0,\tilde{\underline{T}}}$	−3,320.26€	−5,234.89€
$K^u_{0,\tilde{\underline{T}}}$	−2,075.78€	−3,938.75€
$K^l_{1,\tilde{\underline{T}}}$	−3,000.00€	−3,000.00€
$K^u_{1,\tilde{\underline{T}}}$	−3,000.00€	−3,000.00€

The utility value $U_{0,0}$ for a correct choice of H_0 is defined through the cost (2,000€) for the installation of the machine axis. The two utility values $U_{1,0}$ and $U_{1,1}$ lead to a mechanical finishing of the machine axis and therefore to costs of 3,000€. The worst case scenario is a type II error, when the length of the machine axis is not correct but it is classified as belonging to the region of acceptance. The costs for $U_{0,1}$ are 10,000€.

The expected utilities for the two hypotheses are computed with (11.13) and (11.14). The results are given in Table 11.3.

The null hypothesis for the TCA 1101 is rejected in any case. The final test decision for the TCA 2003 depends on the particular situation. The test value is classified as belonging to the region of acceptance, when ϕ_{crit} from (11.16) is larger than 0.257.

6 Conclusions and Outlook

An optimal tests decision requires the consideration of both measurement uncertainties and the consequences of test decisions. In this paper, a strategy is shown which allows handling data which are both randomly varying and imprecise. At present, the final test decision in case of fuzzy-random intervals needs a final expert statement, if the present situation is safety-relevant or not.

However, the presented methods allow improved statements about the influence of uncertainties and consequences in e.g., the significance of measurement results and in the sensitivity of measurement setups.

There is some more work needed concerning the adaptation of the theoretical methods to practical applications. It is certainly worthwhile to extend the methods to the fuzzy-random case with arbitrary membership functions (based on expert opinions about imprecision). The methods must also be extended to test situations, where the pdf of the null and alternative hypothesis is known.

Acknowledgement The paper shows results of the research projects KU 1250/4–1 and 4–2 "Geodätische Deformationsanalysen unter Berücksichtigung von Beobachtungsimpräzision und Objektunschärfe", which was funded by the German Research Foundation (DFG).

References

Dubois DJ, Prade HM (1980) Fuzzy sets and systems: theory and applications. Academic, New York

Hurwicz L (1951) Optimality criteria for decision making under ignorance. Cowles commission, discussion paper, statistics, no. 370

Klir GJ (2006) Uncertainty and information: foundations of generalized information theory. Wiley, Hoboken

Kutterer H, Neumann I (2007) Multidimensional statistical tests for imprecise data. In: Xu P, Liu J, Dermanis A (eds) Proceedings of the 6th Hotine-Marussi-Symposium, IAG, vol 132, pp 232–237

Kwakernaak H (1978) Fuzzy random variables: definitions and theorems. Inf Sci 15:1–29

Luce RD, Raiffa H (1989) Games and decisions: introduction and critical survey. Dover, New York

Möller B, Beer M (2004) Fuzzy randomness – uncertainty in civil engineering and computational mechanics. Springer, Berlin and New York

Neumann I (2009) On the analysis of an extended uncertainty budget in parameter estimation and hypothesis testing (in German). PhD thesis, Leibniz University of Hannover, Germany

Nguyen HT, Kreinovich V (1996) Nested intervals and sets: concepts, relations to fuzzy sets, and applications. In: Kearfott RB, Kreinovich V (eds) Applications of interval computations. Kluwer, Dordrecht, pp 245–290

Zadeh LA (1965) Fuzzy sets. Inf Control 8:338–353

Sequential Monte Carlo Filtering for Nonlinear GNSS Trajectories

12

H. Alkhatib, J.-A. Paffenholz, and H. Kutterer

Abstract

The Kalman filter is supposed to be the optimal analytical closed-form solution for the Bayesian space-state estimation problem, if the state-space system is linear and the system noises are additive Gaussian. Unfortunately, except in the above mentioned cases, there is no closed-form solution to the filtering problem. So it is necessary to adopt alternative techniques in order to solve the Bayesian filtering problem. Sequential Monte Carlo (SMC) filtering – or commonly known as particle filter – is a well known approach that allows to reach this goal numerically, and works properly with nonlinear, non-Gaussian state estimation. However, computational difficulties could occur concerning the sufficient number of particles to be drawn. We present in this paper a more efficient approach, which is based on the combination of SMC filter and the extended Kalman filter. We identified the resulting filter as extended Kalman particle filter (EKPF). This filter is applied to a method for the direct geo-referencing of 3D terrestrial laser scans.

Keywords

Nonlinear state estimation • Bayesian Filtering • Sequential Monte Carlo Filtering • GNSS

1 Introduction

Linear filtering theory according to Kalman and Bucy (1960) is optimal only if the system, which consists of measurement and transition equations, is linear and the error process is Gaussian. Unfortunately, the modeling of reality sometimes differs from these optimal assumptions and nonlinear, non-Gaussian, and non-stationary state estimation should be taken into account. Thus over the years a multitude of approximate nonlinear filters has been proposed; see e.g., Doucet et al. (2001), and Simon (2006). A well known analytical approximation to handle a nonlinear system is to linearize the measurement and the system equations using Taylor series expansions; see e.g., Simon (2006).

However, as pointed out in Doucet et al. (2001) this type of nonlinear filter which includes the first-order

H. Alkhatib (✉) · J.-A. Paffenholz
Geodätisches Institut der Leibniz Universität Hannover,
D-30167 Hannover, Nienburger Str. 1, Germany
e-mail: alkhatib@gih.uni-hannover.de

H. Kutterer
Geodätisches Institut der Leibniz Universität Hannover,
D-30167 Hannover, Nienburger Str. 1, Germany

Bundesamt für Kartographie und Geodäsie
Richard-Strauss-Allee 11, D-60598 Frankfurt am Main

N. Sneeuw et al. (eds.), *VII Hotine-Marussi Symposium on Mathematical Geodesy*, International Association of Geodesy Symposia 137, DOI 10.1007/978-3-642-22078-4_12, © Springer-Verlag Berlin Heidelberg 2012

and the higher-order extended Kalman filter (EKF), is prone to diverge if the system equations are highly nonlinear. This gives us the motivation to use other filter techniques such as the sequential Monte Carlo (SMC) approach in order to take the nonlinearities into account. The SMC filter (also known as particle filter (PF)) is a suboptimal filter for implementing the recursive Bayesian filter by Monte Carlo (MC) techniques; see e.g., Doucet et al. (2001) and Ristic et al. (2004). The main idea behind the SMC filter is to approximate the posterior power density function (PDF) of the state parameters by a set of random samples, which can be generated from a known PDF. By means of the drawn particles the mean as well as the variance-covariance information of the state vector are estimated.

In order to obtain an equivalent representation of the posterior PDF a large number of particles should be drawn. Unfortunately, the high computational cost due to the large number of required particles restricts the use of SMC in many applications. In this paper, two filtering techniques are discussed: first, the generic PF, and, second a filtering technique, which can significantly improve the performance of PF, and which reduces the computational cost of the algorithm.

2 Nonlinear State Estimation

2.1 The Mathematical Model

Before describing the different filter algorithms, we briefly introduce the notation and the terminology used throughout this paper. To define the problem of nonlinear filtering, let us consider the state vector $\mathbf{x}_k \in \mathbb{R}^{n_x}$, where n_x is the dimension of the state vector, and k is the time index. The evolution of the state vector $\mathbf{x}_k$ is described by the dynamic model:

$$\mathbf{x}_{k+1} = f(\mathbf{x}_k, \mathbf{u}_k, \mathbf{w}_k) \tag{12.1}$$

where f is a known, in general nonlinear function of $\mathbf{x}_k$, $\mathbf{u}_k$ the vector of known (deterministic) input, and $\mathbf{w}_k$ is the process noise vector, which is caused by mismodeling effects and other disturbances in the motion model. The main aim of filtering is to estimate the optimal state vector $\mathbf{x}_{k+1}$ from the observations $\mathbf{y}_{k+1} \in \mathbb{R}^{n_y}$ and $\mathbf{x}_k$ where n_y is the dimension of the measurement model:

$$\mathbf{y}_{k+1} = h(\mathbf{x}_{k+1}, \mathbf{v}_{k+1}). \tag{12.2}$$

In (12.2) h is a known, in general nonlinear function, and $\mathbf{v}_{k+1}$ is the measurement noise vector, which obeys a known PDF and is mutually independent with the system noise $\mathbf{w}_k$.

2.2 The Bayes Filter

From a Bayesian perspective, the filtering problem is to estimate the state $\mathbf{x}_{k+1}$ recursively given the data $\mathbf{y}_{1:k+1}$ up to time $k+1$. Thus, it is required to evaluate the joint posterior PDF given the hole data. That is:

$$p(\mathbf{x}_{k+1}|\,\mathbf{y}_{1:k+1}) = \frac{p(\mathbf{y}_{k+1}|\,\mathbf{x}_{k+1}) \cdot p(\mathbf{x}_{k+1}|\,\mathbf{y}_{1:k})}{p(\mathbf{y}_{k+1}|\,\mathbf{y}_{1:k})} \tag{12.3}$$

where the posterior PDF at time k, $p(\mathbf{x}_k|\,\mathbf{y}_{1:k})$, is first projected forward in time in order to calculate the prior PDF at time $k+1$. This is done by using the probabilistic process model (cf. Simon 2006, pp. 464):

$$p(\mathbf{x}_{k+1}|\,\mathbf{y}_{1:k}) = \int p(\mathbf{x}_{k+1}|\,\mathbf{x}_k) \cdot p(\mathbf{x}_k|\,\mathbf{y}_{1:k}) d\mathbf{x}_k. \tag{12.4}$$

The probabilistic model of the state evolution $p(\mathbf{x}_{k+1}|\,\mathbf{x}_k)$ is defined by the system described in (12.1) and the known PDF of the noise vector $\mathbf{w}_k$. The term $p(\mathbf{y}_{k+1}|\,\mathbf{y}_{1:k})$ in (12.3) is a normalizing factor. Figure 12.1 illustrates the kth recursive step of sequential Bayesian filtering, along with inputs and outputs.

Simultaneously with the recursion given jointly by (12.3) and (12.4), we can estimate the current state via a *maximum a posteriori* (MAP) approach (see, for instance, Koch (2007)):

$$\hat{\mathbf{x}}_{k+1} = \max_{\mathbf{x}_{k+1}} p(\mathbf{x}_{k+1}|\,\mathbf{y}_{1:k+1}). \tag{12.5}$$

Note, that a closed-form solution for the filtering problem presented in (12.3) and (12.5) only exists if the system equations presented in (12.1) and (12.2) are linear, and both the system noise and the observation noise are Gaussian. When these conditions are fulfilled, we obtain the known Kalman filter, which is a special sequential Bayesian filter where the posterior density $p(\mathbf{x}_{k+1}|\,\mathbf{y}_{1:k+1})$ also becomes Gaussian, refer to (Arulampalam et al. 2002).

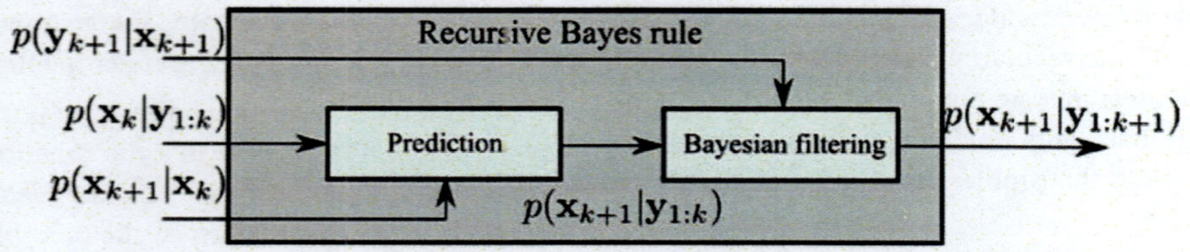

Fig. 12.1 Recursive computation in sequential Bayesian filtering. Each epoch k has two computation steps: the prediction step and the filtering step. These steps are sequential. The prediction unit takes in the motion model density and the posterior density of previous time step k and outputs predicted posterior density $p(\mathbf{x}_{k+1}|\mathbf{y}_{1:k})$. Next the Bayesian filtering unit takes in this predicted posterior density and the likelihood density to estimate the posterior density $p(\mathbf{x}_{k+1}|\mathbf{y}_{1:k+1})$ for the current time step

2.3 The Generic Particle Filter

In this section, we describe how to approximate the optimal Bayesian solution (see Sect. 2.2) if an analytical solution is unsolvable. The PF is a suboptimal solution to approximate the Bayesian estimator given in (12.3) numerically by means of SMC techniques.

The main idea of SMC is based on particle representation of a PDF. The SMC technique is used to determine the components of the state vector in the nonlinear filtering system given by (12.1) and (12.2). The resulting MC algorithm is known as sequential importance sampling (SIS). This method approximates the posterior PDF by a set of M weighted samples of this density without making any explicit assumption about its form and can thus be used in general nonlinear, non-Gaussian systems.

Let the particle set $\left\{\mathbf{x}_{0:k}^{(i)}, w_k^{(i)}\right\}_{i=1}^{M}$ denote a random measure that characterizes the posterior PDF $p(\mathbf{x}_{0:k}|\mathbf{y}_{1:k})$, where $\left\{\mathbf{x}_{0:k}^{(i)}, i = 1,\ldots,M\right\}$ is a set of realization points with associated weights $\left\{w_k^{(i)}, i = 1,\ldots,M\right\}$, and $\{\mathbf{x}_j, 0 = 1,\ldots,k\}$ is the set of all states up to epoch k. One can then approximate the posterior density in epoch k as:

$$p(\mathbf{x}_{0:k}|\mathbf{y}_{1:k}) \approx \sum_{i=1}^{M} w_k^{(i)} \delta\left(\mathbf{x}_{0:k} - \mathbf{x}_{0:k}^{(i)}\right) \tag{12.6}$$

where δ represents the Kronecker Delta function, and the associated weights $w_k^{(i)}$ sum up to unity. The weights $w_k^{(i)}$ in (12.6) are chosen using the principle of importance sampling, cf. Doucet et al. (2001). After a lengthly derivation, which will not be given here due to lack of space, the weights are computed recursively based on the weight update equation (refer to Ristic et al. 2004, pp. 37–39):

$$w_{k+1}^{(i)} = w_k^{(i)} \frac{p\left(\mathbf{y}_{k+1}|\mathbf{x}_{k+1}^{(i)}\right) p\left(\mathbf{x}_{k+1}^{(i)}|\mathbf{x}_k^{(i)}\right)}{\pi\left(\mathbf{x}_{k+1}^{(i)}|\mathbf{x}_{1:k}^{(i)}, \mathbf{y}_{1:k}\right)} \tag{12.7}$$

where $\pi(.)$ is a known PDF from which it is easy to draw samples, $p\left(\mathbf{y}_{k+1}|\mathbf{x}_{k+1}^{(i)}\right)$ the evaluated likelihood PDF for each particle and $p\left(\mathbf{x}_{k+1}^{(i)}|\mathbf{x}_k^{(i)}\right)$ is simply the PDF of the state at epoch $k+1$ given a specific state at previous epoch k for every particle $\mathbf{x}_k^{(i)}$. The weight update equation (12.7) yields a sequential update of the importance weights, given an appropriate choice of the proposal distribution $\pi(.)$. Doucet et al. (2001) show that the selection of the proposal PDF is one of the most critical issues in the SIS algorithm.

The SIS algorithm starts with M initialization values of the state vector $\left(\mathbf{x}_0^{(i)}\right)$ with $i \in 1,\ldots,M$, which can be randomly generated from the initial PDF $\pi(\mathbf{x}_0)$. These particles are then propagated at each epoch $k = 1,2,\ldots,n$ in forward by substitution in the dynamic equation (12.1). In order to distinguish between this drawn particles (they were indicated in (12.7) as $\mathbf{x}_k^{(i)}$) and the resampled particles in the following step, we rename the resulting particles as $\mathbf{x}_{k,-}^{(i)}$. As the current observations $\mathbf{y}_k$ become available, we compute the conditional likelihood of each particle: $p\left(\mathbf{y}_k|\mathbf{x}_{k,-}^{(i)}\right)$. The evaluation of the likelihood is based on the known PDF of the measurement noise and on the nonlinear measurement equation. On the basis of (12.7) we recursively compute the relative weights. Before we evaluate the current state obtained by (12.5) and move to the next time step, the particles are resampeld.

In other words, we randomly generate new particles $\mathbf{x}_{k,+}^{(i)}$ based on the relative weights. Particles with relatively small weights are eliminated. Otherwise, particles with large weights are duplicated. This resampling is used to avoid the problem of degeneracy of drawn particles.

2.4 The Extended Kalman Particle Filter

In this section, an implementation issue to improve the performance of PF presented in Sect. 2.3 is introduced. A shortcoming of the PF algorithm is the computational cost caused by the increase of the generated particles. A large number of samples should be drawn in order to achieve the convergence of the algorithm, and to estimate the desired state vectors and its covariance matrix. In Ristic et al. (2004) and Simon (2006), several implementation issues are considered for improving the PF algorithm, including degeneracy, the selection of the importance density, and particle filters with an improved sample diversity. Due to lack of space we only discuss the developed approach for enhancement of convergence based on combination with the well known Kalman filter such as the extended Kalman filter (EKF). The novelty of the proposed EKPF algorithm is the update of each particle at every time step k using the EKF, when a new measurement $\mathbf{y}_k$ arrives. In other words, we are running an extra EKF step for ith particle $\mathbf{x}_{k+1,-}^{(i)}$ at the epoch $k+1$:

$$\begin{aligned}
\mathbf{P}_{k+1,-}^{(i)} &= \mathbf{F}_k^{(i)}\mathbf{P}_{k,+}^{(i)}\mathbf{F}_k^{(i)T} + \mathbf{Q}_k \\
\mathbf{K}_{k+1}^{(i)} &= \mathbf{P}_{k+1,-}^{(i)}\mathbf{H}_{k+1}^{(i)T}\left(\mathbf{H}_{k+1}^{(i)}\mathbf{P}_{k+1,-}^{(i)}\mathbf{H}_{k+1,i}^{T} + \mathbf{R}_{k+1}\right)^{-1} \\
\mathbf{x}_{k+1,+}^{(i)} &= \mathbf{x}_{k+1,-}^{(i)} + \mathbf{K}_{k+1}^{(i)}\left[\mathbf{y}_{k+1} - \mathbf{h}(\mathbf{x}_{k+1,-}^{(i)})\right] \\
\mathbf{P}_{k+1,+}^{(i)} &= \left(\mathbf{I} - \mathbf{K}_{k+1}^{(i)}\mathbf{H}_{k+1}^{(i)}\right)\mathbf{P}_{k+1,-}^{(i)}.
\end{aligned} \tag{12.8}$$

$\mathbf{K}_{k+1}^{(i)}$ represents the Kalman gain of the ith particle, and $\mathbf{P}_{k+1}^{(i)}$ is the appropriate estimation of the state covariance matrix. $\mathbf{Q}_k$ and $\mathbf{R}_{k+1}$ are the covariance matrices of the process noise vector $\mathbf{w}_k$ and the observation noise vector $\mathbf{v}_{k+1}$, respectively. We distinguish in (12.8) between the prior $\mathbf{P}_{k+1,-}^{(i)}$ and the posterior $\mathbf{P}_{k+1,+}^{(i)}$. The transition and design matrices $\mathbf{F}_k^{(i)}$ and $\mathbf{H}_{k+1}^{(i)}$ in (12.8) are defined as: $\mathbf{F}_k^{(i)} = \frac{\partial f}{\partial \mathbf{x}}\Big|_{\mathbf{x}=\mathbf{x}_{k,+}^{(i)}}$ and $\mathbf{H}_{k+1}^{(i)} = \frac{\partial h}{\partial \mathbf{x}}\Big|_{\mathbf{x}=\mathbf{x}_{k+1,-}^{(i)}}$, respectively. The Taylor series are evaluated for the transition matrix $\mathbf{F}$ for the particle from the previous epoch $\mathbf{x}_{k,+}^{(i)}$ and for the predicted particle from the current epoch $\mathbf{x}_{k+1,-}^{(i)}$ for the matrix $\mathbf{H}$. Please note that the functions $f(.)$ and $h(.)$ are both time invariant. The key idea behind this approach is the substitution of the possibly nonlinear model given by (12.1) and (12.2) with a linearized model to reduce the variance of the drawn particles in order to get short computing times without increasing the number of samples.

The generated prior particles $\mathbf{x}_{k,-}^{(i)}$ would be transformed to a new set of particles $\mathbf{x}_{k,+}^{(i)}$ using the EKF step given by (12.8). Based on the transformed particles $\mathbf{x}_{k,+}^{(i)}$ and their $\mathbf{P}_{k+1,+}^{(i)}$ we generate and propagate a new set of particles using the Gaussian PDF:

$$\mathbf{x}_{k+1}^{(i)} \sim p\left(\mathbf{x}_{k+1}^{(i)}|\mathbf{x}_k^{(i)}, \mathbf{y}_{k+1}\right) \sim \mathcal{N}\left(\mathbf{x}_{k+1,+}^{(i)}, \mathbf{P}_{k+1,+}^{(i)}\right) \tag{12.9}$$

where $\sim$ in (12.9) means that the particles are drawn from a specific PDF. The remaining computational steps of the EKPF are similar to the generic PF.

3 Numerical Study and Results

In this section an application of the algorithms presented in Sects. 2.3 and 2.4 is shown and the results are discussed. The main goal of the numerical investigation is to derive position and orientation parameters for the transformation of a local sensor-defined coordinate system (denoted by upper index L) to an global earth centered, earth fixed coordinate system (denoted by upper index G). This is a typical task within the direct geo-referencing procedure of 3D terrestrial laser scans. For this purpose, an adapted sensor-driven method based on a multi-sensor system (MSS) has been developed at the Geodetic Institute of the Leibniz Universität Hannover (GIH). The MSS is established by a sensor fusion of a phase-based terrestrial laser scanner (TLS) and additional navigation sensors to observe the parameters.

The above mentioned transformation parameters include the position of the MSS, which is equal to the translation vector and a rotation matrix, which contains the orientation of the three axes of the MSS – roll, pitch and yaw angle, known from aeronautics. The mathematical modeling of the MSS in form of a EKF approach is presented in Paffenholz et al. (2009).

This approach uses the constant rotation of the TLS about its local vertical axis (z^L) in combination with kinematic GNSS measurements to estimate four of the six degrees of freedom of the transformation – the position vector as well as the orientation in the horizontal plane. Therefore, one GNSS antenna is mounted eccentrically on the TLS. In order to optimize the direct geo-referencing strategy the MSS is enhanced with additional navigation sensors to estimate the residual spatial rotation angles about the x^L- and y^L-axis.

In this MSS application the trajectory can be described by a circle in 3D space. This parameterization is due to the circular motion of the antenna reference point (ARP) caused by the constant rotation of the TLS about the z^L-axis, as already mentioned. The orientation change of the ARP within two time steps is given by the circular arc segment s divided by the radius r_k.

The state vector is expressed by the components:

$$\mathbf{x}_k^G = \left[\mathbf{X}_k^G \; \alpha_{S,k}^G \; \beta_{S,k}^L \; \gamma_{S,k}^L\right]^T \qquad (12.10)$$

where $\mathbf{X}_k^G$ is the global position of the ARP at the epoch k, $\alpha_{S,k}^G$ describes the azimuthal orientation of the MSS, $\beta_{S,k}^L$ the inclination in scan direction and $\gamma_{S,k}^L$ is perpendicular to the scan direction. The space state model leads to:

$$\mathbf{x}_{k+1} = \begin{bmatrix} \mathbf{X}_k^G + \mathbf{R}_L^G(\lambda,\varphi)\cdot\mathbf{R}_{SN}^{aP}\left(\alpha_{S,k}^G\right)\cdot\boldsymbol{\Delta X}_k^L \\ \alpha_{S,k}^G + \frac{s_k}{r_k} \\ \beta_{S,k}^L \\ \gamma_{S,k}^L \end{bmatrix} + \mathbf{w}_k. \qquad (12.11)$$

The term $\Delta\mathbf{X}_k^L$ in (12.11) is given by:

$$\Delta\mathbf{X}_k^L = \begin{bmatrix} r_k\cdot\cos\left(\gamma_{S,k}^L\right)\cdot\sin\left(\frac{s_k}{r_k}\right) \\ r_k\cdot\cos\left(\gamma_{S,k}^L\right)\cdot\cos\left(\frac{s_k}{r_k}\right) \\ s_k\cdot\sin\left(\beta_{S,k}^L\right) \end{bmatrix} - \boldsymbol{\Delta X}_{SN,k}^{GNSS} \qquad (12.12)$$

where $\boldsymbol{\Delta X}_{SN,k}^{GNSS}$ represents the eccentric position of the GNSS antenna. $\Delta\mathbf{X}_k^L$ in (12.11) is responsible for the high-nonlinearity in the space state model. It should be pointed out, that in Paffenholz et al. (2009) additional adaptive parameters are considered in the space state model. However, the consideration of such adaptive parameters in the EKPF algorithm (refer to Sect. 2.4) needs a significant modification, which will be shown in future works.

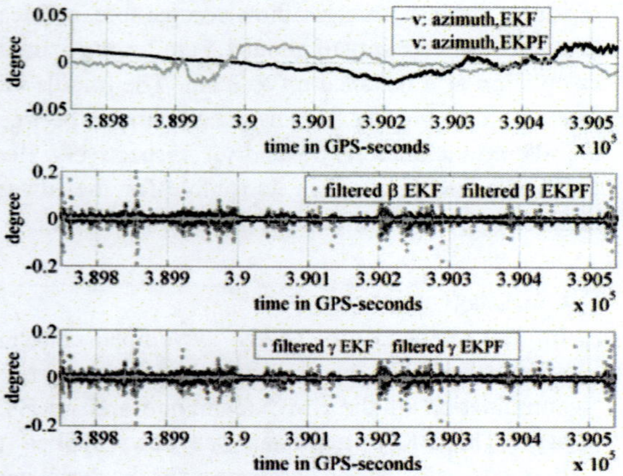

Fig. 12.2 Filtering results of EKF and the EKPF; *top*: the residuals obtained within a linear regression of the orientation α_S^G; *middle* and *bottom*: The filtered inclinations β_S^L and γ_S^L. The EKPF approach shows a significant improvement of the filter effect for the filtered inclination whereas the filtered results of the estimated azimuth are comparable

The measurement model is characterized by the position of the GNSS antenna $\mathbf{X}_k^G$, and the measurements of the inclination sensor $\beta_{S,k}^L$ and $\gamma_{S,k}^L$. That yields:

$$\mathbf{y}_{k+1} = \begin{bmatrix} \mathbf{X}_{k+1}^G \\ \beta_{S,k+1}^L \\ \gamma_{S,k+1}^L \end{bmatrix} = \mathbf{H}_{k+1}\cdot\mathbf{x}_{k+1}^G + \mathbf{v}_{k+1}. \qquad (12.13)$$

As start value for the EKPF approach (see Sect. 2.4) we randomly drawn 500 particles from $\mathcal{N}(\mathbf{x}_0, \mathbf{P}_0)$ with $\mathbf{x}_0 = \mathbf{0}$ and $\mathbf{P}_0$ the initial covariance matrix which has been chosen equally to the noise covariance matrix in Paffenholz et al. (2009).

Figure 12.2 presents a subsample of the estimated state parameters by classical EKF algorithm (black circles) and EKPF approach (gray points). The upper part of this figure shows the residuals obtained within a linear regression of the orientation α_S^G. Due to the constant rotation of the TLS about its vertical axis, we expect a linear relationship between α_S^G and time. Therefore, the residuals are quality indicators. The residuals are in both algorithms comparable, and lead to a metric uncertainty of about 1.5 cm for the azimuth calculation at a distance up to 35 m. The middle and lower part of Fig. 12.2 show a comparison between the filtered inclinations β_S^L and γ_S^L, respectively. Here again, the EKPF effect is noticeable for the filtered inclinations, mainly in case of higher noise level.

Conclusion

In this paper, the newly developed filtering approach EKPF was introduced. It is based on a combination of the SMC technique and an EKF step. The EKPF approach has been applied to derive transformation parameters for the direct geo-referencing of 3D terrestrial laser scans. The results show an improvement of the filter effect. They were compared to the classical EKF approach. The main benefit of the developed approach is the better performance in case of high-nonlinear space state equations. A second important result, which could be not shown in the above example due to lack of space, is the significant decrease of the number of the generated particles compared to the generic PF.

References

Arulampalam MS, Maskell S, Gordon N, Clapp T (2002) A tutorial on particle filters for online nonlinear/non-Gaussian Bayesian tracking. IEEE Trans. Signal Proc., vol 50, pp 174–188

Doucet A, de Freitas N, Gordon N (2001) Sequential Monte Carlo methods in practice. Springer, New York, Berlin

Kalman RE, Bucy RS (1960) New results in linear filtering and prediction theory. J Basic Eng Trans ASME, D 83:95–108

Koch KR (2007) Introduction to Bayesian statistics. 2nd edn. Springer, Berlin

Paffenholz J-A, Alkhatib H, Brieden P, Kutterer H (2009) Optimized direct geo-referencing strategy for a TLS-based Multi-Sensor-System. In: Grün A, Kahmen H (eds) Optical 3D-measurement techniques IX, Vienna, pp 287–292

Ristic B, Arulampalam S, Gordon N (2004) Beyond the Kalman filter, particle filters for tracking applications. Artech House, Boston

Simon D (2006) Optimal state estimation. Wiley, Hoboken

Uncertainty Assessment of Some Data-Adaptive M-Estimators

13

Jan Martin Brockmann and Boris Kargoll

Abstract

In this paper, we review a data-adaptive class of robust estimators consisting of convex combinations of the loss functions with respect to the L_1- and Huber's M-estimator as proposed by Dodge and Jureckova (2000). The great advantage of this approach in comparison to the traditional procedure of applying a single estimator is that the optimal weight factor, representing the data-dependent minimum-variance estimator within that class, may be estimated from the data itself. Depending on the data characteristics, one could obtain pure L_2, L_1 and Huber's estimator, as well as any convex combination between these three. We demonstrate the computational and statistical efficiency of this approach by providing an iteratively reweighted least squares algorithm and Monte Carlo uncertainties of the weight factor.

1 Introduction

The general context of this paper is given by a statistical analysis of data structured in terms of a linear model with independent and homoscedastic error terms, which are assumed to be possibly affected by outliers. Here we adopt the notion that outliers may be viewed as either deterministic, as it would be the case for gross measurement/recording errors or neglected external effects, or as random, due to inherent variability (cf. Barnett and Lewis 1994, p. 42). The statistical methods presented in this paper aim at accommodating for outliers of the latter type; outlier/discordancy tests will not be discussed (see e.g. Kargoll 2005). More specifically, we will assume that the outliers can be explained by the fact that the totality of data to be analyzed do not follow the Gaussian distribution, but some form of outlier distribution. Particular observations which appear as discordant with the others, when looked at through the "Gaussian lense", could then be explained reasonably well in terms of an outlier distribution with thicker tails. We will therefore assume that there exists a particular outlier distribution, which may either be an entirely non-Gaussian type or a contaminated Gaussian distribution, under which all of the given observations appear as concordant. We will focus attention on one particular type of outlier distribution, the family of convex mixtures of the Gaussian and Laplacian distribution.

Recent contributions to the field of mathematical geodesy, addressing the outlier problem, were often focussed on L_1-norm (e.g. Marshall 2002; Junhuan 2005) and M-estimators (e.g. Chang and Guo 2005)

J.M. Brockmann (✉) · B. Kargoll
Institute of Geodesy and Geoinformation, Department of Theoretical Geodesy, University of Bonn, D-53115 Bonn, Nussallee 17, Germany
e-mail: brockmann@geod.uni-bonn.de, kargoll@geod.uni-bonn.de

N. Sneeuw et al. (eds.), *VII Hotine-Marussi Symposium on Mathematical Geodesy*, International Association of Geodesy Symposia 137, DOI 10.1007/978-3-642-22078-4_13, © Springer-Verlag Berlin Heidelberg 2012

individually, whereas this paper aims at demonstrating a way of combining estimators such as L_1-norm, L_2-norm, and Huber's M-estimator according to the approach elaborated in Dodge and Jureckova (2000). We will extend their investigations by evaluating the appropriateness of such combined estimators under the aforementioned two types of outlier distributions in terms of the accuracy and precision with which the weighting factor within the convex combination of estimators and the variance of unit weight can be estimated from the observations. In this context we use the notion of a *data-adaptive* estimator, meaning that the specific form of estimator is estimated from the data itself (cf. Hogg 1974).

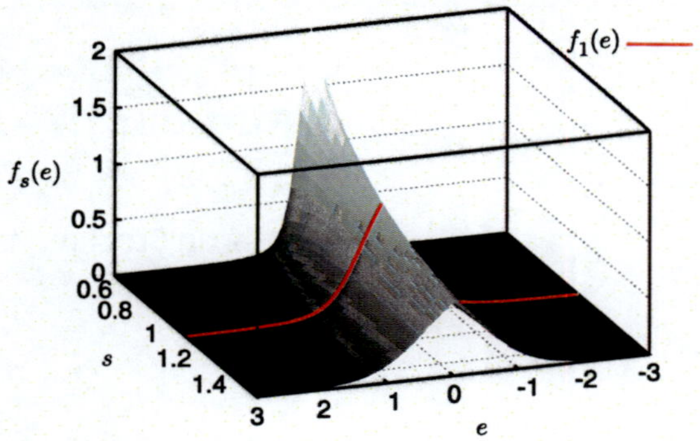

Fig. 13.1 Example for the family of mixture densities, defined by (13.3), for a fixed $\delta = 0.5$. The *highlighted line* ($s = 1$) refers to the basis density of that family

2 Theory

2.1 General Model Assumptions

We will restrict attention to a linear model

$$\ell_i = \mathbf{A}_i \mathbf{x} + e_i, \quad (i = 1, \ldots, n) \tag{13.1}$$

where ℓ_i denotes the ith observation, $\mathbf{A}_i$ the ith row of the $n \times m$ design matrix $\mathbf{A}$ (assumed to be of full rank), $\mathbf{x}$ the unknown parameter vector, and e_i independent and homoscedastic errors with some unknown scalable density function

$$f_s(e) = \frac{1}{s} f_1\left(\frac{e}{s}\right), \quad e \in \mathbb{R} \tag{13.2}$$

with $s > 0$, satisfying the technical assumptions (1) $0 < f_s(0) < \infty$, (2) $0 < \sigma^2 = \int e^2 f_s(e) de < \infty$, (3) $f_s(e) = f_s(-e)$ and (4) $f_1(0) = 1$.

Assumption (4) will be of great importance as it allows one to express the true density f_s in terms of some standardized basis density f_1 and a scale factor s, which will be seen to render the theoretical solution to the estimation problem concerning $\mathbf{x}$ feasible.

2.2 Some Specific Outlier Distributions

In this paper, we will focus attention on the family of Gaussian/Laplacian mixture distributions. If we define the scale factor $s(\sigma) = \sqrt{2\pi\sigma^2}$, then the mixture of Gaussian and Laplacian densities may be expressed as (similarly to the definition provided by Dodge and Jureckova 2000, p. 38)

$$f_{s(\sigma)}(e) = (1-\delta)\frac{1}{\sqrt{2\pi\sigma^2}} \exp\left(-\frac{1}{2}\frac{e^2}{\sigma^2}\right) + \delta \frac{1}{\sqrt{2\pi\sigma^2}} \exp\left(\frac{-2}{\sqrt{2\pi\sigma^2}}|e|\right) \tag{13.3}$$

with weighting factor $0 \le \delta \le 1$ and $\sigma > 0$ (Fig. 13.1). Note that (13.3) becomes a pure Gaussian distribution with standard deviation σ for $\delta = 0$ and a pure Laplacian distribution (with scale $b = \sqrt{2\pi\sigma^2}/2$) for the choice $\delta = 1$. It should be noted that, due to the interdependence of σ and b, the Gaussian and Laplacian parts cannot be scaled independently. If we set $\sigma = 1/\sqrt{2\pi}$ then we see immediately that $s(\sigma) = 1$ and consequently $f_s(0) = f_1(0) = 1$; thus we obtain

$$f_1(e) = (1-\delta)\exp\left(-\pi e^2\right) + \delta \exp(-2|e|), \tag{13.4}$$

which may be used to generate any of the mixture densities (13.3) via (13.2).

We currently investigate the family of rescaled t-distribution as a further potentially useful class of outlier distributions in this context, which we will not explore in the present paper.

2.3 Properties of the Data-Adaptive M-Estimator

To obtain estimates appropriate under such data distributions, Dodge and Jureckova (2000, p. 6) proposed minimization of

$$\sum_{i=1}^{n} \rho_{k,\lambda}\left(\frac{\ell_i - \mathbf{A}_i\mathbf{x}}{s}\right) \tag{13.5}$$

using the convex combination of the loss functions with respect to Huber's M-estimator (cf. Huber 1981) and the L_1-norm (cf. Koch 1999, p. 262) estimator, in terms of the weighting parameter $\lambda \in [0, 1]$, i.e.

$$\rho_{k,\lambda}(e) = (1-\lambda)\begin{Bmatrix} e^2 & \text{if } |e| \le k \\ 2k|e| - k^2 & \text{if } |e| > k \end{Bmatrix} + \lambda|e|. \tag{13.6}$$

The psi-function, defined as $\psi_{k,\lambda} = \rho'_{k,\lambda}$ follows to be

$$\psi_{k,\lambda}(e) = (1-\lambda)\begin{Bmatrix} 2e & \text{if } |e| \le k \\ 2k\text{sign}(e) & \text{if } |e| > k \end{Bmatrix} + \lambda\text{sign}(e). \tag{13.7}$$

As we intend to compute the estimates via the method of iteratively reweighted least squares (IRLS), described in a subsequent section, it will be useful to note that the weight function, defined by $w_{k,\lambda}(e) = \psi_{k,\lambda}(e)/e$, reads

$$w_{k,\lambda}(e) = (1-\lambda)\begin{Bmatrix} 2 & \text{if } |e| \le k \\ 2\frac{k}{|e|} & \text{if } |e| > k \end{Bmatrix} + \lambda\frac{1}{|e|}. \tag{13.8}$$

According to Dodge and Jureckova (2000, p. 81), this M-estimator $\widetilde{\mathcal{X}}$ of the unknown true values ξ asymptotically follows the Gaussian distribution

$$\widetilde{\mathcal{X}} \overset{a}{\sim} N\left(\xi, \sigma^2_{\psi,f}(\mathbf{A}^T\mathbf{A})^{-1}\right), \tag{13.9}$$

where the variance factor $\sigma^2_{\psi,f}$ generally depends on the psi-function through the values of λ and k, on the true data density through the scale factor s and the basis density f_1. For studentized M-estimators, this variance factor is defined as (Dodge and Jureckova 2000, p. 40)

$$\sigma^2_{\psi,f} = \frac{\int \psi^2_{k,\lambda}\left(\frac{e}{s}\right) f_s(e)de}{\left(\int f_s(se)\frac{d\psi_{k,\lambda}(e)}{de}de\right)^2} \tag{13.10}$$

It is easily shown that substitution of (13.2) and (13.7) leads to the expression (similarly to Dodge and Jureckova 2000, p. 78)

$$\sigma^2_{\psi,f} = \frac{s^2}{4}\frac{(1-\lambda)^2\mu_1^2 + 2\lambda(1-\lambda)\vartheta_1 + \lambda^2}{((1-\lambda)\gamma_1 + \lambda)^2}, \tag{13.11}$$

with quantities

$$\gamma_1 := \int_{-k}^{k} f_1(e)de = F_1(k) - F_1(-k), \tag{13.12}$$

$$\vartheta_1 := \int_{-\infty}^{\infty} |\psi_{k,1}(e)| f_1(e)de,$$

$$= 2\int_{-k}^{k} |e| f_1(e)de + 4k\left[1 - F_1(k)\right], \tag{13.13}$$

$$\mu_1^2 := \int_{-\infty}^{\infty} \psi^2_{k,1}(e) f_1(e)de,$$

$$= 4\int_{-k}^{k} e^2 f_1(e)de + 8k^2\left[1 - F_1(k)\right], \tag{13.14}$$

characterizing the data distribution, with respect to Huber's M-estimator $\psi_{k,1}$ and the true basis density f_1 and the corresponding distribution function F_1. It will be convenient to use the quadratic approximation in λ (Dodge and Jureckova 2000, p. 79)

$$\sigma^2_{\psi,f} \approx \frac{s^2}{4}\left((1-\lambda)^2\frac{\mu_1^2}{\gamma_1^2} + 2\lambda(1-\lambda)\frac{\vartheta_1}{\gamma_1} + \lambda^2\right) \tag{13.15}$$

instead as we wish to find the optimal value for $\lambda \in [0, 1]$ which minimizes the variance (a convex or concave function depending on the values for γ_1, ϑ_1 and μ_1). Minimizing (13.15) as a function of λ then yields

$$\lambda_{\text{opt}} = \begin{cases} 0 & \text{if } \mu_1^2 \le \vartheta_1\gamma_1 \text{ and } \mu_1^2 < \gamma_1^2 \\ \dfrac{\mu_1^2 - \vartheta_1\gamma_1}{\mu_1^2 - 2\vartheta_1\gamma_1 + \gamma_1^2} & \text{if } \vartheta_1\gamma_1 \le \gamma_1^2 \text{ and } \vartheta_1\gamma_1 < \mu_1^2. \\ 1 & \text{if } \gamma_1^2 < \vartheta_1\gamma_1 \text{ and } \gamma_1^2 < \mu_1^2 \end{cases} \tag{13.16}$$

We will now demonstrate how λ_{opt} and the corresponding minimum variance $\sigma^2_{\psi,f}$ are estimated from given data by substituting empirical quantities ($\hat{s}, \hat{\gamma}_1, \hat{\vartheta}_1$, and $\hat{\mu}_1^2$) for their theoretical counterparts.

3 Algorithms and Estimation Procedure

3.1 Estimation of the Scale s

Evaluating (13.2) for $e = 0$ and using the fourth technical assumption we see that $f_s(0) = 1/s$ or $s = 1/f_s(0)$. The estimation of the scale parameter is thus reduced essentially to the estimation of $f_s(0)$, which we performed via the well known kernel density estimation (KDE) method (see e.g. Peracchi 2001, p. 447ff). A different, however far more complex, approach based on regression quantiles has been proposed by Dodge and Jureckova (2000, p. 151).

As a starting point for the scale estimation we compute the residuals $\hat{v}_i = \mathbf{A}_i\hat{\mathbf{x}}_{L1} - \ell_i$ using the scale invariant robust L_1-norm estimates. Then the kernel density estimate at a point e_0 is given by

$$\hat{f}_s(e_0) = \frac{1}{n}\sum_{i=1}^{n}\frac{1}{b}K\left(\frac{e_0 - \hat{v}_i}{b}\right) \tag{13.17}$$

where we used a Gaussian kernel for $K(\cdot)$ with bandwidth $b = 1.48 \cdot \text{median}\,(|\hat{\mathbf{v}}|)\left(\frac{4}{3n}\right)^{\frac{1}{5}}$ (cf. Bowman and Azzalini 1997, p. 31). Equation (13.17) now allows one to estimate the scale parameter through the relation

$$\hat{s} = \hat{f}_s(0)^{-1}. \tag{13.18}$$

Other kernels and similar bandwidths could be used without affecting the results much.

3.2 Estimation of γ_1, ϑ_1 and μ_1^2

As γ_1 represents the probability mass under f_1 between $-k$ and k according to (13.12), we arrive at a corresponding empirical measure by determining the fraction of the standardized residuals within that range,[1]

$$\hat{\gamma}_1 = \frac{1}{n}\sum_{i=1}^{n} I\left(\left|\frac{\hat{v}_i}{\hat{s}}\right| \le k\right). \tag{13.19}$$

Similarly, the integrals in (13.13) and (13.14) may be approximated by

$$\hat{\vartheta}_1 = \frac{2}{n-m}\sum_{i=1}^{n}\left|\frac{\hat{v}_i}{\hat{s}}\right| I\left(\left|\frac{\hat{v}_i}{\hat{s}}\right| \le k\right) + 2k(1-\hat{\gamma}_1), \tag{13.20}$$

$$\hat{\mu}_1^2 = \frac{4}{n-m}\sum_{i=1}^{n}\frac{\hat{v}_i^2}{\hat{s}^2} I\left(\left|\frac{\hat{v}_i}{\hat{s}}\right| \le k\right) + 4k^2(1-\hat{\gamma}_1). \tag{13.21}$$

Dodge and Jureckova (2000, p. 80) demonstrate that these estimators converge in probability to their theoretical values. Now these estimates can be substituted into (13.16) and (13.15) to obtain first the empirical weighting factor $\hat{\lambda}_{\text{opt}}$ and subsequently the variance factor $\hat{\sigma}^2$.

3.3 Estimation of the Parameters $\mathbf{x}$

Algorithm 1 demonstrates how the estimates for the parameters $\mathbf{x}$ are computed via iteratively reweighted least squares (IRLS) regarding the data-adaptive M-estimator defined by the loss function (13.6) and the corresponding weight function (13.8). Table 13.1 shows the parameter settings for which the various estimators (L_1, L_2, HUBER, $L_2 + L_1$, HUBER $+ L_1$) can be obtained with this algorithm.

Algorithm 1: Adaptive IRLS M-estimation

Data: $\mathbf{A}_{[n\times m]}$ ... design matrix, $\boldsymbol{\ell}_{[n\times 1]}$ ... observations
Result: $\hat{\mathbf{x}}^{(i_{\max})}_{[m\times 1]}$ ... final solution

estimate L_1-norm solution $\hat{\mathbf{x}}_{L1}$ using $(\mathbf{A}, \boldsymbol{\ell})$
$\hat{\mathbf{v}}^{(0)} = \mathbf{A}\hat{\mathbf{x}}_{L1} - \boldsymbol{\ell}$
estimate $\hat{s}$ // *cf. eq. 13.17 and 13.18*
compute $\hat{\gamma}_1, \hat{\vartheta}_1, \hat{\mu}_1^2$ // *cf. eq. 13.19, 13.20 and 13.21*
compute $\hat{\lambda}_{\text{opt}}$ // *cf. eq. 13.16*
compute $\hat{\sigma}^2$ // *cf. eq. 13.15*
// *IRLS iterations*
for $i = 1$ **to** $i_{\max}$ **do**
  $\mathbf{P}^{(i)}(j,j) = w\left(\hat{v}_j^{(i)}/\hat{s}, \hat{\lambda}, k\right)$ // *cf. eq. 13.8*
  $\hat{\mathbf{x}}^{(i)} = \left(\mathbf{A}^T\mathbf{P}^{(i)}\mathbf{A}\right)^{-1}\mathbf{A}^T\mathbf{P}^{(i)}\boldsymbol{\ell}$
  $\hat{\mathbf{v}}^{(i)} = \mathbf{A}\hat{\mathbf{x}}^{(i)} - \boldsymbol{\ell}$
end
$\widehat{\boldsymbol{\Sigma}}^{(i)}_{\mathcal{X}} = \hat{\sigma}^2\left(\mathbf{A}^T\mathbf{A}\right)^{-1}$ // *cf. eq. 13.9*
return $\hat{\mathbf{x}}^{(i_{\max})}$, $\widehat{\boldsymbol{\Sigma}}^{(i)}_{\mathcal{X}}$

Table 13.1 Parameter configurations for different estimators using algorithm 1

	k	λ	s
L_1	arbitrary	1	1
L_2	$\to\infty$	0	1
HUBER	1.2 ... 1.8	0	1
$L_2 + L_1$	$\to\infty$	estimated	estimated
HUBER+L_1	1.2 ... 1.8	estimated	estimated

[1] $I(\cdot)$ denotes the indicator function, which returns 1 if the statement in the argument is true and 0 otherwise.

4 Numerical Simulations

As far as both the small-sample and the large-sample properties of the presented data-adaptive M-estimator are concerned, the estimation of the scale parameter s, the weighting parameter λ and the variance factor $\sigma^2_{\psi,f}$ is of special interest. In this contribution we will concentrate on the estimation of the mixture parameter and of its accuracy as well as its empirical distribution, the behaviour of which depends on the number of observations n. For this purpose we performed Monte Carlo simulations to determine a reasonable sample size n, for which a reliable estimate for λ is obtained. These simulations were performed using a Fourier series as functional model, which we found suitable to simulate different configurations in terms of numbers of parameters and numbers of observations. The synthetic observations computed from true Fourier coefficients were disturbed by adding white noise generated from the Gauss/Laplace mixture distribution (with $\delta = 0.5$ and $s = 2.5066$ in (13.3)) using the Acceptance-Rejection method for random number generation (cf. Koch 2007, p. 196). These simulations were carried out using different numbers of observations (i.e. 50 ... 100,000), a Fourier series of degree two, and two estimators ($L_2 + L_1$, Huber+L_1). We used a maximum number of 1,000 iteration steps and a termination criteria of $|\mathbf{x}^{(i)} - \mathbf{x}^{(i-1)}| < 10^{-8}$. Then we determined the mean value and the standard deviation of the estimated $\hat{\lambda}_{opt}$ from 100,000 Monte Carlo samples. The estimation of λ is influenced by the scale estimation; to separate the effects, we used the true scale s instead of the estimated scale as in algorithm 1.

The results of the simulations are summarized in Figs. 13.2 and 13.3. It is seen that for sample sizes as

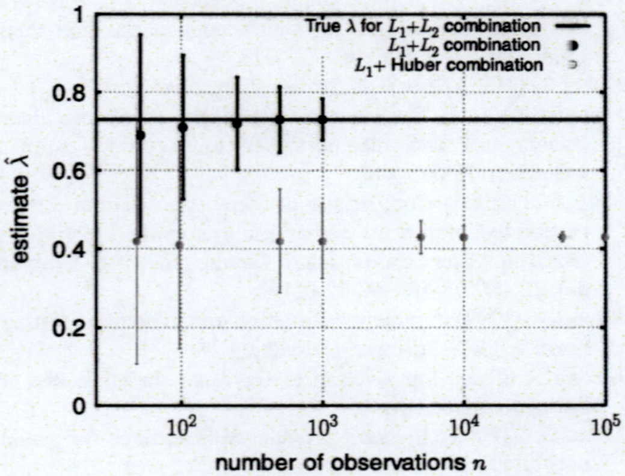

Fig. 13.2 Mean values and standard deviations for the λ-estimation for different sample sizes

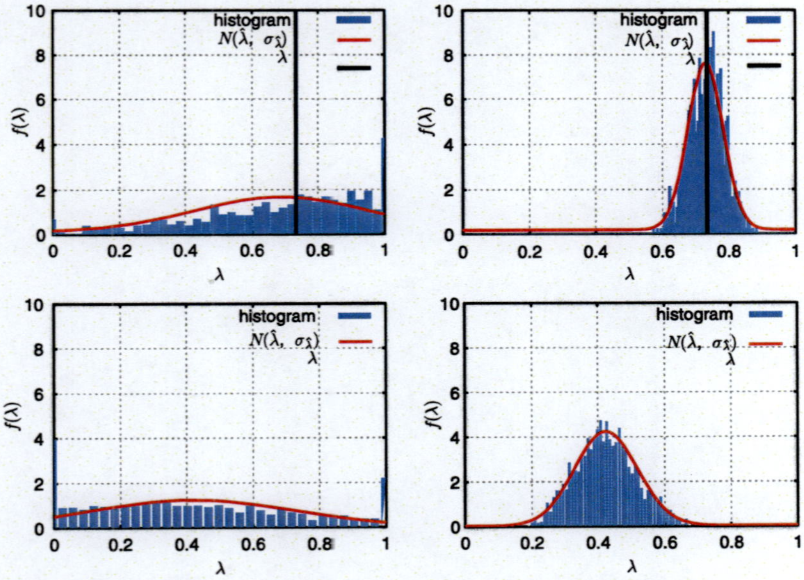

Fig. 13.3 Histogram of λ-estimates with sample size 50 (*left column*) and 1,000 (*right column*) for the $L_1 + L_2$ combination (*upper row*) and the L_1+HUBER combination (*lower row*). The *grey line* shows a normal distribution with the estimated mean and standard deviation. The *black line* shows the true value for λ (not available for L_1+HUBER combination)

low as $n = 50$, bias, standard deviation and the range of λ estimates are generally very large. For $n = 1{,}000$ both the bias and the standard deviation are already relatively low and the histogram of the λ-estimates roughly reflects a Gaussian distribution (Fig. 13.3). For very large n ($n > 50{,}000$) the bias and standard deviation are seen to vanish.

5 Summary and Outlook

We demonstrated the theoretical idea, an algorithm and some empirical small- and large-sample properties of an estimator for the mixture parameter with respect to the Gauss/Laplace mixture distribution in the context of data-adaptive M-estimation, in particular for $L_2 + L_1$ and Huber+L_1 combination. We intend to extend these investigations to various estimators for the scale parameter s (e.g. based on simple KDE) and for the variance/covariance matrix of the estimated parameters.

Acknowledgements The computations were performed on the JUMP supercomputer in Jülich. The computing time was granted by the John von Neumann Institute for Computing (project 1827)

References

Barnett V, Lewis T (1994) Outliers in statistical data, 3rd edn. Wiley, Chichester

Bowman AW, Azzalini A (1997) Applied smoothing techniques for data analysis. Oxford Statistical Science Series, vol 18. Oxford University Press

Chang XW, Guo Y (2005) Huber's M-estimation in relative GPS positioning: computational aspects. J Geodesy 79: 351–362

Dodge Y, J Jureckova (2000) Adaptive regression. Springer, Berlin

Hogg RV (1974) Adaptive robust procedures: a partial review and some suggestions for future applications and theory. J Am Stat Assoc 69(348):909–923

Huber PJ (1981) Robust statistics. Wiley, New York

Junhuan P (2005) The asymptotic variance-covariance matrix, Baarda test and the reliability of L_1-norm estimates. J Geodesy 78:668–682

Kargoll B (2005) Comparison of some robust parameter estimation techniques for outlier analysis applied to simulated GOCE mission data. In: Jekeli C et al. (eds) IAG Symposia, doi: 10.1007/3-540-26932-0_14

Koch KR (1999) Parameter estimation and hypothesis testing in linear models. Springer, Heidelberg

Koch KR (2007) Introduction to Bayesian statistics, 2nd edn. Springer, Heidelberg

Marshall J (2002) L_1-norm pre-analysis measures for geodetic networks. J Geodesy, 76:334–344

Peracchi F (2001) Econometrics. Wiley, New York

Uniqueness and Nonuniqueness of the GNSS Carrier-Phase Compass Readings 14

P.J.G. Teunissen

Abstract

In this contribution we analyse the possible nonuniqueness in the least-squares solution of the GNSS carrier-phase compass model. It is shown that this lack of uniqueness may manifest itself in the fixed baseline estimator and therefore in the GNSS compass readings. We present the conditions under which nonuniqueness occurs and give explicit expressions for these nonunique least-squares solutions.

Keywords

GNSS-compass • Ambiguity resolution • Attitude nonuniqueness

1 Introduction

Global Navigation Satellite System (GNSS) attitude determination is a field with a wide variety of challenging (terrestrial, air and space) applications, see e.g. Cohen (1992), Lu (1995), Tu et al. (1996), Montgomery et al. (1997), Park and Teunissen (2003), Simsky et al. (2005), Kuylen et al. (2006), Teunissen (2006), Hide and Pinchin (2007).

In the present contribution we consider the determination of heading and elevation (or yaw and pitch) and therefore restrict ourselves to the two-antenna, single baseline case. GNSS carrier phase data and integer ambiguity resolution are needed in order to determine the compass parameters with the highest possible precision. Short baseline, epoch-by-epoch, successful ambiguity resolution is however only possible if two or more frequencies are used, but not in the single-frequency case. In the single-frequency case, the underlying model is too weak to ensure a sufficiently high probability of correct integer estimation. Hence, for the single-frequency case, the corresponding GNSS model needs to be strengthened and this can be done by considering the length of the (small) baseline to be known. This model is referred to as the GNSS compass model and it differs from the standard GNSS single baseline model in that the known length of the baseline is added as a (weighted) constraint.

The inclusion of the baseline length constraint strengthens the model, thereby increasing the ambiguity success rates significantly, but at the same time it also complicates the least-squares estimation process. This is particularly true for short to very-short GNSS baselines (less than 1 m), as a reduction in baseline length increases the nonlinearity of the curved manifold. Related to the high-nonlinearity is another potential complication, namely the occurrence of singularities in the solution process. We will show that non-uniqueness in the attitude solutions may indeed

P.J.G. Teunissen (✉)
Delft Institute of Earth Observation and Space Systems, Delft University of Technology, The Netherlands

Department of Spatial Sciences, Curtin University of Technology, Perth, Australia
e-mail: p.j.g.teunissen@tudelft.nl

N. Sneeuw et al. (eds.), *VII Hotine-Marussi Symposium on Mathematical Geodesy*, International Association of Geodesy Symposia 137, DOI 10.1007/978-3-642-22078-4_14, © Springer-Verlag Berlin Heidelberg 2012

occur and we will characterize them for different measurement scenarios. Singularities of different types may occur and we will identify them in relation to various data subspaces. The theoretical analysis presented also improves our understanding of the near-singular situations. Due to a lack of space, the theorems are presented without an extensive proof. These will be published elsewhere.

2 The GNSS Compass Model

In principle all GNSS baseline models can be cast in the following frame of linear(ized) observation equations,

$$\mathrm{E}(y) = Aa + Bb, \quad \mathrm{D}(y) = Q_{yy}, \quad a \in \mathbb{Z}^p, b \in \mathbb{R}^n \tag{14.1}$$

where y is the given GNSS data vector of order m, and a and b are the unknown parameter vectors of order p and n respectively. E(.) and D(.) denote the expectation and dispersion operator, and A and B are the given design matrices that link the data vector to the unknown parameters. Matrix A contains the carrier wavelengths and the geometry matrix B contains the receiver-satellite unit line-of-sight vectors. The variance matrix of y is given by the positive definite matrix Q_{yy}. The data vector y will usually consist of the 'observed minus computed' single- or multi-frequency double-difference (DD) phase and/or pseudorange (code) observations accumulated over all observation epochs. The entries of vector a are then the DD carrier phase ambiguities, expressed in units of cycles rather than range. They are known to be *integers*, $a \in \mathbb{Z}^p$. The entries of the vector b will consist of the remaining unknown parameters, such as baseline components (coordinates) and possibly atmospheric delay parameters (troposphere, ionosphere). They are known to be real-valued, $b \in \mathbb{R}^n$. Vectors a and b are referred to as the *ambiguity* vector and *baseline* vector, respectively.

Since we consider the GNSS-Compass application in the present contribution, we restrict attention to the case of satellite tracking with two near-by antennas. The short distance between the two antennas implies that we may neglect the (differential) atmospheric delays. Thus b consists then only of the three coordinates of the between baseline vector of the two antennas.

If we may assume that the two antennas are firmly attached to the body of the moving platform, the length of the baseline vector may be determined a priori. In that case we can strengthen the GNSS model (14.1) by including the additional observation equation

$$\mathrm{E}(l) = \|b\|, \quad \mathrm{D}(l) = \sigma_l^2 \tag{14.2}$$

The required compass information (e.g. heading and pitch) follows from the baseline solution of the GNSS compass model (14.1) and (14.2). To obtain the most precise compass information, use needs to be made of the very precise carrier phase data. The inclusion of the carrier phase data into the model accounts for the presence of the unknown integer ambiguity vector a in (14.1).

3 The Least Squares Compass Solution

The least-squares (LS) objective function of the GNSS compass model (14.1) and (14.2) is given as $H(a,b) = \|y - Aa - Bb\|^2_{Q_{yy}} + \sigma_l^{-2}(l - \|b\|)^2$, with $a \in \mathbb{Z}^p$, $b \in \mathbb{R}^n$, and $\|.\|^2_{Q_{yy}} = (.)^T Q_{yy}^{-1}(.)$. The LS parameter solution is therefore given by the minimizers

$$\begin{aligned} \check{a} &= \arg\min_{a\in\mathbb{Z}^p}[\min_{b\in\mathbb{R}^n} H(a,b)] \\ \check{b} &= \arg\min_{b\in\mathbb{R}^n} H(\check{a},b) \end{aligned} \tag{14.3}$$

This can be worked out further if we let $\hat{a}$, with variance matrix $Q_{\hat{a}\hat{a}}$, denote the LS ambiguity solution of (14.1) without the integer constraint $a \in \mathbb{Z}^n$, and let $\hat{b}(a)$, with variance matrix $Q_{\hat{b}(a)\hat{b}(a)}$, denote the conditional LS baseline solution of (14.1) assuming a known. Then the LS solution (14.3) can be shown to work out as

$$\begin{aligned} \check{a} &= \arg\min_{a\in\mathbb{Z}^p}\left(\|\hat{a} - a\|^2_{Q_{\hat{a}\hat{a}}} + \min_{b\in\mathbb{R}^n} G(a,b)\right) \\ \check{b} &= \arg\min_{b\in\mathbb{R}^n} G(\check{a},b) \end{aligned} \tag{14.4}$$

where $G(a,b) = \|\hat{b}(a) - b\|^2_{Q_{\hat{b}(a)\hat{b}(a)}} + \sigma_l^{-2}(l - \|b\|)^2$.

Note that (14.4) reduces to the LS parameter solution of the GNSS-baseline model (14.1) in case $\sigma_l^2 = \infty$. Then $\min_{b\in\mathbb{R}^n} G(a,b) = 0$ and $\arg\min_{b\in\mathbb{R}^n} G(a,b) = \hat{b}(a)$, from which it follows that the minimizers of $H(a,b)$ are given as

$$\check{a} = \arg\min_{a\in\mathbb{Z}^n} \|\hat{a} - a\|^2_{Q_{\hat{a}\hat{a}}} \text{ and } \check{b} = \hat{b}(\check{a}) \tag{14.5}$$

This is the commonly used solution for real-time kinematic (RTK) GNSS baseline processing, see e.g. Strang and Borre (1997), Teunissen and Kleusberg (1998), Misra and Enge (2001), Hofmann-Wellenhoff and Lichtenegger (2001), Leick (2003).

The computational complexity of (14.5) resides in the computation of the integer least-squares (ILS) solution $\check{a}$. Its computation is based on an integer search inside an ellipsoidal search space, which can be efficiently executed by means of the standard LAMBDA method, see Teunissen (1994) and Teunissen (1995). The computation of $\check{b} = \hat{b}(\check{a})$, the so-called fixed baseline, is straightforward once $\check{a}$ is known.

In our present case, we have $\sigma_l^2 \neq \infty$. This increases the computational complexity considerably. First, the computation of the fixed baseline vector is more complicated; compare (14.4) with (14.5). Second, the computation of $\check{a}$ is now based on an integer search in a *non-ellipsoidal* search space. An efficient method for this search has been developed, see Teunissen (2006), Buist (2007), Park and Teunissen (2007), Giorgi et al. (2008).

4 Nonuniqueness of Compass Solution

Note that the minimization problem $\min_{b \in \mathbb{R}^n} G(a, b)$ of (14.4), is part of the ambiguity objective function. Thus for every evaluation of the ambiguity objective function, this minimization problem needs to be solved. A proper understanding of this minimization problem is therefore essential for the GNSS-compass ambiguity resolution problem. To simplify notation, we define

$$F(b) = \|b_0 - b\|_Q^2 + \sigma_l^{-2}(l - \|b\|)^2 \qquad (14.6)$$

Then, for $Q = Q_{\hat{b}(a)\hat{b}(a)}$, we have $\min_b G(a, b) = \min_b F(b)$ if $b_0 = \hat{b}(a)$, and $\check{b} = \arg\min_b F(b)$ if $b_0 = \hat{b}(\check{a})$, see (14.3) and (14.4). Since all the properties of the fixed baseline estimator can be derived from $F(b)$, we use from now on the simplified notation of (14.6).

The minimization of (14.6) is a nonlinear least-squares problem of which the manifold is highly curved if Q is large and l is small. This is the typical case for the GNSS compass, where the baseline is very short and the single-epoch solution is determined by the relative poor code data.

The problem of minimizing $F(b)$ can be described in geometric terms as the problem of finding a point of contact between the b_0-centred ellipsoid $\|b_0 - b\|_Q^2 =$ constant and the origin-centred sphere $\|b\|^2 =$ constant. These points of contact are easily determined in case Q is a scaled unit matrix, but not so in the general case.

We have the following theorem.

Theorem 1. *Let* $M_b = Q^{-1} + \sigma_l^{-2}(1 - l/\|b\|)I_n$. *Then* $F(\hat{b}) \leq F(b)$ *for all* $b \in \mathbb{R}^n$ *if and only if* $M_{\hat{b}}\hat{b} = Q^{-1}b_0$ *and* $M_{\hat{b}} \geq 0$.

This theorem formulates necessary and sufficiency conditions for $\hat{b}$ to be a *global* minimizer of $F(b)$. It also provides the conditions for having nonunique minimizers. Note that $M_b = M_{\hat{b}}$ for any $b \neq \hat{b}$ that has the same length as $\hat{b}$, $\|b\| = \|\hat{b}\|$. Thus for nonunique minimizers to exist, it is necessary that $M_{\hat{b}}$ is singular.

It can be shown that M_b is singular if and only if $\|b\| = l/(1 + \sigma_l^2\lambda_1)$, where λ_1 is the smallest eigenvalue of Q^{-1}. Nonunique minimizers, if they exist, lie therefore all on the sphere with radius γ_1.

5 When do the Nonunique Solutions Exist?

Before we can determine the nonunique solutions, we first need to know whether they exist. The *consistency* requirement of the system of equations, $M_b b = Q^{-1}b_0$ and $\|b\| = \gamma_1$, results in two conditions that the data vector b_0 has to satisfy. The first condition is that b_0 must lie in the range space of matrix QM_b, $b_0 \in R(QM_b)$. The second condition comes into play when the first condition is satisfied. This second condition puts restrictions on the length of b_0. Not every b_0 that makes $M_b b = Q^{-1}b_0$ consistent, will namely produce a solution that satisfies $\|b\| = \gamma_1$.

We start with the first condition. Let the orthogonal matrix of eigenvectors of Q^{-1} be partitioned as $U = (U_1, U_2)$, with U_1 containing all eigenvectors having λ_1 as eigenvalue. Then the null space and range space of QM_b are spanned by the columns of U_1 and U_2, respectively. Hence, the first consistency condition can be formulated as

$$U_1^T b_0 = 0 \text{ or } b_0 \in R(U_2) \qquad (14.7)$$

If this condition is satisfied, then $M_b b = Q^{-1}b_0$ is solvable and its solutions can be expressed as

$$b = b_p + U_1\beta_1 \tag{14.8}$$

where $b_p = (QM_b)^{+}b_0$ is a *particular* solution, with $(QM_b)^{+}$ the pseudo inverse of QM_b, and where $U_1\beta_1$ is the *homogeneous* solution, with β_1 still undetermined.

Since b of (14.8) has to satisfy $\|b\| = \gamma_1$, the as yet undetermined β_1 cannot take on values freely, but has to satisfy

$$\|\beta_1\|^2 = \gamma_1^2 - \|(QM_b)^{+}b_0\|^2 \geq 0 \tag{14.9}$$

This shows that b_0 may not have an arbitrary length. Hence, the nonnegativity condition of (14.9) is the second consistency condition that b_0 has to satisfy.

If we take the two conditions, (14.7) and (14.9), together, we may summarize our result as follows.

Theorem 2a. *The function $F(b)$ (cf. (14.6)) has nonunique minimizers if and only if*

$$b_0 \in \mathbb{C} = \{x \in \mathbb{R}^n \mid \|x\|_\Sigma^2 \leq \gamma_1^2,\ x \in R(U_2)\} \tag{14.10}$$

where $\Sigma^{-1} = U_1U_1^T + (QM_b)^{+T}(QM_b)^{+}$.

Thus we now know, if the data vector b_0 lies in the intersection of $R(U_2)$ and the origin-centred ellipsoidal region $\|x\|_\Sigma^2 \leq \gamma_1^2$, that we will have more than one minimizer of $F(b)$.

6 The Nonunique Solutions Determined

It is now not difficult anymore to determine the nonunique solutions. We already know that the nonunique minimizers, if they exist, lie all on the sphere with radius γ_1, denoted as S_{γ_1}. This combined with the general solution of $M_b b = Q^{-1}b_0$ (cf. (14.8)), gives the following result.

Theorem 2b. *The nonunique minimizers of $F(b)$ (cf. (14.6)), if they exist, are given by the solution set*

$$\mathbb{S} = \{b_p + R(U_1)\} \cap S_{\gamma_1} \tag{14.11}$$

where $b_p = (QM_b)^{+}b_0$.

The consistency set $\mathbb{C}$ and solution set $\mathbb{S}$ can both be given a clear geometric interpretation. The set $\mathbb{C}$ describes the two consistency conditions (14.7) and (14.9). Geometrically this set describes the intersection of an origin-centred ellipsoidal region with the linear manifold $R(U_2)$. Since $R(U_1)$ and $R(U_2)$ are each others orthogonal complement, we have $\dim R(U_2) = n - \dim R(U_1)$. Thus if $\dim R(U_1) = n$, then $\dim R(U_2) = 0$ and $\mathbb{C} = \{0\}$, and if $\dim R(U_1) = n-1$, then $\dim R(U_2) = 1$ and $\mathbb{C}$ reduces to an origin centred interval.

The solution set $\mathbb{S}$ is the intersection of the linear manifold $b_p + R(U_1)$ with the sphere S_{γ_1}. It consists of two points if the linear manifold is a straight line ($\dim R(U_1) = 1$) and it forms a circle if the linear manifold is a plane ($\dim R(U_1) = 2$). Since the dimension of the linear manifold is equal to the number of times the eigenvalues of Q^{-1} are equal to λ_1, we have $1 \leq \dim R(U_1) \leq n$. If $\dim R(U_1) = n$, then $b_p + R(U_1) = \mathbb{R}^n$ and $\mathbb{S} = S_{\gamma_1}$. This is the special case when all eigenvalues of Q^{-1} are equal (i.e. Q is a scaled unit matrix).

As an illustration, we now show for the cases $n = 1, 2, 3$ how the sets $\mathbb{C}$ and $\mathbb{S}$ may look like.

Case $n = 1$: If $n = 1$, then $U_1 = 1$, $U_2 = 0$, and $Q = \lambda_1^{-1}$. Therefore $\mathbb{C} = \{0\}$ and $\mathbb{S} = S_{\gamma_1} = \{b \in \mathbb{R} \mid b = \pm\gamma_1\}$. Since $F(b) = \sigma_l^{-2}(1+\sigma_l^2\lambda_1)(b^2 - 2\gamma_1|b| + l\gamma_1)$ for $b_0 = 0$, it is readily verified that $\pm\gamma_1$ are indeed its two minimizers (see Fig. 14.1).

Case $n = 2$: We now have two cases: (a) $U_1 = I_2$, $U_2 = 0$ and (b) $U_1 = u_1$, $U_2 = u_2$. In case (a) we have $\mathbb{C} = \{0\}$ and $\mathbb{S} = S_{\gamma_1}$. In case (b), $\mathbb{C}$ is an origin-centred interval of length $2\gamma_1(1 - \lambda_1/\lambda_2)$ along the u_2 direction and $\mathbb{S}$ consists of the two intersection points of the line $b = b_p + u_1\alpha$ with the circle S_{γ_1} (see Fig. 14.2). Note, if b_0 lies on the edge of $\mathbb{C}$ and thus

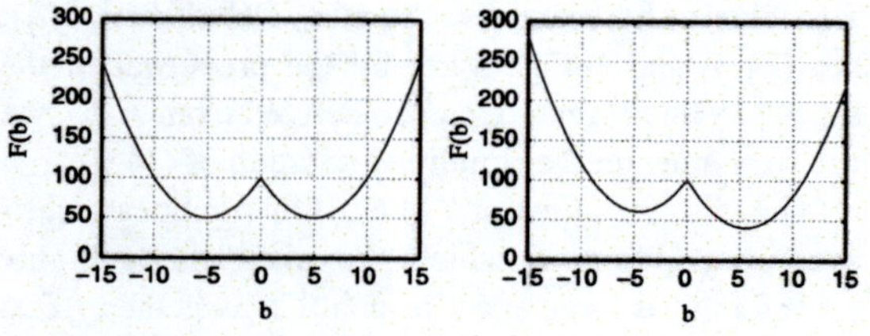

Fig. 14.1 The function $F(b) = \lambda_1(b_0 - b)^2 + \sigma_l^{-2}(l - |b|)^2$ for $b_0 = 0$ (*left*) and $b_0 \neq 0$ (*right*)

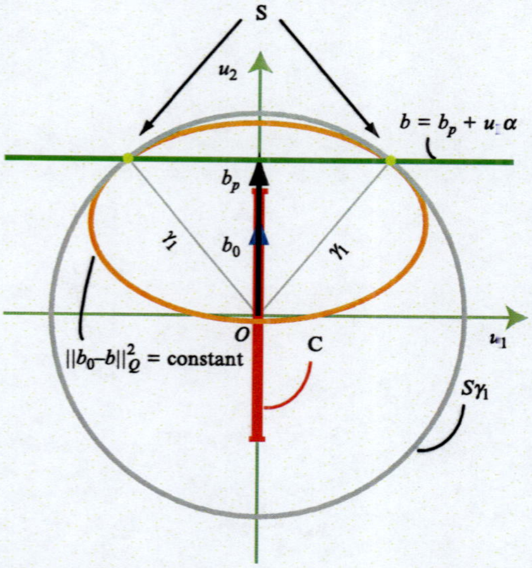

Fig. 14.2 The case $n = 2$ with $U_1 = u_1$, $U_2 = u_2$: Shown are the *circle* S_{γ_1} with axes u_1 and u_2, the interval $\mathbb{C}$, the two solution points of $\mathbb{S}$, the *line* $b = b_p + u_1\alpha$ and the *ellipse* $\|b_0 - b\|_Q^2 =$ constant

$\|b_0\| = \gamma_1(1 - \lambda_1/\lambda_2)$, that $\|b_p\| = \gamma_1$ and the two intersection points coincide in one point.

Case $n = 3$: We have the three cases: (a) $U_1 = I_3$, $U_2 = 0$, (b) $U_1 = (u_1, u_2)$, $U_2 = u_3$, and (c) $U_1 = u_1$, $U_2 = (u_2, u_3)$. In case (a) we have $\mathbb{C} = \{0\}$ and $\mathbb{S} = S_{\gamma_1}$. In case (b), $\mathbb{C}$ is an origin-centred interval of length $2\gamma_1(1 - \lambda_1/\lambda_3)$ along the u_3 direction and the solution space $\mathbb{S}$ is the circle with centre b_p and radius $\sqrt{\gamma_1^2 - \|b_p\|^2}$ that follows from intersecting the plane $b = b_p + u_1\alpha_1 + u_2\alpha_2$ with the sphere S_{γ_1}. In case (c), $\mathbb{C}$ is an origin-centred ellipse in the U_2-plane with principal axes $\gamma_1(1 - \lambda_1/\lambda_2)u_2$ and $\gamma_1(1 - \lambda_1/\lambda_3)u_3$. The solution space $\mathbb{S}$ is then the two point intersection of the line $b = b_p + u_1\alpha$ with the sphere S_{γ_1}.

The latter case ($n = 3$ (c)) is the one that is most likely to occur with GNSS, since the eigenvalues of the variance matrix $Q = Q_{\hat{b}(a)\hat{b}(a)}$ will usually all be different.

Acknowledgements The research of the author has been supported by an Australian Research Council Federation Fellowship (project number FF0883188). This support is gratefully acknowledged.

References

Buist P (2007) The baseline constrained LAMBDA method for single epoch, single frequency attitude determination applications. Proceedings of ION GPS-2007, 2007

Cohen CE (1992) Attitude determination using GPS. PhD Thesis, Stanford University

Giorgi G, Teunissen PJG, Buist P (2008) A search and shrink approach for the baseline constrained LAMBDA method: experimental results. Proceedings International GPS/GNSS Symposium, Tokyo, Japan, Nov. 11–14, 2008

Hide C, Pinchin DJ (2007) Development of a low cost multiple GPS antenna attitude system. Proceedings ION GNSS 2007, 88–95, 2007

Hofmann-Wellenhoff B, H Lichtenegger (2001): Global positioning system: theory and practice, 5nd edn. Springer, Berlin

Kuylen LV, Nemry P, Boon F, Simsky A, Lorga JFM (2006) Comparison of attitude performance for multi-antenna receivers. Eur J Navigation, 4(2)

Leick A (2003) GPS satellite surveying, 3rd edn. Wiley, New York

Lu G (1995) Development of a GPS multi-antenna system for attitude determination. UCGE Reports 20073, Dept. of Geomatics Eng., University of Calgary, 1995

Misra P, Enge P (2001) Global positioning system: signals, measurements, and performance. Ganga-Jamuna Press, Lincoln, MA

Montgomery P, IY Bar-Itzhack, J Garrick (1997): Algorithm for attitude determination using GPS. Proceedings of AIAA Guidance, Navigation and Control Conf., New Orleans, LA, USA. Aug. 1997, Paper no. AIAA 97-3616, 1997.

Park C, Teunissen PJG (2003) A new carrier phase ambiguity estimation for gnss attitude determination systems. Proceedings of International GPS/GNSS Symposium, Tokyo, 2003.

Park C, Teunissen PJG (2007) A baseline constrained lambda method for integer ambiguity resolution of GNSS attitude determination systems. J Control, Robotics Systems 14(6):587–594

Simsky A, Vander Kuylen L, Boon F (2005) Single-board attitude determination system based on the PolaRx2@GPS receiver. Proc ENC GNSS 2005, Munich, Germany, 19–22 July, 2005.

Strang G, Borre K (1997) Linear algebra, geodesy, and GPS. Wellesley-Cambridge Press, Wellesley, MA

Teunissen PJG (1994) A new method for fast carrier phase ambiguity estimation. Proceedings IEEE PLANS'94, Position Location and Navigation Symposium, pp 562–573, 1994

Teunissen PJG (1995) The least-squares ambiguity decorrelation adjustment: a method for fast GPS integer ambiguity estimation. J Geodesy 70:65–82

Teunissen PJG (2006) The LAMBDA method for the GNSS compass. Artificial Satellites, 41(3):89–103

Teunissen PJG, Kleusberg (eds) (1998) GPS for Geodesy, 2nd enlarged (edn) Springer, Berlin

Tu CH, KY Wang, WC Melton (1996) GPS compass: a novel navigation equipment. Proceedings of ION National Technical Meeting, Santa Monica, CA, USA, 1996

The Effect of Reordering Strategies on Rounding Errors in Large, Sparse Equation Systems

15

A. Ernst and W.-D. Schuh

Abstract

The effect of reordering strategies on the rounding errors is considered for the factorization and solution of sparse symmetric systems. On the one hand, a reduction of rounding errors can be expected, because the number of floating point operations decreases. On the other hand, the clustering of neighboring parameters and therefore the fixing of the sequence of parameter elimination may result in numerical instabilities. These effects are demonstrated for sparse covariance matrices in Wiener filtering. In particular Cholesky factorization and profile reordering in conjunction with envelope storage schemes are examined.

1 Introduction

In this work we investigate the hypothesis that reordering the sequence of unknown parameters of a sparse equation system has no negative effect on the rounding errors. In principle the sequence of the elimination of unknowns is subject to an appropriate pivoting strategy to deal with numerical instabilities. Strongly correlated parameters are separated by reordering the sequence of parameter elimination. In contrast to the pivoting strategy the reordering scheme for sparse systems aims at a clustering of neighbored data points. This yields a small profile and only few fill-ins during the solution process (Ernst 2009). From the numerical point of view reordering counteracts pivoting. As a typical and also most critical application we have a look at Wiener filtering and other prediction processes where large covariance matrices are generated. Compactly supported covariance functions in 2D (Sansò and Schuh 1987) and 3D (Gaspari and Cohn 1999; Gaspari et al. 2006; Moreaux 2008) allow for a sparse representation of the covariance information considering the positive definiteness. Naturally, we exploit the sparse structure of the covariance matrices as much as possible by an efficient reordering algorithm (e.g. reversed Cuthill-McKee (Gibbs et al. 1976) or banker's algorithm (Snay 1976)) and an appropriate storage schema (Ernst 2009). As outlined in Schuh (1991) the numerical stability of covariance matrices in prediction procedures is basically influenced by the shape of the covariance function, the variance of the uncorrelated noise and the data distribution. Especially neighboring data points cause numerical problems. To study the numerical behaviour in detail a specific rounding error analysis is necessary. Whereas norm-based perturbation bounds (Stewart 1973) are focused on the global assessment of algorithmic processes and ignore the sparsity of a system, a stochastic approach (Meissl 1980) allows for an individual handling.

The paper is organized as follows. Section 2 defines some fundamental terms concerning rounding error

A. Ernst · W.-D. Schuh (✉)
Institut für Geodäsie und Geoinformation der Universität Bonn, D-53115 Bonn, Nussallee 17, Germany
e-mail: Schuh@uni-bonn.de

N. Sneeuw et al. (eds.), *VII Hotine-Marussi Symposium on Mathematical Geodesy*, International Association of Geodesy Symposia 137, DOI 10.1007/978-3-642-22078-4_15,

analysis. Section 3 presents norm-based rounding error analysis applied to Cholesky's algorithm and introduces the stochastic approach for a precise rounding error analysis and an algorithmic procedure to overcome the recursive variance propagation within Cholesky factorization. Section 4 gives an example. The paper finishes with conclusions.

2 Rounding Error Analysis

For an efficient solution on a computer each floating point number d is represented by its machine representation $\overline{d}$. Today the widely-used IEEE standard 754 defines the representation, rounding algorithms, mathematical operations and exception handling for floating point arithmetics (IEEE 2008). A floating point number d in binary coded 64-bit (double precision) representation consists of

$$\overline{d} = (-1)^s \cdot m \cdot b^q, \tag{15.1}$$

where s denotes a binary digit for the sign of the number, m the mantissa with $\tau = 53$ binary digits, b the basis 2, and q the exponent with 10 binary digits and a given bias. The relative error

$$\left|\frac{\overline{d}-d}{d}\right| \leq \varepsilon_m, \tag{15.2}$$

defines the unit roundoff or machine epsilon ε_m. This quantity depends on the number of digits of the mantissa τ and the rounding procedure. True rounding (rounding to nearest) yields $\varepsilon_m = 2^{-\tau}$. A mapping error occurs also during each arithmetic operation. The computer evaluates the computed function $\overline{f}(\overline{d})$ instead of the mathematical function $f(d)$.

A rounding error analysis provides information about the perturbance measured by the size of $|f(d) - \overline{f}(\overline{d})|$, the difference between the mathematically rigorous result $f(d)$ and the function $\overline{f}(\overline{d})$ evaluated with machine numbers. Expanding this norm by plus minus $f(\overline{d})$ we get the inequality

$$|f(d) - \overline{f}(\overline{d})| \leq |f(d) - f(\overline{d})| + |f(\overline{d}) - \overline{f}(\overline{d})|. \tag{15.3}$$

The first absolute term on the right-hand side of inequality (15.3) characterizes the *stability of the problem* closely connected with the condition of the problem, whereas the second term contains information about the *stability of the algorithm*, where beside the condition also the order and number of operations in the algorithm has to be taken into account (Dahmen and Reusken 2008).

3 Rounding Error Analysis Applied to Cholesky's Algorithm

Without restricting the generality we focus our investigation on the Cholesky solution of an n-dimensional equation system $\mathbf{N}\mathbf{x} = \mathbf{y}$, where the positive definite, symmetric matrix $\mathbf{N}$ is factorized by $\mathbf{N} = \mathbf{R}^T\mathbf{R}$ into a unique upper triangular matrix $\mathbf{R}$ with positive diagonal elements. For a given right hand side $\mathbf{y}$ the unknown parameter vector $\mathbf{x}$ is computed by the solution of two triangular systems. In the forward substitution step $\mathbf{R}^T\mathbf{z} = \mathbf{y}$ the auxiliary vector $\mathbf{z}$ is determined and after this the unknown parameter vector $\mathbf{x}$ results from the backward substitution step $\mathbf{R}\mathbf{x} = \mathbf{z}$.

In general the effect of rounding errors in a triangular factorization process can be measured indirectly by an estimation of the coefficients of the disturbed system $(\mathbf{N} + \Delta\mathbf{N})\overline{\mathbf{x}} = \mathbf{y}$, which are given by

$$|\Delta n_{ij}| \leq (c_1 n + 2c_2 n^2 + c_2^2 n^3 \varepsilon_m) \max_{i,j} |n_{ij}| \; g \; \varepsilon_m \tag{15.4}$$

where c_1 and c_2 are constants of the order unity and g denotes the growth factor, which is defined by half of the magnitude of the largest number occurring during the whole computation divided by the largest absolute value in $\mathbf{N}$ (Stewart 1973, Theorem 5.3, p. 155). Applying the propagation of relative errors in linear equation systems

$$\frac{\|\Delta\mathbf{x}\|}{\|\mathbf{x}\|} \leq \|\mathbf{N}\|\|\mathbf{N}^{-1}\| \left(\frac{\|\Delta\mathbf{N}\|}{\|\mathbf{N}\|} + \frac{\|\Delta\mathbf{y}\|}{\|\mathbf{y}\|}\right) \tag{15.5}$$

(cf. Kreyszig 1993, p. 998) the disturbances in $\Delta\mathbf{N}$ of (15.4) can be propagated to the relative disturbances of the solution vector. Introducing the norm $\|\mathbf{N}\|$ by the *infinity norm* $\|\mathbf{N}\| = n \max_{i,j} |n_{ij}|$ and substitute (15.4) in (15.5) yields

$$\frac{\|\mathbf{x} - \overline{\mathbf{x}}\|}{\|\mathbf{x}\|} \leq \|\mathbf{N}\|\|\mathbf{N}^{-1}\|(c_1 + 2c_2 n + c_2^2 n^2 \varepsilon_m) \; g \; \varepsilon_m. \tag{15.6}$$

In contrast to LU factorization strategies, the growth factor of Cholesky decomposition is not affected by

Table 15.1 Stochastical description for rounding errors of arithmetical operations. γ denotes the smallest integer power satisfying the inequality

Operation	ε	$E\{\varepsilon\}$	$\sigma\{\varepsilon\} = \frac{c}{\sqrt{12}}\varepsilon_m$
Summation	$\varepsilon^{(a)}$	0	$c = 2^\gamma > \max(\|a\|, \|b\|, \|a+b\|)$
Subtraction	$\varepsilon^{(s)}$	0	$c = 2^\gamma > \max(\|a\|, \|b\|, \|a-b\|)$
Multiplication	$\varepsilon^{(m)}$	0	$c = 2^\gamma > \|a \cdot b\|$
Division	$\varepsilon^{(d)}$	0	$c = 2^\gamma > \|a/b\|$
Square root	$\varepsilon^{(sq)}$	0	$c = 2^\gamma > \|\sqrt{a}\|$

the pivoting strategy and is bounded by $g \leq 1$. The influence of rounding errors is dominated by the linear term $2c_2n$. This term is mainly caused by the accumulation of the scalar products, and can be reduced by a higher precision in the computation of the scalar product (Stewart 1973, p. 156). However, also sparsity reduces the number of operations and may have a positive influence on the rounding errors.

To allow for an individual analysis Meissl (1980) introduced a stochastic approach to estimate the rounding error for very large networks in particular for the adjustment of the US ground-control network. The rounding error ε is considered a random variable and defined by its expectation $E\{\varepsilon\}$ and variance $\sigma^2\{\varepsilon\}$. Table 15.1 contains the expectation and variance for the arithmetic operations used in the Cholesky algorithm. The expectation depends on the rounding algorithm. In the IEEE 754 definitions true rounding is implemented, so in this case no bias occurs. The variances $\sigma^2\{\varepsilon\}$ of the individual operations are given in the last column of Table 15.1. The variance depends on the factor c, which is an operation dependent number, and on the machine epsilon ε_m. The factor $\frac{1}{\sqrt{12}}$ is defined by the variance of a uniformly distributed random variable. The factor c characterizes the maximum number of digits that are lost during the operation and depends for the addition/subtraction on the maximum of the input values as well as on the result of the operation. Within the other operations of multiplication, division and square root the factor c depends only on the magnitude of the result.

In contrast to Meissl's approach where a rough estimation of the number of operations and the magnitude of the quantities is used to propagate the rounding error for the large system, we consider each individual computing step. All functional dependencies during the Cholesky decomposition are taken into account and we perform a rigorous variance propagation for the whole solution process. The rounding errors in each operation are modeled individually by the size of the actual operators,

$$\overline{f}(\overline{a},\overline{b}) = f(a+\varepsilon_a, b+\varepsilon_b) + \varepsilon_{f(\overline{a},\overline{b})}. \tag{15.7}$$

Here ε_a and ε_b denotes the perturbance of the coefficients and $\varepsilon_{f(\overline{a},\overline{b})}$ the rounding error during the operation. Applying linear perturbation theory we get

$$\overline{f}(\overline{a},\overline{b}) = f(a,b) + c_1\varepsilon_a + c_2\varepsilon_b + \varepsilon_{f(\overline{a},\overline{b})}. \tag{15.8}$$

Collecting the ε-quantities in the variable $\varepsilon_{f(a,b)}$ yields

$$\overline{f}(\overline{a},\overline{b}) = f(a,b) + \varepsilon_{f(a,b)}. \tag{15.9}$$

The Cholesky factorization is a recursive evaluation process. All elements r_{ij}, r_{ii}, z_i and x_i depend on previous evaluated elements and all these elements are correlated. To show the principle approach we pick out a special operation, the computation of

$$r_{ij} = \left(n_{ij} - \sum_{k=1}^{i-1} r_{ki}r_{kj}\right)/r_{ii}, \; i = 1\ldots j, \; j = 1\ldots n \tag{15.10}$$

(Meissl 1980, eq. 3.31). For the evaluation the disturbed values $\overline{r}_{ij}$ and $\overline{n}_{ij}$ (ref. (15.9)) as well as the basic rounding errors $\varepsilon^{(m)}$, $\varepsilon^{(s)}$ and $\varepsilon^{(d)}$ caused by the arithmetic operations have to be taken into account,

$$\overline{r}_{ij} = \frac{\overline{n}_{ij} - \sum_{k=1}^{i-1}\left(\left(\overline{r}_{ki}\overline{r}_{kj} + \varepsilon_k^{(m)}\right) + \varepsilon_k^{(s)}\right)}{\overline{r}_{ii}} + \varepsilon^{(d)}. \tag{15.11}$$

Applying linear perturbation theory the individual basic errors can be summarized by $\varepsilon_{\varepsilon_{r_{ij}}}$

$$\varepsilon_{\varepsilon_{r_{ij}}} = \frac{1}{\overline{r}_{ii}}\sum_{k=1}^{i-1}\left(\varepsilon_k^{(m)} + \varepsilon_k^{(s)}\right) + \varepsilon^{(d)}. \tag{15.12}$$

It should be mentioned that the order of the computing steps is important because the rounding errors are not commutative as they depend on the size of the result. Taken into account also the disturbances of the input quantities we get

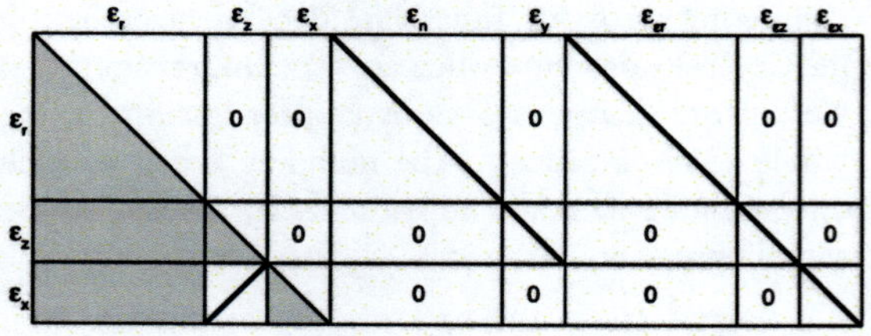

Fig. 15.1 Structure of the implicit error formulation of the complete solution process with Cholesky's algorithm. The *filled parts* depict dense matrix structures, the *fat lines* represent diagonal matrix entries, and the *white parts* contain just zeros

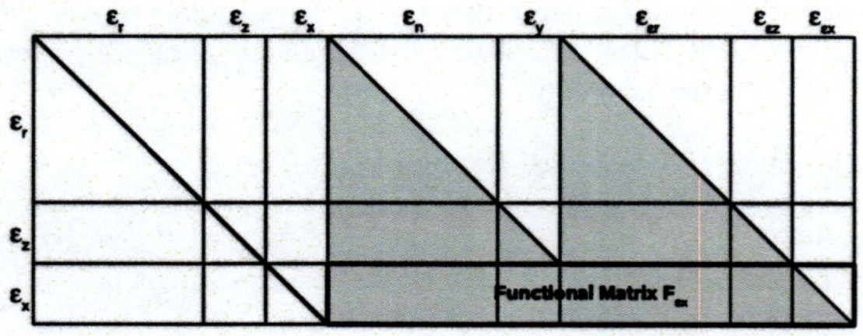

Fig. 15.2 Structure of the explicit rounding error formulation after reduction by the Gauss–Jordan algorithm. Dependencies shifted to the right blocks of the known rounding errors

$$r_{ij} + \varepsilon_{r_{ij}} = \frac{(n_{ij} + \varepsilon_{n_{ij}}) - \sum_{k=1}^{i-1} (r_{ki} + \varepsilon_{r_{ki}})(r_{kj} + \varepsilon_{r_{kj}})}{r_{ii} + \varepsilon_{r_{ii}}} + \varepsilon_{\varepsilon_{r_{ij}}}. \tag{15.13}$$

By expanding this equation and disregarding second order terms of ε we get a linearized construction for the evaluated Cholesky-element $\bar{r}_{ij}$,

$$r_{ij} + \varepsilon_{r_{ij}} = r_{ij} + \frac{1}{r_{ii}}\varepsilon_{n_{ij}} - \sum_{k=1}^{i-1}\left(\frac{r_{kj}}{r_{ii}}\varepsilon_{r_{ki}} + \frac{r_{ki}}{r_{ii}}\varepsilon_{r_{kj}}\right) - \frac{1}{2r_{ii}}\varepsilon_{r_{ii}} + \varepsilon_{\varepsilon_{r_{ij}}}. \tag{15.14}$$

We end up with an implicit formulation of $\varepsilon_{r_{ij}}$, which depends on the already computed quantities $\varepsilon_{n_{ij}}$, $\varepsilon_{r_{ki}}$, $\varepsilon_{r_{kj}}$ and $\varepsilon_{\varepsilon_{r_{ij}}}$. The same approach is also applied to the quantities of the forward and backward substitution step, ε_{z_i} and ε_{x_i}.

The structure of the complete implicit equation system is shown in Fig. 15.1. The system shows the functional dependencies of the rounding errors as they are formulated in (15.14). The system is ordered column wise by the errors of the derived quantities ($\varepsilon_{r_{ij}}$, ε_{z_i}, ε_{x_i}) followed by the input errors ($\varepsilon_{n_{ij}}$, ε_{y_i}) and the individually processed errors of the operations ($\varepsilon_{\varepsilon_{r_{ij}}}$, $\varepsilon_{\varepsilon_{z_i}}$, $\varepsilon_{\varepsilon_{x_i}}$) defined by (15.12). Out of these implicit equations an explicit formulation for the unknown rounding errors is needed. Therefore, the system is factorized by the Gauss–Jordan algorithm, which solves for the dependencies of the Cholesky quantities (see Fig. 15.2).

To compute the rounding error covariance matrix of the unknown parameters **x** the bordered block in Fig. 15.2 is needed. This block is the functional matrix $\mathbf{F}_{\varepsilon_x}$ in the variance propagation $\Sigma\{\varepsilon_x\} = \mathbf{F}_{\varepsilon_x}\Sigma\{\varepsilon_{basic}\}\mathbf{F}_{\varepsilon_x}^T$. $\Sigma\{\varepsilon_{basic}\}$ contains the uncorrelated basic rounding errors that arise during the solution, e.g $\varepsilon_{\varepsilon_{r_{ij}}}$ of (15.12) and the a priori error information of the normal equation system. The result is a full covariance matrix $\Sigma\{\boldsymbol{\varepsilon}_{\mathbf{x}}\}$ where the variances describe the stochastic rounding errors of the solution of the equation system. The covariances also contain information concerning the correlations between the single rounding errors.

4 Simulations

The algorithm outlined in Sect. 3 is tested with a Wiener–Kolmogorov filtering of Bouguer anomalies derived at irregular positions. Figure 15.3 shows the spatial data distribution with the residuals and identified outliers. The measurements are reduced by a polynomial of second order to ensure stationarity. The residual signal **s** is predicted by $\mathbf{s} = \Sigma\{\mathbf{s}, \Delta\mathbf{l}\}\,\Sigma\{\Delta\mathbf{l}\}^{-1}\,\Delta\mathbf{l}$, where $\Delta\mathbf{l}$ denotes the vector with the trend reduced measurements and $\Sigma\{\Delta\mathbf{l}\}$ the covariances. The matrix is deduced from the analytic covariance function, where the empirical covariances are approximated by a Bessel function combined with a compactly supported function (Sansò and Schuh 1987; Moreaux 2008). This leads to a sparse matrix $\Sigma\{\Delta\mathbf{l}\}$. The matrix $\Sigma\{\mathbf{s}, \Delta\mathbf{l}\}$ defines the covariances between the data points and the prediction points.

We focus our attention on the inversion process $\mathbf{w} = \Sigma\{\Delta\mathbf{l}\}^{-1}\,\Delta\mathbf{l}$. This is equivalent to the solution of the linear equation system $\Sigma\{\Delta\mathbf{l}\}\,\mathbf{w} = \Delta\mathbf{l}$. The matrix $\Sigma\{\Delta\mathbf{l}\}$ is reordered with three different numbering schemes. The first scheme is the natural form given

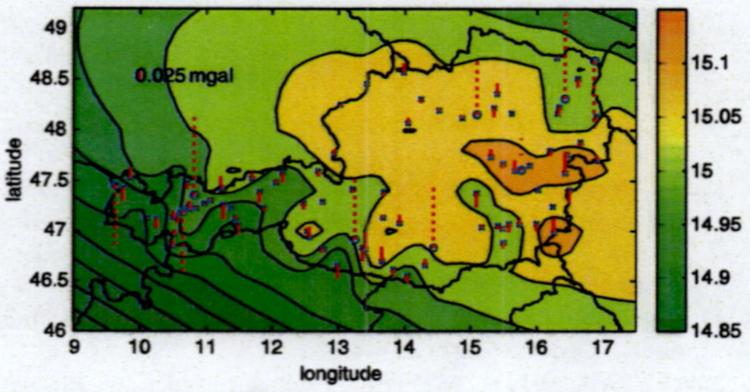

Fig. 15.3 Wiener–Kolmogorov prediction of Austrian Bouguer anomalies (data set Ruess 1986, intern communication). *Red bars* display the residuals. *Red dotted bars* display identified outliers

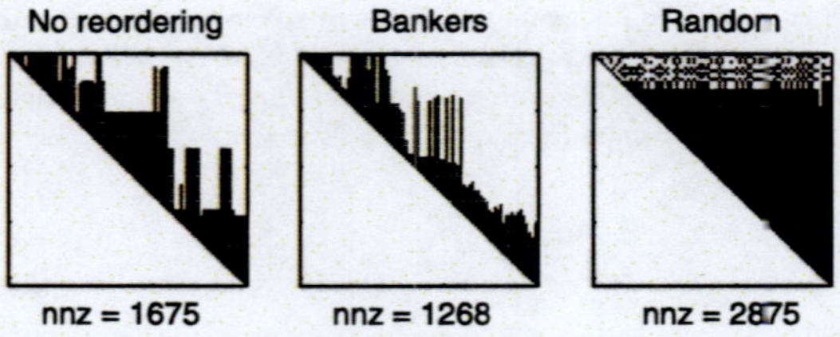

Fig. 15.4 Profiles of test system after Cholesky factorization with different reordering strategies

by the ordering of the data. The second scheme is produced by a reordering with the banker's algorithm (Snay 1976). The third system is generated by a randomized ordering. The three profiles are shown in Fig. 15.4.

To analyze the different rounding errors dependent on the condition of the system we vary the condition of the system by arbitrary choices of the uncorrelated noise in the data points, which is defined by the difference between the empirical and the analytic covariance function at the distance zero. The noise is added on the main diagonal of $\Sigma\{\Delta \mathbf{l}\}$ and stabilizes the system. These systems are tested with the developed algorithm and the covariance matrix of the rounding errors of the solution vector $\mathbf{w}$ is computed. Results from the simulation are shown in Fig. 15.5. The maximum rounding errors are plotted for the different numbering schemes and systems. The errors have almost the same size for the same condition number. They do not differ significantly because of the reordering strategy. The reordering with the banker's algorithm influences the rounding errors positively at higher condition numbers. There the randomized

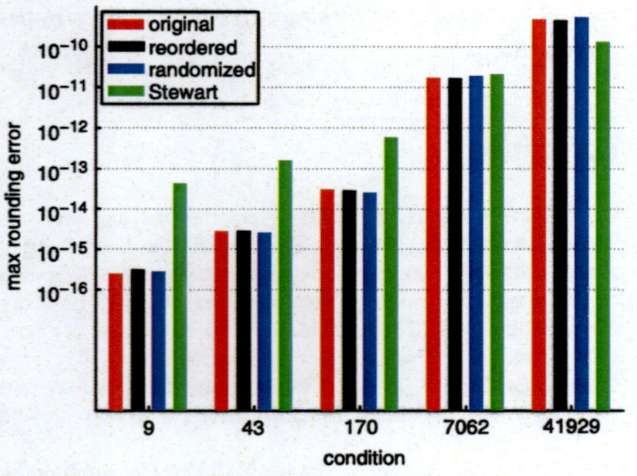

Fig. 15.5 Results of the simulation for various condition numbers with the three numbering schemes and Stewart's rounding error approximation with c_1 and c_2 of the order unity fixed with 1

ordering produces the highest rounding errors. But in general the size of the rounding errors is essentially influenced by the condition number. A clustering of numerical instabilities cannot be observed. For well conditioned systems the algorithm gives a more optimistic approximation of the rounding errors than the formula by Stewart up to the factor 100. For bad conditioned systems both approximations show similar results, but for instable systems Stewart's approximation is smaller than the stochastic errors by a factor of five but this depends basically on the choice of the constants c_1 and c_2 in (15.6). For the stochastic approach besides the absolute rounding errors also individual values including the correlations between the rounding errors can be analyzed. It can be observed that the correlations depend very strongly on the size of the rounding errors. The larger they are the higher they are correlated.

5 Conclusion and Discussion

We investigated the hypothesis that reordering induces a clustering of instabilities. The stochastic approach allows for an individual analysis of rounding errors in evaluation processes. As demonstrated here also complex recursive algorithms can be handled and rigorously computed. With respect to our hypothesis it can be stated that the clustering of numerical instabilities caused by reordering strategy has no negative impact on rounding errors.

References

Dahmen W, Reusken A (2008) Numerics for engineers and natural scientists (German). Springer, Berlin, Heidelberg

Ernst A (2009) Implementation of efficient algorithms to reorder, solve and invert sparse normal equation systems with geodetic applications (German). Master's thesis, Unitversity Bonn, Institute of Geodesy and Geoinformation

Gaspari G, Cohn S, (1999) Construction of correlation functions in two and three dimensions. Q J Roy Meteorol Soc 125(554):723–757

Gaspari G, Cohn S, Guo J, Pawson S (2006) Construction and application of covariance functions with variable length-fields. Q J Roy Meteorol Soc, 132:1815–1838

Gibbs N, Poole W, Stockmeyer P (1976) An algorithm for reducing the bandwidth and profile of a sparse matrix. SIAM J Numer Anal 13:236–250

IEEE Computer Society (2008) IEEE standard for floating point arithmetic. Technical report, The Institute of Electrical and Electronics Engineers, Inc

Kreyszig E (1993) Advanced engineering mathematics, 7th edn. Wiley, New York

Meissl P (1980) A priori prediction of roundoff error accumulation in the solution of a super-large geodetic normal equation system. NOAA/National Ocean Survey's National Geodetic Survey (NGS), Rockville, Md., Professional Paper 12

Moreaux G (2008) Compactly supported radial covariance functions. J Geodesy 82(7):431–443

Sansò F, Schuh W-D (1987) Finite covariance functions. Bulletin Géodésique 61:331–347

Schuh W-D (1991) Numerical behaviour of covariance matrices and their influence on iterative solution techniques. In: Rapp R, Sansò F (eds) Determination of the Geoid – Present and Future, IAG Proceedings, vol 106. Springer, Heidelberg, 432–441

Snay R (1976) Reducing the profile of sparse symmetric matrices. NOAA Technical Memorandum, NOS NGS-4

Stewart G (1973) Introduction to matrix computations. Academic press, New York, San Fancisco, London

Performance Analysis of Isotropic Spherical Harmonic Spectral Windows

16

B. Devaraju and N. Sneeuw

Abstract

Spatial smoothing or spectral filtering using spherical harmonic spectral windows is performed, for example, to reduce noise, or to bring two datasets to the same resolution. Despite a number of spectral windows available, no framework exists to analyse their performance in the spatial domain before a choice can be made. In view of this, a set of parameters is devised to analyse the performance of isotropic spectral windows in the spatial domain. Using these parameters five isotropic spectral windows – Gauss, von Hann, Pellinen, box-car and Butterworth – are analysed in terms of their efficacy in reducing the resolution of a given dataset. The analysis shows that the parameters designed in this contribution provide a new perspective to the qualitative and quantitative analysis of isotropic spectral windows. Further, the analysis points out Butterworth (order 2) and von Hann windows as the appropriate windows for reducing resolution of datasets.

Keywords

Isotropy • Performance analysis • Spectral windows • Smoothing • Filtering

1 Spatial Smoothing

Smoothing the gravity field on the sphere using isotropic filter windows was formalised for physical geodesy by Jekeli (1981) in order *to reduce the resolution of the dataset*. There he adapted the windows that were already in use in Fourier spectral analysis of one-dimensional (1-D) data to the sphere. The interest in such filter windows resurfaced with the advent of GRACE time-variable gravity field data due to the need *for suppressing noisy higher frequencies* in the GRACE dataset (Wahr et al. 1998). In due course, a variety of methods were proposed to smooth the noisy GRACE data, for example, Han et al. (2005); Swenson & Wahr (2006); Kusche (2007); Klees et al. (2008). An attempt was made by Kusche (2007) to derive performance measures for the inter-comparison of these filter kernels. However, a unifying framework to analyse the performance of these filters in both the spatial and spectral domains does not exist. In this contribution, performance measures will be derived and applied to determine the appropriate isotropic spectral window for resolution reduction of a given gravity field. The performance measures are to a larger extent an adaptation of the measures devised by Harris (1978) for windows in 1-D Fourier spectral analysis.

B. Devaraju (✉) · N. Sneeuw
Institute of Geodesy, University of Stuttgart,
Geschwister-Scholl-Str. 24D, D-70174, Stuttgart, Germany
e-mail: devaraju@gis.uni-stuttgart.de

In the rest of this document, the terms *smoothing operator/window* and *filter* will be used inter-changeably.

2 Isotropic Windows

The weights of isotropic windows on the sphere depend only on the spherical distance between the points, which implies that they are independent of the location and direction. They are the simplest class of windows defined on the sphere. Table 16.1 shows a few well-known isotropic windows used in physical geodesy. Following is the harmonic transform pair for an isotropic window.

$$W(\psi) = \sum_{l=0}^{L} \frac{2l+1}{2} w_l P_l(\cos\psi) \qquad (16.1a)$$

$$w_l = \int_0^{\pi} W(\psi) P_l(\cos\psi) \sin\psi \, d\psi \qquad (16.1b)$$

where

$W(\psi)$ – Isotropic function on the unit sphere
$P_l(\cos\psi)$ – Legendre polynomial of degree l
ψ – Spherical distance
w_l – Legendre polynomial coefficients

Table 16.1 Some well-known isotropic windows. All spatial cross-sections shown here are reconstructed/constructed using (16.1a)

Filter	Definition
Gauss	$W(\psi) = b\,\frac{e^{-b(1-\cos\psi)}}{1-e^{-2b}}, \quad b = \frac{\ln(2)}{1-\cos\psi_0}$
von Hann	$W(\psi) = \begin{cases} \frac{1}{2}\left(1+\cos\frac{\pi\psi}{\psi_0}\right), & 0 \le \psi \le \psi_0 \\ 0, & \psi_0 \le \psi \le \pi \end{cases}$
Pellinen	$W(\psi) = \begin{cases} 1\,, 0 \le \psi \le \psi_0 \\ 0\,, \psi_0 \le \psi \le \pi \end{cases}$
Box-car	$w_l = \begin{cases} 1\,, l = 0,\ldots,l_c \\ 0\,, l > l_c \end{cases}$
Butterworth	$w_l = \frac{1}{\sqrt{1+\left(\frac{l}{l_c}\right)^{2k}}}, \quad k = 1,2,3,\ldots$

	Gauss	von Hann	Pellinen	Box-car	Butterworth
Spatial					
Spectral					

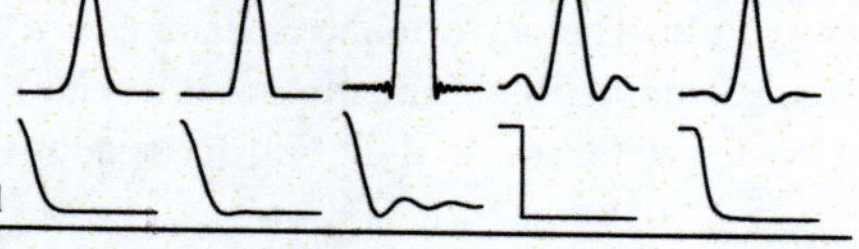

Smoothing a field $f(\theta,\lambda)$ on the sphere with an isotropic window is a convolution in the spatial domain, and a multiplication in the spectral domain.

$$f'(\theta,\lambda) = f(\theta,\lambda) * W(\psi) \qquad (16.2a)$$

$$= \sum_{l=0}^{\infty} w_l \sum_{m=-l}^{l} f_{lm} Y_{lm}(\theta,\lambda) \qquad (16.2b)$$

where θ, λ are the co-latitude and longitude, $Y_{lm}(\cdot,\cdot)$ is the surface spherical harmonic of degree l and order m, and f_{lm} is its coefficient.

3 Performance Measures

The performance measures that are designed here use the energy associated with the filter kernel as the basis. In Fig. 16.1, the definitions of different terms that will be used in the rest of this contribution are illustrated. The magnitude of the filter weights and the energy of the filter will all be measured in units of decibel (dB).

$$W'(\psi)\ [\text{dB}] = 20 \log_{10}\left(\frac{W(\psi)}{W(0)}\right) \qquad (16.3)$$

Processing loss (α_L) A certain amount of the signal is lost when a field is filtered. *Processing loss* computes the amount of signal lost due to filtering.

$$\alpha_L = 1 - \frac{\iint_\Omega f'^2(\theta,\lambda)\,d\Omega}{\iint_\Omega f^2(\theta,\lambda)\,d\Omega} = 1 - \frac{\sum_{l,m} w_l^2 f_{lm}^2}{\sum_{l,m} f_{lm}^2} \qquad (16.4a)$$

$$= 1 - \alpha \qquad (16.4b)$$

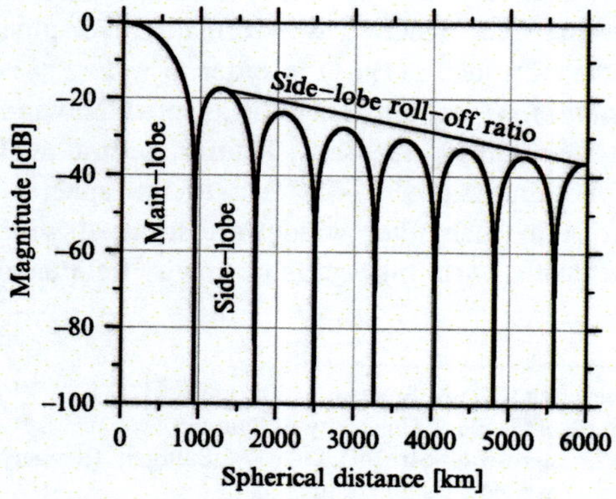

Fig. 16.1 Anatomy of an isotropic filter kernel

$$\alpha_{\text{SNR}} = \frac{\text{SNR}(f'(\Omega))}{\text{SNR}(f(\Omega))}, \text{ where SNR} = \frac{\text{Signal power}}{\text{Noise power}} \quad (16.4c)$$

where α is the damping factor (Kusche 2007). If the isotropic windows are used for reducing noise, then (16.4c) provides better information than (16.4b) as it includes the smoothing applied to the noise as well.

Main-lobe half-width (ψ_M) Main-lobe half-width is the point of first zero-crossing of the isotropic window. This definition is applicable only for filters, whose spatial filter weights oscillate around zero. In order to determine the main-lobe width of filters whose weights decay to zero and do not cross it, a few other definitions are required.

1. *Spatial variance of normalized energy of the spatial filter kernel* For non-zero-crossing windows, the main-lobe half-width is calculated via the *second moments* of the energy function. In order to determine the second moments, the energy function is treated as a probability density function by normalizing it as done by Kusche (2007).

$$\psi_M^2 = \int_0^\pi \psi^2 \frac{W^2(\psi)}{\int_0^\pi W^2(\psi)\sin\psi \mathrm{d}\psi} \sin\psi \mathrm{d}\psi \quad (16.5)$$

2. *Half-width of fraction of the peak* The half-width from kernel location at which the function attains a certain fraction of its peak value.

$$W(\psi_M) = \frac{1}{n} W(0) \quad (16.6)$$

3. *Half-width at fraction of total energy* The half-width at which the filter has accumulated a desired fraction of total energy contained in the function.

$$\int_0^{\psi_M} W^2(\psi)\sin\psi \mathrm{d}\psi = \frac{1}{n}\int_0^\pi W^2(\psi)\sin\psi \mathrm{d}\psi \quad (16.7)$$

Highest side-lobe level It is defined as the peak of the highest side-lobe. This is an important quantity both for noise reduction and resolution reduction as it determines the single largest unwanted contribution from outside the main-lobe. For non-zero-crossing windows, it is the magnitude at main-lobe half-width.

Side-lobe roll-off ratio (ϱ_s) It is defined as the ratio between peak window weights of the first and last side-lobe (cf. Fig. 16.1). This ratio determines how fast the oscillations around the main-lobe subside, i.e., approach zero. For non-zero-crossing windows, it is the ratio between window weights at the main-lobe width and at the end of the window.

Spatial leakage (ξ) Spatial leakage is the fraction of energy contributed to the total energy by the side-lobes. Thus, the total energy contained in the filter is the sum of leakage and main-lobe energy concentration (β).

$$\xi = \frac{\int_{\psi_M}^\pi W^2(\psi)\sin\psi \mathrm{d}\psi}{\int_0^\pi W^2(\psi)\sin\psi \mathrm{d}\psi} \quad (16.8a)$$

$$= 1 - \frac{\int_0^{\psi_M} W^2(\psi)\sin\psi \mathrm{d}\psi}{\int_0^\pi W^2(\psi)\sin\psi \mathrm{d}\psi} = 1 - \beta \quad (16.8b)$$

4 Performance Analysis

The main idea of applying windows in the spectral domain is to reduce the resolution or smooth noise in the dataset. The windows must decay slowly towards zero so that they can be transformed from the spectrum to the space without any artefacts. The artefacts, which are inevitable in the case of discontinuous and rapidly decaying windows, mainly occur as oscillating weights in the transformed window (in space). Therefore, the region under these oscillating weights, called the side-lobes, can be considered unwanted contribution. This implies that the side-lobe levels must be smaller to ensure that the smoothed signal represented at a point in space is mostly represented by the main-lobe.

An ideal filter window applied in the spectral domain for reducing the resolution should have the corresponding main-lobe width without any side-lobes, and should have minimum processing loss. This implies that in the less-ideal cases, as in Fig. 16.2, the filters should have desired main-lobe width, less leakage, lower side-lobe levels, minimum processing loss and faster side-lobe roll-off. Of these, highest side-lobe level and processing loss provide an overview of

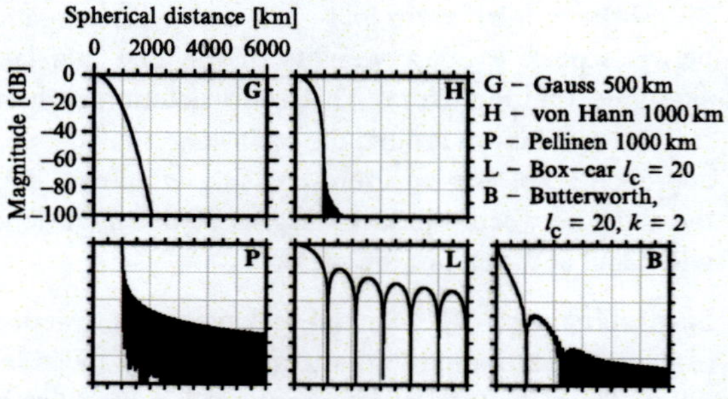

Fig. 16.2 Spatial weights of the filters in decibel

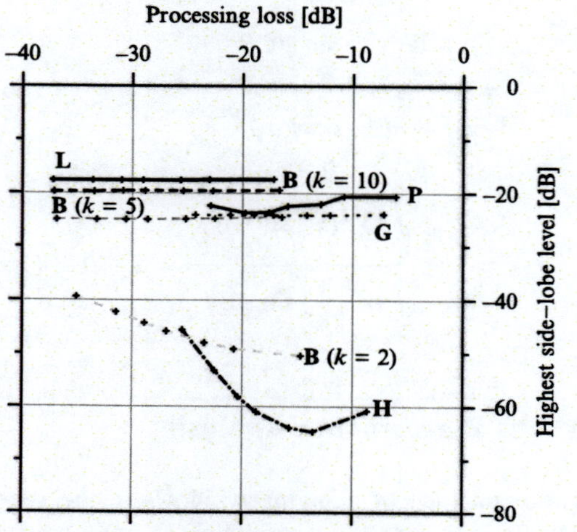

Fig. 16.3 Scatter plot of the highest side-lobe level and processing loss. The filters shown in Fig. 16.2 are compared for seven different smoothing radii: 200, 300, 400, 500, 750, 1,000, and 2,000 km. These are arranged from *left* to *right* for each curve of the filter. Good filters for a given smoothing radii will be in the *lower left corner* of the plot

how much leakage can happen and how much energy will be lost due to the filter, respectively. Figure 16.3 is constructed based on these observations.

Filter kernel comparison Gauss, von Hann, Pellinen, box-car (Shannon window), and Butterworth. All the filters have been transformed into spatial kernels from their spectral coefficients, which means in the case of Gauss, von Hann, and Pellinen filters, the kernels are reconstructions. The spectral windows were all expanded upto degree 360 for seven different radii. Since the five filter kernels are defined in completely different ways (cf. Table 16.1), the following procedure was adopted for comparison.

The smoothing radii for Gauss filter is defined such that the filter weights drop to half at the prescribed smoothing radius. Similarly, the von Hann filter weights reduce to half at half the prescribed smoothing radius, which means that Gauss 500 km and von Hann 1,000 km filters will have the same filter weight at 500 km from the kernel location. This provides a basis for comparing Gauss and von Hann filters. The definition of von Hann filter shows that the prescribed smoothing radius is also a cut-off radius (cf. Table 16.1) and hence, it can be compared directly with a Pellinen filter of the same cut-off radius. Since Box-car and Butterworth filters are based on cut-off degrees, they can be conveniently converted to spatial scales by using half-wavelength rule $\left(\frac{\lambda}{2} \approx \frac{20{,}000}{l}\right)$. Excluding the Gauss filter, the main-lobe width values obtained from the analysis, shown in Table 16.2, is a proof to this procedure. The main-lobe half-width of Gauss filter is the cut-off radius of the comparable von Hann filter.

In Fig. 16.2, the magnitudes of filter weights are shown in decibels for a 1,000 km smoothing radius. The reason for plotting the filter weights in decibels is that the filter structure is far more clearer than the usual cross-section plots. Figure 16.3 shows the scatter plot between highest side-lobe level and processing loss. Kaula's rule $\left(\sigma_l^2 = \frac{160\times10^{-12}}{l^3}\right)$ was used for calculating processing loss as the calculation requires spherical harmonic coefficients of the field. Table 16.2 provides values for the rest of the performance measures. The ideal filter for a given smoothing radius will be in the lower left corner of Fig. 16.3 as it will have a very low side-lobe level and a minimal processing loss.

Butterworth filters of order 2 and von Hann filters come close to satisfying the *good filter* criteria. While Butterworth filters lose less during filtering, they have more leakage than von Hann filters due to their side-lobe levels (cf. Table 16.2). The von Hann filter has a narrow main-lobe width with a very low side-lobe level, in addition to the fact that leakage is nearly non-existent. However, they lose a moderate part of the energy during filtering. It is clear from Fig. 16.3 that von Hann filters have the lowest side-lobe levels, and have processing loss slightly less than Pellinen & Gauss but much more than Butterworth & Box-car filters.

Pellinen filters outperform the box-car filters. They have less leakage, lower side-lobe levels, and narrower main-lobes compared to the box-car filter. However,

Table 16.2 Main-lobe width (ψ_M), spatial leakage (ξ), and side-lobe roll off ratio (ρ_s) are provided for the isotropic filters compared here. ψ_M is the first zero-crossing of the filter weights

Filter	ψ_M [km]	ξ [%]	ρ_s [dB]
Gauss (at $W(\psi) = 0.5$)			
100 km	200	0.39	−106.16
250 km	500	0.39	−147.63
500 km	1,000	0.39	−169.12
von Hann			
200 km	200.37	0.01	−30.86
500 km	498.53	0.00	−32.70
1,000 km	990.05	0.00	−84.04
Pellinen			
200 km	235.44	0.50	−27.96
500 km	533.80	0.15	−26.48
1,000 km	1,033.94	0.08	−23.88
Box-car (l_c)			
100	242.05	16.22	−22.52
40	596.12	16.21	−14.70
20	1,163.78	16.18	−8.94
Butterworth (l_c)			
$k = 2$			
100	193.96	0.33	−22.10
40	522.78	0.10	−19.43
20	1,062.39	0.10	−18.34
$k = 5$			
100	212.60	2.55	−63.39
40	528.39	2.59	−87.49
20	1,044.56	2.65	−105.90
$k = 10$			
100	225.22	6.96	−122.17
40	559.05	7.00	−170.86
20	1,105.27	7.08	−88.98

the performance of the reconstructed Pellinen filter is highly dependent on the number of terms used to reconstruct it.

Butterworth filters of order 2 have relatively lower side-lobe levels, relatively less leakage, and a minimal processing loss, which make them attractive for resolution reduction. The lower order Butterworth filters decay slowly, and so the cut-off degree must be far less than the maximum degree of expansion. If the cut-off degree is closer to the maximum degree of expansion, then there will be truncation effects and that can lead to change in filter characteristics.

Gauss filters are different from other filters compared here as by definition they do not reach zero and therefore, will not have a side-lobe. The side-lobes (not shown here) are an artefact of the truncation of the harmonic expansion in the spectral domain. Since they are comparable to von Hann filters, their main-lobe width can be taken to be twice their prescribed smoothing radius as mentioned previously. It is interesting to note that the leakage remains constant even with a change in the smoothing radius. Other than the steep side-lobe roll-off, there is not much of note about this filter.

Box-car filter performs the worst due to very high side-lobe levels, enormous leakage, and wider than expected main-lobe width. Inspite of having the least processing loss, Box-car filters have a lot of undesirable characteristics, especially their leakage.

5 Summary and Conclusions

A set of measures were developed to analyse the performance of isotropic spectral windows: processing loss, main-lobe half-width, spatial leakage, highest side-lobe level, and side-lobe roll-off ratio. Using these parameters five different filters of varying smoothing radii were analysed for their efficacy in reducing the resolution of a given field.

The performance measures provide deeper insights into the characteristics of the filters, which makes it easier to choose a filter and its smoothing radii depending on the needs of the problem in hand. This was shown for the case of resolution reduction, where the measures point to Butterworth filters of order 2 (if lower processing loss is desired), and von Hann filters (if lower leakage is desired).

Acknowledgements This study is part of *Direct Water Balance* project, funded by DFG SPP1257 "Mass transport and mass distribution in the system Earth" programme. The authors thank two anonymous reviewers for their constructive comments, which helped improve the manuscript.

References

Han SC, Shum CK, Jekeli C et al (2005) Non-isotropic filtering of GRACE temporal gravity for geophysical signal enhancement. Geophys J Int 163:18–25

Harris FJ (1978) On the use of windows for harmonic analysis with the discrete Fourier transform. Proceedings of the IEEE 66(1):51–93

Jekeli C (1981) Alternative methods to smooth the Earth's gravity field. Tech. Rep. 327, Department of Geodetic Science and Surveying, The Ohio State University

Klees R, Revtova EA, Gunter BC et al (2008) The design of an optimal filter for monthly GRACE gravity models. Geophys J Int 175:417–432

Kusche J (2007) Approximate decorrelation and non-isotropic smoothing of time-variable GRACE-type gravity field models. J Geodesy 81:733–749

Swenson S, Wahr J (2006) Post-processing removal of correlated errors in GRACE data. *Geophys Res Lett* 33:L08 402

Wahr J, Molenaar M, Bryan F (1998) Time variability of the Earth's gravity field: Hydrological and oceanic effects and their possible detection using GRACE. J Geophys Res 103(B12):30 205–30 229

Uniqueness Theorems for Inverse Gravimetric Problems

17

D. Sampietro and F. Sansò

Abstract

The inverse gravimetric problem, namely the determination of the internal density distribution of a body from the exterior gravity field, is known to have a very large indeterminacy while it is well identified and described in functional terms. However, when density models are strongly reduced to simple classes, or functional subspaces, the uniqueness property of the inversion is retrieved. Uniqueness theorems are proved for three simple cases in Cartesian approximation:

- The recovery of the interface between two layers of known density
- The recovery of a laterally varying density distribution, in a two layers model, given the geometry of the problem (topography and depth of compensation)
- The recovery of the distribution of the vertical gradient of density, in a two layers model, given the geometry of the problem (topography and depth of compensation) and the density distribution at sea level.

Keywords

Inverse gravimetric problems • Earth's density anomalies • Uniqueness theorems

1 Introduction

The inverse gravimetric problem consists in recovering the Earth's density distribution from observations of functionals of the gravitational potential. This problem, is based on the inversion of Newton's gravitational potential:

$$V(P) = G \int\int\int_B \frac{\rho(Q)}{r_Q(P)} dv(Q) \qquad (17.1)$$

D. Sampietro (✉) · F. Sansò
DIIAR, Politecnico di Milano, Polo Regionale di Como, Via Valleggo 11, 22100 Como, Italy
e-mail: daniele.sampietro@polimi.it

and of its derivatives, which are expressed in term of Fredholm integral equations of the first kind (see for example Lavrent et al. 1986). In literature various publications have studied this inverse gravimetric problem (see, for instance, Nettleton 1939, 1940; Ballani and Stromeyer 1982, 1990; Vaníček and Christou 1994; Michel 1999; Tarantola 2005; and the references therein). Considering (17.1) it turns out that each of Hadamard's criteria for a well-posed problem is violated (see for example Michel 2005), in particular if the problem is solvable, then the space of all solutions corresponding to a fixed potential $V(P)$ is infinite-dimensional. In specific the densities that do not produce an external field, are those that are orthogonal to all (square integrable) harmonic function in B.

N. Sneeuw et al. (eds.), *VII Hotine-Marussi Symposium on Mathematical Geodesy*, International Association of Geodesy Symposia 137, DOI 10.1007/978-3-642-22078-4_17, © Springer-Verlag Berlin Heidelberg 2012

The non-uniqueness can be treated, for instance, by considering hypotheses on the shape of the density discontinuity. In fact, in principle, one can think that it is better to use a rough geophysical hypothesis and to find a unique solution, rather than accepting a solution that can be very far from reality because it corresponds to a purely mathematical criterion (see Sansò et al. 1986). In this work the uniqueness of the solution for three different inverse gravimetric problems under the assumptions of a two layer body in Cartesian approximation is proved. Note that for some geophysical problems such as the study of the Moho topography or the study of mountains roots these approaches are frequently used (see e.g.Gangui 1998 or Lessel 1998).

We underline here that in this work "Cartesian approximation" means that we consider a Cartesian reference frame in which the Z axis is oriented as the prevailing direction of the plumb line in the area.

2 Considered Inverse Problems

The determination of the structure (density and depth of discontinuity) of a two layer body, based on the inversion of Newton's gravitational potential, is an ill-posed problem. Nevertheless under some assumptions the uniqueness of the solution can be retrieved. In this work three cases (fixed geometry, fixed density distribution and vertical gradient), with different initial hypothesis, are considered. To fix the ideas we think of the two layer case as referring, in Cartesian approximation, to the Earth crust and mantle. In this context the three cases can be seen as the estimation of the horizontal density gradient in the Earth crust, the estimation of the Moho depth and the estimation of the crust density gradient in the vertical direction respectively.

In the first case (fixed geometry) the uniqueness of the solution is proved for an inverse problem in which we want to estimate the density distribution from gravimetric observations knowing the geometry of the problem (i.e. the topography and the surface of the discontinuity between crust and mantle). Vice-versa in the second case we suppose to know the value of the density in the two layers and the topography and we want to estimate the depth of compensation. In the last case the uniqueness is proved for the estimation of the vertical gradient of the density from gravity observations, given the geometry and the density distribution

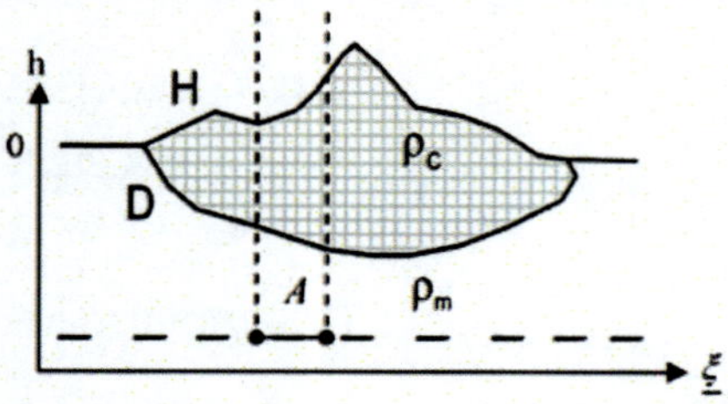

Fig. 17.1 Geometry and the notation used

at a known altitude. The three cases can be summarized as follow (refer to Fig. 17.1 for the notation):

- Fixed geometry:
  - $H(\underline{\xi})$ is the given topography
  - $\rho_c(\underline{\xi})$ is the unknown crust density as function of the planar coordinates only
  - $-D(\underline{\xi})$ is the known geometry of the surface between crust and mantle
- Fixed density distribution:
  - $H(\underline{\xi})$ is the given topography
  - $\rho_c(\underline{\xi})$ is the given crust density
  - $-\bar{D}(\underline{\xi})$ is the unknown depth of compensation
- Vertical gradient:
  - $H(\underline{\xi})$ is the given topography
  - $\rho_c(\underline{\xi})=\rho_0(\underline{\xi})+(h-H)\rho'(\underline{\xi})$ is the linearized, in the vertical direction, crust density
  - $\rho_0(\underline{\xi})$ is the known density at the upper level
  - $\rho'(\underline{\xi})$ is the unknown density vertical gradient
  - $-\bar{D}(\underline{\xi})$ is the known depth of compensation.

In all the cases, the mantle density, is considered as a known constant. The geometry and the notation used in the three cases are described in Fig. 17.1.

To prove the uniqueness of the solution for our inverse problems we consider two bodies, B_1 and B_2 respectively, with the same external surface S and generating the same external (i.e. in $h > H$) gravity field. We restrict here to the case that the perturbing body B has a finite extension, as shown in Fig. 17.1.

Moreover we assume that:

- Both the bodies are constituted by two layers: mantle and crust
- D_1 and D_2 are the regular surfaces between mantle and crust for the two bodies
- ρ_m, is the mantle density for both the bodies while ρ_{c1} ρ_{c2} are the crust densities for B_1 and B_2 respectively
- $\rho_m > \rho_{c1}$ and $\rho_m > \rho_{c2}$

- $V_1 = V_2 = V$ where V_1 is the potential outside S due to B_1 and V_2 is the potential outside S due to B_2.

Naturally this condition guarantees that the potentials V_1 and V_2 do coincide in $\{h \geq H\}$. Let us observe that the solution of the second problem, with some little variations, has already been given in literature (Barzaghi and Sansò 1988, Biagi 1997, Sampietro 2009) and it is presented here only for the sake of completeness and comparison with the solution of the other two cases.

3 Uniqueness Theorems

We want to prove that if we take into account the previous hypotheses the corresponding inverse gravimetric problems admit a unique solution.

1. To demonstrate the thesis for problem 1 consider a third body B, which density is, $\delta\rho = \rho_{c1} - \rho_{c2}$ obtained as the difference between B_1 and B_2 (see Fig. 17.2). Obviously, since $V_1 = V_2$ on S including the whole $\underline{\xi}$ plane, B generates a zero-potential everywhere outside the body:

$$V(\delta\rho) = 0 \, outside \, B. \qquad (17.2)$$

Note that in principle we know that $V = 0$ only on the upper surface (Fig. 17.2), but due to the hypothesis of finite extension of B and thanks to the unique continuation property holding for harmonic functions, we can claim that $V = 0$ everywhere outside B (Moritz 1980).

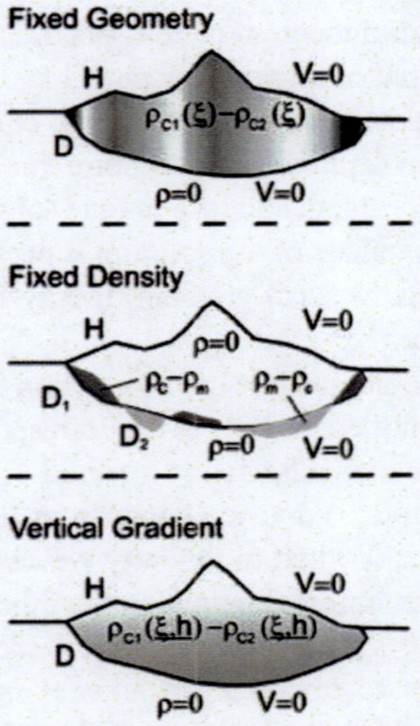

Fig. 17.2 The Body B in the three cases considered

Consider now the classical inner product $<\cdot,\cdot>$ in L^2, since, for (17.1) and (17.2), for P$<\delta\rho, \frac{1}{r_Q(P)}> = 0$ outside B and since the set $\{\delta\rho, \frac{1}{r_Q(P)}\}$ is dense in the subspace of functions harmonic in B (one of the many formulations of Krarup's theorem, see Krarup 1975; Sansò 1982) it holds $<\delta\rho, u> = 0 \;\; \forall u \in L^2(B)$, i.e. belongs to the orthogonal complement of the space of square-integrable functions harmonic in B. A detailed proof of this theorem (theorem of zero-potential distributions) is given in Sansò 1980.

Let's consider now a generic square integrable function u, harmonic in B, since $\delta\rho$ belongs to the orthogonal complement of the space of square-integrable functions harmonic in B (see Sansò 1980) we have that:

$$\int_B \left[\delta\rho\left(\underline{\xi}, h\right)\right] u \, dB = 0 \, \forall u, \; \Delta u = 0 \, in \, B. \qquad (17.3)$$

Take an arbitrary set A on the plane $\underline{\xi}$, contained on the projection of B on this plane (see Fig. 17.1) and define a typical test function $\Phi_A(\underline{\xi})$ smooth and positive in A but null outside A. Now let v be harmonic in B and such that:

$$\begin{cases} v(\underline{\xi}, H) = \Phi_A(\underline{\xi}) \\ v(\underline{\xi}, -D) = 0 \end{cases} \qquad (17.4)$$

furthermore define

$$u = \frac{\partial v}{\partial h}. \qquad (17.5)$$

It is easy to see that v is a smooth function, so that $v \in L^2(B)$. If we think at the fixed geometry model, we have that $D_1(\underline{\xi}) = D_2(\underline{\xi}) = D(\underline{\xi})$ and that since both ρ_{c1} and ρ_{c2} do not depend from h also $\delta\rho$ will not depend from h (i.e. $\delta\rho = \delta\rho(\underline{\xi})$). Substituting (17.5) into integral (17.3) we get:

$$\int_S d\underline{\xi} \int_{-D}^{H} \left[\delta\rho(\underline{\xi})\right] \frac{\partial v}{\partial h} dh = 0. \qquad (17.6)$$

Solving the integral in the vertical direction we obtain:

$$\int_S \left[\delta\rho(\underline{\xi})\right][v(H) - v(-D)]\ d(\underline{\xi}) = 0. \quad (17.7)$$

Taking into account a small two-dimensional set A, in which we know that $\delta\rho$ does not change its sign and considering the properties of v described into (17.4), we have:

$$\int_A \delta\rho(\underline{\xi})\Phi_A(\underline{\xi})d\underline{\xi} = 0. \quad (17.8)$$

This implies $\delta\rho(\underline{\xi}) = 0$ in A and, since the previous equivalence must hold for every set A and for every function v, it must be $\delta\rho(\underline{\xi}) = 0$ i.e. $\rho_{c1}(\underline{\xi}) = \rho_{c2}(\underline{\xi})$.

2. In the case of a fixed, laterally varying density model we assume $\rho_{c1}(\underline{\xi}) = \rho_{c2}(\underline{\xi}) = \rho(\underline{\xi})$ while D_1 and D_2 can be different. Going back to (17.3) and considering the new assumptions we can write:

$$\begin{cases} \int_S d\underline{\xi} \int_{-D_2}^{-D_1} (\rho_m - \rho)\frac{\partial v}{\partial h}dh = 0 \ \ if\ D_2 > D_1 \\ \int_S d\underline{\xi} \int_{-D_1}^{-D_2} (\rho - \rho_m)\frac{\partial v}{\partial h}dh = 0 \ \ if\ D_1 > D_2 \end{cases}. \quad (17.9)$$

The only solution of the previous system is $D_1(\underline{\xi}) = D_2(\underline{\xi})$. In fact if we suppose to take the set A where for example $D_2 > D_1$ we obtain:

$$\int_A (\rho_m - \rho_2)\Phi_A(\underline{\xi})d\underline{\xi} = 0 \quad (17.10)$$

that is not possible since this integral, for the properties of the function Φ_A and because $\rho_m > \rho_2$, is always positive.

3. Consider now the last case (vertical density gradient). This means that by hypothesis the following density model holds:

$$\rho_c(\underline{\xi}) = \rho_0(\underline{\xi}) + (h - H)\rho'(\underline{\xi}) \quad (17.11)$$

where $\rho_0(\underline{\xi})$ is the density on the external surface S, which is assumed to be known, while $\rho'(\underline{\xi})$, the vertical gradient, is unknown. Equation (17.3) still holds and by hypothesis we can write:

$$\delta\rho_c(\underline{\xi}) = (h - H)\delta\rho' \quad (17.12)$$

where $\delta\rho' = \rho_1'(\underline{\xi}) - \rho_2'(\underline{\xi})$.

Furthermore in this case we use again the same function v as defined in (17.5) but we put $u = \frac{\partial^2 v}{\partial h^2}$. Substituting (17.12) into (17.3) we obtain:

$$\int_S d\underline{\xi} \int_{-D}^{H} [(h - H)\delta\rho']\frac{\partial^2 v}{\partial h^2}dh = 0 \quad (17.13)$$

solving the integral in the vertical direction we get:

$$\int_S d\underline{\xi}\delta\rho' \left[v(-D) - v(H) - (D + H)\left.\frac{\partial v}{\partial h}\right|_{-D}\right] = 0. \quad (17.14)$$

Again, considering the properties of the function v, integral (17.14) becomes:

$$\int_A d\underline{\xi}\delta\rho' \left[-(D + H)\left.\frac{\partial v}{\partial h}\right|_{-D} - v(H)\right] = 0. \quad (17.15)$$

In integral (17.15) the integrand in square brackets is always negative since $v(\underline{\xi}, H) = \Phi_A(\underline{\xi}) \geq 0$, and $\frac{\partial v}{\partial h}|_{-D}$ is positive, or better non-negative, because of standard maximum properties of harmonic functions. Therefore it must be $\delta\rho'(\underline{\xi}) = 0$ i.e. $\rho_1'(\underline{\xi}) = \rho_2'(\underline{\xi})$.

Conclusion

The general theorem of zero-potential distributions proves that in a two layer model and under simple assumptions (known topography, constant density or known depth of compensation) the inverse gravitational problem admits a unique solution.

The uniqueness of the solution is proved also when the linear, vertical gradient density distribution is unknown.

Once the uniqueness of the solution is guaranteed, we are entitled to apply to the corresponding inverse problem a regularization method (like Tikhonov regularization) and we know from litterature (e.g. Schock 2005) that in this way we can approximate the true solution, dominating the inherent instabilities. This is the reason why, in inverse problem like this, it is so important to address specifically the question of the uniqueness of the solution, at least for simplified models.

References

Ballani L, Stromeyer D (1982) The inverse gravimetric problem: a Hilbert Space Approach. In: Holota P (ed) Proceedings of the international symposium 'Figure of the Earth, the Moon, and other Planets', 20–25 September 1982, Prague, Czech Republic, pp 359–373

Ballani L, Stromeyer D (1990) On the structure of uniqueness in linear inverse source problems. In: Vogel A et al (eds) Theory and practice of geophysical data inversion, Proceedings of the 8th international mathematical geophysics seminar on model optimization in exploration geophysics. Vieweg-Verlag, Braunschweig

Barzaghi R, Sansò F (1988) Remarks on the inverse gravimetric problem. Geophys J 92:505–511

Biagi L (1997) La stima della Moho mediante il metodo della collocazione. PhD Thesis, Dipertimento di scienza dell'Ingegneria sezione osservatorio geofisico, Università degli Studi di Modena

Gangui AH (1998) A combined structural interpretation based on seismic data and 3-D gravity modeling in the Northern Puna, Eastern Cordillera, Argentina. Dissertation FU Berlin, Berliner Geowissenschaftliche Abhandlungen, Reihe B, Band 27, Berlin

Krarup T (1975) On potential theory. Matematical geodesy part 1 Methode der Verfahren der Mathematischen Physik/ Band 12 Wissenschaftsverlag

Lavrent MM, Romanov VG, Shishatskii SP (1986) Ill-posed problems of mathematical physics and analysis. MS Bookstore

Lessel K (1998) Die Krustenstruktur der Zentralen Anden in Nordchile (21–24°S), abgeleitet aus 3D-Modellierungen refraktionsseismischer Daten. Dissertation FU Berlin, Berliner Geowissenschaftliche Abhandlungen, Reihe B, Band 31, Berlin

Michel V (1999) A multiscale method for the gravimetry problem—theoretical and numerical aspects of harmonic and anharmonic modelling. PhD Thesis, Geomathematics Group, Department of Mathematics, University of Kaiserslautern (Shaker Verlag, Aachen)

Michel V (2005) Regularized wavelet-based multiresolution recovery of the harmonic mass density distribution from data of the Earth's gravitational field at satellite height. Inverse Prob 21:997–1025

Moritz H (1980) Advanced physical geodesy. Abacus Press, Tunbridge

Nettleton LL (1939) Determination of density for reduction of gravity observations. Geophysics 4:176–183

Nettleton LL (1940) Geophysical prospecting for oil. McGraw-Hill, New York

Sampietro D (2009) An inverse gravimetric problem with goce data. PhD Thesis. Doctoral programs in Geodesy and Geomatics. Politecnico di Milano, polo regionale di Como

Sansò F (1980) Internal collocation. Memorie dell'Accademia dei Lincei, Vol. XVI, n. 1

Sansò F (1982) A note on density problems and the Runge-Krarup's theorem. Bollettino di Geodesia e Scienze Affini 41(4):421–447

Sansò F, Barzaghi R, Tscherning CC (1986) Choice of norm for the density distribution of the Earth. Geophys J Royal Astronomical Soc 87(1):123–141

Schock E (2005) On the asymptotic order of accuracy of Tikhonov regularization. J Optim Theory Appl 44:95–104

Tarantola A (2005) Inverse problem theory and methods for model parameter estimation. Soc Ind Appl Math

Vaníček P, Kleusberg A (1985) What an external gravitational potential can really tell us about mass distribution. Bollettino di Geofisica Teorica ed Applicata XXCII(108): 243–250

Vaníček P, Christou NT (1994) Geoid and its geophysical interpretations. CRC, New York

Part IV

Time Series Analysis and Prediction of Multi-Dimensional Signals in Geodesy

Towards a Multi-Scale Representation of Multi-Dimensional Signals

18

Michael Schmidt

Abstract

For analyzing and representing a one-dimensional signal wavelet methods are used for a long time. The basic feature of wavelet analysis is the localization property, i.e. it is – depending on the chosen wavelet – possible to study a signal just in a finite interval. Nowadays a large number of satellite missions allows to monitor various geophysical phenomena. Since often regional phenomena have to be studied, multi-dimensional wavelet methods come into question. In this paper the basic principles of a multi-scale representation of multi-dimensional signals using B-spline wavelets are presented. Finally the procedure is applied to an example of ionosphere research.

Keywords

Series expansions • B-spline functions • Wavelet functions • Multi-scale representation

1 Introduction

Today a large number of satellite missions allows the monitoring of various geophysical phenomena, e.g. for studying climate change. Frequently, scientists are interested in regional processes such as post-glacial rebound or the equatorial anomaly in ionosphere research. For a long time wavelets have been considered as a candidate for regional representations due to their localization feature and their flexible filtering characteristics; see, e.g., Freeden (1999) or Schmidt et al. (2007a). Modern spaceborne observation techniques provide huge data sets. For the evaluation and interpretation of such kind of data representations are required which are characterized by both effective numerical algorithms and an efficient data handling. The *multi-scale representation* (*MSR*) – also known as multi-resolution representation – splits an input signal into a certain number of detail signals by successive low-pass filtering (Schmidt et al. 2007a). Hence, each detail signal is a band-pass filtered version of the input signal, related to a specific frequency band, i.e. resolution level. Numerically this decomposition process is realized by the highly effective pyramid algorithm. In case of a two-dimensional (2-D) MSR for further efficiency reasons so-called *tensor product* scaling and wavelet functions can be introduced. Schmidt (2001) formulated a 2-D MSR generated by orthogonal wavelet functions such as the Haar function using mathematically the Kronecker product. Since in many geodetic and

M. Schmidt (✉)
Deutsches Geodätisches Forschungsinstitut (DGFI), D-80539 München, Alfons-Goppel-Str. 11, Germany
e-mail: schmidt@dgfi.badw.de

N. Sneeuw et al. (eds.), *VII Hotine-Marussi Symposium on Mathematical Geodesy*, International Association of Geodesy Symposia 137, DOI 10.1007/978-3-642-22078-4_18,

geophysical applications spatio-temporal signals are studied, the 2-D MSR based on tensor products is generalized in this paper to the multi-dimensional (M-D) case. B-spline representations of M-D signals are widely used, e.g. in reverse engineering for fitting surfaces to given measurements (Koch 2009). Because of their mathematical properties (e.g. compactness, smoothness) B-splines are taken here to generate the M-D MSR. In summary this paper can be seen as a guide for applying a MSR based on tensor product B-spline wavelet functions to M-D problems.

In the following section the fundamentals of the 1-D MSR based on B-splines are outlined. The results are generalized in the third section to the M-D case. As an example the derived procedure is finally applied to the vertical total electron content within the ionosphere.

2 1-D Multi-Scale B-Spline Representation

As level-j scaling functions $\phi_{j;k}(x)$ of a 1-D MSR with $x \in \mathbb{R}$ the *normalized quadratic B-splines* $N^2_{j;k}(x)$ with $k = 0, \ldots, K_j - 1$ are introduced, i.e. $\phi_{j;k}(x) = N^2_{j;k}(x)$. K_j means the total number of scaling functions of the resolution level (scale) $j \in \mathbb{N}_0$. Assuming that a sequence of non-decreasing values $t^j_0, \ldots, t^j_{K_j+2}$, called knots, is given, the normalized quadratic B-spline is defined via the recursion formula

$$N^m_{j;k}(x) = \frac{x - t^j_k}{t^j_{k+m} - t^j_k} N^{m-1}_{j,k}(x) + \frac{t^j_{k+m+1} - x}{t^j_{k+m+1} - t^j_{k+1}} N^{m-1}_{j,k+1}(x) \qquad (18.1)$$

with $m = 1, 2$ and initial values $N^0_{j;k}(x) = 1$ for $t^j_k \leq x < t^j_{k+1}$ and $N^0_{j;k}(x) = 0$ else; see, e.g. Stollnitz et al. (1995) or Schmidt (2007). Note, in (18.1) a factor is set to zero if the corresponding denominator is equal to zero.

A B-spline is *compactly supported*, i.e. its values are different from zero only in a finite interval on the real axis, mathematically written as $\operatorname{supp} N^2_{j;k} = [t^j_k, t^j_{k+3})$. Since the B-splines shall be used for regional modeling, so-called *endpoint-interpolating normalized quadratic B-splines (ENQ B-splines)* defined on the unit interval $\mathbb{I} = [0, 1]$, i.e. $x \in \mathbb{I}$ are introduced. For that purpose the first three knots are set to the value zero and the last three knots to the value one. Hence, the knot sequence is given as

$$0 = t^j_0 = t^j_1 = t^j_2 < \cdots < t^j_{K_j} = t^j_{K_j+1} = t^j_{K_j+2} = 1 \qquad (18.2)$$

with $t^j_k = (k-2)h_j$, $h_j = 1/(K_j - 2)$ and $K_j = 2^j + 2$. Figures of ENQ B-splines can be found, e.g. in the publications of Schmidt (2007) and Zeilhofer (2008).

The K_j scaling functions $\phi_{j;k}(x) = N^2_{j;k}(x)$ constitute a basis of the so-called *scaling space* V_j. The MSR requires that there exists a sequence of nested subspaces V_j with $j \in \mathbb{N}_0$, such that $V_0 \subset V_1 \subset \cdots \subset V_{j-1} \subset V_j \subset V_{j+1} \subset \cdots \subset L^2(\mathbb{I})$ holds. Hence, it follows that the scaling functions $\phi_{j-1;l}(x)$ with $l = 0, \ldots, K_{j-1}-1$ and $\phi_{j-1;l} \in V_{j-1}$ can be expressed by a linear combination of the scaling functions $\phi_{j;k}(x)$ with $k = 0, \ldots, K_j - 1$ and $\phi_{j;k} \in V_j$, i.e. the *two-scale relation*

$$\phi_{j-1;l}(x) = \sum_{k=2l-(K_j-1)}^{2l} p_{j;k}\phi_{j;2l-k}(x), \qquad (18.3)$$

also known as the *refinement relation* holds. By defining the $K_{j-1} \times 1$ scaling vector

$$\boldsymbol{\phi}_{j-1}(x) = [\phi_{j-1;0}(x), \ldots, \phi_{j-1;K_{j-1}-1}(x)]^T \qquad (18.4)$$

of resolution level $j-1$ and the $K_j \times 1$ scaling vector

$$\boldsymbol{\phi}_j(x) = [\phi_{j;0}(x), \ldots, \phi_{j;K_j-1}(x)]^T \qquad (18.5)$$

of resolution level j the two-scale relation (18.3) can be rewritten as the matrix equation

$$\boldsymbol{\phi}^T_{j-1}(x) = \boldsymbol{\phi}^T_j(x)\mathbf{P}_j. \qquad (18.6)$$

The entries $p_{j;k}$ of the $K_j \times K_{j-1}$ matrix $\mathbf{P}_j$ of the refinement process are given as

$$\mathbf{P}_j = \frac{1}{4}\begin{bmatrix} 4\,0\,0\,0 \cdots 0\,0\,0 \\ 2\,2\,0\,0 \cdots 0\,0\,0 \\ 0\,3\,1\,0 \cdots 0\,0\,0 \\ 0\,1\,3\,0 \cdots 0\,0\,0 \\ 0\,0\,3\,1 \cdots 0\,0\,0 \\ 0\,0\,1\,3 \cdots 0\,0\,0 \\ 0\,0\,0\,3 \cdots 0\,0\,0 \\ 0\,0\,0\,1 \cdots 0\,0\,0 \\ \cdots\cdots \\ 0\,0\,0\,0 \cdots 1\,0\,0 \\ 0\,0\,0\,0 \cdots 3\,0\,0 \\ 0\,0\,0\,0 \cdots 3\,1\,0 \\ 0\,0\,0\,0 \cdots 1\,3\,0 \\ 0\,0\,0\,0 \cdots 0\,2\,2 \\ 0\,0\,0\,0 \cdots 0\,0\,4 \end{bmatrix},$$

see, e.g. Stollnitz et al. (1995) or Lyche and Schumaker (2001).

Next the so-called *detail space* W_{j-1} defined as the orthogonal complement of the scaling space V_{j-1} in the scaling space V_j, i.e. $V_j = V_{j-1} \oplus W_{j-1}$ is introduced. Furthermore, it is assumed that the so-called *wavelet functions* $\psi_{j-1;l}(x)$ with $l = 0, \ldots, L_{j-1} - 1$ and $L_{j-1} = K_j - K_{j-1}$ are base functions of W_{j-1}. Since the relations $W_{j-1} \subset V_j$, $\psi_{j-1;l} \in W_{j-1}$ and $\phi_{j;k} \in V_j$ hold, the series expansion

$$\psi_{j-1;l}(x) = \sum_{k=2l-(K_j-1)}^{2l} q_{j;k}\phi_{j;2l-k}(x) \quad (18.7)$$

is formulated analogously to (18.3) as the *second two-scale relation*; but this time $l = 0, \ldots, L_{j-1} - 1$ holds. With the $L_{j-1} \times 1$ vector

$$\boldsymbol{\psi}_{j-1}(x) = [\psi_{j-1;0}(x), \ldots, \psi_{j-1;L_{j-1}-1}(x)]^T \quad (18.8)$$

of level $j-1$ (18.7) can be rewritten as

$$\boldsymbol{\psi}^T_{j-1}(x) = \boldsymbol{\phi}^T_j(x)\mathbf{Q}_j, \quad (18.9)$$

wherein the $K_j \times L_{j-1}$ matrix $\mathbf{Q}_j$ with elements $q_{j;k}$ can be determined from the given matrix $\mathbf{P}_j$ as will be shown in the following. Since, as mentioned before, the wavelets $\psi_{j-1;l}(x)$ are assumed to be orthogonal to the scaling functions $\phi_{j-1;k}(x)$, the inner product

$$\int_0^1 \psi_{j-1;l}(x)\phi_{j-1;k}(x)dx = \langle \psi_{j-1;l}, \phi_{j-1;k}\rangle_{L^2(\mathbb{I})} = 0 \quad (18.10)$$

vanishes for $l = 0, \ldots, L_{j-1} - 1$ and $k = 0, \ldots, K_{j-1} - 1$. Wavelet functions which fulfill this condition are called *pre-wavelets* or *semi-orthogonal wavelets*. Substituting (18.3) and (18.7) into (18.10) yields $\sum_m \sum_n q_{j;m} p_{j;n} \langle \phi_{j;2l-m}, \phi_{j;2k-n}\rangle_{L^2(\mathbb{I})} = 0$. Writing this equation for each (l,k)-combination gives

$$\mathbf{Q}^T_j \mathbf{G}_j \mathbf{P}_j = \mathbf{0}, \quad (18.11)$$

wherein $\mathbf{0}$ is the $L_{j-1} \times K_{j-1}$ zero matrix. Since the $K_j \times K_j$ matrix $\mathbf{G}_j$ with elements $\langle \phi_{j;2l-m}, \phi_{j;2k-n}\rangle_{L^2(\mathbb{I})}$ is positive definite, the Cholesky factorization can be applied and yields $\mathbf{G}_j = \mathbf{L}_j\mathbf{L}^T_j$, wherein $\mathbf{L}_j$ is an $K_j \times K_j$ lower triangular matrix (Koch 1999). Substituting the Cholesky factorization into (18.11) gives

$$\widetilde{\mathbf{P}}^T_j \widetilde{\mathbf{Q}}_j = \mathbf{0}^T. \quad (18.12)$$

Thus, the columns of the $K_j \times L_{j-1}$ matrix

$$\widetilde{\mathbf{Q}}_j = \mathbf{L}^T_j \mathbf{Q}_j \quad (18.13)$$

are a basis of the null space of the given matrix $\widetilde{\mathbf{P}}^T_j = \mathbf{P}^T_j \mathbf{L}_j$. If $\widetilde{\mathbf{Q}}_j$ is calculated from (18.12) (see, e.g. Koch 1999), the desired matrix $\mathbf{Q}_j$ is obtained from (18.13) and can be introduced into the two-scale relation (18.9). However, with the columns of $\widetilde{\mathbf{Q}}_j$, i.e. with the basis of the null space of $\widetilde{\mathbf{P}}^T_j$, the matrix $\mathbf{Q}_j$ is not unique. Examples for $\mathbf{Q}_j$ are presented by Stollnitz et al. (1995) and Lyche and Schumaker (2001); see also Zeilhofer (2008).

Next, the *decomposition* equation will be derived, which is required for the MSR. Since the relations $\phi_j \in V_j$, $\phi_{j-1} \in V_{j-1}$ and $\psi_{j-1} \in W_{j-1}$ hold, it follows from $V_j = V_{j-1} \oplus W_{j-1}$

$$\boldsymbol{\phi}^T_j(x) = \boldsymbol{\phi}^T_{j-1}(x)\overline{\mathbf{P}}_j + \boldsymbol{\psi}^T_{j-1}(x)\overline{\mathbf{Q}}_j, \quad (18.14)$$

wherein $\overline{\mathbf{P}}_j$ and $\overline{\mathbf{Q}}_j$ are $K_{j-1} \times K_j$ and $L_{j-1} \times K_j$ initially unknown coefficient matrices, respectively. Inserting (18.6) and (18.9) into (18.14) yields $\boldsymbol{\phi}^T_j(x) = \boldsymbol{\phi}^T_j(x)\mathbf{P}_j\overline{\mathbf{P}}_j + \boldsymbol{\phi}^T_j(x)\mathbf{Q}_j\overline{\mathbf{Q}}_j$ and therefore

$$\mathbf{I} = \mathbf{P}_j\overline{\mathbf{P}}_j + \mathbf{Q}_j\overline{\mathbf{Q}}_j = \left[\mathbf{P}_j\mathbf{Q}_j\right]\begin{bmatrix}\overline{\mathbf{P}}_j\\ \overline{\mathbf{Q}}_j\end{bmatrix}. \quad (18.15)$$

Since the $K_j \times K_j$ matrix $\left[\mathbf{P}_j\mathbf{Q}_j\right]$ is due to (18.12) and (18.13) of full rank, i.e. rank $\left[\mathbf{P}_j\mathbf{Q}_j\right] = K_j$, it follows from (18.15)

$$\begin{bmatrix}\overline{\mathbf{P}}_j\\ \overline{\mathbf{Q}}_j\end{bmatrix} = \left[\mathbf{P}_j\mathbf{Q}_j\right]^{-1}. \quad (18.16)$$

Now, both the two-scale relations (18.6) and (18.9) as well as the decomposition equation (18.14) will be used to construct the MSR. It can be achieved in two steps, namely the decomposition of the signal into level-dependent coefficients (*analysis*) and the (re)construction by means of the detail signals (*synthesis*).

Assuming that a signal $f_J(x)$ with $f_J \in V_J$ is given, the decomposition

$$f_J(x) = f_{j'}(x) + \sum_{j=j'}^{J-1} g_j(x) \quad (18.17)$$

can be formulated (see Schmidt 2007), wherein $f_{j'}(x)$ with $f_{j'} \in V_{j'}$ and $j' \in \{0, \ldots, J-1\}$ is defined as

$$f_{j'}(x) = \sum_{k=0}^{K_{j'}-1} d_{j';k}\phi_{j';k}(x) = \boldsymbol{\phi}_{j'}^T(x)\mathbf{d}_{j'} \quad (18.18)$$

and means a *low-pass* filtered, i.e. smoothed version of $f_J(x)$. The $K_{j'} \times 1$ *scaling coefficient vector* $\mathbf{d}_{j'}$ of level j' reads

$$\mathbf{d}_{j'} = \left[d_{j';0}, \ldots, d_{j';K_{j'}-1}\right]^T. \quad (18.19)$$

The *detail signals*

$$g_j(x) = \sum_{k=0}^{L_j-1} c_{j;k}\psi_{j;k}(x) = \boldsymbol{\psi}_j^T(x)\mathbf{c}_j \quad (18.20)$$

with $g_j \in W_j$ and $j = j', \ldots, J-1$ are *band-pass* filtered versions of $f_J(x)$. The $L_j \times 1$ *wavelet coefficient vector* $\mathbf{c}_j$ of level j is defined as

$$\mathbf{c}_j = \left[c_{j;0}, \ldots, c_{j;L_j-1}\right]^T. \quad (18.21)$$

Hence, the decomposition (18.17) is the MSR of the signal $f_J(x)$. For the M-D case, which will be outlined in the next section, (18.17) is rewritten with $j' = J - I$ as

$$f_J(x) = f_{j'}(x) + \sum_{i=1}^{I} g_{J-i}(x). \quad (18.22)$$

It can be seen from (18.20) that the lower the level j is chosen the smaller is the number L_j of addends, i.e. the number of wavelet coefficients $c_{j;k}$ is decreasing with the level value. This is due to the fact that coarse structures, i.e low-frequency signals, are describable by just a few coefficients, whereas fine structures or high frequency signals require a large number of coefficients. In addition the probably most important feature of the MSR presented here is the dependency of the coefficient vectors (18.19) and (18.21) on each other. For the computation of the level–$(j-1)$ vectors $\mathbf{d}_{j-1}$ and $\mathbf{c}_{j-1}$ from the level-j vector $\mathbf{d}_j$ the signal $f_j(x)$ is expressed as $f_j(x) = f_{j-1}(x) + g_{j-1}(x)$ according to (18.17). It follows with (18.18) and (18.20)

$$\boldsymbol{\phi}_j^T(x)\mathbf{d}_j = \boldsymbol{\phi}_{j-1}^T(x)\mathbf{d}_{j-1} + \boldsymbol{\psi}_{j-1}^T(x)\mathbf{c}_{j-1}. \quad (18.23)$$

Substituting the decomposition equation (18.14) for $\boldsymbol{\phi}_j^T(x)$ on the left-hand side of (18.23) yields $\boldsymbol{\phi}_{j-1}^T(x)\overline{\mathbf{P}}_j\mathbf{d}_j + \boldsymbol{\psi}_{j-1}^T(x)\overline{\mathbf{Q}}_j\mathbf{d}_j$. The comparison of this result with the right-hand side of (18.23) gives the desired relations

$$\mathbf{d}_{j-1} = \overline{\mathbf{P}}_j\mathbf{d}_j \quad \text{and} \quad \mathbf{c}_{j-1} = \overline{\mathbf{Q}}_j\mathbf{d}_j \quad (18.24)$$

for $j = j' + 1, \ldots, J$, which connect the scaling and wavelet coefficient vectors recursively. To be more specific, the K_j elements of the level-j scaling coefficient vector $\mathbf{d}_j$ are transformed into the K_{j-1} elements of the level-$(j-1)$ scaling coefficient vector $\mathbf{d}_{j-1}$ and the L_{j-1} components of the level-$(j-1)$ wavelet coefficient vector $\mathbf{c}_{j-1}$ with $K_j = K_{j-1} + L_{j-1}$. Hereby the transformation matrices $\overline{\mathbf{P}}_j$ and $\overline{\mathbf{Q}}_j$, computed from (18.16), provide the *downsampling* from level j to level $j-1$. Hence, (18.24) are rewritten as

$$\begin{bmatrix}\mathbf{d}_{j-1}\\ \mathbf{c}_{j-1}\end{bmatrix} = \begin{bmatrix}\overline{\mathbf{P}}_j\\ \overline{\mathbf{Q}}_j\end{bmatrix}\mathbf{d}_j. \quad (18.25)$$

Solving this equation for $\mathbf{d}_j$ gives under the consideration of (18.16) the inverse relation

$$\mathbf{d}_j = \mathbf{P}_j \mathbf{d}_{j-1} + \mathbf{Q}_j \mathbf{c}_{j-1}. \quad (18.26)$$

Thus, the level-j scaling coefficient vector $\mathbf{d}_j$ is computed from the level-$(j-1)$ scaling and wavelet coefficient vectors $\mathbf{d}_{j-1}$ and $\mathbf{c}_{j-1}$ by means of the transformation matrices $\mathbf{P}_j$ and $\mathbf{Q}_j$.

Now the procedure of decomposition and reconstruction of a signal can be explained as follows. The *initialization step* is started with the expansion (18.18) for altogether P observations $y(x_p) =: y_p$ with $p = 1, \ldots, P$. Note, that the value J has to be chosen such that $K_J = 2^J + 2 < P$ holds. With the measurement error $e(x_p) =: e_p$ the left-hand side of (18.18) with $j' = J$ is replaced by $f_J(x_p) = y(x_p) + e(x_p) = y_p + e_p$. Thus, the *observation equation* reads

$$y_p + e_p = \sum_{k=0}^{K_J-1} d_{J;k}\phi_{J;k}(x_p) = \boldsymbol{\phi}_J^T(x_p)\mathbf{d}_J, \quad (18.27)$$

wherein the $K_J \times 1$ vector $\mathbf{d}_J$ of the level-J scaling coefficients $d_{J;k}$ is unknown. Introducing the $P \times 1$ vectors $\mathbf{y} = (y_p)$ and $\mathbf{e} = (e_p)$ of the observations and the measurement errors, the $P \times K_J$ coefficient matrix $\mathbf{X}_J = [\boldsymbol{\phi}_J(x_1), \ldots, \boldsymbol{\phi}_J(x_P)]^T$ and the $P \times P$ covariance matrix $D(\mathbf{y})$ of the observations, the linear model

$$\mathbf{y} + \mathbf{e} = \mathbf{X}_J \mathbf{d}_J \quad \text{with} \quad D(\mathbf{y}) = \sigma_y^2 \mathbf{P}_y^{-1} \quad (18.28)$$

is established. Herein σ_y^2 and $\mathbf{P}_y$ are denoted as the variance factor and the weight matrix, respectively. Since the model (18.28) means a *Gauss–Markov model*, the *least-squares estimator* under the assumption rank $\mathbf{X}_J = K_J$ is given as

$$\widehat{\mathbf{d}}_J = (\mathbf{X}_J^T \mathbf{P}_y \mathbf{X}_J)^{-1} \mathbf{X}_J^T \mathbf{P}_y \mathbf{y}, \quad (18.29)$$

see, e.g. Koch (1999). The corresponding covariance matrix reads $D(\widehat{\mathbf{d}}_J) = \sigma_y^2 (\mathbf{X}_J^T \mathbf{P}_y \mathbf{X}_J)^{-1}$. An unbiased estimator of the variance factor σ_y^2 is given as $\widehat{\sigma}_y^2 = \frac{\widehat{\mathbf{e}}^T \mathbf{P}_y \widehat{\mathbf{e}}}{P-K_J}$, wherein $\widehat{\mathbf{e}} = \mathbf{X}_J \widehat{\mathbf{d}}_J - \mathbf{y}$ means the residual vector, i.e. the least-squares estimation of the error vector. Thus, the estimated covariance matrix $\widehat{D}(\widehat{\mathbf{d}}_J)$ of the estimation (18.29) follows as

$$\widehat{D}(\widehat{\mathbf{d}}_J) = \widehat{\sigma}_y^2 (\mathbf{X}_J^T \mathbf{P}_y \mathbf{X}_J)^{-1}.$$

Within the *pyramid steps* the estimators $\widehat{\mathbf{d}}_{j-1} = \overline{\mathbf{P}}_j \widehat{\mathbf{d}}_j$ and $\widehat{\mathbf{c}}_{j-1} = \overline{\mathbf{Q}}_j \widehat{\mathbf{d}}_j$ are computed from (18.24) recursively. Applying the law of error propagation to these relations yields the estimated covariance matrices

$$\widehat{D}(\widehat{\mathbf{d}}_{j-1}) = \overline{\mathbf{P}}_j \widehat{D}(\widehat{\mathbf{d}}_j) \overline{\mathbf{P}}_j^T \text{ and}$$

$$\widehat{D}(\widehat{\mathbf{c}}_{j-1}) = \overline{\mathbf{Q}}_j \widehat{D}(\widehat{\mathbf{d}}_j) \overline{\mathbf{Q}}_j^T.$$

A *hypothesis testing* can be used to check the elements $\widehat{c}_{j-1;l}$ with $l = 0, \ldots, L_{j-1} - 1$ on significance; see Koch (1999). Other *data compression techniques* are, e.g. discussed by Ogden (1997). Either with the estimation $\widehat{\mathbf{c}}_{j-1}$ or with a corresponding compressed version the estimated detail signals $\widehat{g}_{j-1}(x) = \boldsymbol{\psi}_{j-1}^T(x)\widehat{\mathbf{c}}_{j-1}$ with variances $\widehat{V}(\widehat{g}_{j-1}(x)) = \boldsymbol{\psi}_{j-1}^T(x)\widehat{D}(\widehat{\mathbf{c}}_{j-1})\boldsymbol{\psi}_{J-1}(x)$ are calculable from (18.20).

In the next section the previous results are generalized to the M-D case. A detailed treatise of the M-D MSR is given by Zeilhofer (2008).

3 Multi-Dimensional B-Spline MSR

First the M-D vector $\mathbf{x} = [x_1, \ldots, x_M]^T$ with $\mathbf{x} \in \mathbb{I}^M$ and $\mathbb{I}^M = \bigotimes_{m=1}^M \mathbb{I}$ of coordinates $x_m \in \mathbb{I}$ with $m = 1, \ldots, M$ is introduced. Assuming that a M-D signal $f_{J_1,\ldots,J_M}(\mathbf{x}) = f_{\mathbf{J}}(\mathbf{x})$ with $f_{\mathbf{J}} \in V_{\mathbf{J}} = V_{J_1,\ldots,J_M}$ and $V_{\mathbf{J}} \subset L^2(\mathbb{I}^M)$ is given, (18.18) can be generalized to

$$f_{\mathbf{J}}(\mathbf{x}) = \sum_{k_1=0}^{K_{J_1}-1} \cdots \sum_{k_M=0}^{K_{J_M}-1} d_{\mathbf{J};k_1,\ldots,k_M} \phi_{\mathbf{J};k_1,\ldots,k_M}(\mathbf{x})$$

$$= \sum_{\mathbf{k}} d_{\mathbf{J};\mathbf{k}} \phi_{\mathbf{J};\mathbf{k}}(\mathbf{x}) = \boldsymbol{\phi}_{\mathbf{J}}^T(\mathbf{x})\mathbf{d}_{\mathbf{J}}, \quad (18.30)$$

wherein $\phi_{\mathbf{J};\mathbf{k}}(\mathbf{x})$ is a M-D *scaling function* of resolution levels $J_1, \ldots, J_M$ w.r.t. the coordinates $x_1, \ldots, x_M$. In the *tensor product approach* $\phi_{\mathbf{J};\mathbf{k}}(\mathbf{x})$ is replaced by the product of M 1-D scaling functions $\phi_{J_m;k_m}(x_m)$ with $m = 1, \ldots, M$, i.e.

$$\phi_{\mathbf{J};\mathbf{k}}(\mathbf{x}) = \prod_{m=1}^M \phi_{J_m;k_m}(x_m). \quad (18.31)$$

Note, that for the M-D B-spline representation the functions $\phi_{J_m;k_m}(x_m)$ are defined as ENQ B-splines. According to

$$\boldsymbol{\phi}_J(\mathbf{x}) = \boldsymbol{\phi}_{J_M}(x_M) \otimes \ldots \otimes \boldsymbol{\phi}_{J_2}(x_2) \otimes \boldsymbol{\phi}_{J_1}(x_1), \quad (18.32)$$

the $K_J (= \prod_{m=1}^{M} K_{J_m}) \times 1$ vector $\boldsymbol{\phi}_J(\mathbf{x})$ can be written as the Kronecker product sum ("$\otimes$"; see, e.g. Koch 1999) of the M vectors $\boldsymbol{\phi}_{J_m}(x_m)$ as introduced in (18.5) with $j = J_m$. The $K_J \times 1$ scaling coefficient vector $\mathbf{d}_J$ reads

$$\mathbf{d}_J = \left[d_{J;0,\ldots,0}, \ldots, d_{J;K_{J_1}-1,\ldots,K_{J_M}-1}\right]^T. \quad (18.33)$$

As generalization of the 1-D MSR (18.22) the M-D MSR of the signal $f_J(\mathbf{x})$ is defined as

$$f_J(\mathbf{x}) = f_{\mathbf{j}'}(\mathbf{x}) + \sum_{i=1}^{I} \sum_{\lambda=1}^{\Lambda} g^{\lambda}_{\mathbf{J}-\mathbf{i}}(\mathbf{x}) \quad (18.34)$$

with $\Lambda = 2^M - 1$ and the two $M \times 1$ vectors $\mathbf{j}' = \left[j'_1 = J_1 - I, \ldots, j'_M = J_M - I\right]^T$, $\mathbf{i} = \left[i, \ldots, i\right]^T$. The level-$\mathbf{j}'$ smoothed version $f_{\mathbf{j}'}(\mathbf{x})$ and the level-$(\mathbf{J} - \mathbf{i})$ detail signal $g^{\lambda}_{\mathbf{J}-\mathbf{i}}(\mathbf{x})$ are defined as

$$f_{\mathbf{j}'}(\mathbf{x}) = \sum_{\mathbf{k}} d_{\mathbf{j}';\mathbf{k}} \phi_{\mathbf{j}';\mathbf{k}}(\mathbf{x}) = \boldsymbol{\phi}^T_{\mathbf{j}'}(\mathbf{x}) \mathbf{d}_{\mathbf{j}'},$$

$$g^{\lambda}_{\mathbf{J}-\mathbf{i}}(\mathbf{x}) = \sum_{\mathbf{k}} c^{\lambda}_{\mathbf{J}-\mathbf{i};\mathbf{k}} \psi^{\lambda}_{\mathbf{J}-\mathbf{i};\mathbf{k}}(\mathbf{x}) = (\boldsymbol{\psi}^{\lambda}_{\mathbf{J}-\mathbf{i}}(\mathbf{x}))^T \mathbf{c}^{\lambda}_{\mathbf{J}-\mathbf{i}}.$$

The altogether Λ level-$(\mathbf{J} - \mathbf{i})$ tensor product wavelet functions $\psi^{\lambda}_{\mathbf{J}-\mathbf{i};\mathbf{k}}(\mathbf{x})$ with $\lambda = 1, \ldots, \Lambda$ are given as

$$\psi^{1}_{\mathbf{J}-\mathbf{i};\mathbf{k}}(\mathbf{x}) = \phi_{J_1-i;k_1}(x_1)\phi_{J_2-i;k_2}(x_2)\ldots$$
$$\ldots\phi_{J_{M-1}-i;k_{M-1}}(x_{M-1})\psi_{J_M-i;k_M}(x_M),$$
$$\psi^{2}_{\mathbf{J}-\mathbf{i};\mathbf{k}}(\mathbf{x}) = \phi_{J_1-i;k_1}(x_1)\phi_{J_2-i;k_2}(x_2)\ldots$$
$$\ldots\psi_{J_{M-1}-i;k_{M-1}}(x_{M-1})\phi_{J_M-i;k_M}(x_M),$$
$$\ldots$$
$$\psi^{M}_{\mathbf{J}-\mathbf{i};\mathbf{k}}(\mathbf{x}) = \psi_{J_1-i;k_1}(x_1)\phi_{J_2-i;k_2}(x_2)\ldots$$
$$\ldots\phi_{J_{M-1}-i;k_{M-1}}(x_{M-1})\phi_{J_M-i;k_M}(x_M),$$
$$\psi^{M+1}_{\mathbf{J}-\mathbf{i};\mathbf{k}}(\mathbf{x}) = \phi_{J_1-i;k_1}(x_1)\phi_{J_2-i;k_2}(x_2)\ldots$$
$$\ldots\psi_{J_{M-1}-i;k_{M-1}}(x_{M-1})\psi_{J_M-i;k_M}(x_M),$$
$$\ldots$$
$$\psi^{\Lambda}_{\mathbf{J}-\mathbf{i};\mathbf{k}}(\mathbf{x}) = \psi_{J_1-i;k_1}(x_1)\psi_{J_2-i;k_2}(x_2)\ldots$$
$$\ldots\psi_{J_{M-1}-i;k_{M-1}}(x_{M-1})\psi_{J_M-i;k_M}(x_M).$$

The decomposition equations between the scaling coefficient vectors $\mathbf{d}_{\mathbf{j}}, \mathbf{d}_{\mathbf{j}-\mathbf{1}}$ and the wavelet coefficient vectors $\mathbf{c}^{\lambda}_{\mathbf{j}-\mathbf{1}}$ with $\lambda = 1, \ldots, \Lambda$ are derived to

$$\mathbf{d}_{\mathbf{j}-\mathbf{1}} = (\overline{\mathbf{P}}_{j_M} \otimes \overline{\mathbf{P}}_{j_{M-1}} \otimes \ldots \otimes \overline{\mathbf{P}}_{j_1})\mathbf{d}_{\mathbf{j}},$$
$$\mathbf{c}^{1}_{\mathbf{j}-\mathbf{1}} = (\overline{\mathbf{Q}}_{j_M} \otimes \overline{\mathbf{P}}_{j_{M-1}} \otimes \ldots \otimes \overline{\mathbf{P}}_{j_1})\mathbf{d}_{\mathbf{j}},$$
$$\mathbf{c}^{2}_{\mathbf{j}-\mathbf{1}} = (\overline{\mathbf{P}}_{j_M} \otimes \overline{\mathbf{Q}}_{j_{M-1}} \otimes \ldots \otimes \overline{\mathbf{P}}_{j_1})\mathbf{d}_{\mathbf{j}},$$
$$\ldots$$
$$\mathbf{c}^{M}_{\mathbf{j}-\mathbf{1}} = (\overline{\mathbf{P}}_{j_M} \otimes \overline{\mathbf{P}}_{j_{M-1}} \otimes \ldots \otimes \overline{\mathbf{Q}}_{j_1})\mathbf{d}_{\mathbf{j}},$$
$$\mathbf{c}^{M+1}_{\mathbf{j}-\mathbf{1}} = (\overline{\mathbf{Q}}_{j_M} \otimes \overline{\mathbf{Q}}_{j_{M-1}} \otimes \ldots \otimes \overline{\mathbf{P}}_{j_1})\mathbf{d}_{\mathbf{j}},$$
$$\ldots$$
$$\mathbf{c}^{\Lambda}_{\mathbf{j}-\mathbf{1}} = (\overline{\mathbf{Q}}_{j_M} \otimes \overline{\mathbf{Q}}_{j_{M-1}} \otimes \ldots \otimes \overline{\mathbf{Q}}_{j_1})\mathbf{d}_{\mathbf{j}} \quad (18.35)$$

are obtained as the generalization of the 1-D relations (18.24); the $K_{j_m-1} \times K_{j_m}$ and $L_{j_m-1} \times K_{j_m}$ matrices $\overline{\mathbf{P}}_{j_m}$ and $\overline{\mathbf{Q}}_{j_m}$ were already introduced in (18.16).

Let $y(\mathbf{x}_p) = y_p$ be a M-D observation with $p = 1, \ldots, P$ and $e(\mathbf{x}_p) = e_p$ the corresponding measurement error, the observation equation follows from (18.30) and reads

$$y_p + e_p = \sum_{\mathbf{k}} d_{\mathbf{J};\mathbf{k}} \phi_{\mathbf{J};\mathbf{k}}(\mathbf{x}_p) = \boldsymbol{\phi}^T_{\mathbf{J}}(\mathbf{x}_p)\mathbf{d}_{\mathbf{J}}, \quad (18.36)$$

herein the vector $\mathbf{d}_J$ as defined in (18.33) has to be determined within the initialization step of the M-D MSR; see also Schmidt et al. (2007b). Introducing the covariance matrix $D(\mathbf{y})$ of the $P \times 1$ observation vector $\mathbf{y}$, the linear adjustment model is established. Consequently, the results presented before for the 1-D case are also valid for the M-D problem. In the pyramid steps of the MSR the relations (18.35) have to be used. Note, if the input data is given on a regular grid, too, the *lofting method* can be applied for solving (18.36) with much less computational complexity; see Koch (2009).

4 Example

With $M = 2$ the 2-D multi-scale B-spline approach is outlined in this example; $J_1 = J_2 = J$ is set for the highest resolution levels w.r.t. the coordinates $x_1 = x$

(geographical normalized longitude) and $x_2 = y$ (geographical normalized latitude) with $\mathbf{x} = [x, y]^T \in \mathbb{I}^2$. Consequently, the representation (18.30) of the 2-D signal $f_J(\mathbf{x}) = f_J(x, y)$ with $f_J \in V_J$ reduces with (18.31) to

$$f_J(x,y) = \sum_{k_1=0}^{K_J-1} \sum_{k_2=0}^{K_J-1} d_{J;k_1,k_2} \phi_{J;k_1}(x) \phi_{J;k_2}(y) \tag{18.37}$$

with $K_{J_1} = K_{J_2} = K_J = 2^J + 2$ as defined before. The $K_J \times 1$ vector $\mathbf{d}_J$ introduced in (18.33) can be rewritten as

$$\mathbf{d}_J = \text{vec}\mathbf{D}_J, \tag{18.38}$$

herein 'vec' means the vec-operator (Koch 1999). The $K_J \times K_J$ scaling coefficient matrix $\mathbf{D}_J$ is defined as

$$\mathbf{D}_J = \begin{bmatrix} d_{J;0,0} & d_{J;0,1} & \dots & d_{J;0,K_J-1} \\ d_{J;1,0} & d_{J;1,1} & \dots & d_{J;1,K_J-1} \\ \dots & \dots & \dots & \dots \\ d_{J;K_J-1,0} & d_{J;K_J-1,1} & \dots & d_{J;K_J-1,K_J-1} \end{bmatrix}. \tag{18.39}$$

With the Kronecker product representation (18.32), (18.37) reads $f_J(x, y) = (\boldsymbol{\phi}_J^T(y) \otimes \boldsymbol{\phi}_J^T(x))\,\text{vec}\mathbf{D}_J$; considering further the computation rules for the Kronecker product (Koch 1999) the matrix equation

$$f_J(x,y) = \boldsymbol{\phi}_J^T(x) \mathbf{D}_J \boldsymbol{\phi}_J(y) \tag{18.40}$$

is obtained. With $j_1' = j_2' = j'$ it follows from (18.34)

$$f_J(x,y) = f_{j'}(x,y) + \sum_{i=1}^{I} \sum_{\lambda=1}^{3} g_{J-i}^{\lambda}(x,y), \tag{18.41}$$

wherein the signals $f_{j'}(x,y) = f_{j'}(x,y)$ and $g_{J-i}^{\lambda}(x,y) = g_{J-i}^{\lambda}(x,y)$ are computable via the relations

$$\begin{aligned} f_{j'}(x,y) &= \boldsymbol{\phi}_{j'}^T(x) \mathbf{D}_{j'} \boldsymbol{\phi}_{j'}(y), \\ g_{J-i}^{1}(x,y) &= \boldsymbol{\phi}_{J-i}^T(x) \mathbf{C}_{J-i}^{1} \boldsymbol{\psi}_{J-i}(y), \\ g_{J-i}^{2}(x,y) &= \boldsymbol{\psi}_{J-i}^T(x) \mathbf{C}_{J-i}^{2} \boldsymbol{\phi}_{J-i}(y), \\ g_{J-i}^{3}(x,y) &= \boldsymbol{\psi}_{J-i}^T(x) \mathbf{C}_{J-i}^{3} \boldsymbol{\psi}_{J-i}(y), \end{aligned} \tag{18.42}$$

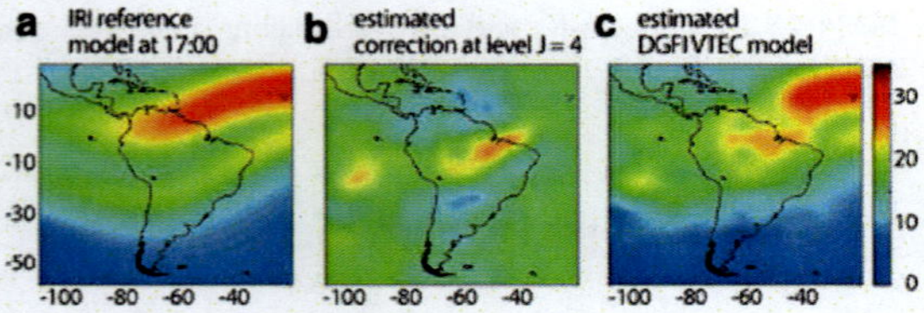

Fig. 18.1 (a) Reference model $IRI(x, y)$ of VTEC at July 21, 2006 at 5:00 p.m. UT; (b) estimated VTEC correction model $\widehat{f}_4(x, y)$ at the same time; (c) the sum of the two other panels, i.e. the estimated VTEC model $IRI(x, y) + \widehat{f}_4(x, y)$; all data in TECU. The colorbar of panel b is the same as in Fig. 18.2

the vectors $\boldsymbol{\psi}_{J-i}(x)$ and $\boldsymbol{\psi}_{J-i}(y)$ were already introduced in (18.8). Furthermore in (18.42) the $K_{J-i} \times L_{J-i}$ matrix $\mathbf{C}_{J-i}^1$, the $L_{J-i} \times K_{J-i}$ matrix $\mathbf{C}_{J-i}^2$ and the $L_{J-i} \times L_{J-i}$ matrix $\mathbf{C}_{J-i}^3$ defined via the relations $\mathbf{c}_{J-i}^{\lambda} = \text{vec}\mathbf{C}_{J-i}^{\lambda}$ with $\lambda = 1, 2, 3$ analogously to (18.38) are used. The corresponding 2-D downsampling equations are following from (18.35) and read

$$\begin{bmatrix} \mathbf{D}_{j-1} & \mathbf{C}_{j-1}^1 \\ \mathbf{C}_{j-1}^2 & \mathbf{C}_{j-1}^3 \end{bmatrix} = \begin{bmatrix} \overline{\mathbf{P}}_j \\ \overline{\mathbf{Q}}_j \end{bmatrix} \mathbf{D}_j \begin{bmatrix} \overline{\mathbf{P}}_j^T & \overline{\mathbf{Q}}_j^T \end{bmatrix} \tag{18.43}$$

for $j = j', \dots, J-1$.

For numerical demonstration of the 2-D MSR terrestrial GNSS data, radio occultation data from the COSMIC/FORMOSAT-3 mission and radar altimetry data from the Jason-1 and Envisat missions are evaluated to derive a correction model of the vertical total electron content (VTEC) w.r.t. the International Reference Ionosphere (IRI) over South and Central America; for more details see Dettmering et al. (2011). Figure 18.1 shows in panel b the signal $\widehat{f}_J(x, y)$ at level $J = 4$, i.e. the estimated level-4 VTEC correction model to IRI (shown in panel a) calculated from the scaling coefficient matrix $\widehat{\mathbf{D}}_4$, which is the result from the least-squares adjustment of the input data mentioned before. Since we chose $K_4 = 2^4 + 2 = 18$, the 18×18 matrix $\widehat{\mathbf{D}}_4$ contains 324 scaling coefficients $\widehat{d}_{4;k_1,k_2}$ with $k_1, k_2 = 0, \dots, 17$; see Table 18.1.

Figure 18.2 displays the MSR of the estimated level-4 VTEC correction model (panel a, corresponds to panel b in Fig. 18.1). The low-pass filtered (smoothed) signals $\widehat{f}_3(x, y)$, $\widehat{f}_2(x, y)$, and $\widehat{f}_1(x, y)$ (panels b–d) are calculated according to the first equation of (18.42). The panels e–g visualize the sums $\widehat{g}_{4-i}(x, y) = \sum_{\lambda=1}^{3} \widehat{g}_{4-i}^{\lambda}(x, y)$ with $i = 1, 2, 3$.

Table 18.1 Numbers K_j and L_j of the scaling and wavelet coefficient matrices $\mathbf{D}_j$ and $\mathbf{C}_j^\lambda$

Level j	K_j	L_j	$\mathbf{D}_j$ $K_j \times K_j$	$\mathbf{C}_j^1$ $K_j \times L_j$	$\mathbf{C}_j^2$ $L_j \times K_j$	$\mathbf{C}_j^3$ $L_j \times L_j$	Total #
4	18	–	324	–	–	–	324
3	10	8	100	80	80	64	324
2	6	4	36	24	24	16	100
1	4	2	16	8	8	4	36
0	3	1	9	3	3	1	16

The last column means the total number of coefficients in level j

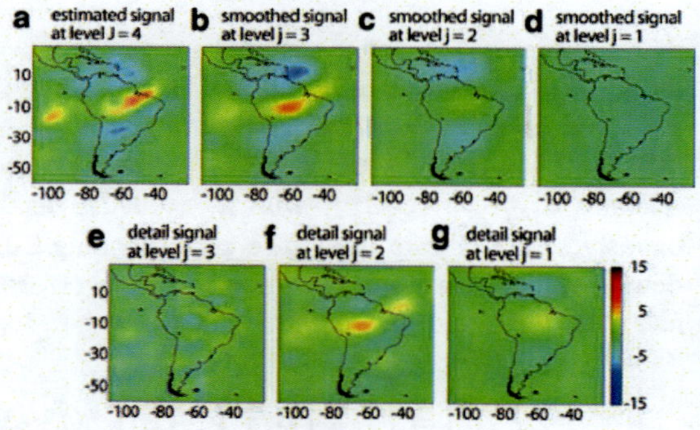

Fig. 18.2 (b)–(d) Estimated low-pass filtered signals $\widehat{f}_3(x,y)$, $\widehat{f}_2(x,y)$, $\widehat{f}_1(x,y)$ of the level-4 VTEC correction model (panel a, corresponds to Fig. 18.1b; (e)–(g) estimated band-pass filtered detail signals $\widehat{g}_3(x,y)$, $\widehat{g}_2(x,y)$, $\widehat{g}_1(x,y)$; all data in TECU.

The signals $\widehat{g}_{4-i}^\lambda(x,y)$ with $\lambda = 1,2,3$ are all computed via the last three equations in (18.42). It is clearly visible that the lower the level value is set the coarser are the extracted structures. Table 18.1 displays the sizes of the scaling and wavelet coefficient matrices for the levels $j = 0,1,2,3,4$ computed from the decomposition scheme (18.43). According to the principle of the MSR the sum of the three detail signals $\widehat{g}_{4-i}(x,y)$, $i = 1,2,3$ and the smoothed signal $\widehat{f}_1(x,y)$ yields the estimated VTEC correction model $\widehat{f}_4(x,y)$. The application of data compression techniques is discussed by Schmidt (2007) and Zeilhofer (2008).

5 Summary and Outlook

In this paper it was demonstrated how a MSR based on ENQ B-spline wavelets and tensor products can be applied to M-D signals. Since the derived approach is based on Euclidean theory, it is restricted to regional or local areas. The basic feature of the MSR is the decomposition of a signal into band-pass filtered detail signals; as already standard in digital image processing the reconstruction step should include efficient data compression techniques. This way just the significant information of the signal is stored. Furthermore, Kalman filtering can be applied efficiently to the MSR. In ionosphere research the presented procedure can, e.g. be applied to model the 4-D electron density (Zeilhofer et al. 2009). The author will present these issues in an upcoming paper.

References

Dettmering D, Schmidt M, Heinkelmann R, Seitz M (2011) Combination of different satellite observation data for regional ionosphere modeling. J Geodesy, doi:10.1007/s00190-010-0423-1

Freeden W (1999) Multiscale modelling of spaceborne geodata. Teubner, Leipzig

Koch KR (1999) Parameter estimation and hypothesis testing in linear models. Springer, Berlin

Koch KR (2009) Identity of simultaneous estimates of control points and of their estimates by the lofting method for NURBS surface fitting. Int J Adv Manuf Technol. doi:10.1007/s00170-009-1934-x

Lyche T, Schumaker LL (2001) A multiresolution tensor spline method for fitting functions on the sphere. SIAM J Sci Comput 22(2):724–746

Ogden RT (1997) Essential wavelets for statistical applications and data analysis. Birkhäuser, Boston.

Schmidt M (2001) Basic principles of wavelet analysis and applications in geodesy (in German). Post Doctoral Thesis, Shaker, Aachen

Schmidt M (2007) Wavelet modeling in support of IRI. J Adv Space Res 39:932–940

Schmidt M, Fengler M, Mayer-Gürr T, Eicker A, Kusche J, Sánchez L, Han S-C (2007a) Regional gravity modeling in terms of spherical base functions. J Geodesy 81:17–38

Schmidt M, Bilitza D, Shum CK, Zeilhofer C (2007b) Regional 4-D modeling of the ionospheric electron content. J Adv Space Res 42:787–790

Stollnitz EJ, DeRose TD, Salesin DH (1995) Wavelets for computer graphics: a primer, Part I. IEEE Comput Graph Appl 15(3):76–84; Part II, IEEE Comput Graph Appl 15(4) 75–85

Zeilhofer C (2008) Multi-dimensional B-spline modeling of spatio-temporal ionospheric signals. German Geodetic Commission, Series A, 123, Munich, Germany.

Zeilhofer C, Schmidt M, Bilitza D, Shum CK (2009) Regional 4-D modeling of the ionospheric electron density from satellite data and IRI. J Adv Space Res 43:1669–1675

Analysing Time Series of GNSS Residuals by Means of AR(I)MA Processes

19

X. Luo, M. Mayer, and B. Heck

Abstract

The classical least-squares (LS) algorithm is widely applied in processing data from Global Navigation Satellite Systems (GNSS). However, some limiting factors impacting the accuracy measures of unknown parameters such as temporal correlations of observational data are neglected in most GNSS processing software products. In order to study the temporal correlation characteristics of GNSS observations, this paper introduces autoregressive (integrated) moving average (AR(I)MA) processes to analyse residual time series resulting from the LS evaluation. Based on a representative data base the influences of various factors, like baseline length, multipath effects, observation weighting, atmospheric conditions on ARIMA identification are investigated. Additionally, different temporal correlation models, for example first-order AR processes, ARMA processes, and empirically determined analytical autocorrelation functions are compared with respect to model appropriateness and efficiency.

Keywords

GNSS • Stochastic model • Temporal correlations • Time series analysis • AR(I)MA processes

1 Introduction

Accompanying the modernisation and completion of Global Navigation Satellite Systems (GNSS), the requirements on positioning accuracy and reliability increase in a wide range of geodetic applications, for example highly sensitive deformation monitoring. To meet the rising demands on accurate positions as well as realistic quality measures, improvements not only in hardware but also in mathematical modelling within GNSS data processing are necessary. The mathematical models consist of the functional and the stochastic model. In contrast to the intensively investigated functional model, the stochastic model formulating the observations' statistical properties is still under development. One essential deficiency of the stochastic model is caused by neglecting temporal correlations of GNSS observations, which leads to over-optimistic accuracy estimates. The temporal correlations originating from turbulent irregularities in the Earth's lower atmosphere (Wheelon 2001) are affected, for instance, by multipath effects and receiver-dependent signal processing techniques (Tiberius et al. 1999).

X. Luo (✉) · M. Mayer · B. Heck
Geodetic Institute, Karlsruhe Institute of Technology (KIT), D-76131 Karlsruhe, Englerstr. 7, Germany
e-mail: luo@kit.edu

N. Sneeuw et al. (eds.), *VII Hotine-Marussi Symposium on Mathematical Geodesy*, International Association of Geodesy Symposia 137, DOI 10.1007/978-3-642-22078-4_19,

Up to now, different approaches for modelling temporal correlations of GNSS observations have been proposed. Tiberius and Kenselaar (2003) applied variance component estimation to construct an appropriate variance-covariance model (VCM) for GNSS data processing. Furthermore, Howind (2005) used an empirically determined analytical autocorrelation function (ACF) and found significant variations of up to 2 cm in estimated site coordinates, in particular for long baselines. Based on the atmospheric turbulence theory Schön and Brunner (2008) suggested a fully-populated VCM which enables the understanding of physical processes correlating and decorrelating GNSS phase observations. Applying this advanced VCM, more realistic formal coordinate variances can be obtained despite highly redundant observation data. Although being different in modelling theories, all the aforementioned approaches aim at realistic quality measures of the unknown parameters, such as site coordinates, ambiguities, and site-specific troposphere parameters.

This paper presents an innovative procedure for modelling temporal correlations of GNSS observations using AR(I)MA processes. Section 2 gives a brief introduction to AR(I)MA processes and ARMA modelling. In Sect. 3 the impacts of different factors on model identification are empirically investigated based on representative data. Section 4 outlines an advanced approach for modelling trends of GNSS residuals. Finally, Sect. 5 compares three temporal correlation models concerning model appropriateness and efficiency.

2 AR(I)MA Processes

As an important parametric family of stationary time series, ARMA processes play a key role in modelling time series data. In the context of geodesy, Li et al. (2000) employed ARMA processes to analyse crustal deformations in central Japan. Wang et al. (2002) investigated the noise characteristics of GNSS carrier phase measurements using first-order autoregressive (AR(1)) processes. In connection with the determination of the Earth's gravity field, Klees et al. (2003) exploited ARMA representation of coloured observation noise in large LS problems to achieve a faster solution of a Toeplitz system of linear equations. In the following text, AR(I)MA processes and the associated procedures for parameter estimation are briefly described.

The time series $\{Y_t\}$ is an ARMA(p, q) process if $\{Y_t\}$ is stationary and if for each time index t,

$$Y_t - \sum_{i=1}^{p} \phi_i Y_{t-i} = Z_t + \sum_{j=1}^{q} \theta_j Z_{t-j}, \qquad (19.1)$$

where $\{Z_t\}$ is referred to as a white noise (WN) process of random variables, each with zero-mean and variance σ^2 (see, e.g., Brockwell and Davis 2002). The integer numbers p and q denote the ARMA orders. The terms $\boldsymbol{\phi} = (\phi_1, \ldots, \phi_p)^T$ and $\boldsymbol{\theta} = (\theta_1, \ldots, \theta_q)^T$ are the model coefficients. In particular, $\{Y_t\}$ is an autoregressive process of order p (AR(p)) if $\theta_j = 0$ for $j = 1, \ldots, q$, and a moving average process of order q (MA(q)) if $\phi_i = 0$ for $i = 1, \ldots, p$. The corresponding pth- and qth-degree characteristic polynomials are

$$\phi(z) = 1 - \phi_1 z - \cdots - \phi_p z^p, \qquad (19.2)$$

$$\theta(z) = 1 + \theta_1 z + \cdots + \theta_q z^q. \qquad (19.3)$$

A unique stationary solution $\{Y_t\}$ of (19.1) exists only if $\phi(z) \neq 0$ for all $|z| = 1$. An ARMA(p, q) process is causal (invertible) if $\phi(z) \neq 0$ ($\theta(z) \neq 0$) for all $|z| \leq 1$.

ARMA modelling aims at the determination of an appropriate ARMA(p, q) model to represent an observed stationary time series. It involves a number of interrelated problems, such as order selection, parameter estimation, etc. Using the software ITSM2000-V.7.1 provided by Brockwell and Davis (2002), ARMA modelling is performed in three steps.

In the first step the given data are transformed into continuous and stationary time series. There are two general approaches to handle the trends. One is to model the trend, then to subtract it from the data. The other is to eliminate the trend by differencing. If a differenced time series with the order of differencing d is an ARMA(p, q) process, the original undifferenced time series is denoted as an ARIMA(p, d, q) process. The stationarity of detrended data can be verified using variance homogeneity tests (e.g., Teusch 2006), sign tests for trend (e.g., Hartung et al. 2005), and unit root tests (e.g., Said and Dickey 1984).

The second step deals with model identification and parameter estimation. The order selection is carried out based on the so-called AICC (Akaike

Information Criterion with small sample Correction) statistic

$$\text{AICC} = -2\ln L(\boldsymbol{\phi}, \boldsymbol{\theta}, \sigma^2) + 2\frac{(p+q+1)n}{n-p-q-2} \quad (19.4)$$

proposed by Hurvich and Tsai (1989), where L denotes the Gaussian likelihood function and n the sample size. All ARMA (p, q) models with p and q in the specified limits are considered and the pair (p, q) holding the smallest AICC value is selected. Besides the AICC criterion there are other criteria for order selection, like FPE and BIC statistics (see, e.g., Brockwell and Davis 2002). Following the model identification, the initial values of $\boldsymbol{\phi}$, $\boldsymbol{\theta}$, and σ^2 can be computed, for example using the Hannen–Rissanen algorithm (Hannen and Rissanen 1982). The final parameters are maximum likelihood estimators, whereas a non-linear optimisation of the initial approximation is undertaken numerically to maximise the Gaussian likelihood function.

In the last step the model validity is judged by analysing the sample ACF of residuals which represent the discrepancies between the given data and the fitted model, as well as by employing statistical hypothesis tests for residual randomness.

Table 19.1 Parameter settings of the GNSS data processing

Observations	1-Hz GPS phase double differences
Processing interval	DOY2007: 161–181, UT: 15–18 h
Observation weighting	ELV: $\sin^2 e$, SNR: $f(SNR)$ e: satellite elevation angle SNR: signal-to-noise ratio
Elevation cut-off	3°
Satellite orbits	Precise IGS products
Earth rotation parameters	Precise IGS products
Ionospheric model	Precise CODE products
Tropospheric model	Niell_{dry} (a priori model)
Mapping function	Niell_{wet} (Niell 1996)
Troposphere parameters	15 min (time window)
Ambiguity resolution	SIGMA strategy (L5, L3)
Antenna correction	Individual absolute calibration

Table 19.2 Analysed factors impacting ARIMA identification

Factor	Comparison between	#SDDR
Multipath impact (MP)	HEDA (MP: strong, 54.1 km)	60
	TAAF (MP: weak, 53.7 km)	62
Baseline length (BL)	RATA (MP: weak, 203.7 km)	56
	SIBI (MP: weak, 42.5 km)	63
Observation weighting (WGT)	ELV: Dach et al. (2007)	285
	SNR: Luo et al. (2008)	285
Relative humidity (RH)	Wet days (mean: 75.8%)	55
	Dry days (mean: 46.0%)	55
Wind velocity (WV)	Windy days (mean: 5 m/s)	51
	Calm days (mean: 2 m/s)	50

3 ARIMA Identification

In order to investigate the influences of different factors impacting GNSS positioning quality on model identification, 21-day 1-Hz GPS phase observations from the SA*POS*®(Satellite Positioning Service of the German State Survey) network are processed with the Bernese GPS Software 5.0 (Dach et al. 2007) in post-processing mode considering multipath impact, baseline length, observation weighting model, and atmospheric conditions based on meteorological surface data provided by the German Meteorological Service (DWD). Table 19.1 gives selected important parameter settings of the GNSS data processing.

A total set of 285 studentised double difference residual (SDDR) time series of identical length (3,600 values) and relating to sidereal time are obtained. The use of SDDR rather than LS residuals is due to the more homogeneous variances of SDDR without attenuation of temporal correlations (Howind 2005). Considering the results of the unit root tests indicating difference-stationary behaviour and regarding the specific irregular data properties, the SDDR time series are primarily modelled by means of ARIMA$(p, 1, q)$ processes. Applying first-order differencing, the apparent trends are sufficiently eliminated and the achieved stationarity is proved by the tests for stationarity mentioned in Sect. 2. Table 19.2 provides an overview of the considered impacting factors and the corresponding numbers of the used SDDR time series.

The influences of these factors on model identification are investigated based on the corresponding empirical cumulative distribution functions of the sum of the selected order parameters p and q. Figure 19.1 visualises the quantile differences of $p + q$ under different aspects. Compared to other analysed factors, multipath impact illustrates the most significant effect on order selection. The α-quantile values of $p + q$ rise with increased multipath, and fall with stronger wind. Thus, under strong multipath conditions, higher ARMA orders are necessary for sufficient data characterisation.

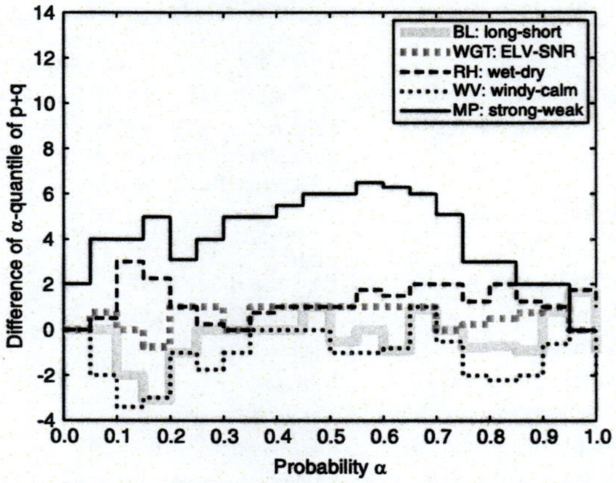

Fig. 19.1 Influences of different factors on model identification

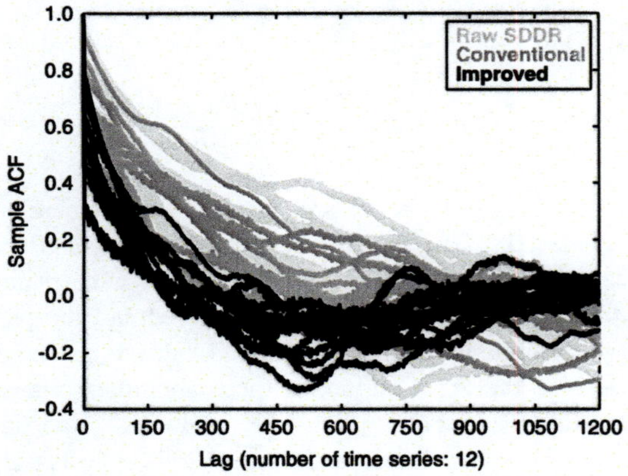

Fig. 19.2 Comparison of sample ACFs with respect to the applied approaches for modelling residual trends (SDDR: SIBI0917)

Utilising ARIMA processes, trends are effectively eliminated by differencing. However, the temporal correlation characteristics existing in the original SDDR can not be retrieved from the differenced series. The use of ARMA processes to model undifferenced data requires appropriate strategies for trend modelling.

4 Modelling Trends of GNSS Residuals

Benefiting from the repetition of GPS satellite geometry and assuming invariable observation environments, a conventional approach for modelling residual trends of GNSS observations is to calculate the epoch-wise arithmetic means of the residual time series relating to sidereal time and being available on different days (e.g., Howind 2005; Ragheb et al. 2007). However, due to variable atmospheric conditions the performance of the conventional trend modelling degrades if a large number of sidereal days are involved in calculating the epoch-wise mean values.

Within this study an improved iterative procedure for modelling GNSS residual trends has been developed. At each iteration, the three (empirically determined) most similar SDDR time series holding the highest correlation coefficients are chosen to calculate the epoch-wise mean values which are then subtracted from the selected data. Subsequently, a second degree polynomial fitting is carried out in order to correct the remaining long-periodic effects. At the end of each iteration, the processed series are removed from the residual data base.

Applying this improved trend modelling procedure, the computed epoch-wise mean values are generally more appropriate to reflect the data characteristics. As a result, the detrended SDDR exhibit smaller variation ranges and more homogenous variances. Comparing the sample ACFs of a representative example displayed in Fig. 19.2, the conventional approach leads to merely insignificant improvements in correlation structure, while using the advanced one, the corresponding sample ACFs illustrate a considerably faster decay and a clearly smaller variation range. These enhancements emphasise that appropriate trend modelling is mandatory for analysing temporal correlations of GNSS observations based on residual sample ACFs.

5 Comparing Temporal Correlation Models

Following the advanced trend modelling, the detrended SDDR time series are analysed using AR(1) processes, ARMA(p, q) processes, and an empirically derived analytical ACF suggested by Howind (2005). The model ACF of an AR(1) process ($Y_t - \phi Y_{t-1} = Z_t$ with $|\phi| < 1$) has the form

$$\mathrm{ACF}_{\mathrm{AR(1)}}(h) = \phi^h \ (h \geq 0), \tag{19.5}$$

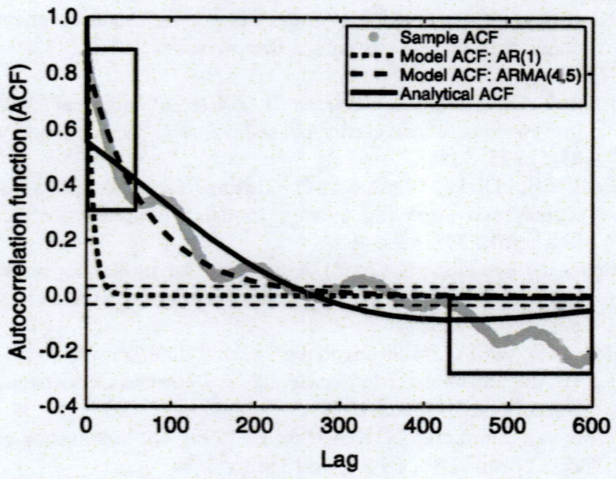

Fig. 19.3 Comparison of different temporal correlation models under strong multipath conditions (SDDR: HEDA2922177)

and it decreases exponentially with increasing epoch distance h (lag). The model ACF of an ARMA(p, q) process can be obtained, for example by solving a set of homogenous linear difference equations after successfully determining the unknown model parameters within ARMA modelling (see, e.g., Brockwell and Davis 2002, Chap. 3.2). The analytical ACF consists of an exponential function and a cosine oscillation component

$$\mathrm{ACF}_{\mathrm{LS}}(h) = C \cdot \exp\left(-\frac{h}{ND_1}\right) \cos\left(\frac{2\pi \cdot h}{ND_1 \cdot T}\right) \tag{19.6}$$

for $h \geq 1$, where ND_1 denotes the smallest lag value at which the sample ACF falls below zero. ND_1 providing valuable information about correlation length is determined numerically a priori, while the parameters C (scaling factor) and T (oscillation period) are estimated using LS regression on sample ACF values.

The temporal correlation models are compared based on the associated ACFs. Under normal observation conditions, both ARMA processes and the applied analytical ACF are capable of modelling the temporal correlation behaviour of GNSS observations. In the presence of strong multipath effects, as exemplarily visualised in Fig. 19.3, ARMA models show large deviations mainly at high lag values because of their short-memory nature, while the estimated analytical ACFs differ from the sample ACFs also within the low-valued lag areas as a result of the global minimisation using LS regression. Under both circumstances, AR(1) processes are obviously insufficient to reflect the observations' temporal correlation characteristics due to the extremely rapid decay of the corresponding model ACFs. Moreover, considering the arithmetic mean of ND_1 of approx. 250 s in this case study, residual time series of between 600 and 1,000 s appear to be efficient for temporal correlation analysis.

6 Conclusions and Outlook

In this paper autoregressive (integrated) moving average (AR(I)MA) processes are investigated to analyse time series of GNSS residuals with the final purpose to propose a reliable and practicable temporal correlation model for GNSS data processing. Based on a representative data base the investigation results show that multipath effects and wind velocity considerably affect ARIMA identification. Applying an improved approach for modelling residual trends, the determined mean correlation length of GPS observations amounts to approx. 250 s. Furthermore, first-order autoregressive processes are clearly insufficient to characterise the temporal correlation behaviour of GNSS observations, while both ARMA processes and the employed analytical ACF turn out to be applicable under normal observation conditions. Compared to estimating analytical ACFs, fitting ARMA processes is more time-consuming and computational-intensive.

The future work will concentrate on the verification of the impact of atmospheric conditions on ARIMA identification. Additionally, other criteria for model identification, for example applied by Klees et al. (2003), will be utilised. Finally, the performance of the advanced procedure for modelling trends of GNSS residuals has to be assessed.

Acknowledgements The Deutsche Forschungsgemeinschaft (DFG, German Research Foundation) is gratefully acknowledged for supporting this research work. We also thank two anonymous reviewers for their valuable comments.

References

Brockwell PJ, Davis RA (2002) Introduction to time series and forecasting, 2nd edn. Springer, New York

Dach R, Hugentobler U, Fridez P, Meindl M (2007) Bernese GPS Software Version 5.0. Astronomical Institute, University of Bern, Bern

Hannen EJ, Rissanen J (1982) Recursive estimation of mixed autoregressive moving-average order. Biometrika 69(1): 81–94

Hartung J, Elpelt B, Klösener KH (2005) Statistik: Lehr- und Handbuch der angewandten Statistik, 14th edn. Oldenbourg Wissenschaftsverlag, Munich

Howind J (2005) Analyse des stochastischen Modells von GPS-Trägerphasenbeobachtungen. Deutsche Geodätische Kommission, Munich

Hurvich CM, Tsai CL (1989) Regression and time series model selection in small samples. Biometrika 76(2): 297–307

Klees R, Ditmar P, Broersen P (2003) How to handle colored observation noise in large least-squares problems. J Geodesy 76(11–12):629–640

Li J, Miyashita K, Kato T, Miyazaki S (2000) GPS time series modeling by autoregressive moving average method: application to the crustal deformation in central Japan. Earth Planets Space 52(3):155–162

Luo X, Mayer M, Heck B (2008) Improving the stochastic model of GNSS observations by means of SNR-based weighting. In: Sideris MG (ed) Observing our changing Earth. Proceedings of the 2007 IAG general assembly, 02–13 July 2007, Perugia, Italy, IAG Symposia, vol 133, pp 725–734

Niell AE (1996) Global mapping functions for the atmosphere delay at radio wavelengths. J Geophys Res 101(B2):3227–3246

Ragheb AE, Clarke PJ, Edwards SJ (2007) GPS sidereal filtering: coordinate- and carrier-phase-level strategies. J Geodesy 81(5):325–335

Said SE, Dickey DA (1984) Testing for unit roots in autoregressive-moving average models of unknown order. Biometrika 71(3):599–607

Schön S, Brunner FK (2008) A proposal for modelling physical correlations of GPS phase observations. J Geodesy 82(10):601–612

Teusch A (2006) Einführung in die Spektral- und Zeitreihenanalyse mit Beispielen aus der Geodäsie. Deutsche Geodätische Kommission, Munich

Tiberius C, Jonkman N, Kenselaar F (1999) The stochastics of GPS observables. GPS World 10(2):49–54

Tiberius C, Kenselaar, F (2003) Variance component estimation and precise GPS positioning: case study. J Surv Eng 129(1):11–18

Wang J, Satirapod C, Rizos C (2002) Stochastic assessment of GPS carrier phase measurements for precise static relative positioning. J Geodesy 76(2):95–104

Wheelon AD (2001) Electromagnetic scintillation: I. Geometrical optics. Cambridge University Press, Cambridge

Discontinuity Detection and Removal from Data Time Series

20

M. Roggero

Abstract

The aim of time series analysis is to distinguish between stochastic and deterministic signals, which are generated by different sources and mixed in the data time series. Before analyzing long term linear trend and periodic effects, it is necessary to detect and remove time series discontinuities, often undocumented. Discontinuities can occur in the case of hardware change, data model change or even signal source and environmental variations.

A data time series can be interpreted as a stochastic process plus a step function that represents the time series discontinuities or jumps. Modeling the process as a discrete-time linear system, it can be described by a finite state vector evolving with known dynamics, and by constant biases. The constant biases are described by a matrix of zeroes and ones, but generally the number and the position of jumps are unknown, and it cannot be defined univocally.

Since it is not possible to build a bias model a priori, the null hypothesis H_0 with no jump can be tested against a certain number of alternative hypotheses H_A, with a jump in a given epoch. An alternative hypothesis can be formulated for each observation epoch. The adequacy of the model can be verified using the ratio test, which is known to have the χ^2 distribution. After detecting the jumps, they can be estimated and removed. Simulated and real data examples will be given.

Keywords

Time series • Least mean squares • DIA

1 Introduction

The analysis of observations time series is applied in many problems of space geodesy. The same definition of a geodetic reference frame also includes the temporal coordinates, and its realization involves the use of observations time series. Moreover, time series analysis is an important aid in the analysis of deformations, in the case of landslides, of crustal deformations or geodynamic continental drift. Time series analysis can finally provide a direct estimation of the observation accuracies (repeatability).

GNSS coordinates time series are complex processes generated by the sum of different effects that can be modeled by a functional plus a stochastic

M. Roggero (✉)
Dipartimento di scienze e tecnologie per i processi di INSEdiamento, Viale Mattioli 39, 10125, Torino, Italy
e-mail: marco.roggero@polito.it

N. Sneeuw et al. (eds.), *VII Hotine-Marussi Symposium on Mathematical Geodesy*, International Association of Geodesy Symposia 137, DOI 10.1007/978-3-642-22078-4_20, © Springer-Verlag Berlin Heidelberg 2012

model. Generally the functional model is interpreted as the sum of a long term linear trend, a step function and cyclical components, as in (Perfetti 2006) and (Ostini et al. 2008). The stochastic model is a cyclo-stationary stochastic process, whose statistical properties vary periodically. Colored noise, non constant observation noise (heteroskedasticity) and data gaps must be taken into account. Discontinuities and data gaps must be localized, estimated and removed, before fitting one or more linear models to remove the trend and analyzing the time series in the frequency domain.

All behavior which is a result of abrupt changes in coordinates are identified as "discontinuities of degree zero", level shifts, or simply jumps. Jumps can be caused by antenna or hardware change, site effects or also earthquakes.

The changes in velocity described by multi linear models, and usually caused by earthquakes, are identified as "discontinuity of degree one".

A full time series analysis can be roughly divided into five steps:

1. Detection and removal of the level shifts (discontinuities of degree zero)
2. Detection of the velocity changes (discontinuities of degree one)
3. Fitting of one or more linear models to remove the trend
4. Analysis in the frequency domain and removal of the cyclical components
5. Noise analysis

The presented approach focuses on step 1 and is intended to detect, estimate and remove the level shifts, performing iteratively the so called detection, identification and adaptation procedure (DIA), presented in (Teunissen 1998), as applied in (Perfetti 2006). However, instead of assuming an a priori functional model, the station motion is represented as a discrete-time Markov process. The state vector can be designed in 3D, taking into account coordinate cross correlations, and it is estimated by least squares, constraining the system dynamic as in (Albertella et al. 2005) and (Roggero 2006). This approach also makes it possible to consider documented and undocumented jumps, to predict the station coordinates in data gaps (2.4), and to correctly represent the pre-seismic and the post-seismic deformations or other non-linear behaviors, as will be shown in 3.2.

2 Algorithms

2.1 Constrain Dynamics in Least Squares

Let us consider a discrete time linear system described by a state vector x and by a constant bias vector b; the system evolves with known dynamic through the epochs t ($\in t[1,n]$) with system noise ν (with variance-covariance matrix $R_{\nu\nu}$) and observation noise ε (with variance-covariance matrix $R_{\varepsilon\varepsilon}$):

$$\begin{cases} x_{t+1} = T_{t+1}x_t + B_{t+1}b_t + \nu_{t+1} \\ y_{t+1} = H_{t+1}x_{t+1} + C_{t+1}b_{t+1} + \varepsilon_{t+1} \\ b_{t+1} = b_t \end{cases} \tag{20.1}$$

where b is a bias vector linked to the system dynamic and to the observations by the matrices B and C. The transition matrix T describes the equations of the motion and H is the matrix of the coefficients that links the unknown parameters x to the observations y. The system (20.1) usually leads to a quite large normal matrix to be inverted, taking advantage of its sparse structure. The structure of the system (20.1) is described in detail in (Roggero 2006). Note that partitioning the normal matrix and using the Schur domain decomposition, the bias vector can be estimated independently by the whole state vector, as already shown in (Roggero 2006).

In coordinate time series analysis, both the state vector x and the observations vector y contain coordinates and have the same number of elements. The observed coordinates y are derived by GNSS observations, e.g. resulting from a network compensation, while the estimated coordinates x take into account a functional and a stochastic model of the system dynamics.

The special case of coordinate time series with jumps can be described by a constant velocity model in T, and by the unknown matrices B and C with $B = C$.

$$\begin{aligned} p_{t+1} &= p_t + v_t \cdot \delta t \\ v_{t+1} &= v_t \end{aligned} \qquad T = \begin{bmatrix} 1 & \delta t \\ & 1 \end{bmatrix} \tag{20.2}$$

where p and v are positions and velocities. The positions of the jumps are represented by the matrices B and C, whose elements are 0 or 1, as shown in

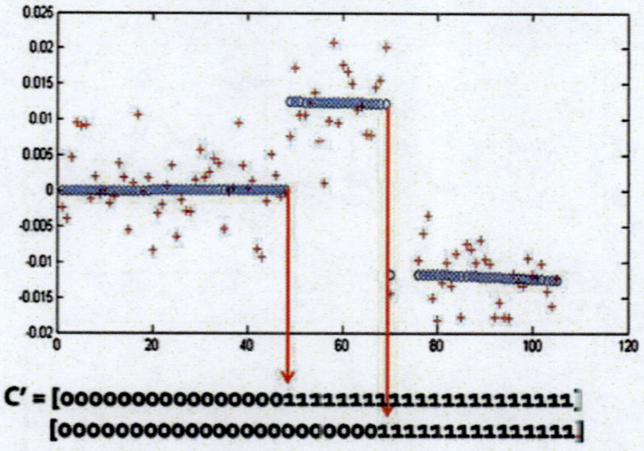

Fig. 20.1 The bias matrix C: + observed coordinates y, ○ estimated coordinates x. The observed coordinates are synthetic (observation noise $\sigma = 5\,\text{mm}$, power of the test $\alpha = 5\%$)

Fig. 20.1. The number of rows is equal to the number of observation epochs, while the number of columns is equal to the unknown number of jumps to be estimated. These matrices can be known a priori in the case of documented jumps, or determined by means of the algorithm described in 2.2 for undocumented jumps.

The matrix B has the function of removing the effect of the estimated biases on the estimated parameters x, and can therefore alternatively be assumed equal to zero (the level shifts are not removed), or equal to C (adaptation, the level shifts are removed).

2.2 Jump Detection and Estimation

Assuming that the observation vector y is normally distributed and has variance-covariance matrix Q_y, we can make the null hypothesis H_0, in the absence of discontinuities and a number of alternative hypotheses H_A in presence of discontinuities. Under these assumptions the deterministic $\Delta(y)$ and the stochastic $\Sigma(y)$ models take the form:

$$H_0: \begin{cases} \Delta(y) = Hx \\ \Sigma(y) = Q_y \end{cases}; \quad H_A: \begin{cases} \Delta(y) = Hx + Cb \\ \Sigma(y) = Q_y \end{cases} \tag{20.3}$$

The estimate of these two cases will also provide the respective variances of unit weight $\hat{\sigma}_0(0)$ and $\hat{\sigma}_0(A)$. The variance $\hat{\sigma}_0(0)$ can be used as an a priori value $\sigma_0(0)$ in the case A. We can make an alternative hypothesis for each observation epoch, if we wish to localize the discontinuities with full time resolution,

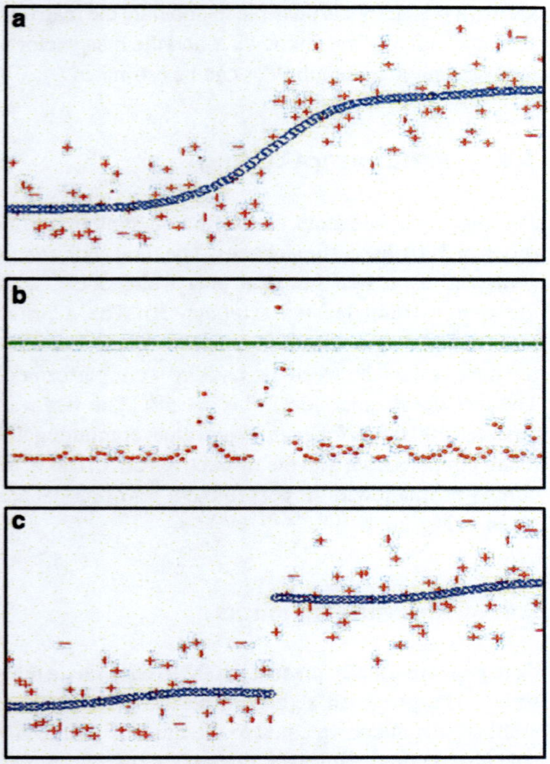

Fig. 20.2 Series of 100 synthetic height coordinates of a permanent GNSS station, with $\sigma_h = 0.005\,\text{m}$. A discontinuity of 0.015 m is located at day 50. Legend: + observed coordinates y, ○ estimated coordinates x. (**a**) Null hypothesis H_0, no discontinuities; estimated coordinates are smoothed. (**b**) Ratio test: the upper limit is exceeded in two contiguous cases; the discontinuity is located at the maximum between the two. (**c**) Alternative hypothesis H_A, estimation in the presence of the identified discontinuity

and verify the adequacy of the model by means of a test on the variable ratio, that has a χ^2 distribution:

$$\frac{\chi^2_{\left(\frac{\alpha}{2}\right)}}{n-r} < \frac{\hat{\sigma}_0^2(A)}{\sigma_0^2(0)} < \frac{\chi^2_{\left(1-\frac{\alpha}{2}\right)}}{n-r} \tag{20.4}$$

If the ratio exceeds one of the two thresholds (Fig. 20.2b), the initial hypothesis (Fig. 20.2a) must be discarded, and the alternative hypothesis of the presence of a jump in the considered observation epoch accepted (Fig. 20.2c). This test shall be carried out for each observation epoch, and requires the estimation of as many alternative models. Terminated this phase of testing, the final alternative model will include all the detected jumps and the documented discontinuities known a priori, so it is possible to populate the matrices B and C. Finally the unknown x and the bias vector b representing jump amplitudes can be estimated.

2.3 Performance Testing

The proposed algorithm was tested on synthetic time series of 100 observation epochs. The time series were generated by a step function plus white noise, with the step position located at epoch 50. The a priori observation σ was assumed equal to 4 mm, and tests for jump values between 1σ and 5σ were performed. The test significance level is $\alpha = 5\%$. The test was repeated 100 times for each jump value, computing the mean jump epoch t_j and the respective RMS. The % of correct estimatiowith respect to the a priori model are given in Fig. 20.3.

2.4 Algorithm Remarks

Data gaps are usually present in GNSS coordinate time series. The proposed algorithm basically estimates a number x of unknown equal to the number y of available observations. However, this causes discontinuities in the system dynamics in the presence of data gaps. The solution consists of predicting the system dynamic in the data gap, estimating a number of unknown x equal to the number of possible observations y:

$$y = t(n) - t(1) + 1 \tag{20.5}$$

where $t(1)$ and $t(n)$ are the epochs of the first and last observations. The matrix B must be modified to be coherent (Fig. 20.4).

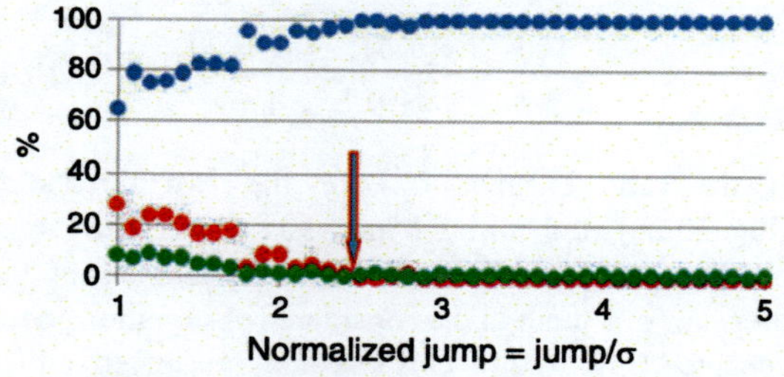

Fig. 20.3 Performance test: + correct estimations, • false jumps, o jump estimation errors. False jumps are detected for normalized jump values lower than 2.5

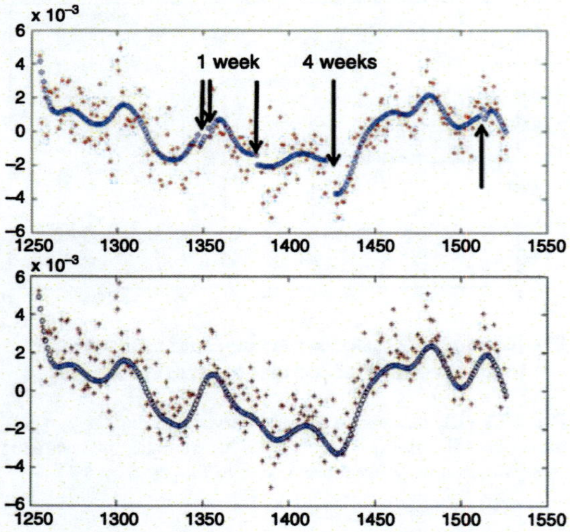

Fig. 20.4 Effect of estimating the system dynamics in the data gaps. + observed coordinates y, o estimated coordinates x

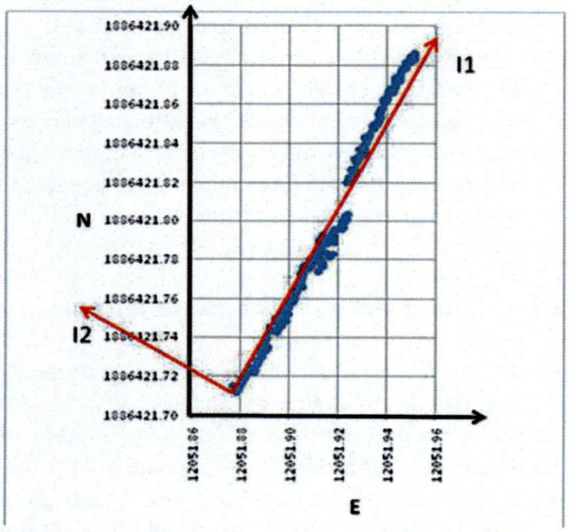

Fig. 20.5 Decorrelation of the horizontal components. The map coordinates (E, N) and the inertial ones (I_1, I_2) are represented

In GNSS coordinate time series can also be useful to uncorrelate horizontal components by principal component analysis. It is necessary to transform the horizontal coordinates in the principal system instead of in the cartographic system as shown in Fig. 20.5, and to estimate the jumps of the components I_1 and I_2 in the time domain.

Finally, the effect of uncertainties in the stochastic model must be underlined. For GNSS coordinate time series it is quite well known, but the noise stochastic properties varies with time and are different for different stations, depending on hardware, antenna and site effects. Autocorrelation violates the ordinary least squares assumption that the error terms are time uncorrelated. While it does not bias the least square parameter estimates, the standard errors tend to be underestimated, when the autocorrelations of the errors at low lags are positive.

3 Real Data Examples

3.1 Analysis of IGS Time Series

Different IGS permanent stations coordinate time series have been analyzed to test the algorithm. The case of Wroclaw station (WROC) can be taken as an example. 5 years of weekly solutions were used, in which there is a documented jump in week 1400, corresponding to the IGb00-IGS05 transition. The documented jump can be included in the null hypothesis. Moreover, two undocumented jumps were detected in the North component. The first one also was detected in the East component, the last one in height component as shown in Table 20.1.

Removing the jumps and the linear trend from the East and North component, the periodic effects became clearly visible (see Fig. 20.6).

3.2 L'Aquila Earthquake Time Series

The 2009 L'Aquila earthquake occurred in the region of Abruzzo, in central Italy, in 2009 (Fig. 20.7). The main shock occurred on 6 April, and was rated 5.8 on the Richter scale; its epicenter was near L'Aquila, the capital of the Abruzzo, which together with surrounding villages suffered most damage. There were several thousand foreshocks and aftershocks from December 2008 to July 2009, more than thirty of which had a Richter magnitude of over 3.5.

Table 20.1 Detected jumps and estimated level shifts

	Week	Jump (m)
E	1,389	−0.003
N	1,389	−0.016
N	1,400	0.006
N	1,426	0.015
h	1,426	0.014

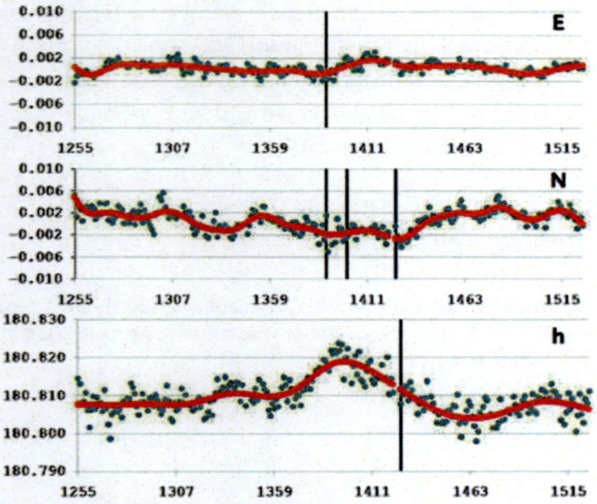

Fig. 20.6 WROC station cleaned time series. The vertical lines mark the position of detected and removed jumps. The linear trend has been removed from E and N components

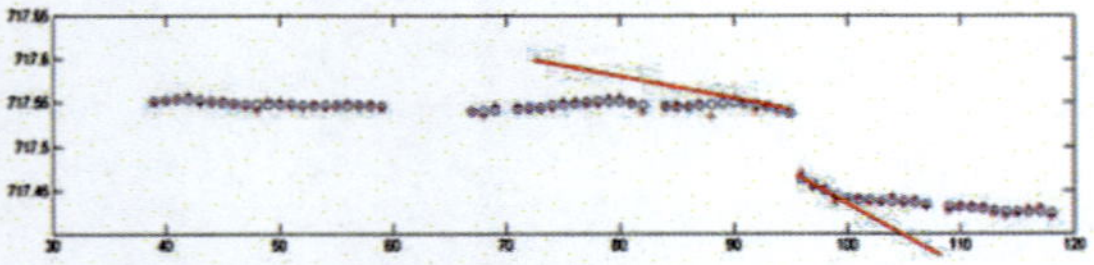

Fig. 20.7 Abruzzo earthquake: up component in the Paganica station (PAGA). Are evidenced the pre-seismic and the post-seismic velocities. The co-seismic deformation is about 7 cm down, while the post-seismic deformation is about 5 cm in 22 days. + observed coordinates y, ○ estimated coordinates x

Conclusion

The time series analysis allows the identification of undocumented anomalies, which if not removed may lead to incorrect interpretations and, therefore, would compromise the proper estimation of the estimated station velocities. Also in the case of documented discontinuities, the magnitude of the jump remains not always easily deductible and must be estimated.

The sampling of the time series is usually non uniform due to data gaps, so that it is not possible to perform the frequency analysis using the FFT. Operating on the estimated positions x, it is possible. However, in case of non uniform data sampling other algorithms can be suitable, such as FAMOUS – Frequency Analysis Mapping On Unusual Sampling (Mignard 2003).

The L'Aquila earthquake GNSS time series were estimated by Eng. Stefano Caldera, whom I thank for giving me the results.

References

Albertella A, Betti B, Sansò F, Tornatore V (2005) Real time and batch navigation solutions: Alternative approaches. Bull SIFET n. 4, pp. 82–99

Mignard F (2003) FAMOUS, Frequency analysis mapping on unusual sampling, (OCA Cassiopee), Software

Ostini L, Dach R, Meindl M, Schaer S, Hugentobler U (2008) FODITS: A new tool of the Bernese GPS software to analyze time series. EUREF 2008 Symposium, Brussels

Perfetti N (2006) Detection of station coordinates discontinuities within the Italian GPS Fiducial network. J Geodes ISSN 0949–7714 (Print) 1432–1394 (Online)

Roggero M (2006) Kinematic GPS batch processing, a source for large sparse problems. VI Hotine Marussi Symposium on Theoretical and Computational Geodesy, Wuhan

Teunissen P (1998) Quality control and GPS. In: Teunissen PJG, Kleusberg A (ed) GPS for geodesy. Springer, Berlin. ISBN 3-540-63661-7

G-MoDe Detection of Small and Rapid Movements by a Single GPS Carrier Phase Receiver

21

Sébastien Guillaume, Prof. Dr. Alain Geiger, and Fabio Forrer

Abstract

This paper presents a new method to detect small and rapid movements in real-time with a single L1-GPS receiver. The method is based on the prediction of single differences between satellites by appropriate Filter methods. The movements' detection is based, on the one hand, on displacements computed with multiple prediction lengths, and on the other hand, on the time series of displacements computed with the previous epochs. The accuracy of this method is demonstrated by real 10-[Hz] observations with two L1-receivers. One antenna remained completely static, while characteristic movements were applied on the other. This allow us to compare on the one hand the displacements of the displaced antenna estimated by G-MoDe with standard differential kinematic processing and on the other hand the resolution and the noise behavior of G-MoDe applied on the data of the static antenna. The results indicate that, in good conditions, horizontal movements and of short duration and oscillatory movements above 5 mm amplitude are significantly detected (95%).

Keywords

Displacement Detection • GNSS Signal Processing • Kalman Filter • L1 Carrier Phase Receiver • Real-Time algorithm

1 Introduction

The advent of GPS in the 1980s was a revolution in many domains, notably in geodesy. In this field, the majority of coordinates' determinations are based on carrier phase measurements. Because of systematic errors and ambiguities contained in phase measurement, it is necessary to observe the same satellites with two or more receivers simultaneously. After suitable treatments, this method makes it possible to determine the vector between antennas very accurately. Precision ranges from millimeter in static mode, to centimeter in the kinematic mode (Hugentobler et al. 2007; Leick 2004). An other technique that uses a single receiver (to determine absolute positions with a precision of a few millimeters in static mode and a few centimeters in the kinematic one) is known as Precise Point Positioning (PPP) (Shen et al. 2002).

The phenomena we want to investigate are dominated by rapid deformations on the order of a few millimeters. A data processing in static mode

S. Guillaume (✉) · Prof. Dr. A. Geiger · F. Forrer
Institute of Geodesy and Photogrammetry, CH-8093 Zurich, Schafmattstrasse 34, Switzerland
e-mail: alain.geiger@geod.baug.ethz.ch

N. Sneeuw et al. (eds.), *VII Hotine-Marussi Symposium on Mathematical Geodesy*, International Association of Geodesy Symposia 137, DOI 10.1007/978-3-642-22078-4_21,

is very accurate and powerful for long duration and continuous deformations but unsuccessful for small and quick displacements which can occur at any moment. In fact, it is necessary to measure the same points during a few hours to obtain highly accurate and reliable positions [mm] in static mode. A solution would be to work in kinematic mode to obtain positions at a high sampling rate. However, the gain in the sampling rate leads to the loss of precision. In addition, the noise of the obtained time series is in the range of the centimeter and remains highly coloured. Therefore, this method is not adequate for high accurate applications.

In this paper, an alternative method is presented (G-MoDe©, GNSS-Movement Detection). It is based on the filtering of the carrier phase observations of a single receiver. The main goal is to detect and measure the rapid movements of a few millimeters. The presented algorithm allows to analyse at these small displacements in real time and at a high sampling rate. It avoids complicated processing like ambiguity resolution and additional parameter estimation for error modelling: like precise satellites orbits, tropospheric and ionospheric models, phase center offset variation, multipath effects, clock errors etc. The basic principle is to assume that all effects which affect the carrier phase observations vary continuously; therefore, they are short time predictable. Movements of the antenna engender signals in phase measurements which can be detected and used to reconstruct the real displacements.

2 Basic Principle

Our goal is not to determine the coordinates of the receiver but their displacements. Basically, GPS is based on differences of range measurements between satellites and receiver antenna and by assuming that satellites and receiver positions are known, the theoretical distance to each satellite should be equal to the measured range provided all error sources are negligible. If the receiver is displaced, the difference between the theoretical and the measured range to every satellite will equal the orthogonal projection of the displacement's vector onto the receiver-satellite direction.

Regrettably, we cannot measure the geometrical range and calculate the position of satellites accurately enough to easily determine the displacements. Nevertheless, the basic principle that the ranges are modified by the orthogonal projection of the displacement will be used for the detection of movement.

Carrier phase measurements show a time-dependent variation which is mainly due to the satellites' movements along their orbits. It is obvious that numerous effects will also impact the measurements, e.g. change of the slant path delay. However, at the sub-second resolution these time varying effects may be considered to be slow. If on the contrary rapid displacements have to be detected, see Fig. 21.1, an appropriate filter model could help to discover the movement.

In fact, if a displacement occurs between two epochs in the direction of the satellite-receiver direction, the corresponding time series contains

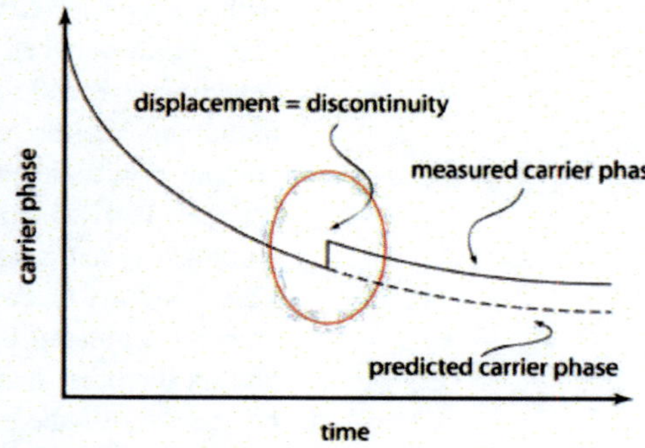

Fig. 21.1 Effect of a rapid displacement on a carrier phase time series. At the time when a displacement occurs, a discontinuity in the time series can be determined with appropriate carrier phase prediction techniques

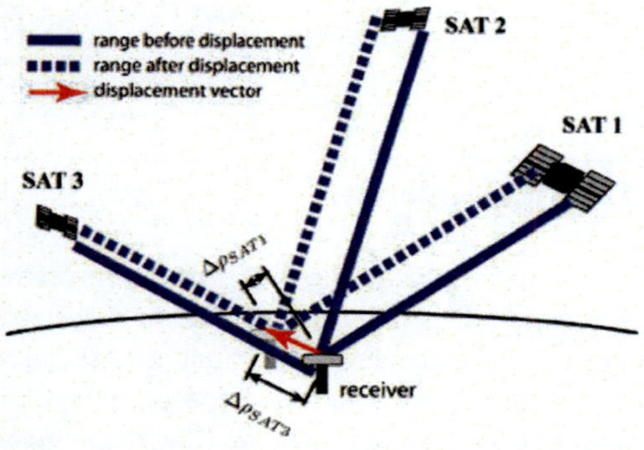

Fig. 21.2 A displacement of the antenna produces a range anomaly equal to the orthogonal projection of the displacement on the receiver-satellite direction. In this case, only the ranges on satellite 1 and 3 are influenced by the displacement. The range of the satellite 2 is invariant because its direction is perpendicular to the displacement

a discontinuity – or a jump – identical to the displacement that should be detectable with a certain accuracy, see Fig. 21.2. Then, if we do that for all visible satellites, the distinction between noise and real displacement can be improved and the 3D displacement vector determined.

3 From the Carrier Phase to the Displacement Vector

After this short introduction on the basic principle, this section explains how it is possible, with real data, to compute the displacement vector from the carrier phase observations measured with a single receiver. As explained in Chap. 2, the goal is to be able to track and to predict carrier phase measurements. That is only possible if it is enough regular and deterministic for the required prediction time span. However, analysing the zero-differences more carefully, it is seen that these conditions are not present at the accuracy level required. In fact, the receiver clock error has a dramatic effect on the regularity of the time series, it engenders jumps which are quasi unpredictable. For this reason, the tracking is done on carrier phase differences between two different satellites, thus, the receiver clock error is completely eliminated.

3.1 Single Difference Tracking

The simplified fundamental carrier phase zero-difference observation equation for L1 is:

$$\Phi^i(t) + \varepsilon^i(t) = \rho^i(t) - \lambda N^i + c[\delta T(t) - \delta t^i(t)] \underbrace{+ d^i_{trop}(t) + d^i_{iono}(t) + \dots}_{d^i_{div}(t)} \quad (21.1)$$

and the effect of a displacement $\mathbf{d}$ added:

$$\Phi^i(t) + \varepsilon^i(t) = -\mathbf{d}(t) \cdot \mathbf{e}_i(t) + \rho^i(t) - \lambda N^i + c[\delta T(t) - \delta t^i(t)] + d^i_{div}(t) \quad (21.2)$$

and forming the single difference between the satellites i and j:

$$\nabla\Phi^{ij}(t) + \nabla\varepsilon^{ij}(t) = -\mathbf{d}(t) \cdot [\mathbf{e}_i(t) - \mathbf{e}_j(t)] \underbrace{+ \nabla\rho^{ij}(t) - \lambda\nabla N^{ij} - c\nabla\delta t^{ij}(t) + \nabla d^{ij}_{div}(t)}_{=f_k(t)} \quad (21.3)$$

where:

i, j = satellite i respectively j

t = GPS time [s]

Φ = carrier phase received [m]

λ = wavelength of L1 [m]

ρ = range between the receiver and a satellite [m]

N = carrier ambiguity [cycles (integer)]

c = speed of light in vacuum [$\frac{m}{sec}$]

δT = receiver clock bias [s]

δt = satellite clock bias [s]

d_{trop} = tropospheric delay [m]

d_{iono} = ionospheric effect [m]

$\mathbf{d}$ = displacement vector [m]

$\mathbf{e}$ = receiver to satellite unit vector [-]

ε = carrier phase random noise [m]

$\nabla .^{ij}$ = $.^i - .^j$, single difference between satellites i and j

f_k = displacement-free single difference k carrier phase function [m].

Without displacement, the function $f_k(t)$ is equal to the observed single difference $\nabla\Phi^{ij}(t)$. If we are able to track and to predict this function for the next epochs, it might become possible to detect and to determine a possible movement of the receiver. The quality of determination of $f_k(t)$ is therefore, crucial for successful prediction of unbiased single differences.

Many kind of function approximation algorithms with specific capabilities are known, but some necessary conditions restrict the choice of them significantly. The algorithm must run real-time and should automatically adapt itself to the varying non-stationary properties of the tracked signal. In our application, we applied the Kalman Filter (Gelb 1988):

$$\begin{cases} \dot{\mathbf{x}}(t) = \mathbf{F}(t)\mathbf{x}(t) + \mathbf{w}(t) \; where : \mathbf{w}(t) \sim N(0, \mathbf{Q}(t)) \\ \mathbf{z}(t) = \mathbf{H}(t)\mathbf{x}(t) + \mathbf{v}(t) \; where : \mathbf{v}(t) \sim N(0, \mathbf{R}(t)) \end{cases} \tag{21.4}$$

The first equation describes the dynamic behavior of the state vector $\mathbf{x}(t) = \left(f_k, \dot{f}_k, \ddot{f}_k, \dddot{f}_k,\right)^T$ to be estimated. The second equation links the observations $\mathbf{z}(t) = \left(\nabla\Phi^{ij}, -\nabla\dot{\Phi}^{ij}\right)^T$ with $\mathbf{x}(t)$ where $\dot{\Phi}$ is the observed Doppler converted in $[\frac{m}{sec}]$. With the matrices F(t) and H(t):

$$\mathbf{F}(t) = \begin{pmatrix} 0\,1\,0\,0 \\ 0\,0\,1\,0 \\ 0\,0\,0\,1 \\ 0\,0\,0\,0 \end{pmatrix}; \qquad \mathbf{H}(t) = \begin{pmatrix} 1\,0\,0\,0 \\ 0\,1\,0\,0 \end{pmatrix}$$

The stochastic model for the observations represented by the matrix $\mathbf{R}(t)$ is a diagonal matrix with 2 [mm] and 30 $[\frac{mm}{sec}]$ standard deviation for the single difference and the Doppler measurements, respectively.

The system noise characterised by the covariance matrix $\mathbf{Q}(t)$ is estimated empirically for each single difference k in the initialization process of the filter. To obtain this covariance matrix, a collocational model is adjusted over an initialization time (~30 [s]), see Fig. 21.3, thus, estimating $f_k(t)$ in an optimal way and whitening the residual noise. Then, at each initialization epoch, the state vector is computed on the one hand by numerical derivation of the collocated function and on the other hand by the step by step resolution of the dynamic system (21.4). Finally, $\mathbf{Q}(t)$ can be estimated with these two states series.

If $\mathbf{D} = (\delta\mathbf{x}_1 - \mu_{\delta\mathbf{x}}, \delta\mathbf{x}_2 - \mu_{\delta\mathbf{x}}, \ldots \delta\mathbf{x}_n - \mu_{\delta\mathbf{x}})^T$ is the matrix which groups the n state vector residuals $\delta\mathbf{x}_i$ ($\mu_{\delta\mathbf{x}}$ is the mean of $\delta\mathbf{x}_i$) of each initialization epoch i, $\mathbf{Q}(t)$ can be empirically computed by:

$$\mathbf{Q} = \frac{1}{n-1} \cdot \mathbf{D}^T\mathbf{D} \tag{21.5}$$

3.2 Estimation of the Displacement

After the initialization and the system identification process, all available single differences are tracked and it is assumed that all $f_k(t)$ are known. From (21.3) we have:

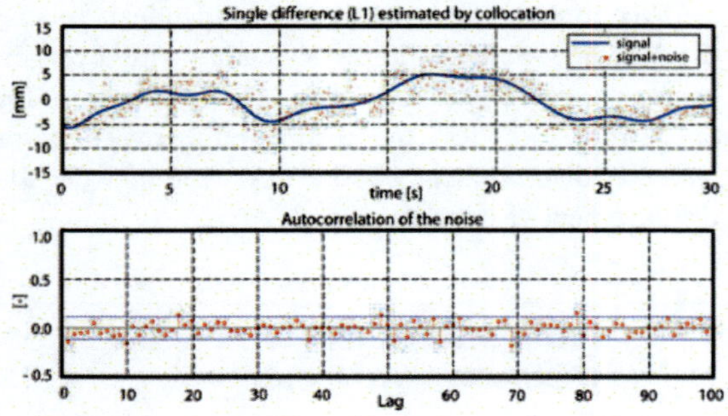

Fig. 21.3 Collocation of a single difference time series of 30 [s] (carrier phase observations acquired at 10 [Hz]). The covariance function is adapted to obtain a white residual noise. Only the signal and the noise components are shown (*upper*). The autocorrelation function proves that the residual noise is close to a white noise process (*lower*)

$$\nabla\Phi^{ij}(t) + \nabla\varepsilon^{ij}(t) = -\mathbf{d}(t) \cdot [\mathbf{e_i}(t) - \mathbf{e_j}(t)] + f_k(t) \tag{21.6}$$

and can be rewritten:

$$\underbrace{\nabla\Phi^{ij}(t) - f_k(t)}_{\mathbf{l}_k} + \underbrace{\nabla\varepsilon^{ij}(t)}_{\mathbf{v}_k} = \underbrace{-[\mathbf{e_i}(t) - \mathbf{e_j}(t)]^T}_{\mathbf{A}_k} \cdot \underbrace{\mathbf{d}(t)}_{\mathbf{x}} \tag{21.7}$$

With all available single differences, the unknown displacement can be estimated :

$$\hat{\mathbf{x}} = (\mathbf{A}^T\mathbf{Q}_{ll}^{-1}\mathbf{A})^{-1}\mathbf{A}^T\mathbf{Q}_{ll}^{-1}\mathbf{l} \tag{21.8}$$

The covariance matrix $\sigma_0^2\mathbf{Q}_{ll}$ is computed by taking into account the correlation due to forming the single differences. A displacement is significantly detected if both the global test (F-Test) of the estimation and the non-zero displacement hypothesis test (congruence test) at a chosen confidence level are successful (Welsch et al. 2000).

4 Test on Real Measurements

To illustrate the algorithm and assess the real accuracy of this method, we have applied the algorithm to observations carried out by two Leica 500 receivers (2 [m] baseline) at an data acquisition rate of 10 [Hz]

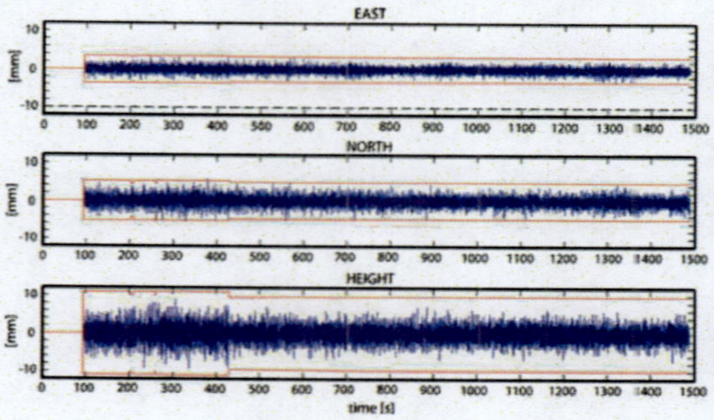

Fig. 21.4 East, North and Height components of the displacement estimated with G-MoDe algorithm in real-time on the static receiver. The a priori 3σ confidence intervals are drawn in red on the time series

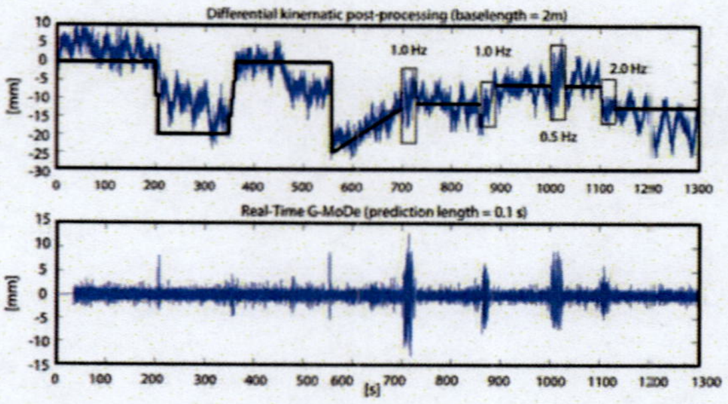

Fig. 21.5 Time series of the east component of the displacements computed with standard kinematic differential processing and G-MoDe algorithm

during 25 [min]. The first was static and different kind of displacement where applied on the second one. This allow us to compare on the one hand the displacements of the displaced antenna estimated by G-MoDe with standard differential kinematic processing and on the other hand the resolution and the noise behavior of G-MoDe applied on the data of the static antenna.

In the Fig. 21.4 it is shown the East, North and Height components of displacements calculated by the filtering and extraction algorithm of the static receiver which give an idea of the accuracy of the filtering. The displacements are computed with a prediction length of one time step. Only L1 carrier phase observations measured on nine satellites are used. The empirical standard deviations are 0.8, 1.3 and 2.1 [mm] for the east, north, and high components respectively.

In the upper Fig. 21.5 the time series of the east component of a standard kinematic processing is shown. The black line represents the movement which were applied. In the first part, rapid displacement in the order of 5–20 [mm] were produced. In the second part, oscillating movement with frequencies from 0.5 to 2.0 [Hz] with amplitudes of 5–20 [mm] were applied.

In the lower Fig. 21.5 the time series of the east component of G-MoDe processing is shown. All rapid and oscillating displacement are well detected.

Conclusion

A new method to detect small and rapid movements in real-time with a single L1-GPS carrier phase receiver has been presented. In good conditions, quick movements above 5 [mm] in the horizontal and 10 [mm] in height can be significantly (95%) detected instantaneously. There are multiple advantages of this technique, compared to the conventional differential processing:

First, it is based on one stand-alone single receiver. No reference station is needed. The processing can be carried out at the measurement site without the need of communication with other stations. In addition, the precision is not dependent on any baseline length.

Secondly, it is not necessary to use complicated models for satellite ephemeris, tropospheric correction etc. The presented method reduces the noise to an almost white spectrum. Finally, low-cost single-frequency receivers which are able to measure the carrier phase with an acquisition rate equal or higher to 10 [Hz] can be used.

References

Gelb, A (1988) Applied optimal estimation. The analytic sciences corporation, MIT Press, Massachusetts

Hugentobler H et al (2005) Bernese GPS Software 5.0. University of Bern, Switzerland

Leick A (2004) GPS satellite surveying. John Wiley, New York, ISBN 0-471-05930-7

Shen X, Gao Y (2002) Kinematic processing analysis of carrier phase based precise point positioning. University of Calgary, Canada

Welsch W et al (2000) Auswertung geodätischer Ueberwachungsmessungen. H. Wichmann Verlag, Heidelberg

Bayesian and Variational Methods for Discontinuity Detection: Theory Overview and Performance Comparison

22

Battista Benciolini, Mirko Reguzzoni, Giovanna Venuti, and Alfonso Vitti

Abstract

Discontinuity detection is of great relevance at different stages of the processing and analysis of geodetic time-series of data. This paper is essentially a review of two possible methods. The first method follows a stochastic approach and exploits the Bayesian theory to compute the posterior distributions of the discontinuity parameters. The epoch and the amplitude of the discontinuity are then selected as maximum a posteriori (MAP). The second method follows a variational approach based on the Mumford and Shah functional to segment the time-series and to detect the discontinuities. Whereas the original formulation was developed in a continuous form, discrete approaches are also available presenting some interesting connections with robust regressions. Both the methods have been applied to identify the occurrence of cycle-slips in GNSS phase measurements. Simulated and real data have been processed to compare the performance and to evaluate pros and cons of the two approaches. Results clearly show that both the methods can successfully identify cycle-slips.

Keywords

Time-series discontinuities • GNSS cycle-slips • Bayesian method • Variational methods

1 Introduction

The occurrence of discontinuities in geodetic data is very common, discontinuity detection is a well known problem and a key task in many applications. Sometimes, one wants to detect and then remove discontinuities in signals so that subsequent analyses or models can be correctly applied on a discontinuity-free signal. Sometimes, one is mainly interested in the discontinuities themselves and in the event epochs. Jumps in coordinates time-series are unwanted events when working on the realization of Reference Systems (Perfetti 2006; Altamimi et al.

B. Benciolini
DICA – Università di Trento,
via Mesiano, 77 – 38123 Trento, Italy

M. Reguzzoni
Department of Geophysics of the Lithosphere – OGS,
c/o Politecnico di Milano – Polo Regionale di Como,
via Valleggio, 11 – 22100 Como, Italy

G. Venuti
DIIAR - Politecnico di Milano – Polo Regionale di Como
via Valleggio, 11 – 22100 Como, Italy

A. Vitti (✉)
DICA – Università di Trento,
via Mesiano, 77 – 38123 Trento, Italy
e-mail: alfonso.vitti@ing.unitn.it

N. Sneeuw et al. (eds.), *VII Hotine-Marussi Symposium on Mathematical Geodesy*, International Association of Geodesy Symposia 137, DOI 10.1007/978-3-642-22078-4_22, © Springer-Verlag Berlin Heidelberg 2012

2007), likewise, cycle-slips have to be detected and possibly corrected when dealing with GNSS phase measurements (Lichtenegger and Hofmann-Wellenhof 1989; Teunissen and Kleusberg 1998; Gao and Li 1999). Again, discontinuities are of great interest in the analysis of velocity fields and coordinate time-series in geophysical applications (Albertella et al. 2007; Dermanis and Kotsakis 2007; Serpelloni et al. 2005), just to mention some examples.

This paper describes two techniques suitable for discontinuity detection. The first method exploits the Bayesian theory to detect jumps in a time-series of data modeled by a smooth regression (de Lacy et al. 2008). Epoch and amplitude of the jumps are described by means of probability distributions. The second method is based on the minimum problem proposed by Mumford and Shah (1989) to solve the signal segmentation problem. Segmentation can be considered here as the partitioning of the data into disjoint and homogeneous regions by smoothing the data and simultaneously locating the region boundaries without smoothing them out. The boundaries of the homogeneous regions are the points where the data present discontinuities.

The main theoretical elements of the two methods are just introduced herein for a matter of space limitation. The treatment of the mathematical subjects is not exhaustive and the description of the methods is introductory and short as well. Regarding the Mumford and Shah model, the primary aim is to show the complex and long way connecting the original formulation and its practical implementation by means of known results.

The two methods have been implemented numerically and used to detect cycle-slips in GNSS phase measurements. The Bayesian method was developed and applied in a previous work by de Lacy et al. (2008) to face this specific problem. Being the effectiveness of theory-based Bayesian models widely proved (see e.g., Koch 2007), the results of the Bayesian method are here used as a reference to assess the performance of the Mumford and Shah model. In particular, to evaluate and compare the different features of the two methods, tests on simulated and real data have been performed obtaining good and consistent results.

2 The Bayesian Method

When a time-series of data is expected to be smooth in time, a polynomial regression can be used to model the data. If a single discontinuity occurs at epoch τ, one can write the following observation equation:

$$\underline{y}_0 = A\underline{x} + k\underline{h}_\tau + \underline{\nu}, \tag{22.1}$$

where $\underline{y}_0$ is the vector of observations; the base functions of the regression are the elements of the design matrix A; $\underline{x}$ is the vector of the unknown regression parameters; τ is the discontinuity epoch; k is the discontinuity amplitude; $\underline{h}_\tau$ is the Heaviside function; $\underline{\nu}$ is the observation noise vector. The observations, the unknown parameters, the discontinuity epoch, the discontinuity amplitude and the variance of the observation noise are all considered as a priori stochastically independent random variables. For each parameter, a probability distribution (prior) is introduced and the Bayesian theorem is applied to compute the joint posterior distribution of the parameters as:

$$p(\underline{x}, \tau, k, \sigma_0^2 \mid \underline{y}_0) = \frac{p(\underline{y}_0 \mid \underline{x}, \tau, k, \sigma_0^2) \cdot p(\underline{x}, \tau, k, \sigma_0^2)}{p(\underline{y}_0)}. \tag{22.2}$$

In the Bayesian approach, the priors can be used to force some specific constraints on the parameters. The constraints are derived from the a priori knowledge about the parameters features, e.g., cycle-slips in some GNSS data combinations take only integer values, hence an integer condition can be enforced on the prior of the amplitude parameter. In the model presented by de Lacy et al. (2008) non-informative priors (Box and Tiao 1992) were used.

The discontinuity is detected by first computing the marginal posterior distribution of the discontinuity epoch $p(\tau \mid \underline{y}_0)$ and then selecting the epoch $\overline{\tau}$ with the maximum a posteriori probability (MAP). The estimate of the discontinuity amplitude is obtained from the conditional posterior distribution $p(k \mid \overline{\tau}, \underline{y}_0)$ exploiting again the MAP principle.

In the model (22.1) only one discontinuity has been considered. To handle correctly the general case with more than one discontinuity, the conditional posterior distribution of the observation noise variance $p(\sigma_0^2 \mid \overline{\tau}, \underline{y}_0)$ is used in a test on the model accuracy to divide the original data set into intervals containing at most one discontinuity.

3 The Mumford and Shah Method

Mumford and Shah (1989) proposed one of the most known models in image segmentation. The model seeks a smooth approximation u of the data g, at the same time the model detects and preserves from the smoothing the discontinuities of g. In two dimensions the Mumford and Shah functional is:

$$MS(u, K) = \int_{\Omega\setminus K} \left[(u-g)^2 + \lambda\,|\nabla u|^2\right] \mathrm{dx} + \alpha\mathscr{H}^1(K), \tag{22.3}$$

where Ω is a bounded set in $\mathbb{R}^2$; $K \subset \overline{\Omega}$ is a compact set representing the contours reconstructed from the discontinuities of g and $\overline{\Omega}$ is the closure of Ω. The function $u \in W^{1,2}(\Omega\setminus K)$ is a smooth approximation of the data g outside K and $W^{1,2}$ is a Sobolev space; $\mathscr{H}^1$ is the 1-dimensional Hausdorff measure; α and λ are positive constants. In three dimensions analogous models are widely adopted in Mathematical Physics, e.g. fracture mechanics, plasticity, static theory of liquid crystals. In one dimension, the counting measure $\#(K)$ replaces the $\mathscr{H}^1(K)$ measure, and u' replaces ∇u.

The first term in (22.3) measures the distance between the solution u and the data g, the second term measures the variations of u, and the third term handles the discontinuities of g. The problem is to find a pair (u, K) that minimizes the functional (22.3) so that the solution u is forced to be close to the data g, strong variations of u are penalized, and the measure of the discontinuity set K is kept as small as possible in order to avoid over-segmented solutions.

The third term in (22.3) misses some good mathematical properties that would ensure the existence of the minimum. To overcome this difficulty, De Giorgi proposed a weak formulation of (22.3) in the space of special functions of bounded variation $SBV(\Omega)$ (see Ambrosio et al. 2000). The relaxed functional in $SBV(\Omega)$ proposed by De Giorgi is:

$$MS_w(u) = \int_{\Omega} \left[(u-g)^2 + \lambda\,|\nabla u|^2\right] \mathrm{dx} + \alpha\mathscr{H}^1(S_u). \tag{22.4}$$

This functional depends only on the function u: the set K has been replaced by the set S_u of the discontinuities of u. In $SBV(\Omega)$, the measure of S_u is negligible (roughly speaking) and hence the integral terms in (22.4) are now defined over the entire domain Ω. De Giorgi et al. (1989) proved the existence of minimizers of (22.4) using the direct methods of the Calculus of Variations and then showed that minimizers of (22.4) provide also minimizers for the Mumford and Shah functional (22.3).

The Mumford and Shah problem belongs to a class of variational problems known as *free discontinuity problems*. Energy terms of different dimension compete in this kind of minimum problem, as in (22.4). The lower dimension energy, e.g., the third term in (22.4), is concentrated on a set that can be recovered by the *discontinuity* set of a suitable function, as done in (22.4). This unknown discontinuity set has an important role and it is not fixed a priori, i.e., it is *free*.

Finding a suitable numerical method to compute a minimizing pair of (22.4) is not trivial because the numerical treatment of $\mathscr{H}^1(S_u)$ is very difficult. The first and most used result is due to Ambrosio and Tortorelli (1992), they proved the variational approximation of the functional (22.4) with a new sequence of functionals which are numerically more tractable. The Ambrosio and Tortorelli functionals are:

$$AT_\varepsilon(u, z) = \int_{\Omega} \left[(u-g)^2 + \lambda z^2\,|\nabla u|^2\right] \mathrm{dx} + \alpha \int_{\Omega} \left[\varepsilon\,|\nabla z|^2 + \frac{(1-z)^2}{\varepsilon}\right] \mathrm{dx}, \tag{22.5}$$

where $z \in W^{1,2}$, $0 \le z \le 1$, and $u \in W^{1,2}(\Omega)$. The variational approximation was proved exploiting the theory of the Γ-Convergence (De Giorgi and Franzoni 1975) and an important result by Modica and Mortola (1977) who proved that $\mathscr{H}^1(S_u)$ can be approximated by the third integral in (22.5). In (22.5) ε is the Γ-Convergence parameter, and the approximation holds when $\varepsilon \to 0$. The functionals (22.5) present only integral terms defined on the entire domain Ω. The auxiliary function z, that mimics the characteristic function of u, is introduced to control strong variations of u. From the practical point of view, a standard finite element method can be used to approximate numerically the functionals (22.5). A finite difference discretization of the Euler equation associated to (22.5) is also possible. Different types of variational approximations exist (see Braides 1998). We mention only a discrete approximation proposed by Chambolle (1995), that is:

$$C(u^h) = h^2 \sum_{i,j} \left(u^h_{i,j} - g^h_{i,j}\right) + \lambda \left[h^2 \sum_{i,j} W_h \left(\frac{u^h_{i+1,j} - u^h_{i,j}}{h} \right) + h^2 \sum_{i,j} W_h \left(\frac{u^h_{i,j+1} - u^h_{i,j}}{h} \right) \right], \quad (22.6)$$

where h is the mesh size and $W_h(t) = \min(t^2, \alpha/h)$, this function is very similar to the "weight" function used in robust regressions.

Blake and Zisserman (1987) proposed an extension of the functional (22.3) where the smoothness of u is controlled by a second order term, e.g., the second derivative of u in 1-D, and where also the set $S_{u'}$ of the discontinuity points of the first derivative of u is explicitly handled. The weak formulation of the Blake and Zisserman functional in one dimension is:

$$BZ_w(u) = \int_\Omega \left[(u-g)^2 + \lambda z^2 (u'')^2\right] \mathrm{d}x + \alpha \#(S_u) + \beta \#(S_{u'}). \quad (22.7)$$

Some variational approximation is still needed to implement the functional (22.7) numerically.

4 Cycle-Slips Detection on Linear Combinations of Undifferenced GPS Observations

The geometry-free linear combination of phase observations from a single satellite to a single receiver depends only on the ionospheric effect, on the initial integer ambiguities and on the electronic biases. This combination presents a noise ranging from 3 to 4 mm. The wavelengths of the two GPS carriers L1 and L2 are two orders of magnitude larger than the noise of their combination $L1-L2$. This means that even cycle-slips with amplitude of one cycle, occurring in one of the two carriers, can be easily identified. This is not the case for some couples of simultaneous cycle-slips on the two carriers with the amplitude of the combination smaller than the $L1 - L2$ noise level. Anyway, these cycle-slips that cannot be detected in the $L1 - L2$ combination have an amplitude that is always 2 or 3 times larger than the $P1 - L1$ noise, ranging from 20 to 30 cm (see de Lacy et al. 2008).

The performance of the variational method has been assessed on the basis of the results by de Lacy et al. (2008) where the Bayesian method was compared with the BERNESE 5.0 scientific software. The comparison has been performed on a simulated time-series of data with known cycle-slips and on real data with simulated cycle-slips. The data are those used by de Lacy et al. (2008). The simulated data set is composed of 500 observations with a noise variance of $\sigma_0^2 = 1$. A first jump of amplitude $k = 3\sigma_0$ is present at epoch $\tau = 200$, a second jump of amplitude $k = 5\sigma_0$ is present at epoch $\tau = 300$ (see de Lacy et al. 2008).

Figures 22.1 and 22.2 show the detection of the first jump by the Bayesian method: the interpolation and the posteriors of the parameters τ and k are plotted. The conditional posterior distribution of the discontinuity amplitude k has been obtained by enforcing an integer condition on the prior of k. The variational segmentation u of the data and the auxiliary function z that "sees" the discontinuities of u are shown in Fig. 22.3. An estimate of the discontinuity amplitudes can be computed as the difference between the values the approximating function u presents before and after the discontinuity epoch ($\bar{k}_{200} = 2.7$, $\bar{k}_{300} = 5.8$). The variational method has been also applied to a subset of

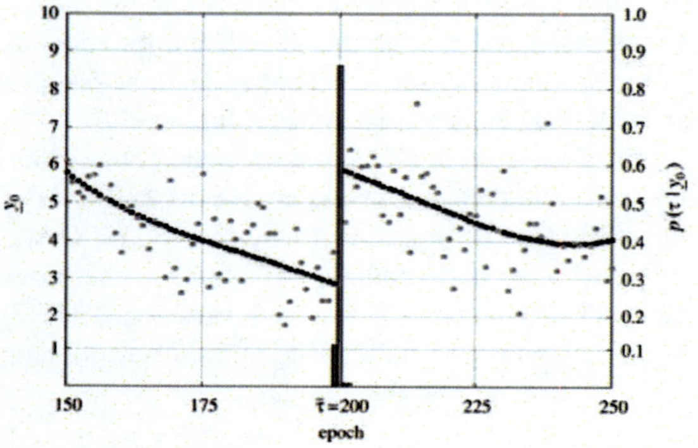

Fig. 22.1 Bayesian method: the most likely interpolation (*black dots*) of the data (*gray dots*) and the marginal posterior of the discontinuity epoch τ given the data (*gray bars*)

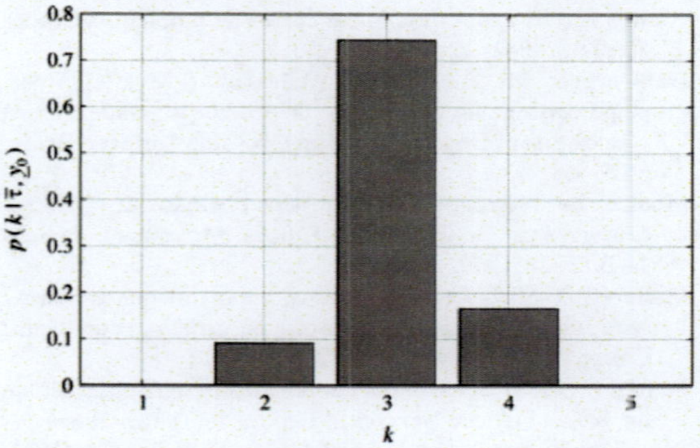

Fig. 22.2 Bayesian method: the conditional posterior of the amplitude k given the data and the estimate of the discontinuity epoch

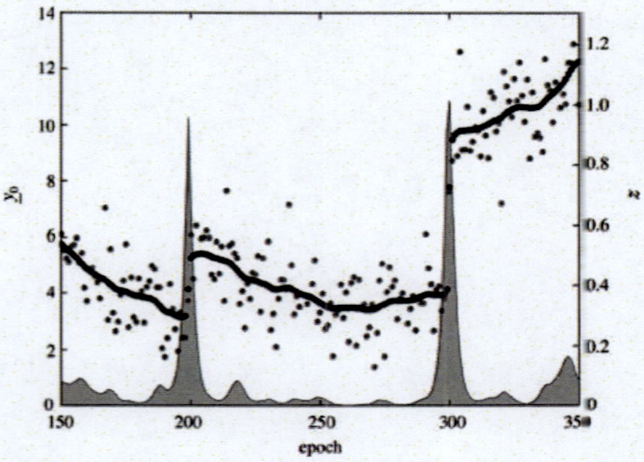

Fig. 22.3 Variational method: the segmentation u (*black dots*) of the data (*gray dots*) and the auxiliary function z (*gray filled*)

the real 30 s GPS data with simulated jumps studied in de Lacy et al. (2008), all the jumps have been correctly detected.

Conclusion

The use of the Mumford and Shah variational model has produced good results, consistent with the reference results of the Bayesian method developed and applied by de Lacy et al. (2008) to study the general problem of detecting discontinuities on a smooth signal. Here the Bayesian and the variational models have been successfully applied to detect cycle-slips, but they could be also used to detect discontinuities in other geodetic applications or in Geophysics.

With the Bayesian approach, significant constraints can be enforced in the priors of the parameters in order to make the discontinuity detection more effective and to achieve results consistent with the a priori knowledge of the features of the unknown parameters. On the other hand, the test on the conditional posterior distribution of the noise variance depends on the knowledge of an a priori value of the noise level which may not always be available.

The Mumford and Shah model has an intrinsic "multi-scale" nature, the ratio between the parameters α and λ is somehow related to the size of the smaller discontinuity that the model can detect and the noise variance of the signal, for details see (Blake and Zisserman 1987). Despite this underlying connection, the choice of proper ratios is far from being an easy task, even when an a priori estimate of the noise level is known.

To exploit the best of the two methods, a sequential application is under investigation. The variational method could be used in a first step with "high sensitive" parameters to detect possible jumps, i.e., real jumps and noise compatible variations. On a second step, the Bayesian model could be used to detect and hence select just the real jumps, and to estimate the corresponding amplitudes.

References

Albertella A, Cazzaniga N, Crespi M, Luzietti L, Sacerdote F, Sansò F (2007) Deformations detection by a Bayesian approach: prior information representation and testing criteria definition. In: International Association of Geodesy Symposia, vol 131, Springer, Berlin, pp 30–37.

Altamimi Z, Collilieux X, Legrand J, Garayt B, Boucher C (2007) ITRF2005: A new release of the International Terrestrial Reference Frame based on time series of stations positions and Earth orientation parameters. J Geophys Res 112, B09401

Ambrosio L, Fusco N, Pallara D (2000) Functions of bounded variations and free discontinuity problems. Oxford University Press, Oxford

Ambrosio L, Tortorelli VM (1992) On the approximation of free discontinuity problems. Bollettino dell'Unione Matematica Italiana 6-B (7):105–123

Blake A, Zisserman A (1987) Visual Reconstruction. The MIT Press, Cambridge

Box GEP, Tiao GC (1992) Bayesian inference in statistical analysis. Wiley, New York

Braides A (1998) Approximation of Free-Discontinuity Problems. Lecture Notes in Mathematics 1694. Springer, Berlin

Chambolle A (1995) Image segmentation by variational methods: Mumford and Shah functional and the discrete approximations. SIAM J Appl Math 55(3):827–863

Dermanis A, Kotsakis C (2007) Estimating crustal deformation parameters from geodetic data: review of existing methodologies, open problems and new challenges. In: International Association of Geodesy Symposia, vol 131, Springer, Berlin, pp 7–18

Gao Y, Li Z (1999) Cycle slip detection and ambiguity resolution algorithm for dual-frequency GPS data processing. Mar Geodes 22(3):169–181

De Giorgi E, Carriero M, Leaci A (1989) Existence theorem for a minimum problem with free discontinuity set. Arch Rational Mech Anal 108:195–218

De Giorgi E, Franzoni T (1975) Su un tipo di convergenza variazionale. Atti Accademia Nazionale dei Lincei - Rend Cl Sci 58:842–850

Koch KR (2007) Introduction to Bayesian Statistics, 2nd edn. Springer, Berlin

de Lacy MC, Reguzzoni M, Sansò F, Venuti G (2008) The Bayesian detection of discontinuities in a polynomial regression and its application to the cycle-slip problem. J Geodes 82(9):527–542

Lichtenegger H, B Hofmann-Wellenhof (1989) GPS-data preprocessing for cycle-slip detection. In: International Association of Geodesy Symposia, vol 102, Springer, Berlin, pp 57–68

Modica L, Mortola S (1977) Un esempio di Gamma-convergenza. Bollettino dell'Unione Matematica Italiana 14-B (5):285–299

Mumford D, Shah J (1989) Optimal approximation by piecewise smooth functions and associated variational problems. Comm Pure Appl Math 42:577–685

Perfetti N (2006) Detection of coordinate discontinuities within the Italian Fiducial Network. J Geodes 80(7):381–396

Serpelloni E, Anzidei M, Baldi P, Casula G, Galvani A (2005) Crustal velocity and strain-rate fields in Italy and surrounding regions: new results from the analysis of permanent and non-permanent GPS networks. Geophys J Int 161(3): 861–880

Teunissen PJ, Kleusberg A (1998) GPS for Geodesy, 2nd edn. Springer, Berlin

Prediction Analysis of UT1-UTC Time Series by Combination of the Least-Squares and Multivariate Autoregressive Method

23

Tomasz Niedzielski and Wiesław Kosek

Abstract

The objective of this paper is to extensively discuss the theory behind the multivariate autoregressive prediction technique used elsewhere for forecasting Universal Time (UT1-UTC) and to characterise its performance depending on input geodetic and geophysical data. This method uses the bivariate time series comprising length-of-day and the axial component of atmospheric angular momentum data and needs to be combined with a least-squares extrapolation of a polynomial-harmonic model. Two daily length-of-day time series, i.e. EOPC04 and EOPC04_05 spanning the time interval from 04.01.1962 to 02.05.2007, are utilised. These time series are corrected for tidal effects following the IERS Conventions model. The data on the axial component of atmospheric angular momentum are processed to gain the 1-day sampling interval and cover the time span listed above. The superior performance of the multivariate autoregressive prediction in comparison to autoregressive forecasting is noticed, in particular during El Niño and La Niña events. However, the accuracy of the multivariate predictions depends on a particular solution of input length-of-day time series. Indeed, for EOPC04-based analysis the multivariate autoregressive predictions are more accurate than for EOPC04_05-based one. This finding can be interpreted as the meaningful influence of smoothing on forecasting performance.

Keywords

Atmospheric angular momentum • El Niño/Southern Ociallation • Length of day • Multivariate autoregressive model • Prediction

T. Niedzielski (✉)
Space Research Centre, Polish Academy of Sciences, ul. Bartycka 18A, 00-716 Warsaw, Poland

Institute of Geography and Regional Development, University of Wrocław, pl. Uniwersytecki 1, 50-137 Wrocław, Poland

Oceanlab, University of Aberdeen, Main Street, Newburgh, Aberdeenshire, AB41 6AA, UK
e-mail: niedzielski@cbk.waw.pl

W. Kosek
Space Research Centre, Polish Academy of Sciences, ul. Bartycka 18A, 00-716 Warsaw, Poland

Department of Land Surveying, University of Agriculture in Kraków, ul. Balicka 253c, 30-198 Kraków, Poland

N. Sneeuw et al. (eds.), *VII Hotine-Marussi Symposium on Mathematical Geodesy*, International Association of Geodesy Symposia 137, DOI 10.1007/978-3-642-22078-4_23,

1 Introduction

Universal Time (UT1-UTC) quantitatively describes the Earth's rotation rate and is one of five Earth Orientation Parameters (EOPs). EOPs are intrinsic to practically perform the time-varying transformation between the International Celestial Reference Frame (ICRF) and the International Terrestrial Reference Frame (ITRF). The knowledge about this transformation and its accuracy is crucial for tracking and navigation of objects in space.

The importance of forecasting UT1-UTC as well as its derivative length-of-day (LOD or Δ) is stressed by many authors (e.g. Schuh et al. 2002; Johnson et al. 2005; Niedzielski and Kosek 2008). The short-term predictions of EOPs are essential to enhance the time-keeping, communication, and navigation in space. The long-term UT1-UTC forecasts, however, can be used for monitoring and prediction of El Niño/Southern Oscillation (ENSO). Indeed, UT1-UTC and LOD data comprise El Niño and La Niña signals (e.g. Rosen et al. 1984; Gross et al. 1996; Abarca del Rio et al. 2000) and hence their prediction can serve well a purpose of ENSO diagnosis (Niedzielski and Kosek 2008; Zhao and Han 2008).

There are several methods suitable for forecasting UT1-UTC and LOD time series. They are usually based on empirical data processing. Probably one of the simplest methods to forecast UT1-UTC and LOD is extrapolation of a polynomial-harmonic least-squares model. The UT1-UTC and LOD time series can also be forecasted using the autoregressive technique (Kosek 1992). To determine the predictions of Earth's rotation rate, Kosek et al. (1998) utilised the autocovariance technique and Gross et al. (1998) applied the Kalman filter. Schuh et al. (2002), Kalarus and Kosek (2004) as well as Kosek et al. (2005) showed that artificial neural networks can be successfully applied to forecast LOD time series. Akyilmaz and Kutterer (2004) found that the fuzzy inference system serves well a purpose of LOD prediction. Following the concept by Freedman et al. (1994), Johnson et al. (2005) utilized the axial component (χ_3) of atmospheric angular momentum (AAM) to support very accurate forecasting technique for UT1-UTC. More recently, Niedzielski and Kosek (2008) applied a multivariate autoregressive model comprising LOD and AAM χ_3 time series and gained the improvement of UT1-UTC predictions during ENSO events.

This paper presents a detailed theory behind the multivariate autoregressive technique and extends the previous study by Niedzielski and Kosek (2008). The impact of different input LOD data on the accuracy of multivariate autoregressive predictions is discussed. Two LOD solutions, EOPC04 and EOPC04_05 spanning the time interval from 04.01.1962 to 02.05.2007, are selected. The differences between the results are interpreted in terms of data smoothing.

2 Multivariate Autoregressive Model

Following Niedzielski and Kosek (2008), three prediction methods are applied: extrapolation of the least-squares polynomial-harmonic model (LS), combination of LS with autoregressive forecasting (AR) denoted by LS+AR, and combination of LS with multivariate autoregressive forecasting (MAR) referred to as LS+MAR. For the sake of brevity, LS and LS+AR methods are not presented here but an extensive presentation of MAR method is given instead. The theory on MAR technique is presented following the paper by Neumaier and Schneider (2001). The application of LS and AR estimations for modelling UT1-UTC and LOD is presented by Niedzielski and Kosek (2008).

2.1 Modelling

The m-variate time series $\mathbf{X}(m,n)$ comprising n vectors is given by (before subtracting LS model):

$$\mathbf{X}(m,n) = \left\{ \begin{bmatrix} X_1^{(1)} \\ X_1^{(2)} \\ \vdots \\ X_1^{(m)} \end{bmatrix}, \begin{bmatrix} X_2^{(1)} \\ X_2^{(2)} \\ \vdots \\ X_2^{(m)} \end{bmatrix}, \ldots, \begin{bmatrix} X_n^{(1)} \\ X_n^{(2)} \\ \vdots \\ X_n^{(m)} \end{bmatrix} \right\}. \tag{23.1}$$

A residual time series $\mathbf{Y}(m,n)$ can be obtained as a difference between the data and the LS model (for each component) and may be denoted by:

$$\mathbf{Y}(m,n) = \left\{ \begin{bmatrix} Y_1^{(1)} \\ Y_1^{(2)} \\ \vdots \\ Y_1^{(m)} \end{bmatrix}, \begin{bmatrix} Y_2^{(1)} \\ Y_2^{(2)} \\ \vdots \\ Y_2^{(m)} \end{bmatrix}, \ldots, \begin{bmatrix} Y_n^{(1)} \\ Y_n^{(2)} \\ \vdots \\ Y_n^{(m)} \end{bmatrix} \right\}. \tag{23.2}$$

A multivariate autoregressive process of order p (MAR(p)), $\mathbf{Y}_t$, is stationary and the following condition holds:

$$\mathbf{Y}_t = \mathbf{w} + \sum_{i=1}^{p} \mathbf{A}_i \mathbf{Y}_{t-i} + \mathbf{E}_t, \tag{23.3}$$

where $\mathbf{Y}_t$ is a random vector of residuals indexed by discrete time t, $\mathbf{A}_i$, $i = 1, \ldots, p$ are autoregressive coefficient matrices, $\mathbf{w}$ is an intercept term, $\mathbf{E}_t$ is a white noise vector with mean zero and a covariance matrix $\mathbf{C}$. Selection of an order p can be done by the Schwarz Bayesian Criterion (SBC) defined as (Schwarz 1978; Neumaier and Schneider 2001):

$$SBC(p) = \frac{l_p}{m} - \left(1 - \frac{n_p}{N}\right) \log N, \tag{23.4}$$

where $l_p = \log\det[(N - n_p)\hat{\mathbf{C}}]$, $n_p = mp + 1$, N is a number of state vectors, m is a dimension of the model, and $\hat{\mathbf{C}}$ is a covariance matrix estimate of the noise vector. Optimal p is an integer, for which SBC attains a minimum value. Estimation of $\mathbf{A}_1, \ldots, \mathbf{A}_p$, $\mathbf{w}$, and $\mathbf{C}$ is performed for a fixed p using a least-squares procedure. A MAR(p) model can also be denoted as:

$$\mathbf{Y}_t = \mathbf{B} \begin{bmatrix} 1 \\ \mathbf{Y}_{t-1} \\ \vdots \\ \mathbf{Y}_{t-p} \end{bmatrix} + \mathbf{E}_t, \tag{23.5}$$

where $\mathbf{B} = [\mathbf{w}\ \mathbf{A}_1\ \ldots\ \mathbf{A}_p]$ is matrix of parameters to be estimated. An estimate of $\mathbf{B}$ can be obtained by:

$$\hat{\mathbf{B}} = \mathbf{W}\mathbf{U}^{-1}, \tag{23.6}$$

where

$$\mathbf{W} = \sum_{i=1}^{N} \left\{ \mathbf{Y}_i \begin{bmatrix} 1 \\ \mathbf{Y}_{i-1} \\ \vdots \\ \mathbf{Y}_{i-p} \end{bmatrix}^T \right\} \tag{23.7}$$

and

$$\mathbf{U} = \sum_{i=1}^{N} \left\{ \begin{bmatrix} 1 \\ \mathbf{Y}_{i-1} \\ \vdots \\ \mathbf{Y}_{i-p} \end{bmatrix} \begin{bmatrix} 1 \\ \mathbf{Y}_{i-1} \\ \vdots \\ \mathbf{Y}_{i-p} \end{bmatrix}^T \right\}. \tag{23.8}$$

In order to estimate $\hat{\mathbf{B}}$ one may apply the Cholesky factorization as:

$$\hat{\mathbf{B}} = (\mathbf{R}_{11}^{-1}\mathbf{R}_{12})^T, \tag{23.9}$$

where $\mathbf{R}_{11}$ and $\mathbf{R}_{12}$ can be obtained from:

$$\begin{bmatrix} \mathbf{U} & \mathbf{W}^T \\ \mathbf{W} & \mathbf{V} \end{bmatrix} = \begin{bmatrix} \mathbf{R}_{11}^T\mathbf{R}_{11} & \mathbf{R}_{11}^T\mathbf{R}_{12} \\ \mathbf{R}_{12}^T\mathbf{R}_{11} & \mathbf{R}_{12}^T\mathbf{R}_{12} + \mathbf{R}_{22}^T\mathbf{R}_{22} \end{bmatrix}, \tag{23.10}$$

where $\mathbf{V} = \sum_{i=1}^{N} \mathbf{Y}_i\mathbf{Y}_i^T$. An estimate of $\mathbf{C}$ can be determined using the following expression:

$$\hat{\mathbf{C}} = \frac{1}{N - n_p} \mathbf{R}_{22}^T\mathbf{R}_{22}. \tag{23.11}$$

2.2 Prediction

A 1-step prediction of a MAR(p) model is given by:

$$P_1\mathbf{Y}_{s+1} = \hat{\mathbf{w}} + \sum_{i=1}^{p} \hat{\mathbf{A}}_i \mathbf{Y}_{s+1-i}, \tag{23.12}$$

where $P_1\mathbf{Y}_{s+1}$ is the prediction vector for time $s+1$ determined at time s; $\hat{\mathbf{A}}_i$, $i = 1, \ldots, p$, are already estimated autoregressive matrices. The 1-step prediction vector is thus defined as:

$$P_1\mathbf{Y}_{s+1} = \begin{bmatrix} P_1 Y_{s+1}^{(1)} \\ \vdots \\ P_1 Y_{s+1}^{(m)} \end{bmatrix}, \tag{23.13}$$

where $P_1 Y_{s+1}^{(j)}$ is the 1-step prediction for j-th component of a multivariate time series $\mathbf{Y}(m,s)$. A k-step prediction, $P_k\mathbf{Y}_{s+k}$, is determined in a stepwise way. First, $P_1\mathbf{Y}_{s+1}$ is computed. Second,

$P_1\mathbf{Y}_{s+1}$ is attached to the time series $\mathbf{Y}(m,s)$ in the following way:

$$\widetilde{\mathbf{Y}}(m,s+1) = \left\{ \begin{bmatrix} Y_1^{(1)} \\ Y_1^{(2)} \\ \vdots \\ Y_1^{(m)} \end{bmatrix}, \ldots, \begin{bmatrix} Y_s^{(1)} \\ Y_s^{(2)} \\ \vdots \\ Y_s^{(m)} \end{bmatrix}, \begin{bmatrix} P_1 Y_{s+1}^{(1)} \\ P_1 Y_{s+1}^{(2)} \\ \vdots \\ P_1 Y_{s+1}^{(m)} \end{bmatrix} \right\}. \quad (23.14)$$

A 2-step prediction, $P_2\mathbf{Y}_{s+2}$, is based on the computation of the 1-step prediction of the time series $\widetilde{\mathbf{Y}}(m,s+1)$. A k-step prediction, $P_k\mathbf{Y}_{s+k}$, can be computed by repeating this procedure k times.

2.3 Prediction Accuracy

Root mean square error (RMSE) of the 1-step prediction can be obtained using the equation:

$$RMSE = \sqrt{\frac{1}{d}\sum_{h=1}^{d}(P_1 X_{s+h}^{(j)} - X_{s+h}^{(j)})^2}, \quad (23.15)$$

where $X_{s+h}^{(j)}$ is j-th component of a vector $\mathbf{X}_{s+h}$; $P_1 X_{s+h}^{(j)}$ is the 1-step prediction of j-th component of a multivariate time series $\mathbf{X}(m, s+h-1)$; d is a number of 1-step predictions; $1 \leq d \leq n-s$. RMSE for k-step predictions is computed in a similar way.

3 Data

For the purpose of the study, three time series are selected. They span the time interval from 04.01.1962 to 02.05.2007.

Two of them are different Δ solutions (EOPC04 and EOPC04_05) used here to carry out the comparative study highlighted in the Introduction. The daily time series Δ are retrieved from EOPC04 and EOPC04_05 data sets. In both cases, Δ data are corrected for tidal effects using the IERS Conventions tidal model $\delta\Delta$ (McCarthy and Petit 2004). The non-tidal length-of-day signal can thus be denoted as $(\Delta - \delta\Delta)_{\mathrm{EOPC04}}$ and $(\Delta - \delta\Delta)_{\mathrm{EOPC04\_05}}$, respectively for EOPC04 and EOPC04_05 solutions. The difference between these time series is depicted in Fig. 23.1. The characterisation of $(\Delta - \delta\Delta)_{\mathrm{EOPC04\_05}}$ data is provided by Niedzielski et al. (2009).

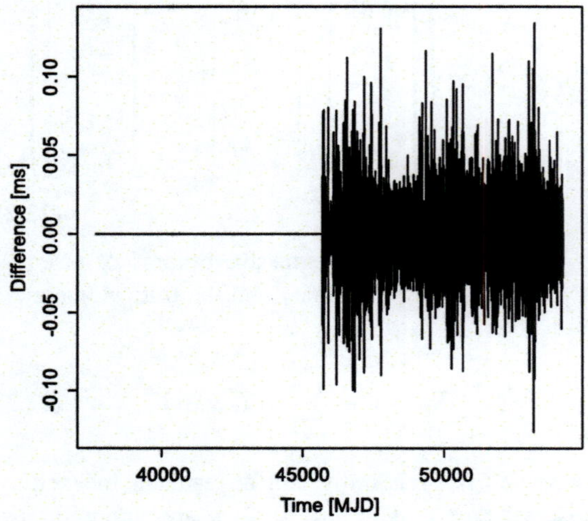

Fig. 23.1 The difference between $(\Delta - \delta\Delta)_{\mathrm{EOPC04}}$ and $(\Delta - \delta\Delta)_{\mathrm{EOPC04\_05}}$ time series

The third data set is AAM χ_3 time series, the sum of wind (motion) and pressure (mass) terms corrected for inverse barometer effect (Kalnay et al. 1996). The data were interpolated to get the 1-day sampling interval.

4 Predictions of UT1-UTC

The analysis indicates that the performance of LS+MAR technique in forecasting LOD and UT1-UTC time series is better than the performance of LS and LS+AR methods. This is particularly well seen during El Niño and La Niña events (Niedzielski and Kosek 2008). One should note that the superior performance may be partially driven either by selecting model orders with dissimilar statistics or by calibrating models with different techniques. Figure 23.2 shows that RMSE values of long-term predictions are the lowest for LS+MAR approach based on LOD input data from EOPC04. The corresponding predictions determined using LOD from EOPC04_05 are less accurate, however, they are still better than those calculated by LS or LS+AR techniques.

It is difficult to address the issue of potential causes of the above-mentioned discrepancy between accuracies of predictions derived using two EOPC solutions. The probable explanation may be related to smoothing as it is usually easier to successfully predict filtered data. It is interesting, however, that the similar difference cannot be observed in the case of LS+AR predictions based on EOPC04 and EOPC04_05 data.

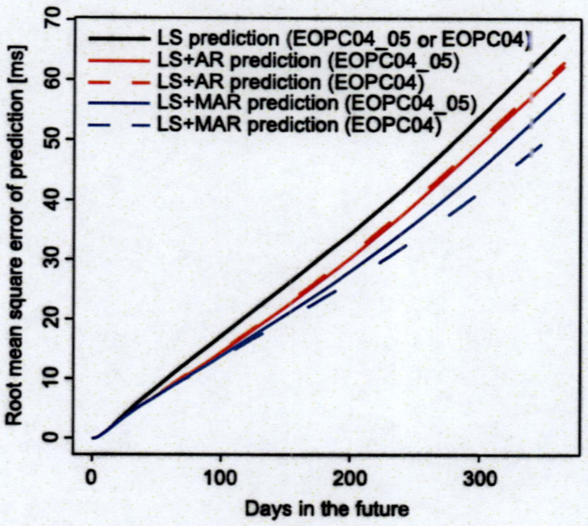

Fig. 23.2 RMSE of predictions based on the LS+MAR method for different input data

Conclusion

In this paper, LS+MAR technique for UT1-UTC forecasting is discussed and shown to be an efficient method in comparison to LS and LS+AR approaches. For long-term predictions, which are crucial for ENSO investigations, the superior performance is provided by LS+MAR method based on the input data from EOPC04 solution. If EOPC04 input data are replaced with EOPC04_05, the EOPC04-based computation still remains to be the most accurate amongst the predictions under study. The potential explanation may be linked to smoothing.

Acknowledgements The research was financed from the Polish science funds for the period of 2009-2011 provided by Polish Ministry of Science and Higher Education through the grant no. N N526 160136 under leadership of Dr Tomasz Niedzielski at the Space Research Centre of Polish Academy of Sciences. The first author was also supported by EU EuroSITES project. The authors of R 2.9.0 – A Language and Environment and additional packages are acknowledged.

References

Abarca del Rio A, Gambis D, Salstein DA (2000) Interannual signals in length of day and atmospheric angular momentum. Annales Geophysicae 18:347–364

Akyilmaz O, Kutt erer H (2004) Prediction of Earth rotation parameters by fuzzy inference systems. J Geodes 78:82–93

Freedman AP, Steppe JA, Dickey JO, Eubanks TM, Sung LY (1994) The short-term prediction of universal time and length of day using atmospheric angular momentum. J Geophys Res 99(B4):6981–6996

Gross RS, Eubanks TM, Steppe JA, Freedman AP, Dickey JO, Runge TF (1998) A Kalman filter-based approach to combining independent Earth-orientation series. J Geodes 72:215–235

Gross RS, Marcus SL, Eubanks TM, Dickey JO, Keppenne CL (1996) Detection of an ENSO signal in seasonal length-of-day variations. Geophys Res Lett 23:3373–3376

Johnson T, Luzum BJ, Ray JR (2005) Improved near-term Earth rotation predictions using atmospheric angular momentum analysis and forecasts. J Geodyn 39:209–221

Kalarus M, Kosek W (2004) Prediction of Earth orientation parameters by artificial neural networks. Artificial Satellites 39:175–184

Kalnay E, Kanamitsu M, Kistler R, Collins W, Deaven D, Gandin L, Iredell M, Saha S, White G, Woollen J, Zhu Y, Leetmaa A, Reynolds B, Chelliah M, Ebisuzaki W, Higgins W, Janowiak J, Mo KC, Ropelewski C, Wang J, Jenne R, Joseph D (1996) The NCEP/NCAR 40-year reanalysis project. Bull Am Meteorol Soc 77:437–471

Kosek W (1992) Short periodic autoregressive prediction of the Earth rotation parameters. Artificial Satellites 27:9–17

Kosek W, Kalarus M, Johnson TJ, Wooden WH, McCarthy DD, Popiński W(2005) A comparison of LOD and UT1-UTC forecasts by different combination prediction techniques. Artificial Satellites 40:119–125

Kosek W, McCarthy DD, Luzum BJ (1998) Possible improvement of Earth orientation forecast using autocovariance prediction procedures. J Geodes 72:189–199

McCarthy DD, Petit G (eds) (2004) IERS Conventions 2003 IERS Technical Note No. 32, Verlag des Bundesamts für Kartographie und Geodäsie, Frankfurt am Main

Neumaier A, Schneider T (2001) Estimation of parameters and eigenmodes of multivariate autoregressive models. ACM Trans Math Software 27:27–57

Niedzielski T, Kosek W (2008) Prediction of UT1-UTC, LOD and AAM χ_3 by combination of least-squares and multivariate stochastic methods. J Geodes 82:83–92

Niedzielski T, Sen AK, Kosek W (2009) On the probability distribution of Earth orientation parameters data. Artficial Satellites 44:33–41

Rosen RD, Salstein DA, Eubanks TM, Dickey JO, Steppe JA (1984) An El Niño signal in atmoshperic angular momentum and Earth rotation. Science 27:411–414

Schuh H, Ulrich M, Egger D, Müller J, Schwegmann W (2002) Prediction of Earth orientation parameters by artificial neural networks. J Geodes 76:247–258

Schwarz G (1978) Estimating the dimension of a model. Ann Stat 6:461–464

Zhao J, Han Y (2008) The relationship between the interannual variation of Earth's rotation and El Niño events. Pure Appl Geophys 165:1435–1443

Next Generation GNSS Single Receiver Cycle Slip Reliability

24

P.J.G. Teunissen and P.F. de Bakker

Abstract

In this contribution we study the multi-frequency, carrier-phase slip detection capabilities of a single receiver. Our analysis is based on an analytical expression that we present for the multi-frequency minimal detectable carrier phase cycle slip.

Keywords

GNSS Cycle Slips • Minimal Detectable Bias (MDB) • Multi-Frequency Receivers

1 Introduction

In this contribution we will study the Global Navigation Satellite System (GNSS) reliability of multi-frequency single-receiver, single-satellite code- and carrier phase time series. Examples of such studies for the single-baseline GNSS models can be found in Teunissen (1998), De Jong (2000), De Jong and Teunissen (2000). There are several advantages to single-receiver, single-satellite data validation. First, it can be executed in real-time inside the receiver and thus enables early quality control on the raw data. Second, the geometry-free single-satellite approach has the advantage that no satellite positions need to be known beforehand and thus no complete navigation messages need to be read and used. Moreover, this approach also makes the method very flexible for processing data from any (future) GNSS in a simple way, like e.g. (modernized) GPS (USA), Galileo (EU), Glonass (Russia), and Compass (China).

Our study of the single-receiver, single-satellite reliability will be analytical and supported with numerical results. As reliability measure we focus on the Minimal Detectable Biases (MDBs). The MDB is a measure for the size of model errors that can be detected with a certain power and a certain probability of false alarm. The MDB can be determined from the functional and stochastic model and is therefore a useful tool to assess how well certain model errors can be detected. We formulate alternative hypotheses for model errors like outliers in de code data on different frequencies, cycle slips in the carrier phase data on different frequencies, potential loss of lock, and ionospheric disturbances. The closed form formulas that will be presented are applicable to any GNSS with an arbitrary number of frequencies and include also the ionosphere-weighted case. Due to lack of space, we only work out the single- and multi-frequency MDBs for cycle slips. However, the same approach can be followed for the other type of model errors as well.

P.J.G. Teunissen (✉)
Delft Institute of Earth Observation and Space Systems, Delft University of Technology, Delft, The Netherlands

Department of Spatial Sciences, Curtin University of Technology, Perth, Australia
e-mail: p.j.g.teunissen@tudelft.nl

P.F. de Bakker
Delft Institute of Earth Observation and Space Systems, Delft University of Technology, Delft, The Netherlands

N. Sneeuw et al. (eds.), *VII Hotine-Marussi Symposium on Mathematical Geodesy*, International Association of Geodesy Symposia 137, DOI 10.1007/978-3-642-22078-4_24,

We emphasize the results for (modernized) GPS and Galileo.

2 The Multi-frequency Single-Receiver Geometry-Free Model

Null Hypothesis: The carrier phase and pseudo range observation equations of a single receiver that tracks a single satellite on frequency f_j $(j = 1, \ldots, n)$ at time instant t $(t = 1, \ldots, k)$, see e.g., Teunissen and Kleusberg (1998), Misra and Enge (2001), Hofmann-Wellenhoff and Lichtenegger (2001), Leick (2003), are given as

$$\begin{aligned} \phi_j(t) &= \rho^*(t) - \mu_j \mathscr{I}(t) + b_{\phi_j} + n_{\phi_j}(t) \\ p_j(t) &= \rho^*(t) + \mu_j \mathscr{I}(t) + b_{p_j} + n_{p_j}(t) \end{aligned} \tag{24.1}$$

where $\phi_j(t)$ and $p_j(t)$ denote the observed carrier phase and pseudo range, respectively, with corresponding zero mean noise terms $n_{\phi_j}(t)$ and $n_{p_j}(t)$. The unknown parameters are $\rho^*(t)$, $\mathscr{I}(t)$, b_{ϕ_j} and b_{p_j}. The lumped parameter $\rho^*(t) = \rho(t) + c\delta t_r(t) - c\delta t^s(t) + T(t)$ is formed from the receiver-satellite range $\rho(t)$, the receiver and satellite clock errors, $c\delta t_r(t)$ and $c\delta t^s(t)$, respectively, and the tropospheric delay $T(t)$. The parameter $\mathscr{I}(t)$ denotes the ionospheric delay expressed in units of range with respect to the *first* frequency. Thus for the f_j-frequency pseudo range observable its coefficient is given as $\mu_j = f_1^2/f_j^2$. The parameters b_{ϕ_j} and b_{p_j} are the phase bias and the instrumental code delay, respectively. The phase bias is the sum of the initial phase, the phase ambiguity and the instrumental phase delay.

Both b_{ϕ_j} and b_{p_j} are assumed to be time-invariant. This is allowed for relatively short time spans, in which the instrumental delays remain sufficiently constant. The time-invariance of b_{ϕ_j} and b_{p_j} implies that only time-differences of $\rho^*(t)$ and $\mathscr{I}(t)$ are estimable. We may therefore just as well formulate the observation equations in time-differenced form. Then the parameters b_{ϕ_j} and b_{p_j} get eliminated and we obtain

$$\begin{aligned} \phi_j(t,s) &= \rho^*(t,s) - \mu_j \mathscr{I}(t,s) + n_{\phi_j}(t,s) \\ p_j(t,s) &= \rho^*(t,s) + \mu_j \mathscr{I}(t,s) + n_{p_j}(t,s) \end{aligned} \tag{24.2}$$

where $\phi_j(t,s) = \phi_j(t) - \phi_j(s)$, with a similar notation for the time-difference of the other variates.

Would we have a priori information available about the ionospheric delays, we could model this through the use of additional observation equations. In our case, we do not assume information about the *absolute* ionospheric delays, but rather on the *relative*, time-differenced, ionospheric delays. We therefore have the additional (pseudo) observation equation

$$\mathscr{I}_o(t,s) = \mathscr{I}(t,s) + n_{\mathscr{I}}(t,s) \tag{24.3}$$

with the (pseudo) ionospheric observable $\mathscr{I}_o(t,s)$. The sample value of $\mathscr{I}_o(t,s)$ is usually taken to be zero.

If we define $\phi(t) = (\phi_1(t), \ldots, \phi_n(t))^T$, $p(t) = (p_1(t), \ldots, p_n(t))^T$, $y(t) = (\phi(t)^T, p(t)^T, \mathscr{I}_o(t))^T$, $g(t) = (\rho^*(t), \mathscr{I}(t))^T$, $\mu = (\mu_1, \ldots, \mu_n)^T$, $y(t,s) = y(t) - y(s)$ and $g(t,s) = g(t) - g(s)$, then the expectation E(.) of the $2n+1$ observation equations of (24.2) and (24.3) can be written in the compact vector-matrix form

$$\mathrm{E}(y(t,s)) = Gg(t,s) \tag{24.4}$$

where

$$G = \begin{bmatrix} e_n & -\mu \\ e_n & +\mu \\ 0 & 1 \end{bmatrix} \tag{24.5}$$

This two-epoch model can be extended to an arbitrary number of epochs. Let $y = (y(1)^T, \ldots, y(k)^T)^T$ and $g = (g(1)^T, \ldots, g(k)^T)^T$, and let D_k be a full rank $k \times (k-1)$ matrix of which the columns span the orthogonal complement of e_k, $D_k^T e_k = 0$ (recall that e_k is a k-vector of 1's). Then $\Delta y = (D_k^T \otimes I_{2n+1})y$ and $\Delta g = (D_k^T \otimes I_2)g$ are the time-differenced vectors of the observables and parameters, respectively, and the k-epoch version of (24.4) can be written as

$$\mathscr{H}_0 : \mathrm{E}(\Delta y) = (I_{k-1} \otimes G)\Delta g \tag{24.6}$$

where $\otimes$ denotes the Kronecker product. Model (24.6), or its two-epoch variant (24.4), will be referred to as our null hypothesis $\mathscr{H}_0$.

Alternative Hypotheses: The data collected by a single GNSS receiver can be corrupted by many different errors. The errors that we consider are the ones that can be modelled as a shift in the mean of the data vector, $\mathrm{E}(\Delta y \mid \mathcal{H}_a) = \mathrm{E}(\Delta y \mid \mathcal{H}_0) + \text{shift}$. Modelling errors of this kind are outliers in the pseudo range data, cycle slips in the carrier phase data, ionospheric disturbances and loss-of-lock. To accomodate these model biases, the alternative hypotheses are formulated as

$$\mathcal{H}_a : \mathrm{E}(\Delta y) = (I_{k-1} \otimes G)\Delta g + (D_k^T s_l \otimes H)b \tag{24.7}$$

where

$$\underset{(2n+1)\times q}{H} = \begin{cases} (I_n, 0, 0)^T & \text{(phase loss of lock)} \\ (\delta_j^T, 0, 0)^T & \text{(carrier phase)} \\ (0, \delta_j^T, 0)^T & \text{(pseudo range)} \\ (0, 0, 1)^T & \text{(ionosphere)} \end{cases} \tag{24.8}$$

and

$$\underset{k\times 1}{s_l} = \begin{cases} (\overset{1}{0}, \ldots, 0, \overset{l}{1}, 0, \ldots, \overset{k}{0})^T & \text{(spike)} \\ (0, \ldots, 0, 1, 1, \ldots, 1)^T & \text{(slip)} \end{cases} \tag{24.9}$$

The n-vector δ_j denotes the unit vector having a 1 as its jth entry.

Stochastic model: With the time-invariant variance matrices of the undifferenced carrier phase and (code) pseudo range observables $\phi(t)$ and $p(t)$ denoted as $Q_{\phi\phi}$ and Q_{pp}, respectively, the dispersion of the two-epoch model (24.4) is assumed to be given as

$$\mathrm{D}\left(y(t,s)\right) = \mathrm{blockdiag}(2Q_{\phi\phi}, 2Q_{pp}, \sigma^2_{\Delta\mathcal{I}}) \tag{24.10}$$

where the scalar $\sigma^2_{\Delta\mathcal{I}}$ denotes the variance of the time-differenced ionospheric delay.

If we assume that the time series of the absolute ionospheric delays can be modelled as a *first-order autoregressive* stochastic process ($\sigma^2_{\mathcal{I}}\beta^{|t-s|}$, with $0 \le \beta \le 1$), the variance of the time-differenced ionospheric delay works out as

$$\sigma^2_{\Delta\mathcal{I}} = 2\sigma^2_{\mathcal{I}}(1 - \beta^{|t-s|}) \tag{24.11}$$

For two successive epochs we have $2\sigma^2_{\mathcal{I}}(1-\beta)$, while for larger time-differences the variance will tend to the white-noise value $2\sigma^2_{\mathcal{I}}$ if $\beta < 1$. Thus $\sigma^2_{\mathcal{I}}$ and β can be used to model the level and smoothness of the noise in the ionospheric delays.

For the measurement precision of the multi-frequency GNSS signals, we assume $Q_{\phi\phi} = \sigma^2_{\phi} I_{2n}$ and $Q_{pp} = \sigma^2_p I_{2n}$, where we used the values as given by De Wilde et al. (2006), see also De Bakker et al. (2009). These zenith-referenced values are summarized in Table 24.1. To obtain the standard deviations for an arbitrary elevation, these values still need to be multiplied with an elevation dependent function. In practice one often uses an exponential or cosecant function, see Fig. 24.1. For these functions, the function values are between 3 and 4 at 15° elevation and approach the minimum of 1 at 90° elevation.

Table 24.1 Standard deviations of undifferenced GPS and Galileo observables (Simsky et al., 2006)

	L1	L2	L5	E1	E5a	E5b	E5	E6
Code (cm)	15	15	3.9	6.1	3.9	3.7	0.9	4.4
Phase (mm)	1.0	1.3	1.3	1.0	1.3	1.3	1.3	1.2

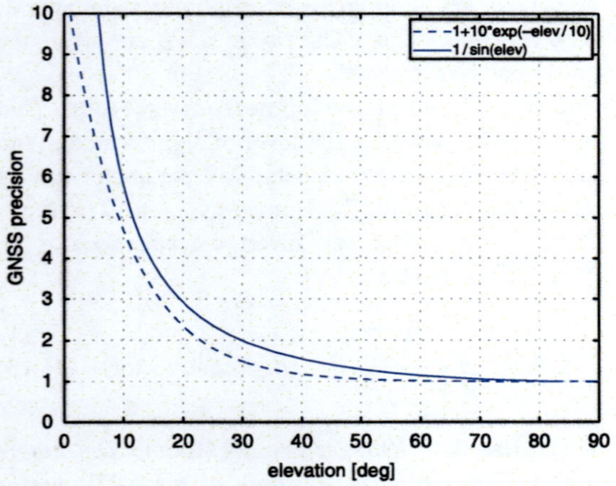

Fig. 24.1 The exponential $1+10\exp(-E/10)$ and the cosecant function $1/sin(E)$

3 Testing and Reliability

In order to test $\mathcal{H}_0$ against $\mathcal{H}_a$, we make use of the *uniformly most powerful invariant* (UMPI) test, see e.g., Arnold (1981), Koch (1999), Teunissen (2006). The UMPI test rejects the null hypothesis $\mathcal{H}_0$ in favour of the alternative hypothesis $\mathcal{H}_a$, if

$$T_q = \hat{b}^T Q_{\hat{b}\hat{b}}^{-1} \hat{b} > \chi^2_\alpha(q, 0) \tag{24.12}$$

where $\hat{b}$, with variance matrix $Q_{\hat{b}\hat{b}}$, is the least-squares estimator of b under $\mathcal{H}_a$, and $\chi^2_\alpha(q,0)$ is the α-level critical value. The UMPI-test statistic T_q is distributed as $T_q \overset{\mathscr{H}_0}{\sim} \chi^2(q,0)$ and $T_q \overset{\mathscr{H}_a}{\sim} \chi^2(q,\lambda)$, respectively, where $\lambda = b^T Q_{\hat{b}\hat{b}}^{-1} b$ is the *noncentrality parameter*.

The power of the test, denoted as γ, is defined as the probability of correctly rejecting $\mathscr{H}_0$, thus $\gamma = \mathrm{P}[T_q > \chi^2_\alpha(q,0) \mid \mathscr{H}_a]$. It depends on q (the dimension of b, a.k.a. degrees of freedom of test), α (level of significance), and through the noncentrality parameter λ, on b (the bias vector). Once q, α and b are given, the power can be computed.

One can however also follow the inverse route. That is, given the power γ, the level of significance α and the dimension q, the noncentrality parameter can be computed, symbolically denoted as $\lambda_0 = \lambda(\alpha, q, \gamma)$. With λ_0 given, one can invert the equation $\lambda_0 = b^T Q_{\hat{b}\hat{b}}^{-1} b$ and obtain

$$\mathrm{MDB} = \sqrt{\frac{\lambda_0}{d^T Q_{\hat{b}\hat{b}}^{-1} d}}\, d \quad (d = \text{unit vector}) \qquad (24.13)$$

This is Baarda's (1968) celebrated *Minimal Detectable Bias* (MDB) vector. The length of the MDB vector is the smallest size of bias vector that can be found with probability γ in the direction d with test (24.12). By letting d vary over the unit sphere in $\mathbb{R}^q$ one obtains the whole range of MDBs that can be detected with probability γ with test (24.12). The MDB can be computed once λ_0 and $Q_{\hat{b}\hat{b}}$ are known. The value of λ_0 depends on q, α and γ. For later use, we have shown the dependence in Fig. 24.2 of $\sqrt{\lambda_0}$ on γ for different values of q and α.

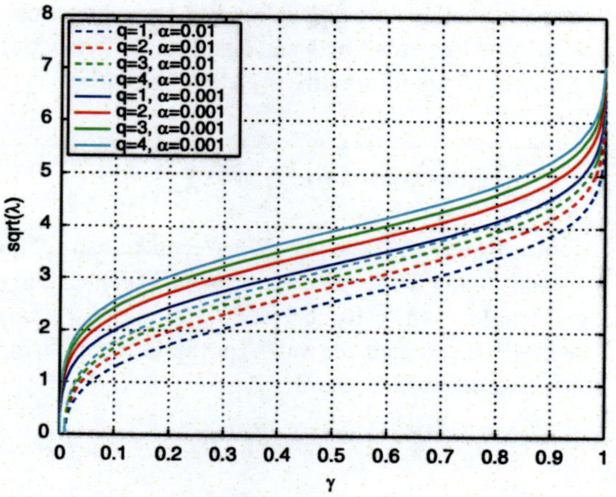

Fig. 24.2 Square root of noncentrality parameter λ_0 as function of power γ for degrees of freedom $q = 1, 2, 3, 4$ and levels of significance $\alpha = 0.01, 0.001$

We now present an analytical expression for the MDB of the single-receiver carrier-phase slip. First we consider the single-frequency receiver, then the multi-frequency GNSS receiver.

4 Single Frequency Receiver MDB-Slip

The two-epoch, single-frequency receiver MDB for a carrier-phase slip can be shown to read as

$$\mathrm{MDB}_{\phi_j} = \sqrt{2(\sigma^2_{\phi_j} + \sigma^2_{p_j} + 4\mu_j^2 \sigma_I^2)\lambda_0} \qquad (24.14)$$

where $\sigma_I = \sigma_{\mathscr{I}} \sqrt{1 - \beta^{|t-s|}}$. This expression clearly shows how the detectability is affected by the measurement precision $(\sigma_{\phi_j}, \sigma_{p_j})$, the signal frequency (μ_j), and the time-smoothness of the ionosphere (σ_I).

In Fig. 24.3 we show the single-frequency phase-slip MDB_{ϕ_j}s for GPS and Galileo as function of σ_I. For Fig. 24.3 we used the frequencies of Tables 24.2 and 24.3, and the standard deviations of Table 24.1. The figure clearly shows the effects of (code) measurement precision and frequency. For small values of σ_I, the effect of (code) measurement precision dominates, while for larger values, the frequency effect starts to be felt.

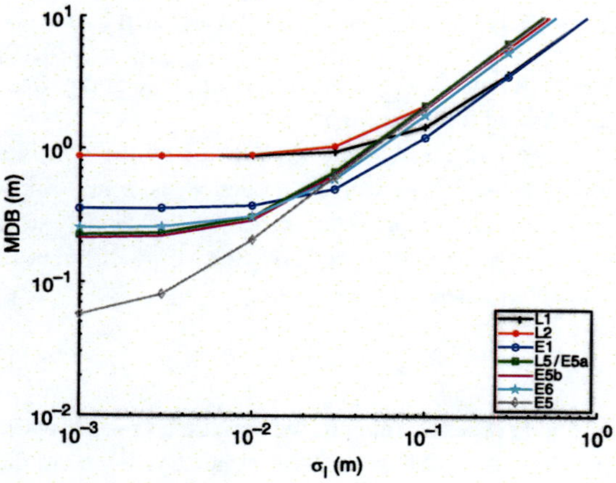

Fig. 24.3 Single-frequency phase-slip MDBs as functie of $\sigma_I = \sigma_{\mathscr{I}} \sqrt{1 - \beta^{|t-s|}}$. The MDBs are shown for the GPS frequencies L1, L2, and L5, and the Galileo frequencies E1, E5, E5a, E5b, E6 (NB: $k = 2$, $\alpha = 0.001$, $\gamma = 0.80$ and $\mathscr{I}(t)$ is defined w.r.t. L1 frequency)

Table 24.2 GPS frequencies and wave lengths

	L1	L2	L5
Frequency (MHz)	1575.42	1227.60	1176.45
Wave length (cm)	19.0	24.4	25.5

Table 24.3 Galileo frequencies and wave lengths

	E1	E5a	E5b	E5	E6
Freq (MHz)	1575.420	1176.450	1207.140	1191.795	1278.750
λ (cm)	19.0	25.5	24.8	25.2	23.4

Since all MDBs, except the E5-MDB, are larger than 20 cm, one can not expect a single-frequency receiver to perform well on these frequencies as fas as cycle slip detection is concerned. Even for those that have their MDB around their wave length – like L5, E5a and E5b – one should keep in mind that these values will become larger for lower elevations.

Cycle slip detection on the E5 frequency does however have a good chance of performing well. The zenith-referenced E5-MDB is about 8 cm for $\sigma_I = 3$ mm. Since this value will have to be multiplied by about 3 to get the 20° elevation MDB, the result still stays below the E5 wave length of 25.2 cm.

For the other frequencies, single-frequency cycle slip detection will be difficult when using the single-receiver, single-satellite geometry-free model.

5 Multi Frequency Receiver MDB-Slips

The two-epoch, multi-frequency carrier phase slip MDB can be shown to read as

$$\mathrm{MDB}_{\phi_j} = \sigma_\phi \sqrt{\left(\frac{2}{1-\frac{1}{n^*}}\right)} \lambda_0 \qquad (24.15)$$

where

$$\frac{1}{n^*} = \frac{1}{n}\frac{1}{1+\varepsilon}\left(1 + \frac{(\mu_j - \frac{1-\varepsilon}{1+\varepsilon}\bar{\mu})^2}{\frac{1}{n}\sum_{i=1}^n \mu_i^2 - \left(\frac{1-\varepsilon}{1+\varepsilon}\right)^2 \bar{\mu}^2 + \frac{\sigma_\phi^2/\sigma_p^2}{n(1+\varepsilon)}}\right) \qquad (24.16)$$

with the phase-code variance ratio $\varepsilon = \sigma_\phi^2/\sigma_p^2$ and the average $\bar{\mu} = \frac{1}{n}\sum_{j=1}^n \mu_j$.

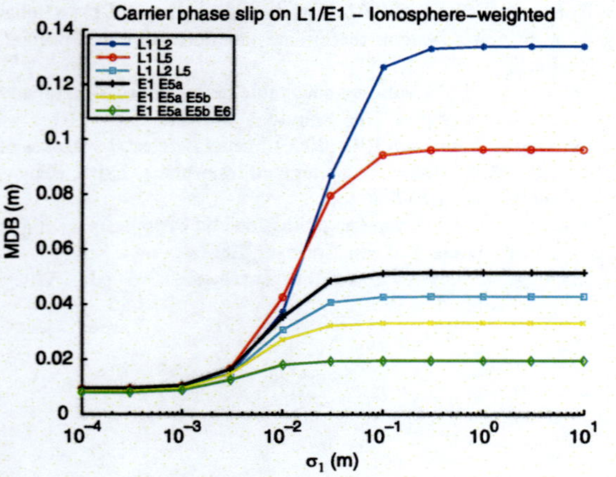

Fig. 24.4 Multi-frequency phase-slip MDBs as functie of $\sigma_I = \sigma_{\mathcal{I}}\sqrt{1-\beta^{|t-s|}}$. The MDBs are shown for dual- and triple-frequency GPS, and for dual- triple- and quadruple-frequency Galileo. (NB: $k = 2$, $\alpha = 0.001$, $\gamma = 0.80$ and $\mathcal{I}(t)$ is defined w.r.t. L1 frequency)

The dual-, triple- and quadruple-frequency MDB phase-slips for GPS and Galileo are shown in Fig. 24.4. The quadruple-frequency Galileo-case performs best, while the dual-frequency GPS-case performs poorest. In all cases however, the MDBs are below the 1-cycle level, even below the 5 cm if $\sigma_I \leq 1$ cm. Thus single-receiver, multi-frequency cycle slip detection will be possible for such ionospheric conditions. For the more extreme case that σ_I is several cm, GPS dual-frequency (e.g. L_1L_2, L_1L_5) cycle slip detection will become problematic for lower elevations.

Acknowledgements The research of the first author has been supported by an Australian Research Council Federation Fellowship (project number FF0883188).

References

Arnold SF (1981) The theory of linear models and multivariate analysis. Wiley, New York

Baarda W (1968) A testing procedure for use in geodetic networks. Netherlands geodetic commission, Publ Geodes, New Series, 2(5)

De Bakker PF, van der Marel H, Tiberius CCJM(2009) Geometry-free undifferenced, single and double differenced analysis of single frequency GPS, EGNOS and GIOVE-A/B measurements. GPS Solut 13(4): 305–314

Hofmann-Wellenhoff B, Lichtenegger H (2001) Global positioning system: theory and practice, 5th edn. Springer, Berlin

De Jong K (2000) Minimal detectable biases of cross-correlated GPS observations. GPS Solutions 3:12–18.

De Jong K, Teunissen PJG (2000) Minimal detectable biases of GPS observations for a weighted ionosphere. Earth, Planets and Space 52:857–862

Koch KR (1999) Parameter estimation and hypothesis testing in linear models, 2nd edn. Springer, Berlin

Leick A (2003) GPS satellite surveying, 3rd edn. Wiley, New York

Misra P, Enge P (2001) Global positioning system: signals, measurements, and performance. Ganga-Jamuna Press, Lincoln MA

Teunissen PJG (1998) Minimal detectable biases of GPS data. J Geodes 72:236–244

Teunissen PJG (2006) Testing theory; an introduction, 2nd edn. Delft VSSD

Teunissen PJG and Kleusberg A (eds) (1998) GPS for geodesy, 2nd enlarged edn., Springer, Berlin

De Wilde W, Wilms F, Simsky A, Sleewaegen J (2006). Early performance results for new Galileo and GPS signals-in-space. In: Proceedings of ENC GNSS, 2006. Manchester

A Constrained Quadratic Programming Technique for Data-Adaptive Design of Decorrelation Filters

25

Lutz Roese-Koerner, Ina Krasbutter, and Wolf-Dieter Schuh

Abstract

Signals from sensors with high sampling rates are often highly correlated. For the decorrelation of such data, which is often applied for the efficient estimation of parametric data models, discrete filters have proven to be both highly flexible and numerically efficient. Standard filter techniques are, however, often not suitable for eliminating strong local fluctuations or trends present in the noise spectral density. Therefore we propose a constrained least-squares filter design method. The spectral features to be filtered out are specified through inequality constraints regarding the noise spectral density. To solve for the optimal filter parameters under such inequality constraints, we review and apply the Active Set Method, a quadratic programming technique. Results are validated by statistical tests. The proposed filter design algorithm is applied to GOCE gradiometer signals to analyze its numerical behaviour and efficiency for a realistic and complex application.

Keywords

Active Set Method • Adjustment with inequality constraints • Decorrelation • Filter

1 Introduction

The most popular method for estimating parameters is ordinary least-squares adjustment, which could easily be extended to handle equality constraints (Koch 1999, pp. 170–177). However, in order to deal with inequality constraints, a transformation into a quadratic program (Gill et al. 1981, pp 177–186), a linear complementarity problem (Cottle et al. 1992) or a least distance program (Lawson and Hanson 1974, pp 158–173) is needed. Roese-Koerner (2009) provides a detailed discussion on these methods.

All these problems could be solved by methods of convex optimization (optimization of a convex objective function subject to constraints, which form a convex set). This paper is focused on the Active Set Method (Gill et al. 1981, pp 199–203) – a very stable quadratic programming approach, which is more memory efficient than e.g. Dantzig's Simplex Algorithm for Quadratic Programs (Dantzig 1998, pp 490–498).

Introducing inequality constraints is helpful in many applications, for example in the design of geodetic networks, to reshape error ellipses (Koch

L. Roese-Koerner (✉) · I. Krasbutter · W.-D. Schuh
Department of Theoretical Geodesy, University of Bonn, Nussallee 17, D-53115 Bonn, Germany
e-mail: roese-koerner@geod.uni-bonn.de

N. Sneeuw et al. (eds.), *VII Hotine-Marussi Symposium on Mathematical Geodesy*, International Association of Geodesy Symposia 137, DOI 10.1007/978-3-642-22078-4_25, © Springer-Verlag Berlin Heidelberg 2012

1982) or to introduce geometric constraints (Wölle 1988). Koch (2006) used inequality constraints in the context of semantic integration of GIS data. Applications can be found also in signal processing, as constraints allow for flexible filter design (Fritsch 1985; Schaffrin 1981).

In the following, we focus on the design of decorrelation filters. As the spectral density of residuals often contains strong local fluctuations or trends, which need special treatment, design of decorrelation filters in the spectral domain is considered. Therefore, we use constraints yielding flexible filter design and an optimal decorrelation. The capability of this approach is shown by decorrelating two different time series. The first one consists of randomly generated colored noise, the second, more sophisticated one, of residuals obtained from adjusted GOCE (Gravity field and steady-state Ocean Circulation Explorer) satellite gravity gradiometry (SGG) data (ESA 1999).

2 Adjustment with Inequality Constraints

The usual way of estimating the parameter vector $\mathbf{x}$ of an overdetermined linear model

$$\mathbf{A}\mathbf{x} = \boldsymbol{\ell} + \mathbf{v}, \tag{25.1}$$

with $(n \times m)$ design matrix $\mathbf{A}$, observation vector $\boldsymbol{\ell}$ and residual vector $\mathbf{v}$, is to minimize the (weighted) sum of squared residuals

$$\Phi\,(\mathbf{v}) = \mathbf{v}^T \boldsymbol{\Sigma}^{-1}\mathbf{v}, \tag{25.2}$$

where $\boldsymbol{\Sigma}$ is the data covariance matrix. This leads to the well-known least-squares adjustment. As it is often reasonable to restrict the parameters to an interval (e.g. positivity, resource limits, budgets), the objective function (25.2) may have to be minimized with respect to p linear inequality constraints

$$b_{j,1}x_1 + \ldots + b_{j,m}x_m \leq b_j \quad \Leftrightarrow \quad \mathbf{B}^T\mathbf{x} \leq \mathbf{b}, \tag{25.3}$$

with constant matrix $\mathbf{B}$, vector $\mathbf{b}$ and $j = 1 \ldots p$. Greater-than-or-equal constraints can be transformed into less-than-or-equal constraints by multiplying the whole equation by minus one. Due to the fact that this problem could not be solved with ordinary least-squares adjustment, we reformulate it as a quadratic program (QP). Once the problem is transformed thusly, there are algorithms capable of dealing with the inequality constraints.

2.1 Quadratic Program

A quadratic program consists of a quadratic objective function, which is to be minimized with respect to linear (inequality) constraints:

constraints:	$\mathbf{B}^T\mathbf{x} \leq \mathbf{b}$
parameters:	$\mathbf{x} \in \mathbb{R}^m$
objective function:	$\gamma_1 \mathbf{x}^T\mathbf{C}\mathbf{x} + \gamma_2 \mathbf{c}^T\mathbf{x} \ldots \text{Min.}$

$\mathbf{C}$ is a constant, symmetric and positive definite matrix, $\mathbf{c}$ a constant vector, and γ_1 and γ_2 are given scalars.

In order to transform the minimization problem (25.2) subject to (25.3) into a quadratic program, solely the objective function has to be reformulated:

$$\begin{aligned}
\Phi(\mathbf{x}) &= \mathbf{v}^T\boldsymbol{\Sigma}^{-1}\mathbf{v} \\
&= (\mathbf{A}\mathbf{x} - \boldsymbol{\ell})^T\boldsymbol{\Sigma}^{-1}(\mathbf{A}\mathbf{x} - \boldsymbol{\ell}) \\
&= \mathbf{x}^T\mathbf{A}^T\boldsymbol{\Sigma}^{-1}\mathbf{A}\mathbf{x} - 2\,\mathbf{x}^T\mathbf{A}^T\boldsymbol{\Sigma}^{-1}\boldsymbol{\ell} + \boldsymbol{\ell}^T\boldsymbol{\Sigma}^{-1}\boldsymbol{\ell} \\
&= \gamma_1\mathbf{x}^T\mathbf{C}\mathbf{x} + \gamma_2\mathbf{c}^T\mathbf{x} \ldots \text{Min.}
\end{aligned} \tag{25.4}$$

by using the substitutions

$$\mathbf{C} = 2\mathbf{A}^T\boldsymbol{\Sigma}^{-1}\mathbf{A}, \quad \mathbf{c} = -2\mathbf{A}^T\boldsymbol{\Sigma}^{-1}\boldsymbol{\ell}, \quad \gamma_1 = \frac{1}{2}, \quad \gamma_2 = 1$$

and neglecting the constant term $\boldsymbol{\ell}^T\boldsymbol{\Sigma}^{-1}\boldsymbol{\ell}$, which is irrelevant to the minimization problem. As a method for solving quadratic programs, we will focus on the Active Set Method (Gill et al. 1981, pp 199–203) for the above mentioned reasons (stability and memory efficiency).

2.2 Active Set Method

The idea behind this iterative algorithm is to start from an (arbitrary) feasible point – i.e. a point that satisfies all inequality constraints (25.3) – and follow the boundary of the feasible set until the minimal

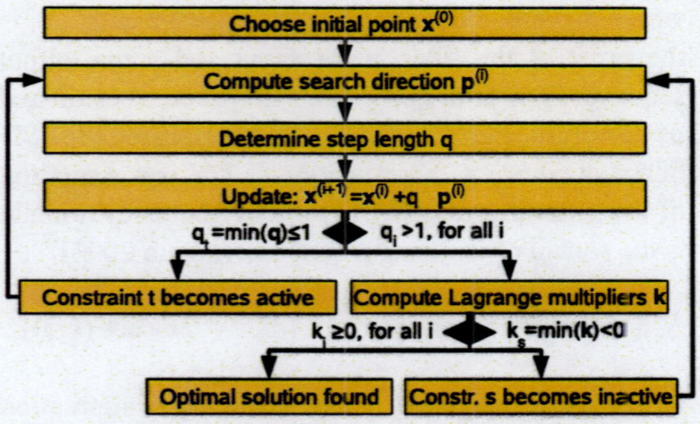

Fig. 25.1 Flowchart of the Active Set Algorithm

objective value is reached. After choosing an initial point $\mathbf{x}^{(0)}$ – e.g. with the "Big-M-Method" (Dantzig and Thapa 2003, pp 115–116) – a search direction $\mathbf{p}^{(0)}$ and an appropriate step length $q^{(0)}$ are computed (superscript numbers denote the iteration steps). Figure 25.1 presents a brief overview of the algorithm. The first step in calculating the search direction is to solve the unconstrained minimization problem (25.2):

$$\hat{\mathbf{x}} = \left(\mathbf{A}^T\boldsymbol{\Sigma}^{-1}\mathbf{A}\right)^{-1}\mathbf{A}^T\boldsymbol{\Sigma}^{-1}\boldsymbol{\ell} = -\mathbf{C}^{-1}\mathbf{c}. \tag{25.5}$$

The vector $\mathbf{p}^{*(0)}$ pointing from the initial point $\mathbf{x}^{(0)}$ to the minimum $\hat{\mathbf{x}}$ of the unconstrained problem is given by their difference

$$\mathbf{p}^{*(0)} = \hat{\mathbf{x}} - \mathbf{x}^{(0)}. \tag{25.6}$$

Next, the so-called Newton direction $\mathbf{p}^{*(0)}$ is projected onto the boundary of the feasible set, i.e. onto the active constraints (cf. Fig. 25.2). A constraint is called "active" if the inequality is exactly satisfied

$$b_{j,1}x_1^{(i)} + b_{j,2}x_2^{(i)} + \ldots + b_{j,m}x_m^{(i)} = b_j. \tag{25.7}$$

At iteration i all active constraints from $\mathbf{B}$ and $\mathbf{b}$ are summarized in the active set matrix $\mathbf{W}^{(i)}$ and the vector $\mathbf{w}^{(i)}$:

$$\mathbf{W}^{(i)T}\mathbf{x}^{(i)} = \mathbf{w}^{(i)}.$$

Due to the metric, which is defined by the unconstrained least-squares objective function, $\mathbf{W}^{(i)}$ must be rescaled. This can be done by multiplying $\mathbf{W}^{(i)}$ from the left with $\mathbf{C}^{-1}$

$$\overline{\mathbf{W}}^{(i)} = \mathbf{C}^{-1}\mathbf{W}^{(i)}, \quad \text{with} \quad \mathbf{C}^{-1} = (2\mathbf{A}^T\boldsymbol{\Sigma}^{-1}\mathbf{A})^{-1}$$

(Roese-Koerner 2009, p 22). Now the search direction $\mathbf{p}^*$ can be projected into the left null space of the matrix $\overline{\mathbf{W}}^{(i)}$ of active constraints

$$\begin{aligned} \mathbf{p} &= <\boldsymbol{\Pi}^{\mathbf{C}}_{S\perp(\overline{\mathbf{W}})}, \mathbf{p}^* >_{\mathbf{C}} \\ &= \boldsymbol{\Pi}^{\mathbf{C}}_{S\perp(\overline{\mathbf{W}})}\mathbf{C}\mathbf{p}^* \\ &= (\mathbf{C}^{-1} - \overline{\mathbf{W}}(\overline{\mathbf{W}}^T\mathbf{C}\overline{\mathbf{W}})^{-1}\overline{\mathbf{W}}^T)\mathbf{C}\mathbf{p}^* \end{aligned}$$

(iteration indices neglected). $\boldsymbol{\Pi}^{\mathbf{C}}_{S\perp(\overline{\mathbf{W}})}$ denotes the projection matrix under the metric $\mathbf{C}$ (Koch 1999, pp 64–66). Thereby, no active constraint is violated by a step in direction $\mathbf{p}^{(i)}$. However, this is not true for inactive constraints. Hence the maximum step length $q_{max}^{(i)}$ has to be computed as the distance to the next inactive constraint along direction $\mathbf{p}^{(i)}$. Here, two scenarios are possible (cf. Fig. 25.1).

If $q_{max}^{(i)} \leq 1$, we set $q^{(i)} = q_{max}^{(i)}$. After updating the parameters

$$\mathbf{x}^{(i+1)} = \mathbf{x}^{(i)} + q^{(i)}\mathbf{p}^{(i)}, \tag{25.8}$$

a new constraint will become active and a new iteration is computed, starting with the determination of a new search direction according to (25.6). Otherwise, if $q_{max} > 1$, it is possible to take a step of optimal length (cf. Fig. 25.2) without violating any constraints, and $q^{(i)} = 1$ is chosen. After computing the update step

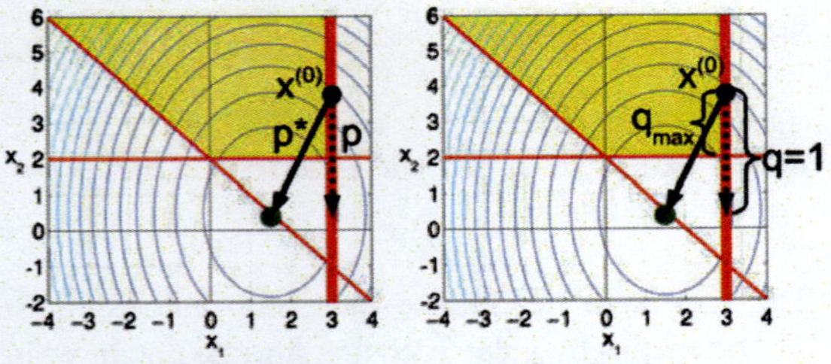

(a) Search direction p. (b) Step length q_{max}.

Fig. 25.2 Search direction $\mathbf{p}$ and step length q_{max} of the Active Set Algorithm. In Fig. 25.2a $\mathbf{p}^*$ denotes the vector to the unconstrained minimum (gray point), the gray lines symbolize constraints ($x_1 \leq 3$, $x_2 \geq 2$, $x_1 + x_2 \geq 2$) and the feasible set is shaded. The active (i.e. exactly satisfied) constraint is highlighted. In Fig. 25.2b the optimal step length $q = 1$ is illustrated, which could not be chosen, because it violates an inactive constraint

(25.8), the objective function (25.4) is extended to the Lagrange function

$$\Phi^* (\mathbf{x}, \mathbf{k}) = \frac{1}{2}\mathbf{x}^T \mathbf{C}\mathbf{x} + \mathbf{c}^T \mathbf{x} + \mathbf{k}^T (\mathbf{B}^T \mathbf{x} - \mathbf{b}), \quad (25.9)$$

as described in Wölle (1988). He proved, that the solution is optimal if and only if all Lagrange multipliers $\mathbf{k}^{(i+1)}$ related to active constraints are positive and that $\mathbf{p}^{*(i+1)}$ is in the column space of $\overline{\mathbf{W}}^{(i+1)}$ and can therefore be expressed as the linear combination

$$\mathbf{p}^{*(i+1)} = \overline{\mathbf{W}}^{(i+1)} \mathbf{k}^{(i+1)}, \quad (25.10)$$

with the Lagrange multipliers as weighting factors. Hence $\mathbf{k}^{(i+1)}$ could be computed by solving (25.10). If all Lagrange multipliers linked with active constraints are positive, the optimal solution is found. Otherwise all constraints with negative Lagrange multipliers are deactivated (Gill et al. 1981, p 170), i.e. removed from the active set $\mathbf{W}$, and the algorithm is started again.

3 Applications in Filter Design

With constrained quadratic programming techniques more flexibility is gained in many applications. Fritsch (1985) and Schaffrin (1981), for instance, applied Lemke's algorithm to design filters in the frequency domain using the transfer function. Thereby, one can control the spectral behavior of the filters by using inequality constraints. In the following we will use the power spectral density (PSD) to estimate decorrelation filters.

3.1 Design of Simple Decorrelation Filters

The aim is to decorrelate data with colored noise characteristics by a decorrelation filter to obtain white noise. An algebraically simple approach is to estimate the coefficients of a non-recursive filter, such as of a symmetric moving-average filter:

$$y_i = \sum_{k=-N}^{N} \beta_k u_{i-k}, \quad \text{with} \quad \beta_{-k} = \beta_k, \quad (25.11)$$

where β_k are the unknown filter coefficients, N is the order of the filter, $\mathbf{u}$ the input and $\mathbf{y}$ the output sequence. The filter coefficients may then be estimated by constrained least-squares adjustment using the PSD of the residuals as observations $\boldsymbol{\ell}$. The representation of the filter equation (25.11) in the frequency domain is the square of its transfer function (Schuh 2003)

$$S(\boldsymbol{\beta}, \omega) = (\beta_0 + 2\beta_1 \cos(\omega) + \ldots + 2\beta_k \cos(k\omega))^2,$$

with angular frequency ω. In a numerical simulation the Active Set Method is used to decorrelate a randomly generated colored noise sequence, solving the quadratic program

constraints:	$\tilde{\ell}_j = \frac{1}{S(\boldsymbol{\beta},\omega_j)} \geq \ell_j, \quad j = 1, \ldots, p$
parameters:	$\boldsymbol{\beta} \in \mathbb{R}^m$
obj. function:	$\sum_{i=1}^{n} \left(\tilde{\ell}_i - \ell_i\right)^2 \ldots \text{Min.}$

where p inverse, adjusted observations $\tilde{\ell}_j$ have to be greater than or equal to the original observations ℓ_j. Figure 25.3 shows the values $\boldsymbol{\ell}$ of the data PSD (dark gray) and the inverse values of the estimated filter $\tilde{\boldsymbol{\ell}}$ (light gray) with $N = 50$. The filter illustrated in Fig. 25.3a is determined with unconstrained least-squares adjustment. The second filter, shown in Fig. 25.3b, has a constraint at angular frequency $\omega = 0.1$ Hz, which is badly denoised in the unconstrained case.

As described in Krasbutter (2009), the estimated decorrelation filters may be verified by statistical tests. One possible choice is the test of autocorrelation (Schlittgen and Streitberg 2001, pp 243-246). Results of the test are illustrated in Fig. 25.4. The horizontal gray lines are boundaries of the confidence interval for type-I-error $\alpha = 0.05$, i.e. for white noise, 95% of the autocorrelation coefficients must lie within this interval.

Using the unconstrained filter, the boundaries were violated by 26.0% of the autocorrelation coefficients of the adjusted observations (see Fig. 25.4). Introducing only one constraint improves the decorrelation process significantly; then only 5.6% of the autocorrelation coefficients are outside the confidence interval. The five percent limit could be reached by applying more constraints.

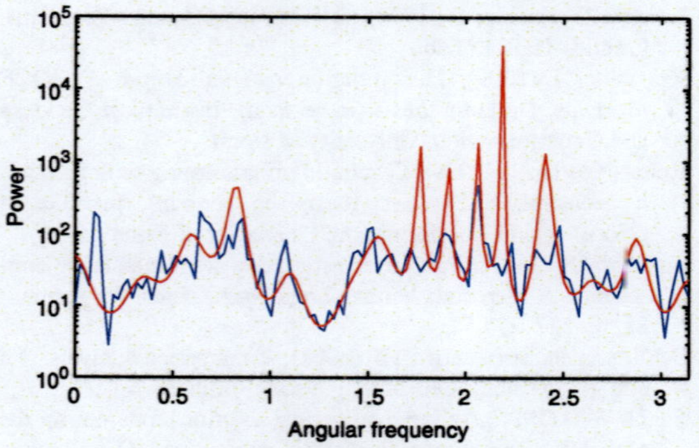

(a) No constraint.

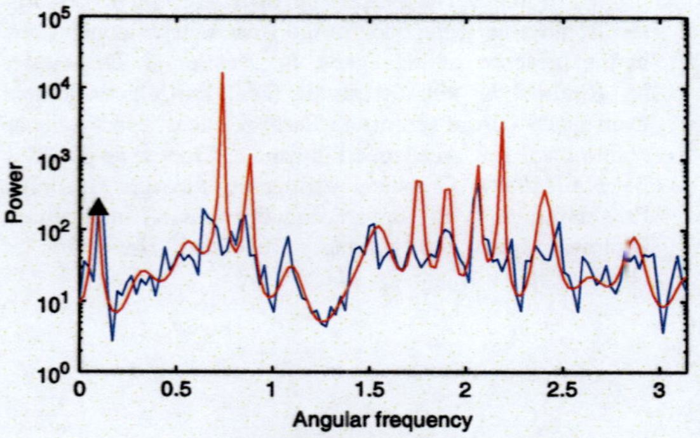

(b) Constraint at $\omega = 0.1$ Hz.

Fig. 25.3 Power spectral density of colored noise (*dark gray*) and the estimated reciprocal filter (*light gray*). The constraint is indicated by a black triangle.

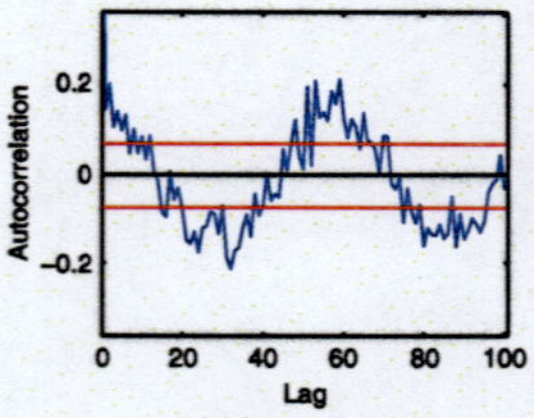

(a) No constraint.

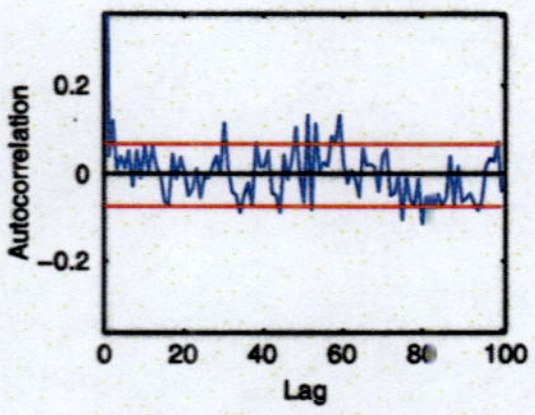

(b) Constraint at $\omega = 0.1$ Hz.

Fig. 25.4 Test of autocorrelation. For white noise, 95% of the autocorrelation coefficients (*gray*) of different lags have to be in the confidence interval (*horizontal gray lines*). (**a**) No constraint (**b**) Constraint at $\omega = 0.1$ Hz

3.2 GOCE Decorrelation Filters

Decorrelation filters are also used for GOCE data analysis as the observations of the gravity gradiometer are highly correlated (Schuh 1996). As described in Schuh

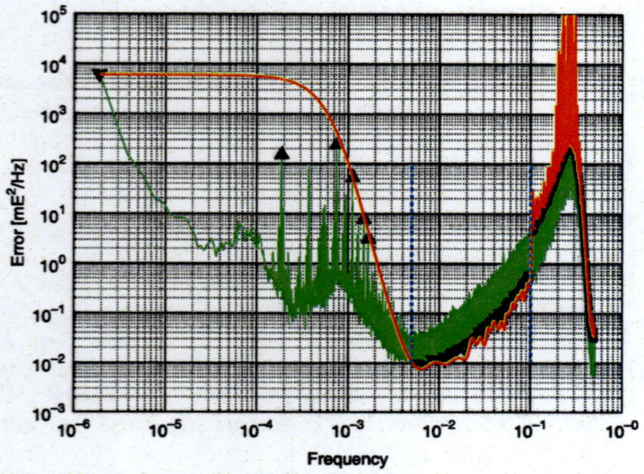

Fig. 25.5 PSD of simulated GOCE residuals (zz-component, *light gray*) and estimated inverse filter (*dark gray*). Constraints are indicated by black triangles. The *dotted vertical lines* denote the measurement bandwidth between 0.005 Hz and 0.1 Hz

(2003) especially symmetric moving-average filters allow for an efficient computation of the parameters within iterative approaches.

The PSD of simulated, highpass-filtered gradiometer residuals (zz-component) and the inverse of the estimated filter are presented in Fig. 25.5. The gradiometer is very sensitive in the bandwidth from 0.005 Hz to 0.1 Hz. Due to that fact, the signal outside the measurement bandwidth contains less information and should therefore be eliminated, while the signal inside should be preserved. This goal could be reached using inequality constraints as illustrated in Fig. 25.3.

Conclusion

Inequality-constrained least-squares problems can easily be transformed into quadratic programs and be solved with the Active Set Method. Concerning the design of decorrelation filters, constrained quadratic programming techniques were demonstrated to lead to improved flexibility (cf. (3.2)) and better rewhitening (cf. (3.1)). One drawback of quadratic programming algorithms is that they yield no stochastic information about the estimated parameters. One possibility to overcome this, is to combine them with Monte Carlo methods in the future.

Acknowledgements This work was financially supported by the BMBF Geotechnologien program REAL GOCE and the ESA contract No. 18308/04/NL/MM.

References

Cottle RW, Pang J-S, Stone RE (1992). The linear complementarity problem. Computer science and scientific computing. Academic Press, San Diego

Dantzig GB (1998). Linear programming and extensions. Princeton University Press, New Jersey

Dantzig GB, Thapa MN (2003). Linear programming 2: theory and extensions. Springer, Berlin

ESA (1999). The four candidate Earth explorer core missions - gravity field and steady-state ocean circulation mission. ESA Report SP-1233(1), Granada

Fritsch D (1985). Some additional informations on the capacity of the linear complementarity algorithm. In: Grafarend EW, Sansò F (eds) Optimization and design of geodetic networks. Springer, Berlin, pp 169–184

Gill P, Murray W, Wright M (1981). Practical optimization. Academic Press, San Diego

Koch K-R (1982). Optimization of the configuration of geodetic networks. In: Proceedings of the international symposium on geodetic networks and computations, Munich

Koch K-R (1999). Parameter estimation and hypothesis testing in linear models. Springer, Berlin

Koch A (2006). Semantic integration of two-dimensional GIS data and digital terrain models. PhD thesis (in german), DGK series C, No. 6, University of Hannover

Lawson C, Hanson R (1974). Solving least squares problems. Prentice-Hall, London

Krasbutter I (2009). Decorrelation and validation of GOCE residuals. Diploma thesis (in german), Institute of Geodesy and Geoinformation, University of Bonn

Roese-Koerner L (2009). Quadratic programming with inequality constraints. Diploma thesis (in german), Institute of Geodesy and Geoinformation, University of Bonn

Schaffrin B (1981). Parameter estimation with inequality constraints. Allgemeine Vermessungs-Nachrichten (in german) 88, pp 227–238

Schlittgen R, Streitberg BHJ (2001). Time series analysis. 9th edn. (in german), Oldenbourg Verlag, Munich

Schuh W-D (1996). Tailored numerical solution strategies for the global determination of the earth's gravity field. Mitteilungen der Geodätischen Institute der TU, No. 8, Graz

Schuh W-D (2003). The processing of band-limited measurements; filtering techniques in the least squares context and in the presence of data gaps. In: Beutler G, Drinkwater M, Rummel R, von Steiger R (eds) Earth gravity field from space – from sensors to Earth sciences, space science reviews, vol 108. Academic Publishers, Kluwer, pp 67–78

Wölle NA (1988). Graphical estimation through Quadratic Programming with linear equalities and inequalities. Diploma thesis (in german), Technical University of Graz

Part V

Geodetic Boundary Value Problems and CM-Geoid Computational Methods

The Geoid Today: Still a Problem of Theory and Practice

26

F. Sansò, R. Barzaghi, and D. Carrion

Abstract

The whole process of the approximation of the anomalous gravity potential is reviewed from the point of view of available geodetic data and their primary combination to define a mathematical standard model, i.e. a suitable boundary value problem.

The construction of global models is then outlined, showing where recent theorems of existence, uniqueness and stability play a role. The local modelling is also summarized, clarifying the conditions under which it has meaning. Finally, some proposal to follow a slightly different line in global modelling are put forward, while open problems in local modelling that need to be further studied are highlighted.

1 Introduction

The main problem of physical geodesy is to establish a model approximating the anomalous gravity potential $T(\underline{x})$, on the Earth's surface and in outer space, to the extent made possible by existing data, mathematical analysis tools and numerical computational tools. On the other hand, close to the Earth's surface, the linearized Bruns relation holds, relating the height anomaly ζ to the anomalous potential T,

$$\zeta = \frac{T}{\gamma_Q} \tag{26.1}$$

with γ_Q the modulus of normal gravity, so that approximating T on some $\widetilde{S}$ (e.g. with $\widetilde{S}$ the telluroid) is the same as approximating ζ on $\widetilde{S}$, i.e. computing the separation between telluroid and the actual Earth's surface. When S refers to the ocean, i.e. on 2/3 of the Earth's surface, ζ becomes the same as the geoid undulation N, hence the title of this paper, where the word *geoid* is used more for reasons of attractiveness than for scientific rigor.

At present, very large improvement is verified in all three fields: data, mathematical models, numerical analysis, so that it is time to rethink the actual situation as for the adequacy of our results with respect to what we have available.

A warning is that in this discussion we shall consider only a stationary picture of the Earth's gravity, i.e. all time variation are assumed to be perturbations known and eliminated from our data.

The target of this paper is to highlight recent achievements on the item and to try to formulate in clear mathematical terms what open problems and future challenges we believe will have to be faced.

F. Sansò (✉) · R. Barzaghi · D. Carrion
DIIAR – Sezione Rilevamento, Politecnico di Milano, P.zza Leonardo da Vinci 32, I-20133 Milano, Italy
e-mail: fernando.sanso@polimi.it

N. Sneeuw et al. (eds.), *VII Hotine-Marussi Symposium on Mathematical Geodesy*, International Association of Geodesy Symposia 137, DOI 10.1007/978-3-642-22078-4_26, © Springer-Verlag Berlin Heidelberg 2012

2 The Data

In this review, we consider as *geodetic data*, for the purposes of physical geodesy, the following data sets:

- Spatial observations: Satellite tracking (ST) Satellite to satellite tracking (SST) Satellite accelerometry (SAC)/gradiometry (SGG), Satellite altimetry (SA), Radar imaging of continenatal areas (SAR)
- Ground observations (gravimetry, relative and absolute, pointwise heights, digital elevation models (DEM))
- Ocean observations (marine gravity, marine positioning, bathymetry).

Of course, there are many more observations that could be ascribed to the geodetic family. In particular, we want to mention aerial gravimetry and gradiometry. Yet this is so close to the surface, be it ground or ocean, that by a small downward continuation we take it as g_G or g_O. Also the deflection of the vertical are traditional geodetic observations, but overall they constitute a small data set. So we better take them as a validation tool, rather than a primary information for the determination of T. In order to have an idea of the usability of the various data sets, we try to give a rough indication of their resolution and accuracy. As for the first three data sets, since they are used to construct a model of the low-degree spherical harmonic representation of T, $\{T_{\ell m}, \ell = 2 - L, |m| \leq \ell\}$, we simply give the maximum degree L which is on the same time an indication of resolution (roughly this is $\sim\frac{180°}{L}$ in degrees on the Earth's sphere) and of accuracy, since at degree L typically commission and omission error balance each other.

We have ESA (1999) for ST, $L \sim 50$; for SST+SA, $L \sim 100$; for SST+SA+SGG, $L \sim 200$.

Moreover we have Andersen and Knudsen (1998) for SA a resolution of ~1 km and an accuracy of ~2 cm; for SAR a standard resolution is 100 m and the accuracy ~10 m Farr et al. (2007).

We notice that the resolution of SA is more than enough since the (stationary) surface of the ocean is naturally smooth. As for the SAR data, the resolution of the surface is enough even in rough areas and the accuracy too, given that a $\sigma_H \sim 10$ m can produce an error up to 2 mGal in gravity corrections (at 2σ level), which is also an accuracy frequently met in the gravity material (see below).

- For g_G we have a resolution which is quite variable from area to area Briais et al. (2008); here we mention only that it is enough to support the creation of a uniform $5' \times 5'$ grid FA gravity anomalies at the ellipsoid level, though the data are really poor in mountainous areas or on continents like Africa.

This means around $9 \cdot 10^6$ data, which is an order of magnitude compatible with the real number in the original data set. As for its accuracy it is also quite varied, ranging from 0.1 mGal to 5 mGal, for data derived from aerial gravimetry,

- For the pointwise height, its resolution is by definition the same as that of g_G, while its accuracy is generally in the range of 1 decimeter
- For DEMs Berry (1999), this can easily reach a resolution of 10 m and an accuracy of 1 m, but this data set is available only on part of the continental area, so we consider it as a tool for an improved approximation of T at a local level,
- As for g_O we have an accuracy ranging from 1 to 10 mGal Wessel and Watts (1988). This data set is ancillary to the creation of a marine gravity data set based on SA and oceanographic information,
- For the pointwise marine positioning we have an accuracy typical of GNSS, in navigation mode, namely in the decimetric range,
- For bathymetry Berry (1999), we can say that the global data set has a resolution of $1' \times 1'$ and an accuracy in the range between one and several hundreds meters, at least, far from coastal areas. Such a resolution and accuracy is sufficient for geodetic purposes, since its corresponding gravity signal is smoothed when propagated from sea floor to sea surface.

This is the picture of the situation that we could summarize by saying that we have global and local geodetic data sets, which include:

- Globally, the vector $\underline{T}$ of harmonic coefficients (up to some degree L) with their covariance matrix; a data set $\{g_G\}$ and $\{h_G\}$ on ground and a data set $\{h_O\}$ on ocean, accompanied by an oceanographic estimate of dynamic height $\{h_D\}$, the set $\{b\}$ of bathymetry,
- Locally, a set of values $\{g_G\}$ with higher resolution than the global, a DEM $\{h_G\}$ with higher resolution than the global; or a set of marine gravity values $\{g_0\}$ and a locally improved bathymetry $\{b\}$.

What is typical of geodesy is that such data sets are primarily combined in order to formulate the problem of determining T in a more treatable form.

3 The Combination of Data Sets

This phase of modelling the problem of the determination of T is critical because on our approach depends the type of problem we will have to solve in the end.

We recall in this section the approach that has been used in the past and still is used today, for global as well as for local modelling. We send to a final paragraph the discussion of a possible alternative approach.

The global modelling is done of several steps that we summarize, warning the reader that here we do not take into account the problem of the unification of the height datum. Also the line presented below might not be universal, but it covers the appraoch used for EGM08 Pavlis (1997).

GLOBAL-STEP 1: Rummel (1993) let $\overline{W} = \overline{U}$ be the value of the gravity potential on the geoid as well as the normal gravity value on the ellipsoid. Then we have (using O as an index to mean on the ocean)

$$W_O - \overline{W} \cong -\gamma(h_0 - N) = -\gamma h_D \quad (26.2)$$

$$W_O - \overline{W} = U_O + T_O - \overline{U} = -\gamma h_0 + T_O \quad (26.3)$$

The combination of (26.2), (26.3) gives

$$T_O = \gamma(h_0 - h_D), \quad (26.4)$$

meaning that we can assume to know T on the ocean, from SA, marine positioning and the model of $\{h_D\}$.

GLOBAL-STEP 2: from $g_{G,P}$ and H_P we compute, with several approximations, Albertella and Sansò (1997), $W_{G,P}$ so that the normal height h_P^* is also available and we can compute the free air anomaly $\Delta g_P = g_{G,P} - \gamma(h_P^*)$.

Note that once H_P was known from leveling but nowadays it is much simpler to determine h_P from GNSS observations. This would allow the computation, at P, of the gravity disturbance $\delta g = g_G - \gamma$.

On the contrary, to compute Δg from h_P, a first *guess* geoid $\widetilde{N}$ has to be used to give $H_P = h_P - \widetilde{N}$.

GLOBAL-STEP 3: the two data sets $\{T_O\}$ and $\{\Delta g_P\}$ are block averaged and then merged with $\{\underline{T}\}$ by least squares to produce a new intermediate model up to a maximum degree L, say $L = 100$. This step is a great contribution of the Ohio State University School Rapp (1997) and it has been accomplished once in preparation of the OSU91 gravity model, up to $L = 360$. Once it was done, its result has been cast into the above model and it has not been necessary to repeat it, for the generation of the new EGM08, up to degree 2160.

GLOBAL-STEP 4: with the help of the intermediate model one can reduce the data set $\{T_O\}$ to another one with a much shorter correlation length. This opens the way to the so-called inverse Stokes theory Barzaghi et al. (1993), namely the prediction of $\{\Delta g_P\}$ on the ocean from (reduced) $\{T_O\}$

GLOBAL-STEP 5: the ground gravity data set $\{\Delta g_P\}$ is manipulated to produce averaged block values $\{\Delta \overline{g}_P\}$ at surface altitude. This is done very much in the same way as it will be described for the local modelling.

GLOBAL-STEP 6: the data set $\{\Delta \overline{g}_P\}$ is downward continued to the ellipsoid, by using Molodensky's theory in the version of Pellinen, Pellinen (1962).

GLOBAL-STEP 7: a final boundary value problem is solved for a regular grid of Δg that now covers the whole ellipsoid (both land and sea).

This is done for instance by least squares Pavlis (1997), what is feasible even to very high degrees like $L = 2160$, because the corresponding normal system becomes block diagonal, due to the regularity of the grid on the ellipsoid.

Despite the difficulty of step 4 and step 6, involving the solution of improperly posed problems, and a certain ambiguity in step 5, where the knowledge of the surface, with respect to which the local terrain correction is done, gets lost, the final result of the whole procedure is spectacular, as the testing of the EGM08 model has proved, Newton's Bulletin (2009).

As for the local modelling of the *geoid*, many (but not all) current approaches can be described as follows:

LOCAL-STEP 1: prepare a local data set $\{\Delta g_P\}$ which has to include more data (higher resolution) than those used for global modelling and subtract the corresponding Δg_P^M computed from a global model; this has the effect of shortening very much the correlation length of the gravity anomaly,

LOCAL-STEP 2: prepare the local DEM, which has to be finer than the global one and implement a residual terrain correction Forsberg (1997) to Δg_P; this implies the definition of some smooth reference DEM and it has a non-univocal but clear impact in smoothing the gravity anomaly data,

LOCAL-STEP 3: switch from the residual Δg to the residual T by one of the local solution methods; one commonly applied theory is that of least squares collocation Moritz (1980), but Stokes formula with FFT is too, Schwarz et al. (1990),

LOCAL-STEP 4: restore the contributions of the global model and of the residual terrain correction, adding them to the residual T computed in step 3. It has to be underlined here that the global and local steps outlined in the paper are so to say along the concepts of Molodensky's theory. There are other approaches where the downward continuation is preceded by a full removal of the intermediate masses. In this respect we would like to mention Martinec (1998), Heck (2003), Sjöberg (2000). This line of thought we call Helmert's line. Although the authors are less confident in this approach, because of the uncertainty of the density of removed masses, it has to be acknowledged that it is capable of producing very good results on a practical ground, as proved e.g. in the Alvernia test promoted by IGeS in 2008 (http://www.iges.polimi.it/).

In this way the local modelling is by now widely applied with different strategies and has allowed the production of many high resolution "local patches of geoid" with accuracies ranging from 1 to 5 cm, IGeS (1995). It is interesting though to observe that such geoid results, in most cases require a *calibration* on a number of known geoid undulations, like those derived from GPS-leveling, beyond the usual adjustment for the vertical datum; otherwise there was often in the solution a long wavelength error with an amplitude in the range of 10 cm, pointing at a certain degree of ambiguity or incompatibility in gluing local to global models which however has been damped in the most recent global model (e.g. EGM08).

Moreover, the appearance of the EGM08 model, with its very high resolution, has put into question the possibility of a local improvement. Of course in principle the addition of new local data should always allow for a better prediction, specially of the gradients of T, so the present situation is just that the local modelling has to be reviewed carefully trying to push further the methods for a better local solution.

4 Global Models and the Analysis of Geodetic BVP's

We like to see a global model, of the form

$$T_{(M)} = \left(\frac{GM}{R}\right)\sum_{\ell=2}^{M}\sum_{m=-\ell}^{\ell} T_{\ell m} \times \left(\frac{SR}{r}\right)^{\ell+1} Y_{\ell m}(\sigma) \tag{26.5}$$

$$\sigma = (\vartheta, \lambda),$$

$$Y_{\ell m} = \text{normalized spherical harmonics}$$

as an approximation of the true anomalous potential of the gravity field. As explained in §3, after a number of manipulation of the data, in principle the determination of T is reduced to the classical scalar linearized Molodensky BVP, namely to find T such that

$$\begin{cases} \Delta T = 0 & \text{in } \widetilde{\Omega}(\partial\widetilde{\Omega} = \widetilde{S}) \\ \underline{e}_\gamma \cdot \nabla T + \frac{\gamma'}{\gamma} T = \Delta g & \text{on } \widetilde{S}(\text{telluroid}) \\ T = O\left(\frac{1}{r^3}\right) & \text{for } r \to \infty \end{cases} \tag{26.6}$$

where

$$\underline{e}_\gamma(P) = \frac{\underline{\gamma}(P)}{\gamma(P)} \cong -\underline{\nu}(P) \tag{26.7}$$

$$\gamma' = \frac{\partial \gamma}{\partial h}. \tag{26.8}$$

The conditions for the existence, uniqueness and stability of solutions of (26.6) has been an intriguing problem keeping a few geodesists busy for some years. One of the most recent result is Sansò and Venuti (2008) and it can be stated as:

Proposition 1. *Assume harmonic coefficients of T are known and fixed up to a degree $L = 25$; assume further the telluroid $\tilde{S} \equiv \{r = \widetilde{R}(\sigma)\}$ has maximum inclination I with respect to the radial direction $\underline{e}_r$,*

$$|I| \leq 60°, \tag{26.9}$$

then for every $\Delta g \in L^2(\sigma)$*, i.e. such that*

$$\|R\Delta g\|^2_{L^2(\sigma)} = \frac{1}{4\pi}\int \Delta g(\sigma)^2 R^3(\sigma) d\sigma < +\infty, \quad (26.10)$$

there is one and only one function T *harmonic in* Ω *and constants* $a_{\ell m}$ *such that*

$$\left. \underline{a}_\gamma \cdot \nabla T + \frac{\gamma'}{\gamma} T \right|_{\widetilde{S}} = \Delta g(\sigma) + \left. \sum_{\ell=0}^{L} \sum_{m=-\ell}^{\ell} a_{\ell m} Z_{\ell m} \right|_{\widetilde{S}} \quad (26.11)$$

and in addition

$$\left| T - \sum_{\ell=0}^{L} \sum_{m=-\ell}^{\ell} T_{\ell m} \left(\frac{R}{r}\right)^{\ell+1} Y_{\ell m}(\sigma) \right| = O\left(\frac{1}{r^{L+2}}\right) \quad n \to \infty. \quad (26.12)$$

The functions $\{Z_{\ell m}(r,\sigma)\}$ *in (26.11) are auxiliary functions implicitly defined by the relations*

$$\frac{1}{4\pi}\int Z_{\ell m}(\widetilde{R}(\sigma),\sigma) Y_{jk}(\sigma) \left[\frac{R}{\widetilde{R}(\sigma)}\right]^{j+1} d\sigma = \delta_{\ell j}\delta_{mk}\left(\frac{R}{\overline{R}}\right)^{\ell+1}, \quad (26.13)$$

with $\overline{R}$ *any fixed radius such that* $\overline{R} > \sup \widetilde{R}(\sigma)$*.*

The function T *so found has a gradient in* $L^2(\sigma)$*, namely it belongs to the space* $HH_1(\widetilde{S})$ *and furthermore there is a constant c such that*

$$\frac{1}{4\pi}\int |\nabla T|^2 R(\sigma)^3 d\sigma < c\frac{1}{4\pi}\int \Delta g^2 R^3(\sigma) d\sigma. \quad (26.14)$$

This gives us (almost) all what we need to know about a reasonable theoretical solution of the BVP. Next, however, we would need to know whether the basis of spherical harmonics, i.e.

$$\left\{ S_{\ell m}(r,\sigma) = \left(\frac{R}{r}\right)^{\ell+1} Y_{\ell m}(\sigma) \right\},$$

is good or not to perform an approximation procedure.

This is true if $\{S_{\ell m}\}$ happens to be a basis of the solution space HH_1. Since the traces on S of $T \in HH_1$ are functions belonging to the classical Sobolev space $H^{1,2}(\widetilde{S})$ and vice versa (at least under regularity conditions on $\widetilde{S}$), implying that $T \in H^{s,2}(\Omega)$ with $s = 3/2$ (cf. Sacerdote and Sansò (2008)), we may conclude that $\{S_{\ell m}\}$ is a basis in HH_1 because this is true for every exterior Sobolev space, as claimed by the Runge – Krarup theorem, Krarup (2006). What Proposition 1 says is that, under the mentioned conditions, we expect that

$$\lim_{M\to\infty} \| T - T_{(M)} \|_{HH_1} = 0, \quad (26.15)$$

i.e. $T_{(M)}$ is an approximation in the sense of HH_1 to T.

This is true if we have a data set Δg given over the whole boundary $\widetilde{S}$, without errors. Since the norm in the data side is $L^2(\widetilde{S})$ we can even say that the solution is stable with respect to "small" errors in $L^2(\widetilde{S})$.

Two comments are in order at this point:

Remark 4.1. Although (26.15) says that there is no limit in principle in the application of global models' theory as for the maximum degree M, or better no other limit than the one imposed by the finiteness of real data, this does not mean that the numerical approach followed until now will be in future the most convenient. In fact, though the numerical solution for a gridded data set on the ellipsoid could be pushed even beyond the present degree 2160, all the previous manipulation, and in particular the downward continuation, becomes questionable for very high degrees. In any case, if not for theoretical reasons at least for numerical convenience when we will try to go directly to the actual boundary or a surface close by, like the telluroid, we shall have to consider the use of a multi-resolution basis, as described e.g. in Freeden et al. (1998), Keller (2004). True it is that the remove-restore method mediated by local solutions is a kind of a naïve multi-resolution approach developed by the geodetic community, as it aims at removing separately the contributions of signals with different spectral signature.

Remark 4.2. Although the relation (26.15) implies that the individual sequence $\{T_{\ell m}\}^{(M)}$, i.e. the coefficient of degree ℓ, order m for the model with maximum degree M, admits a limit, for ℓ, m fixed,

$$\lim_{M\to\infty} T^{(M)}_{\ell m} = T_{\ell m} \quad (26.16)$$

and the coefficients $T_{\ell m}$ so found are the true coefficients of T providing a harmonic series converging outside a Brillouin radius $\overline{R}$

$$\overline{R} \geq \max_{\sigma} \widetilde{R}(\sigma), \tag{26.17}$$

this does not mean that the potential T can be represented by this series down to the level of $\widetilde{S}$. Counter examples are well known in literature Moritz (1980).

5 Local Models and *Local BVP's*

The title of this section is purposely paradoxical. There are no local BVP's for the Laplace operator, or more generally for elliptic problems. At least not properly posed *local* problems. So we think that one should stay within the geodetic tradition and accept that there can be only a local refinement of the approximation of T. This indeed calls immediately for the use of some local basis function. In fact the plane use of spherical harmonics tends to spread the local behavior everywhere on the sphere, very much like to what happens in 1D, if we want to use a Fourier's basis on a small interval to approximate a function on the whole $[0, 2\pi]$ interval.

So local splines of the kind already considered in literature, e.g. in Freeden et al. (1998), Albertella et al. (1999) have to be applied. However, one thing has to be clearly stated on a logical ground: already the idea of performing a local improvement of the approximation of the gravity field requires that the spatial correlation between the gravity signal without and within the local area, should never be strong, otherwise a local improvement of the interpolation would correspond to a global worsening of the approximation. This explains why statistics has entered so naturally into gravity approximation theory. At the same time, the assumption above is realistic only if gross features effects are previously subtracted from the data, i.e. a step of removing a global model has first been applied. This, together with the residual terrain correction removal, usually leaves a smooth, homogenous gravity signal at least for patches of the surface with a diameter of several degrees. When this happens, the theory of homogeneous, isotropic random fields can be used, resulting in a choice of the local spline given by the local covariance function. This approach, known as LSC, Tscherning (1997), can be viewed as a process of choosing the base function for a local approximation by means of a covariance, that automatically incorporates statistical features of the signal. Today the large step forward performed in global modelling with a spatial resolution of ~10 km and an overall accuracy of ~30 cm is strongly pushing the research for a more fitting local approximation method, resolving knots left unsolved by LSC theory.

Indeed we already mentioned other methods applied for the local modelling, providing excellent results too on a numerical ground Schwarz et al. (1990), Sideris (1995), Martinec (1998), Sjöberg (2000).

6 Open Issues and Proposals

In this section we try to summarize our opinions on issues open to discussion suggesting some proposals.

(A) Global modelling:

A1 The downward continuation from the surface to the ellipsoid. This is an improperly posed problem, stabilized implicitly by the computation procedure, though we do not have a clear theoretical control on it. The proposal here could be to define previously an approximate surface $\widetilde{S}$ not too far from the actual Earth's surface S, but smooth enough to form meaningful area averages of gravity anomalies. This has the advantage that a precise residual terrain correction procedure can be set up and no ambiguity is met in the phase of restoring, and on the same time data need to be moved, in computing block averages, only for a very short distance, reducing instability effects. Naturally this requires the study of appropriate numerical methods, for instance a suitable version of Galerkin's method, an important field of current research, Holota and Nesvadba (2008),

A2 The formulation of the geodetic BVP should be reviewed. Ideal would be the implementation of the original altimetry gravimetry problem, where on $\widetilde{S}$ we give directly block averages of T on the ocean and some kind of gravity anomaly on land. Nevertheless as a first step we propose that the linear Molodensky problem for the free air gravity anomaly Δg be substituted by the fixed boundary BVP with the gravity disturbance δg as a datum. We believe that the present knowledge of the geometry of S is sufficient to support this formulation, which on the other hand has much better stability properties than Molodensky's formulation.

The analogous of Proposition 1 for the fixed BVP: assume that δg is square integrable on $\widetilde{S}$ and that the inclination of $\widetilde{S}$ on $\underline{e}_r$ is less than 89°.6, then there is one and only one solution of the fixed BVP, with a gradient square integrable on $\widetilde{S}$.

(B) Local modelling by LSC

B1 Local solutions, in particular LSC, should be studied, adapted to the ellipsoidal geometry. When this problem will be solved we will be able to glue local to a global solution without changing the harmonicity domain, as it is still done today.

B2 We have to study a reasonable probabilistic model for LSC, including a consistent definition of local covariance functions as particular realizations of a unique non-homogeneous covariance model. Again, the solution of this problem will allow patching local solutions without discontinuities.

Last, but not least, the authors think that, since there are at least two approaches to local geoid modelling (Molodensky and Helmert) providing many times solutions of equivalent accuracy, it would be nice to have a clear study showing their almost equivalence, under suitable conditions, on a theoretical ground.

References

Albertella A, Sansò F (1997) A compendium of physical geodesy. In: Lecture Notes of the International School for the Determination and Use of the Geoid. International Geoid Service DIIAR, Politecnico di Milano

Albertella A, Sansò F, Sneeuw N (1999) Band-limited functions on a bounded spherical domain, The Slepian problem on the sphere. J of Geod 73:436–447

Andersen O, Knudsen P (1998) Global marine gravity gield from the ERS1 and Geosat geodetic mission altimetry. J Geoph Res 103:8129–8137

Barzaghi R, Fermi A, Tarantola S, Sansò F (1993) Spectral techniques in inverse Stokes' and overdetermined problems. Surv Geophys 14:4–5

Berry P (1999) Global Digital Elevation Models. Facts or Fiction? Astron Geophys 40:3

Briais A, Bouvalat S, Sarrailh M (2008) The new world gravity map project: a tool for geodynamic studies. Proceedings of the 33rd IGC, Oslo

ESA (1999) Gravity field and steady-state ocean circulation mission. ESA SP-1233(1). ESA Publication Division, c/o ESTEC, Noordwijk, Netherlands

Farr TG et al. (2007) The Shuttle radar topography mission, Rev Geoph 45, RG 2004

Forbserg R (1997) Terrain effects in geoid computations. In: Lecture Notes of the International School for the Determination and Use of the Geoid. International Geoid Service DIIAR, Politecnico di Milano

Freeden W, Gerrens T, Schreiner M (1998) Constructive approximation on the sphere, Oxford Science Publications, New York

Heck B (2003) On Helmert's methods of condensation J of Geod 77:155–170

Holota P (1997) Coerciveness of the linear gravimetric boundary-value problem and a geometrical interpretation J of Geod 71:640–651

Holota P, Nesvadba O (2008) Model Refinements and Numerical Solutions of Weakly Formulated Boundary-Value Problems in Physical Geodesy IAG Symposia 132 In: Proc VI Hotine-Marussi Symposium on Theoretical and Computational Geodesy pp. 320–326, Springer, Berlin

IGeS (1995) New geoids in the world IGeS Bulletin, International Geoid Service DIIAR, Politecnico di Milano

Keller W (2004) Wavelets in Geodesy and Geodynamics. Walter de Gruyter

Krarup T (2006) Mathematical foundations of Geodesy. Selected papers of Torben Krarup. Kai Borre ed, Springer

Martinec Z (1998) Boundary-value problems for gravimetric determination of a precise geoid. In: Lecture Notes in Earth Sciences, N73, Springer

Moritz H (1980) Advanced physical geodesy. Wichmann

External quality evaluation reports of EGM08. Newton's Bulletin 4

Pavlis NK (1997) Development and application of geopotential models. In: Lecture Notes of the International School for the Determination and Use of the Geoid. International Geoid Service at DIIAR, Politecnico di Milano

Pellinen LP (1962) Accounting for topography in the calculation of quasi-geoidal heights and plumb-line deflection from gravity anomalies, Bull Geod, 63

Rapp R (1997) Global models for the one centimeter geoid. Present status and near-term prospects. In: Lecture Notes in Earth Sciences N65 Geodetic boundary value problems in view of the one centimeter geoid, Springer

Rummel R (1993) Principle of satellite altimetry and elimination of radial orbital errors. In: Lecture Notes in Earth Sciences N50 Satellite altimetry in geodesy and oceanography, Springer

Sacerdote F and Sansò F (2008) Least squares, Galerkin and boundary value problems. Proceedings, International Symposium on Gravity, Geoid and Earth Observations, Chania

Sansò F, Venuti G (2008) On the explicit determination of stability constants for linearized geodetic boundary value problems. J of Geod 82:909–916

Schwarz KP, Sideris MG, Forsberg R (1990) The use of FFT techniques in physical geodesy. Geophys J Int 100:485–514

Sideris MG (1995) Fourier geoid determination with irregular data. J of Geod 70:2–12

Sjöberg LE (2000) Topographic effects by the Stokes–Helmert method of geoid and quasi-geoid determinations. J of Geod 74:255–268

Tscherning CC (1997) Geoid determination by least squares collocation using Gravsoft. In: Lecture Notes of the International School for the Determination and Use of the Geoid. International Geoid Service DIIAR, Politecnico di Milano

Wessel P, Watts AB (1988) On the accuracy of marine gravity measurements. J of Geod 93:393–413

Omission Error, Data Requirements, and the Fractal Dimension of the Geoid

27

Christopher Jekeli

Abstract

The newest global geopotential model, EGM08, yields significantly improved height anomaly (and geoid undulation) estimates, but not yet at the level of 1 cm accuracy. Achieving this goal requires higher resolution gravimetric data (among other advancements, both theoretical and numerical). To determine the necessary data resolution, a statistical approach using the power spectral density (psd) of the height anomaly may be used to relate resolution to standard deviation in omission error. Kaula's rule was the first such relationship based on a power-law approximation to the psd. It is shown that the Earth's topography, whose fractal nature implies a power-law attenuation of its psd, and which in many cases is linearly correlated with the gravity anomaly on the basis of Airy's isostatic assumption, can be used to design approximations to the psd of the local height anomaly, thus leading to estimates of the data resolution required to support the 1 cm accuracy level.

Keywords

Omission error • Geoid undulation • Power spectral density • Fractal dimension

1 Introduction

The last decade has witnessed a significant improvement in global geopotential models, not only the most obvious, being the model(s) derived from the Gravity Recovery and Climate Experiment (GRACE) satellite mission launched in March 2002, but also the Earth Gravitational Model 2008 (EGM08, Pavlis et al. 2008) developed from a much expanded and enhanced set of terrestrial measurements, including land and marine gravimetry, satellite altimetry of the oceans, airborne gravimetry over the Arctic and other regions of the world, and topographic data from the Shuttle Radar Topographic Mission (SRTM). While GRACE has yielded orders of magnitude in increased accuracy at lower resolution, the compilation of world-wide high resolution data has produced what may be considered the new standard of high-degree spherical harmonic combination geopotential models. With spherical harmonics complete to degree and order 2160 EGM08 represents a model having 10 km resolution or better (5 arcmin) anywhere in the world, leading one to ask whether or how much more gravity data are needed to produce a corresponding geoid model to a given accuracy.

C. Jekeli (✉)
Division of Geodeticl Science, School of Earth Sciences, Ohio State University, 125 South Oval Mall, Columbus, OH, USA
e-mail: jekeli.1@osu.edu

N. Sneeuw et al. (eds.), *VII Hotine-Marussi Symposium on Mathematical Geodesy*, International Association of Geodesy Symposia 137, DOI 10.1007/978-3-642-22078-4_27,

The answer to this question depends on the relationship between the accuracy and data resolution. In this connection, one might also ask, what is the next goal for a global spherical harmonic expansion of the geopotential. How high in harmonic degree and order should the next model be developed to satisfy geodetic needs? Finally, what do these global models tell us about the Earth's gravity field? What can they tell us about the local or regional field? These are questions that the paper attempts to address, not exhaustively, but tentatively, although specific conclusions are obtained.

We start with the definition of the disturbing potential, T, which is the difference between the actual gravity potential of the Earth, W, and the gravity potential, U, of the normal ellipsoid that rotates with the Earth, has its mass, M (including that of the atmosphere) and whose center is at Earth's center of mass, as well as at the origin of the coordinate system. Then, the spherical harmonic expansion of T in terms of spherical polar coordinates, (r, θ, λ), is:

$$T(r,\theta,\lambda) = \frac{GM}{a}\sum_{n=2}^{\infty}\sum_{m=-n}^{n}\left(\frac{a}{r}\right)^{n+1} C_{nm}\bar{Y}_{nm}(\theta,\lambda), \tag{27.1}$$

where G is Newton's gravitational constant, a is the radius of the bounding (Brillouin) sphere (often approximated by the semi-major axis of the mean-Earth normal ellipsoid), the C_{nm} are the (unit-less) Stokes coefficients, and the $\bar{Y}_{nm}(\theta,\lambda)$ are surface spherical harmonics defined by

$$\bar{Y}_{nm}(\theta,\lambda) = \bar{P}_{n,|m|}(\cos\theta)\begin{cases}\cos m\lambda, & m \ge 0\\ \sin|m|\lambda, & m < 0\end{cases} \tag{27.2}$$

The functions, $\bar{P}_{n,|m|}$, are associated Legendre functions, fully normalized so that

$$\frac{1}{4\pi}\iint_{\sigma}(\bar{Y}_{nm}(\theta,\lambda))^2 d\sigma = 1 \quad \text{for all } n, m, \tag{27.3}$$

where σ represents the unit sphere. The series converges outside the Brillouin sphere and corresponding partial (truncated) series of T are valid approximations only in free space, that is, where T is harmonic (satisfying Laplace's equation); however, see also Sect. 3.

For an arbitrary point, x, on the Earth's surface, we approximate the height anomaly as (Bruns's equation)

$$\zeta(x) = \frac{1}{\gamma(x_0)}T(x), \tag{27.4}$$

where γ is normal gravity evaluated at the telluroid point, x_0, defined according to the Molodensky mapping (Heck 1997). If x is on the geoid, then the height anomaly is also known at the *geoid undulation*, N. Most analyses of the geopotential are performed in terms of the geoid undulation (or, height anomaly) through the proportionality given by (27.4). Some care is required for precise work when using spherical harmonic expansions of T at points on the geoid that are below the Earth's surface, where the truncated series technically is not valid.

2 Degree Variances and Power Spectral Density

A particular spherical harmonic model of the geopotential is given by (27.1) truncated at degree, $n_{\max}$. Each model has two types of error, the commission error committed by using inaccurate data to determine the coefficients (and that, in turn, leads to errors in derived quantities, like the geoid undulation), and the omission error due to the finite maximum degree and order of the model, which defines its maximum resolution. If εC_{nm} denotes the error in model coefficients, then the variance of the error per degree (the error degree variance) is

$$\sigma_n^2(\varepsilon C_{nm}) = \sum_{m=-n}^{n}\varepsilon C_{nm}^2, \qquad 2 \le n \le n_{\max}, \tag{27.5}$$

provided the errors for different orders within a degree are independent, which can happen only if the data that are used to determine the coefficients have errors uncorrelated with respect to longitude (Colombo 1981; Lemoine et al. 1998). Similarly, the degree variances of the field coefficients, themselves, are defined as:

$$\sigma_n^2(C_{nm}) = \sum_{m=-n}^{n} C_{nm}^2, \qquad n \ge 2, \tag{27.6}$$

and this may be understood as a statistical variance under the usual stochastic interpretation of the geopotential field (Moritz 1980).

Our focus is on the omission error, given by

$$\varepsilon T_{\text{omission}} = -\frac{GM}{a}\sum_{n=n_{\max}+1}^{\infty}\sum_{m=-n}^{n}\left(\frac{a}{r}\right)^{n+1} C_{nm}\bar{Y}_{nm}(\theta,\lambda). \tag{27.7}$$

It represents an operational limit in the achievable accuracy of the model – no matter how accurate the coefficients of the model are, the accuracy of the model in its totality is limited by its maximum degree, which ultimately is equivalent (or, at least analogous) to the maximum data resolution. We assume that the omission and commission errors are independent, although this is not strictly true since aliasing errors in the model spectrum result if the data have resolution higher than dictated by $n_{\max}$ (see Jekeli 1996). On the other hand, because of the orthogonality of spherical harmonics, the harmonics in the omission error are statistically independent, and the variance of the total (unit-less) omission error for some fixed r is given by

$$\sigma^2_{\text{omission}}(T) = \left(\frac{GM}{a}\right)^2 \sum_{n=n_{\max}+1}^{\infty} \left(\frac{a}{r}\right)^{2n+2} \sigma_n^2(C_{nm}). \tag{27.8}$$

The height anomaly degree variance on the sphere of radius, a, is defined according to its functional relationship to the disturbing potential. From (27.1), (27.4), and (27.6), we have

$$\sigma_n^2(\zeta) = a^2 \sigma_n^2(C_{nm}), \tag{27.9}$$

where the approximation, $\gamma \approx GM/a^2$, was used. The corresponding omission error variance is the sum of these degree variances with degrees greater than $n_{\max}$. Approximating the geoid by a sphere of radius, $R = 6371$ km, and neglecting the harmonicity error (evaluating T at points where it is not harmonic), we have

$$\sigma_n^2(N) \approx R^2 \left(\frac{a}{R}\right)^{2n+2} \sigma_n^2(C_{nm}). \tag{27.10}$$

As seen in the next section, the amplification factor, $(a/R)^{2n+2}$ will lead to problems at the very high degrees. Thus, it is preferable to consider an alternative representation at local scales.

Instead of a spherical harmonic series, the local (planar) spectrum of the field may be described by a discrete Fourier series, if based on discrete data regularly spaced on the plane. Typical data are the gravity anomaly, Δg, related in planar approximation to the disturbing potential according to

$$\Delta g \approx -\frac{\partial T}{\partial z}, \tag{27.11}$$

where z is vertical distance, positive upwards. Given a set of discrete values of the gravity anomaly, $\Delta g_{\ell_1,\ell_2}$, in a region, their discrete Fourier transform is given by

$$G_{k_1,k_2} = \Delta x_1 \Delta x_2 \sum_{\ell_1=0}^{N_1-1} \sum_{\ell_2=0}^{N_2-1} \Delta g_{\ell_1,\ell_2} e^{-i2\pi(k_1\ell_1+k_2\ell_2)},$$
$$k_p = 0,\ldots,N_p-1, \quad p = 1,2, \tag{27.12}$$

where the Cartesian coordinates of the Δg-values in the plane are $x_1 = \ell_1 \Delta x_1$, $x_2 = \ell_2 \Delta x_2$; and the corresponding cyclical frequencies are

$$(f_1)_{k_1} = \frac{k_1 - N_1/2}{N_1 \Delta x_1}, \qquad (f_2)_{k_2} = \frac{k_2 - N_2/2}{N_2 \Delta x_2}. \tag{27.13}$$

The psd of the gravity anomaly can be estimated by averaging the corresponding periodogram (the square of the spectrum) over all directions of the frequencies:

$$\Phi_{\Delta g}(f_k) = \frac{1}{M_k} \sum_{f_k \approx \sqrt{(f_1)^2_{k_1} + (f_2)^2_{k_2}}} \frac{1}{\Delta x_1 \Delta x_2} |G_{k_1,k_2}|^2, \tag{27.14}$$

where M_k is the number of frequencies, $\sqrt{(f_1)^2_{k_1} + (f_2)^2_{k_2}}$, that satisfy the summation criterion in (27.14).

Using (27.4) for the geoid undulation, the relationship between its spectrum, U_{k_1,k_2}, and that of the gravity anomaly is

$$G_{k_1,k_2} = \left(2\pi \sqrt{(f_1)^2_{k_1} + (f_2)^2_{k_2}}\right) \gamma\, U_{k_1,k_2}, \tag{27.15}$$

where the parenthetical factor accounts for the vertical derivative (27.11). From (27.14), the psd of the geoid undulation is then

$$\Phi_N(f_k) = \frac{1}{(2\pi f_k)^2 \gamma^2} \Phi_{\Delta g}(f_k). \tag{27.16}$$

At the planar level of approximation, we need not distinguish between the geoid undulation and the height anomaly (on the Earth's surface), and the same psd holds for both.

It can be shown (Jekeli 2003) that the psd and degree variances (for any quantity) are related according to

$$\Phi(f_n) = \frac{2\pi R^2}{n} \sigma_n^2, \tag{27.17}$$

where $f_{sub_n} = n/(2 * \pi * R)$.

3 Global PSD Models

Kaula promoted and developed the idea of a stochastic interpretation of the spherical spectrum of the gravitational field; and, an often-used rule-of-thumb (Kaula 1966, p.98), bearing his name, provides a rough, global estimate of the standard deviation of the geoid undulation omission error. His model for the degree variances of the geoid undulation is

$$\sigma_n^2(N) = R^2\,(2n+1)\,\frac{10^{-10}}{n^4}[\mathrm{m}^2]. \tag{27.18}$$

Summing this as in (27.8) yields

$$\sigma_{\mathrm{omission}}(N) = R{\cdot}10^{-5}\sqrt{\sum_{n=n_{\max}+1}^{\infty}\frac{2n+1}{n^4}} \approx \frac{64}{n_{\max}}[\mathrm{m}]. \tag{27.19}$$

Hence, for $n_{\max} = 360$, $\sigma_{\mathrm{omission}}(N)|_{n_{\max}=360} = 18\,\mathrm{cm}$. Many other researchers, notably Rapp (1979), have devised refinements of this model based on improved global geopotential models.

Figure 27.1 shows the power spectral density (using (27.9) and (27.6)) of the height anomaly according to the EGM96 (Lemoine et al. 1998) and EGM08 models, as well as Kaula's rule. Here, we note that coefficients of the EGMs refer to a sphere of radius, $a = 6378136.46\,\mathrm{m}$, which is close to the Brillouin sphere and thus tends to understate the power at the very high frequencies (degrees) on the Earth's surface.

It has been suggested that the Earth's gravitational potential, like its topography and that of other planets, behaves like a fractal (e.g., Turcotte 1987). That is, their psd's obey power laws, essentially of the form:

$$\Phi(f) = bf^{-\alpha}, \qquad f \geq f_0, \tag{27.20}$$

with constants, α, b, where α is related to the dimension of the fractal. Indeed, Kaula's rule, (27.18), is such a power law (compare (27.17) and (27.18) with (27.20)). Fitting this model to the high frequencies of the EGM08 psd, one obtains

$$\begin{aligned}\Phi_\zeta(f) &= bf^{-\alpha}\\ &= 1.611\times10^{-19}f^{-5.306}[\mathrm{m}^2/(\mathrm{cy/m})^2],\\ &\quad f \text{ has units [cy/m]}\end{aligned} \tag{27.21}$$

and integration over frequencies ($f_{\max} < f < \infty$) with $f_{\max} = n_{\max}/(2\pi R)$ yields the following corresponding standard deviation for the omission error (for the height anomaly):

$$\sigma_{\mathrm{omission}}(\zeta) = \frac{2043}{n_{\max}^{1.653}}[\mathrm{m}]. \tag{27.22}$$

For example, if $n_{\max} = 360$, then this model predicts a standard deviation of omission error equal to $\sigma_{\mathrm{omission}}(\zeta)|_{n_{\max}=360} = 12\mathrm{cm}$.

This estimate refers to height anomalies on the Brillouin sphere; whereas, one typically would like the omission error at the Earth's surface (or, for the geoid undulation, on the geoid). Using (27.10) for the degree variances of the geoid undulation (or, also the height anomaly at the surface), we notice an unnatural upward turn from a power-law attenuation at very high frequencies, $f \geq 3\times10^{-5}\,\mathrm{cy/m}$, or degrees, $n \geq 1200$ (see Fig. 27.1). One may conclude that the evaluation of EGM08 near the Earth's surface, even if still in free space, could be in error by several centimeters, especially at the higher latitudes, just due to the apparent beginnings of the divergence of the series below the Brillouin sphere. For example, consider a more natural power-law extension of EGM08 at $R = 6371\,\mathrm{km}$ (not shown, but similar to Kaula's rule):

$$\Phi_N(f) = 4.044\times10^{-12}f^{-3.898}[\mathrm{m}^2/(\mathrm{cy/m})^2]. \tag{27.23}$$

The difference between the cumulative degree variances of EGM08 from degree 1,200 to 2,160 and the

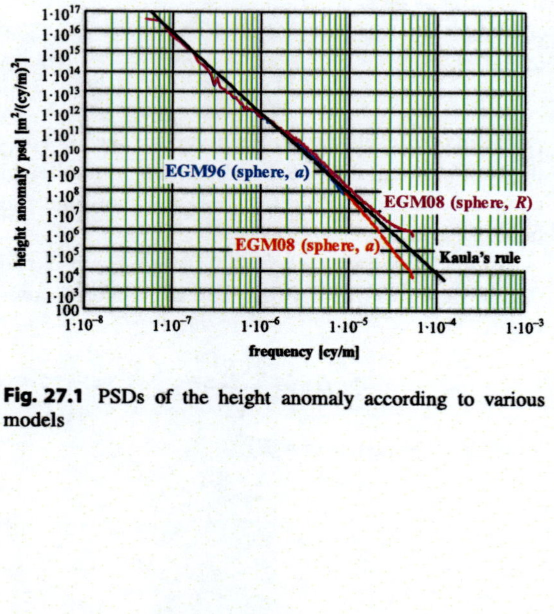

Fig. 27.1 PSDs of the height anomaly according to various models

integral of this power law over the same range of frequencies is

$$R^2 \sum_{n=1200}^{2160} \left(\frac{a}{R}\right)^{2n+2} \sigma_n^2 (C_{nm}) - 2\pi \int_{3\times10^{-5}}^{5.4\times10^{-5}} f \Phi_N(f)\, df$$
$$= 4.6 \times 10^{-3}\text{m}^2 \qquad (27.24)$$

which is about 6.8 cm in terms of standard deviation.

4 Local PSD Models

A model such as (27.22) gives a global average of the omission error. As demonstrated above, it is not particularly useful for local or regional analyses. Therefore, locally one may consider the planar psd of the gravity anomaly and derive a similar power-law model. Three areas in the USA, shown in Fig. 27.2, were evaluated in this manner. Areas 1 and 2 represent relatively smooth topography (and gravity), while Area 3 is rougher due to the Rocky Mountains. The corresponding psd's of the gravity anomaly are shown in Fig. 27.3. Corresponding power-law models of the form of (27.20) are shown in Fig. 27.4 for the derived height anomaly psd's (27.16) in Areas 1 and 3; and Table 27.1 lists the associated parameters, (b, α), as well as the formulas for the standard deviations of the omission errors and the implied required data resolutions for a precision (omission error, only) of 0.7 cm in the geoid undulation (or height anomaly).

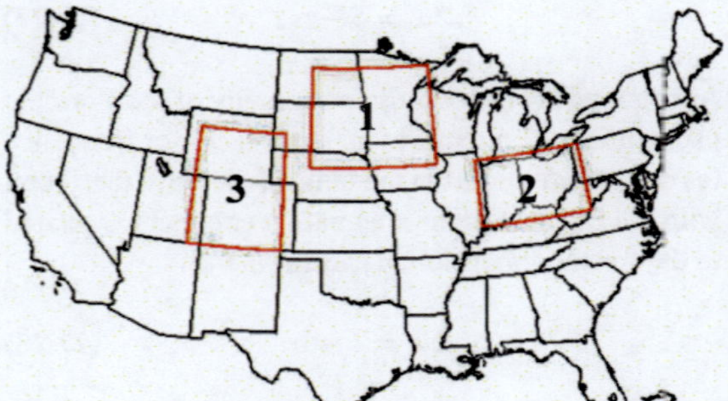

Fig. 27.2 Areas analyzed for the psd's of the gravity anomaly and the topography

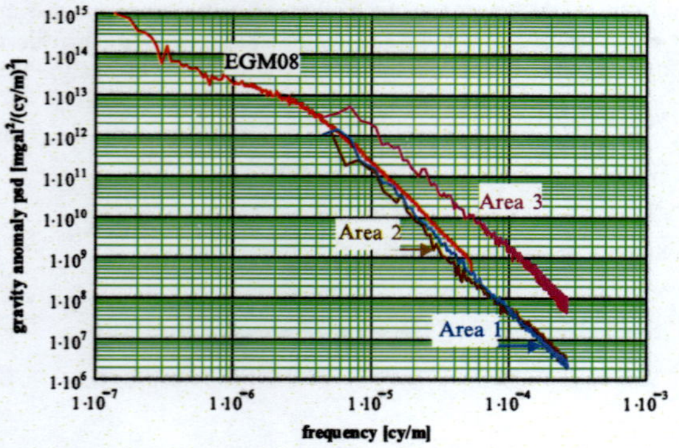

Fig. 27.3 PSDs of the gravity anomaly for the areas shown in Fig. 27.2

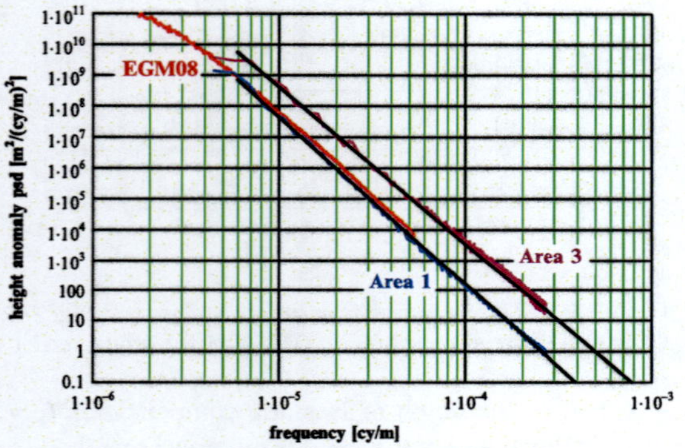

Fig. 27.4 Power-law approximations (straight lines) for the height anomaly psd's in two areas of the USA (Fig. 27.2)

5 PSDs of Topography

In practice, the region of interest likely does not have a sufficiently dense grid of gravity anomalies from which to infer the power law that is needed to determine the required data resolution. In this case, one may appeal to the known correlation between free-air gravity anomalies and topographic heights. Using Helmert condensation layers for the topography and its isostatic compensation according to Airy's model, Jekeli et al. (2009) derived this relationship in terms of Fourier transforms on the plane (see also Forsberg 1985):

$$\mathcal{F}(\Delta g) = 2\pi G\rho \mathcal{F}(\bar{H}) \left(1 - e^{-2\pi\omega D}\right), \qquad (27.25)$$

Table 27.1 Power law parameters and corresponding omission error standard deviations and required data resolution to achieve 0.7 cm precision in the height anomaly (geoid anomaly). The power law parameters refer to the height anomaly psd with units of $[m^2/(cy/m)^2]$

Power law	Omiss. error st. dev	Implied required res.	
Global: $b = 1.611 \times 10^{-19}$, $\alpha = 5.306$	$(\sigma_N)_{omission} = \dfrac{2043}{n_{max}^{1.653}}$ [m]	$n_{max} = 2024$,	5.3 arcmin
Smooth Area 1: $b = 2.802 \times 10^{-20}$, $\alpha = 5.434$	$(\sigma_N)_{omission} = \dfrac{2567}{n_{max}^{1.717}}$ [m]	$n_{max} = 1740$,	6.2 arcmin
Rough Area 3: $b = 1.753 \times 10^{-17}$, $\alpha = 5.073$	$(\sigma_N)_{omission} = \dfrac{2883}{n_{max}^{1.537}}$ [m]	$n_{max} = 4500$,	2.4 arcmin

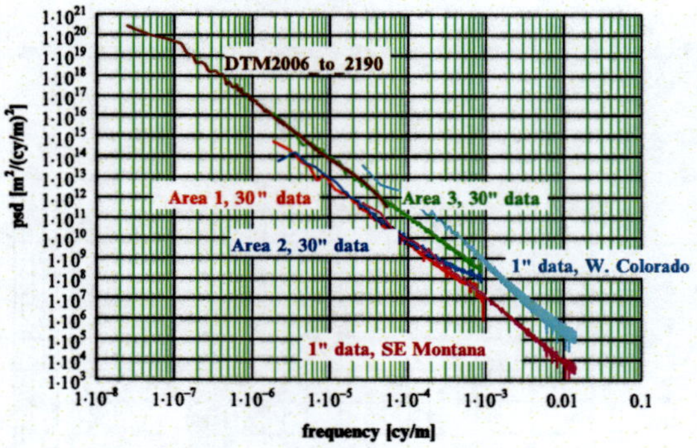

Fig. 27.5 PSDs of topography in various regions (see Fig. 27.2) using data of various resolutions. DTM2006 is a global model

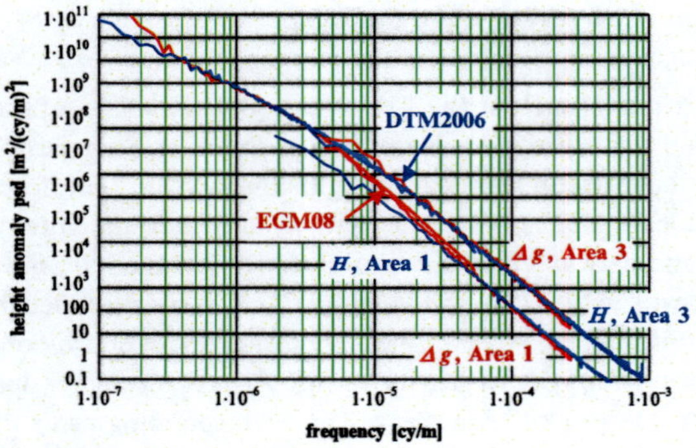

Fig. 27.6 PSDs of the gravity anomaly and the topographic height, scaled to an equivalent height anomaly (geoid undulation) psd using (27.16) and (27.25)

where D is the depth of isostatic compensation (e.g., $D = 30$ km); where the heights are given by

$$\bar{H}(Q) = \begin{cases} H(Q), & Q \in \text{land} \\ -\left(1 - \dfrac{\rho_w}{\rho}\right) B(Q), & Q \in \text{ocean} \end{cases} \tag{27.26}$$

with H the orthometric height and B the (positive) bathymetric depth, ρ the density of the crust and ρ_w the density of sea water; and where $\omega = \sqrt{\mu^2 + \nu^2}$ and μ, ν are spatial frequencies in the cardinal directions of the plane.

Figure 27.5 shows the psd's of the topography for various regions in the USA, all demonstrating essentially a power-law attenuation with frequency. Each of these local psd's is determined from the Fourier spectrum of the topography using the periodogram method, with a subsequent azimuthal averaging to yield an isotropic model (as in (27.14)). The global model (DTM2006) was obtained from the degree variances of a spherical harmonic expansion of heights and ocean depths (Pavlis et al. 2008).

The power-law attenuation of these psd's agrees with the known fractal nature of the topography (Mandelbrot 1983). Indeed, the fractal dimension, d, of a topographic profile is related to its psd attenuation, α', according to (Turcotte 1987):

$$2d = 5 - \alpha'. \tag{27.27}$$

Accounting for the two-dimensionality of the isotropic psd's in Fig. 27.6 requires an extra factor of frequency (as can be inferred from the units of the psd); therefore, with $\alpha = \alpha' + 1$, their attenuation exponent is related to the fractal dimension, d, according to

$$2d = 6 - \alpha. \tag{27.28}$$

From Fig. 27.6, we find $\alpha \approx 2.9$, implying that $d \approx 1.55$, in reasonable agreement with Turcotte's $d = 1.5$.

In the areas of Fig. 27.2, we find strong correlation at the higher frequencies between the gravity anomaly and the topography according to (27.25) (see Fig. 27.6). Thus, the gravity anomaly also has similar fractal characteristics, which could even be conferred onto the geoid (at the high frequencies). The fractal dimension of the gravity anomaly profile, therefore, is also $d = 1.55$; while, for the geoid undulation, where $\alpha \approx 5.3$ (Table 27.1), we have $d = 0.35$. Whether this qualifies the geoid (profile) as a fractal may be debated, considering Mandelbrot's (1983) definition that a fractal dimension should exceed its topological dimension (equal to 1 for the profile).

In any case, Fig. 27.6 shows that it is appropriate to deduce a power-law model for the gravity anomaly and the geoid undulation (height anomaly) psd from topographic data for the purpose of quantifying the omission error and determining corresponding data resolution requirements.

6 Summary Discussion

One of the first tasks in attempting to improve regional geoid (or, quasi-geoid) models is to determine the required gravimetric data resolution for a particular goal in accuracy. Such information then determines the characteristics of the gravity surveys (e.g., the track spacing of an airborne mission) and depends on the roughness of the geopotential field. An old idea, introduced by Kaula, has been refined here whereby a model of the power spectral density of the height anomaly is used to infer the standard deviation of the omission error due to the lack of sufficient resolution in the data. It is shown that a global model, such as EGM08, can yield quite erroneous psd estimates at high frequency near the Earth's surface.

Instead, a proper estimate of the psd at high resolution requires a local or regional approach. In this respect, topographic data can serve as an appropriate proxy, in place of gravimetric data, under the usual isostatic assumption (Airy's model). For some regions in USA with distinctive gravitational characteristics, it is verified that the topography psd in each case follows a power-law attenuation in agreement with its known fractal dimension. This power-law model transfers to the psd's of the gravity anomaly and the height anomaly (or geoid undulation) and provides a simple way to estimate the standard deviation of the omission error and consequently the needed data resolution for a particular accuracy level of the height anomaly in a particular region. It is noted that where the correlation between the gravity anomaly and the topography is not strong (e.g., in tectonic rift and subsidence regions), the gravity anomaly psd may need to be obtained by other means.

Acknowledgements This research was supported by a grant (07KLSGC02), funded through the University of Seoul by the Ministry of Land, Transport and Maritime Affairs, Seoul, Korea.

References

Colombo OL (1981) Numerical methods for harmonic analysis on the sphere. Report no. 310, Department of Geodetic Science and Surveying, Ohio State University, Columbus, Ohio, 1981

Forsberg R (1985) Gravity field terrain effect computations by FFT. Bull Géodés 59(4):342–360

Heck B (1997) Formulation and linearization of boundary value problems: From observables to a mathematical model. In: Sanso F, Rummel R (eds) Geodetic boundary value problems in view of the one centimeter geoid, Springer, Berlin

Jekeli C (1996) Spherical harmonic analysis, aliasing, and filtering. J Geodes 70(4):214–223

Jekeli C (2003) Statistical analysis of moving-base gravimetry and gravity gradiometry. Report No.466, Laboratory for space geodesy and remote sensing research, geodetic science, Ohio State University, Columbus, Ohio, 2003

Jekeli C, Yang HJ, Kwon JH (2009) Using gravity and topography-implied anomalies to assess data requirements for precise geoid computation. J Geodes. doi:10.1007/s00190-009-0337-y

Kaula WM (1966) Theory of Satellite Geodesy. Blaisdell, Waltham

Lemoine FG, Kenyon SC, Factor JK, Trimmer RG, Pavlis NK, Chinn DS, Cox CM, Klosko SM, Luthcke SB, Torrence MH, Wang YM, Williamson RG, Pavlis EC, Rapp RH, Olson TR (1998) The development of the joint NASA GSFC and the National Imagery and Mapping Agency (NIMA) geopotential model EGM96. NASA Technical Paper NASA/TP-1998–206861, Goddard Space Flight Center, Greenbelt, Maryland, 1998

Mandelbrot B (1983) The Fractal Geometry of Nature. Freeman, San Francisco.

Moritz H (1980) Advanced Physical Geodesy. Abacus, Tunbridge Wells, Kent

Pavlis NK, Holmes SA, Kenyon SC, Factor JK (2008) An earth gravitational model to degree 2,160: EGM2008. Presented at the General Assembly of the European Geosciences Union, Vienna, Austria, 13–18 April 2008

Rapp RH (1979) Potential coefficient and anomaly degree variance modeling revisited. Report no.293, geodetic science, Ohio State University, Columbus, Ohio, 1979

Turcotte DL (1987) A fractal interpretation of topography and geoid spectra on the Earth, Moon, Venus, and Mars. J Geophys Res 92(B4):E597–E601

Method of Successive Approximations in Solving Geodetic Boundary Value Problems: Analysis and Numerical Experiments

28

P. Holota and O. Nesvadba

Abstract

After an introductory note reviewing the role and the treatment of boundary problems in physical geodesy, the explanation rests on the concept of the weak solution. The focus is on the linear gravimetric boundary value problem. In this case, however, an oblique derivative in the boundary condition and the need for a numerical integration over the whole and complicated surface of the Earth make the numerical implementation of the concept rather demanding. The intention is to reduce the complexity by means of successive approximations and step by step to take into account effects caused by the obliqueness of the derivative and by the departure of the boundary from a more regular surface. The possibility to use a sphere or an ellipsoid of revolution as an approximation surface is discussed with the aim to simplify the bilinear form that defines the problem under consideration and to justify the use of an approximation of Galerkin's matrix. The discussion is added of extensive numerical simulations and tests using the ETOPO5 boundary for the surface of the Earth and gravity data derived from the EGM96 model of the Earth's gravity field.

Keywords

Earth's gravity field • Geodetic boundary-value problems • Variational methods • Weak solution • Numerical methods • Galerkin's approximations • Iteration and convergence problems

1 Introduction

In refined studies on gravity potential and figure of the Earth terrestrial gravity data play an essential role. Mathematically, one has to deal with boundary value problems. The concept has an important position in physical geodesy.

In our considerations W and U will stand for the gravity and the normal (standard) potential of the Earth. Hence $T = W - U$ is the disturbing potential,

P. Holota (✉)
Research Institute of Geodesy, Topography and Cartography, 250 66 Zdiby 98, Praha-východ, Czech Republic
e-mail: holota@pecny.asu.cas.cz

O. Nesvadba
Land Survey Office, Pod Sídlištěm 9, 182 11 Praha 8, Czech Republic

N. Sneeuw et al. (eds.), *VII Hotine-Marussi Symposium on Mathematical Geodesy*, International Association of Geodesy Symposia 137, DOI 10.1007/978-3-642-22078-4_28, © Springer-Verlag Berlin Heidelberg 2012

$\boldsymbol{g} = \boldsymbol{grad}\, W$, $\boldsymbol{\gamma} = \boldsymbol{grad}\, U$, $g = |\boldsymbol{g}|$ is the measured gravity and $\gamma = |\boldsymbol{\gamma}|$ is the normal gravity. Clearly, $|\cdot|$ means the magnitude of the respective vector.

In practice the determination of T from terrestrial data usually rests on the solution of Molodensky's problem. Nevertheless, in this paper we will follow a somewhat different concept. Recently, one often speaks about the so-called gravimetric boundary value problem. The interest is primarily driven by considerable advances in determining the geometry of the Earth's surface through methods of satellite geodesy. In a linear setting the problem is to find T such that

$$\Delta T = 0 \text{ in } \Omega \tag{28.1}$$

and

$$\langle \boldsymbol{s}, \boldsymbol{grad}\, T\rangle = \partial T/\partial s = -\delta g \text{ on } \partial\Omega \tag{28.2}$$

where $\boldsymbol{s} = -(1/\gamma)\boldsymbol{grad}\, U$, $\delta g = g - \gamma$ is the gravity disturbance, $\langle .,. \rangle$ means the inner product of two vectors in Euclidean three-dimensional space $\boldsymbol{R}^3$, Ω is the exterior of the Earth, $\partial\Omega$ is the boundary of Ω and T is assumed regular at infinity. In particular $T(\boldsymbol{x}) = \mathcal{O}(|\boldsymbol{x}|^{-1})$ as $\boldsymbol{x} \to \infty$, where $\mathcal{O}$ means Landau's symbol, $|\boldsymbol{x}| = (\Sigma_{i=1}^3 x_i^2)^{1/2}$ and x_i, $i = 1, 2, 3$, are rectangular Cartesian co-ordinates. Note that δg represents the input data corrected for the gravitational interaction with the Moon, the Sun and the planets, for the precession and nutation and so on.

In solving boundary value problems of physical geodesy the integral equation method is often applied. The approach is usually combined with the method of a small parameter, which leads to a series representation of the solution and in principle gives a way how to solve the problem for boundaries of a more complex geometry.

Another approach, which in a mathematical sense yields the classical solution, is Green's function method. Green's function can easily be constructed for solution domains of elementary shape. In more general cases additional tools have to be applied, e.g. a transformation of coordinates as in Holota (1991). The solution then leads to successive approximations.

Both the method can be applied to the linear gravimetric boundary value problem too. Nevertheless, we will follow another concept, which is frequently used in many branches of technique and physics for its flexibility. It rests on the notion of the weak solution and the paper ties to results in Nesvadba et al. (2007) and Holota and Nesvadba (2008).

We will look for a measurable function satisfying an integral identity connected with the boundary-value problem in question. As regards function spaces, we will work with Sobolev's weighted space $W_2^{(1)}(\Omega)$ endowed with inner product

$$(u, v)_1 \equiv \int_\Omega \frac{uv}{|\boldsymbol{x}|^2}\, dx + \sum_{i=1}^{3} \int_\Omega \frac{\partial u}{\partial x_i}\frac{\partial v}{\partial x_i} dx \tag{28.3}$$

Also the boundary $\partial\Omega$ of the domain Ω is supposed to have a certain degree of regularity. Putting $\Omega' = \boldsymbol{R}^3 - \Omega \cup \partial\Omega$, we will assume that Ω' is a domain with Lipschitz' boundary.

Example 1. (Neumann's problem). Put

$$A(u, v) = \sum_{i=1}^{3} \int_\Omega \frac{\partial u}{\partial x_i}\frac{\partial v}{\partial x_i} dx \tag{28.4}$$

on the Cartesian product $W_2^{(1)}(\Omega) \times W_2^{(1)}(\Omega)$ and consider an integral identity

$$A(u, v) = \int_{\partial\Omega} v f\, dS \tag{28.5}$$

where $f \in L_2(\partial\Omega)$ and $L_2(\partial\Omega)$ denotes a space of square integrable functions on $\partial\Omega$. In case that (28.5) holds for all $v \in W_2^{(1)}(\Omega)$ the well-known point is that the identity defines the function u uniquely, as an element of $W_2^{(1)}(\Omega)$. The claim rests on the Lax–Milgram theorem, see Nečas (1967) and Rektorys (1977) or Bers et al. (1964). Note that the formulation has also a classical interpretation. Under some regularity assumptions one can apply Green's identity and show that u has to be a solution of Neumann's (exterior) problem, i.e.,

$$\Delta u = 0 \quad \text{in} \quad \Omega \quad \text{and} \quad \frac{\partial u}{\partial n} = -f \quad \text{on} \quad \partial\Omega \tag{28.6}$$

where Δ means Laplace's operator and $\partial/\partial n$ denotes the derivative in the direction of the unit normal $\boldsymbol{n}$ of $\partial\Omega$, see also Holota (2004).

2 An Oblique Derivative Problem

In order to formulate the linear gravimetric boundary value problem in terms of an integral identity similar to (28.5) we have to give the bilinear form $A(u,v)$ a more complex structure. In particular we have to put $A(u,v) = A'(u,v) - A''(u,v)$ where

$$A'(u,v) = \int_{\Omega} \langle \mathbf{grad}\, u, \mathbf{grad}\, v \rangle dx \tag{28.7}$$

$$A''(u,v) = \int_{\Omega} \langle \mathbf{grad}\, v, \mathbf{a} \times \mathbf{grad}\, u \rangle dx + \int_{\Omega} v \langle \mathbf{curl}\, \mathbf{a}, \mathbf{grad}\, u \rangle dx \tag{28.8}$$

and $\boldsymbol{a} = (a_1, a_2, a_3)$ is a vector field such that a_i and also $|\boldsymbol{x}|(\mathbf{curl}\, \boldsymbol{a})_i$, $i = 1, 2, 3$, are Lebesgue measurable functions defined and bounded almost everywhere on Ω, see Holota (1997, 1999, 2000, 2001a,b, 2004, 2005).

Moreover, we have to assume that on the boundary $\partial\Omega$ the vector $\boldsymbol{\sigma} = \boldsymbol{s}/\langle \boldsymbol{s}, \boldsymbol{n} \rangle$ and the field $\mathbf{a}$ are coupled so that $\boldsymbol{\sigma} = \boldsymbol{n} + \mathbf{a} \times \boldsymbol{n}$. Note also that the tie to the classical formulation of the problem [(28.1) and (28.2)] requires that $f = -\gamma(\partial U/\partial n)^{-1}\delta g = \langle \boldsymbol{s}, \boldsymbol{n} \rangle^{-1}\delta g$.

Clearly, the form $A''(u,v)$ represents the oblique derivative in the boundary condition. It has been shown in Holota (2000) that the problem may be solved by means of successive approximations, i.e. we may use a sequence of functions $[u_m]_{m=0}^{\infty}$ defined by

$$A'(u_{m+1}, v) = \int_{\partial\Omega} vf\, dS + A''(u_m, v) \tag{28.9}$$

which is assumed to hold for all $v \in W_2^{(1)}(\Omega)$ and $m = 0, 1, \ldots, \infty$. Conformably to Holota (2000) $[u_m]_{m=0}^{\infty}$ is a Cauchy sequence in $W_2^{(1)}(\Omega)$ and in the norm given by $\|.\|_1 = (\cdot,\cdot)_1^{1/2}$ it converges to a function u, which belongs to $W_2^{(1)}(\Omega)$ and is also the solution of our weakly formulated problem. Nevertheless, we will integrate this iteration process with the treatment of another small effect in the next sections.

3 Modification of the Bilinear Form $A'(u,v)$

Our aim is to follow the concept of direct methods, i.e. to express our problem in terms of a system of linear equations. This also means to compute the entries of the respective Galerkin matrix. They in principle represent values of the bilinear form $A(u,v)$ for u and v equal to members of the chosen function basis. However, the computation may be rather demanding in practice, since the boundary $\partial\Omega$ of the domain Ω will not be of an elementary figure.

It is also worth mentioning that the weak solution is not defined too efficiently in our particular case. $W_2^{(1)}(\Omega)$ is too large. Indeed, the solution, we are looking for, is a harmonic (and therefore an analytic) function, whereas functions from $W_2^{(1)}(\Omega)$ and thus also the base functions have not such a high degree of regularity in general. Inspecting the problem, we can see that it does not mean a principal obstacle. We may consider just a space $H_2^{(1)}(\Omega)$ of those functions from $W_2^{(1)}(\Omega)$ which are harmonic in Ω and reformulate the problem, i.e. to look for $T \in H_2^{(1)}(\Omega)$ such that

$$A(T,v) = \int_{\partial\Omega} vf\, dS \tag{28.10}$$

holds for all $v \in H_2^{(1)}(\Omega)$, cf. Holota (1999, 2000, 2001a,b, 2004, 2005). In addition, for reasons that will be clear in the sequel, we also rearrange the identity (28.10) in the following way

$$A'(T,v) = \int_{\partial\Omega} vf\, dS + A''(T,v) \tag{28.11}$$

It will be our starting point for an iteration solution.

Recall now, that the analyticity of functions from $H_2^{(1)}(\Omega)$ gives us a possibility to apply Runge's property of Laplace's equation [see e.g. Bers et al. (1964) and Krarup (1969)] and to extend all the functions considered to a domain Ω^*, that has a "simpler boundary" and contains the original domain Ω, i.e., $\Omega \subseteq \Omega^*$. This is of practical importance. Indeed, we can modify the bilinear form $A'(u,v)$ slightly. In particular we put

$$A^*(u,v) = \int_{\Omega^*} \langle \mathbf{grad}\, u, \mathbf{grad}\, v \rangle dx \qquad (28.12)$$

for all $u, v \in H_2^{(1)}(\Omega^*)$ and our problem will be to find $T \in H_2^{(1)}(\Omega^*)$ such that

$$A^*(T,v) = \int_{\partial\Omega} vf\, dS + A''(T,v) + F(T,v) \quad (28.13)$$

holds for all $v \in H_2^{(1)}(\Omega^*)$, whereas

$$F(T,v) = A^*(T,v) - A'(T,v) = \int_{D} \langle \mathbf{grad}\, T, \mathbf{grad}\, v \rangle dx \qquad (28.14)$$

and $D = \Omega^* - \Omega$. Recall in this connection that (under our assumptions concerning the regularity of $\partial\Omega$) we also have that the restriction to Ω of the family functions of $H_2^{(1)}(\Omega^*)$ is dense in $H_2^{(1)}(\Omega)$, so that a basis in $H_2^{(1)}(\Omega^*)$, restricted to Ω is a basis of $H_2^{(1)}(\Omega)$.

4 Successive Approximations

Our aim is now to solve the integral identity (28.13) by means of successive approximations. We will consider a sequence of functions $[T_k]_{k=0}^{\infty}$ defined by

$$A^*(T_{k+1}, v) = \int_{\partial\Omega} vf\, dS + A''(T_k, v) + F(T_k, v) \qquad (28.15)$$

valid for all $v \in H(\Omega^*)$ and $k = 1, \ldots, \infty$. For this purpose we first deduce from (28.15) that

$$A^*(T_{k+2} - T_{k+1}, v) = A''(T_{k+1} - T_k, v) + F(T_{k+1} - T_k, v) \qquad (28.16)$$

holds for all $v \in H(\Omega^*)$ and note that on the right hand side $A''(T_{k+1} - T_k, v)$ and $F(T_{k+1} - T_k, v)$, for $T_{k+1} - T_k$ fixed, are bounded linear functionals of the variable v. Indeed, denoting by $||A''||$ the norm of A'', we can find in Holota (2000) that

$$||A''|| \le C||T_{k+1} - T_k||_{H_2^{(1)}(\Omega)} \qquad (28.17)$$

where C is a constant. Clearly, this also yields that

$$||A''|| \le C||T_{k+1} - T_k||_{H_2^{(1)}(\Omega^*)} \qquad (28.18)$$

since $\Omega \subseteq \Omega^*$.

Similarly, denoting by $||F||$ the norm of F we can easily deduce that

$$||F|| \le ||T_{k+1} - T_k||_{H_2^{(1)}(\Omega^*)} \qquad (28.19)$$

Moreover, one can even show that there exists a constant $\mu < 1$ such that

$$||F|| \le \mu||T_{k+1} - T_k||_{H_2^{(1)}(\Omega^*)} \qquad (28.20)$$

The proof has been given in Holota (2005) for functions of equally limited spectrum, but it can be easily generalized for harmonic functions, i.e. for $H_2^{(1)}(\Omega^*)$.

Return now to the bilinear form $A^*(u,v)$ and assume that $\Omega' = R^3 - \bar{\Omega}^*$ is a star-shaped domain with respect to the origin (and has Lipschitz' boundary). Under this assumption

$$||v||^2_{W_2^{(1)}(\Omega^*)} \le \alpha \int_{\Omega^*} |\mathbf{grad}\, v|^2 dx \qquad (28.21)$$

for all $v \in W_2^{(1)}(\Omega^*)$ with $\alpha = 5$. The proof can be found in Holota (1997). Hence

$$A^*(u,u) \ge \frac{1}{\alpha}||u||^2_{H_2^{(1)}(\Omega^*)} \qquad (28.22)$$

holds for all $u \in H_2^{(1)}(\Omega^*)$, i.e., $A^*(u,v)$ is an elliptic bilinear form. In addition, one can easily verify that

$$A^*(u,v) \le ||u||_{H_2^{(1)}(\Omega^*)}||v||_{H_2^{(1)}(\Omega^*)} \qquad (28.23)$$

The properties of $A^*(u,v)$ and the boundedness of the functionals on the right hand side of (28.16) make it now possible to apply Lax–Milgram's theorem, see Nečas (1967), Rektorys (1974) or Bers et al. (1964). It allows us to deduce that

$$||T_{k+2} - T_{k+1}||_{H_2^{(1)}(\Omega^*)} \le \alpha(||A''|| + ||F||) \le c||T_{k+1} - T_k||_{H_2^{(1)}(\Omega^*)} \qquad (28.24)$$

where $c = \alpha(C + \mu)$. Hence for any integer $p > 0$

$$||T_{k+p} - T_{k+1}||_{H_2^{(1)}(\Omega^*)} \leq \frac{c^{k+1}}{1-c}||T_1 - T_0||_{H_2^{(1)}(\Omega^*)} \tag{28.25}$$

which yields $||T_{k+p} - T_{k+1}||_{H_2^{(1)}(\Omega^*)} \to 0$ for $k \to \infty$, provided that $c < 1$. Under this assumption $[T_k]_{k=0}^{\infty}$ is a Cauchy sequence in $H_2^{(1)}(\Omega^*)$ and converges to a function $T \in H_2^{(1)}(\Omega^*)$.

Obviously, the problem, which needs discussion, is the magnitude of the parameters α, C and μ in combination with the condition $c = \alpha(C + \mu) < 1$. Indeed, the analysis above rests but only on "a priory" estimates. As known, they usually are rather pessimistic and here also associated with certain restrictions on the shape of the domain Ω. For these reasons numerical tests, we will add in the sequel, will have a better instructive value.

5 Interpretation in Terms of Bases

Prior to numerical computations we rearrange the identity (28.15) slightly, using the fact that T_k and v are function from $H_2^{(1)}(\Omega^*)$, thus sufficiently regular in the closure of the domain Ω, i.e. in $\bar{\Omega} = \Omega \cup \partial\Omega$. Referring to Holota (1997, 2000), we can write that

$$A''(T_k, v) = \int_{\partial\Omega} v\langle\sigma - \boldsymbol{n}, \boldsymbol{grad}\, T_k\rangle dS \tag{28.26}$$

Moreover, recalling (28.14), we easily obtain

$$A^*(T_{k+1} - T_k, v) = \int_{\partial\Omega} v[f + \langle\sigma, \boldsymbol{grad}\, T_k\rangle]dS \tag{28.27}$$

since

$$A'(T_k, v) = \int_{\Omega} \langle \boldsymbol{grad}\, T_k, \mathbf{grad}\, v\rangle dx = -\int_{\partial\Omega} v\langle\boldsymbol{n}, \boldsymbol{grad}\, T_k\rangle dS \tag{28.28}$$

Alternatively, identity (28.27) can be also written as

$$A^*(T_{k+1} - T_k, v) = \int_{\partial\Omega} v\frac{1}{\langle\boldsymbol{s}, \boldsymbol{n}\rangle}[\delta g + \langle\boldsymbol{s}, \boldsymbol{grad}\, T_k\rangle]dS \tag{28.29}$$

The meaning of the right hand side is obvious. Considering the structure of the boundary condition for T, we can see that, successively, the term $\langle\boldsymbol{s}, \boldsymbol{grad}\, T_k\rangle$ generates the opposite of the measured gravity disturbance δg, provided that $[T_k]_{k=0}^{\infty}$ is a convergent series.

Finally, the identity (28.29), valid for all $v \in H(\Omega^*)$ and $k = 1, \ldots, \infty$, can be taken as a natural starting point of a numerical solution. Indeed, one can approximate T_k by means of a linear combination

$$T_k^{(n)} = \sum_{i=1}^{n} c_{i,k}^{(n)} v_i \tag{28.30}$$

where v_i are members of a function basis of $H_2^{(1)}(\Omega^*)$ that generate an n–dimensional subspace $H_n(\Omega^*) = \mathrm{span}\{v_i, i = 1, \ldots, n\}$ and $c_{i,k}^{(n)}$ are scalar coefficients,

$$c_{i,k}^{(n)} = c_{i,0}^{(n)} + \sum_{j=1}^{k} (\Delta c_{i,j}^{(n)}) \tag{28.31}$$

with increments $\Delta c_{i,j}^{(n)}$

$$\Delta c_{i,j}^{(n)} = c_{i,j}^{(n)} - c_{i,j-1}^{(n)} \tag{28.32}$$

that can be successively obtained for $j = 1, \ldots, k$ from the following system of linear equations

$$\sum_{i=1}^{n} \Delta c_{i,j+1}^{(n)} A^*(v_i, v_m) = \int_{\partial\Omega} v_m \frac{1}{\langle\boldsymbol{s}, \boldsymbol{n}\rangle}[\delta g + \langle\boldsymbol{s}, \boldsymbol{grad}\, T_j^{(n)}\rangle]dS \tag{28.33}$$

where $m = 1, \ldots, n$.

6 Approximation by a Sphere: Test

The exterior of the ETOPO5 surface was used for the domain Ω. The real gravity potential W was simulated by means of the Earth's gravity field model EGM96, see Lemoine et al. (1998). The input gravity disturbances δg on the ETOPO5 boundary, see Fig. 28.1, were referred to the normal gravity potential U with parameters given by the GRS80, see Moritz (2000).

For the numerical integration on the right hand side of (28.33) vertices of an icosahedron were

Fig. 28.1 Simulated gravity disturbances δg

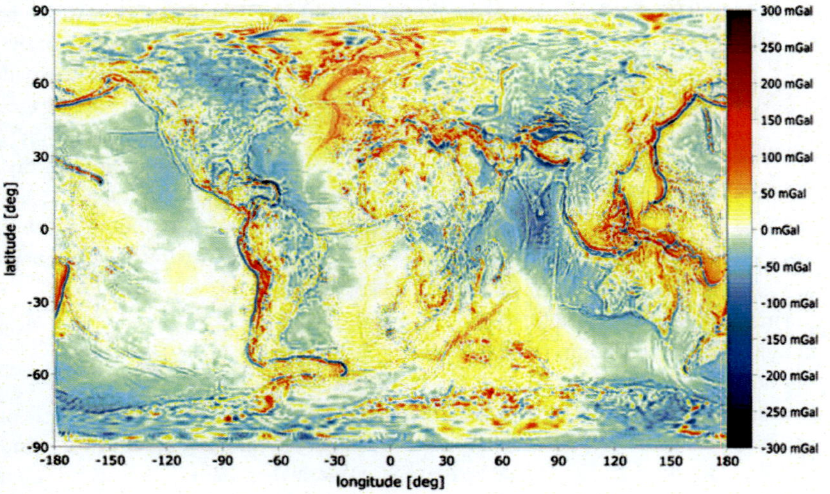

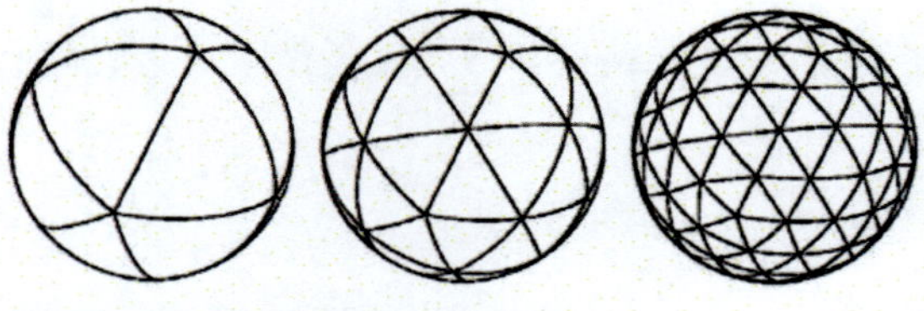

Fig. 28.2 Icosahedron refinement

projected onto the ETOPO5 boundary surface. The triangles obtained in this way were used for generating hierarchical triangulation of the boundary, see Fig.28.2.

In the integration Romberg's method was exploited together with Richardson's extrapolation, which increases the accuracy and provides feedback in the control of the integration error.

For Ω^* the exterior of a sphere of radius R was taken. In this case one can easily compute the reproducing kernel $K(\mathbf{x},\mathbf{y})$ of the space $H_2^{(1)}(\Omega^*)$. Indeed, referring to Holota (2004), for $|\mathbf{x}| \geq R$ and $|\mathbf{y}| > R$ (or $|\mathbf{x}| > R$ and $|\mathbf{y}| \geq R$) we have

$$K(\mathbf{x},\mathbf{y}) = \frac{1}{4\pi R}\sum_{n=0}^{\infty}\frac{2n+1}{n+1}z^{n+1}P_n(\cos\psi) \quad (28.34)$$

where $z = R^2/|\mathbf{x}||\mathbf{y}|$ and ψ is the angle between $\mathbf{y}$ and $\mathbf{x}$. Moreover, we find the closed form

$$K(\mathbf{x},\mathbf{y}) = \frac{1}{4\pi R}\left(\frac{2z}{L} - \ln\frac{L+z-\cos\psi}{1-\cos\psi}\right), \quad (28.35)$$

where $L = (1 - 2z\cos\psi + z^2)^{1/2}$, cf. Tscherning (1975) and Neyman (1979). Recall in particular that the reproducing kernel $K(\mathbf{x},\mathbf{y})$ generates a function basis

$$v_i(\mathbf{x}) = K(\mathbf{x},\mathbf{y}_i), \quad i = 1,2,\ldots,n,\ldots, \quad (28.36)$$

in the space $H_2^{(1)}(\Omega^*)$, see Sansò (1986) and that the functions given by (28.36) have some advantages. In (28.33) they yield the entries of matrix $[A^*]$ in a straightforward way. Indeed, $A^*(v_i, v_m) = K(\mathbf{y}_i, \mathbf{y}_m)$.

Note finally that in our experiment the parking grid of the points $\mathbf{y}_i$, $i = 1,2,\ldots,n$, was given by vertices of the 6th level of the icosahedron refinement (see Fig. 28.2), so that the dimension n of the approximation space $H_n(\Omega^*)$ equals 40962. The radius R agreed with to the minor axis of the GRS80 reference ellipsoid and the height of the point $\mathbf{y}_i$ was ca. 93 km. The structure of the matrix $A^*(v_i, v_j)$ is shown in Fig. 28.3.

Return now to the disturbing potential $T = W - U$. Its values simulated on the ETOPO5 boundary surface can be seen from Fig. 28.4. In the sequel the potential T is used as an "exact" solution and is confronted with the outcome of the individual iteration steps expressed by (28.29)–(28.33).

In the first iteration step $T_0^{(n)} = 0$ was taken for an initial approximation. Subsequently $T_1^{(n)}$ was obtained in the approximation space $H_{40962}(\Omega^*)$. The

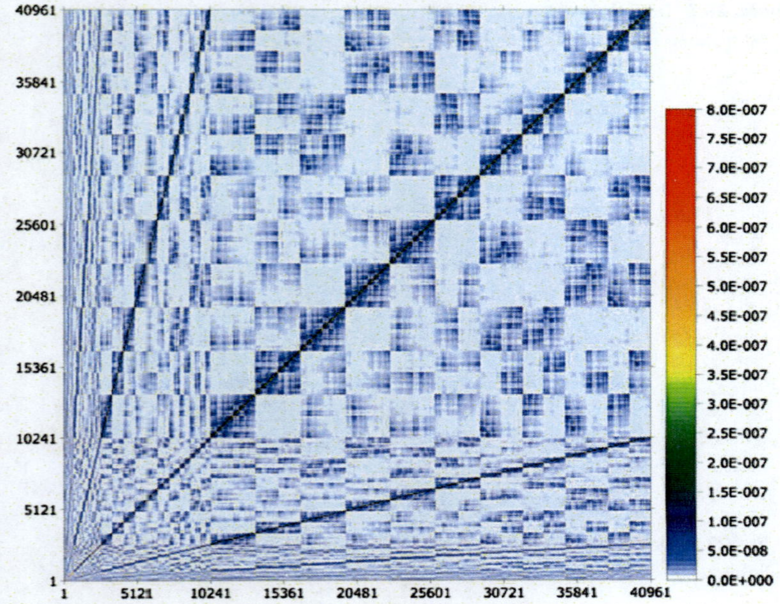

Fig. 28.3 Structure of the matrix $[A^*]$ for $n = 40962$

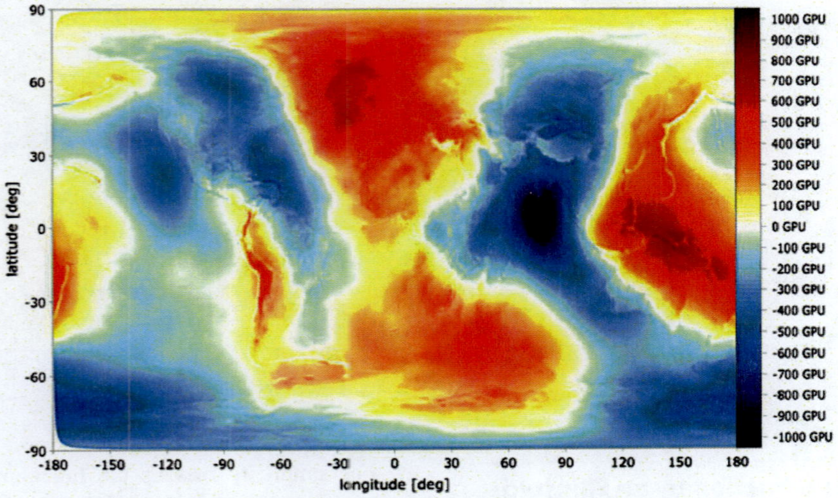

Fig. 28.4 Disturbing potential T on the ETOPO5 boundary surface

deviations of $T_1^{(40962)}$ from T are illustrated in Fig. 28.5 and globally may be characterized by the following triad of values ~ min. : $-51 GPU$, max.: 89 GPU, rms: 6.1 GPU (GPU – geopotential unit, $1\,GPU \equiv 1\,m^2s^{-2}$, roughly corresponding to 10 cm of geoidal height).

Subsequently, a few initial iteration steps indicated a slight improvement. In particular deviations of $T_3^{(40962)}$ from T, as illustrated in Fig. 28.6, may be added of the following global characteristics ~ min.: $-49\,GPU$, max.: 62 GPU, rms: 3.7 GPU. However, a continuation of the iteration procedure step by step demonstrated divergence, especially in polar regions. The result indicates that in general a sphere is too far from the boundary of the domain Ω, i.e. too elementary to be taken in quality of the domain Ω^*. This motivated the following experiment, which is discussed in the next paragraph.

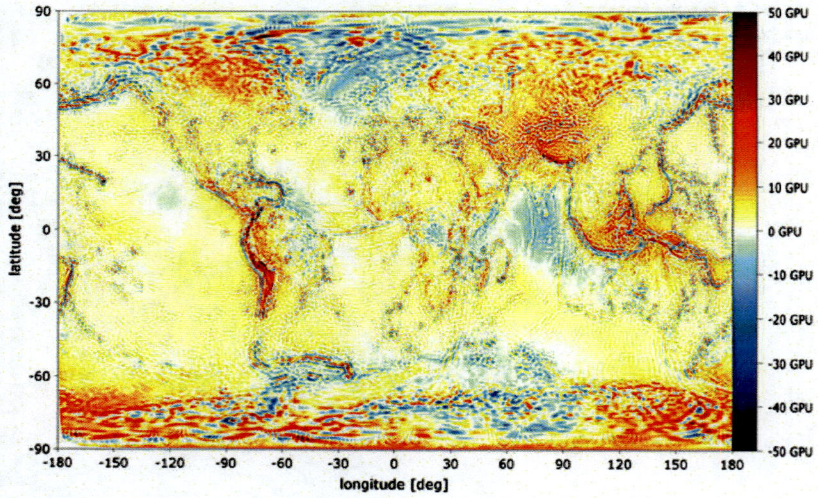

Fig. 28.5 Deviations of the 1st iteration $T_1^{(40962)}$ from T

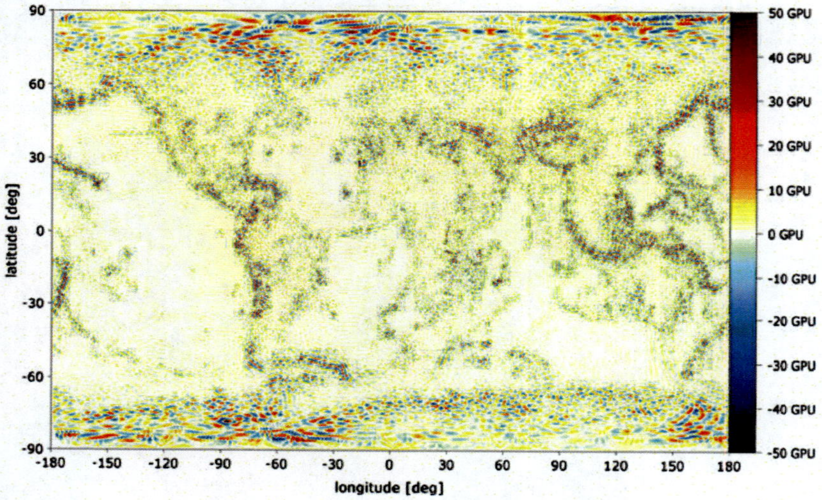

Fig. 28.6 Deviations of the 3rd iteration $T_3^{(40962)}$ from T

7 Approximation by an Ellipsoid: Test and Concluding Remarks

Considering the conclusions above we have reasons to assume that, possibly, the exterior of an ellipsoid of revolution, if used in quality of Ω^*, could lead to better results. This, however, is associated with more complicated formulas for the computation of entries in the matrix $[A^*]$. Unfortunately, they in general do not offer closed form expressions, but only series, in contrast to (28.35).

For simplicity reasons the function basis of the approximation space $H_n(\Omega^*)$ has been generated by the reciprocal distance, i.e.,

$$v_i(\mathbf{x}) = \frac{1}{|\mathbf{x} - \mathbf{y}_i|}, \quad i = 1, 2, \ldots, n \tag{28.37}$$

and for the computation of the entries formulas derived in Holota (2001a, b) have been used. Subsequently, we took $n = 163842$ as the dimension of the approximation space $H_n(\Omega^*)$, which resulted from the 7th level of the icosahedron refinement.

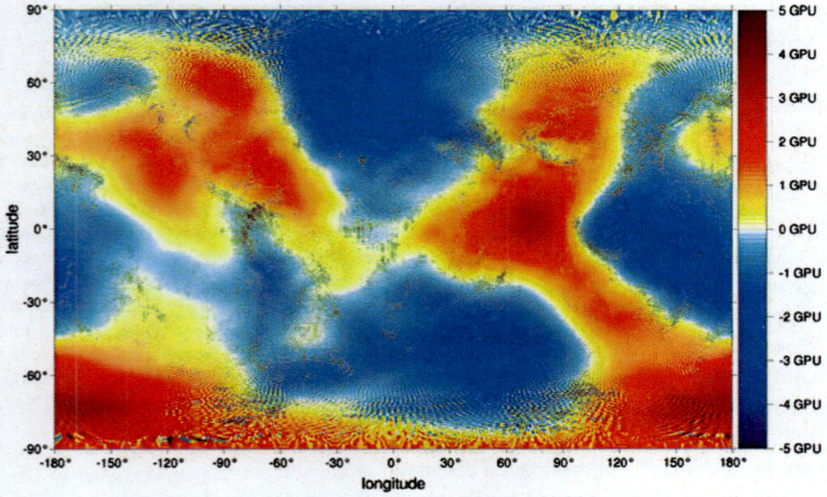

Fig. 28.7 Deviations of the 1st iteration $T_1^{(163842)}$ from T

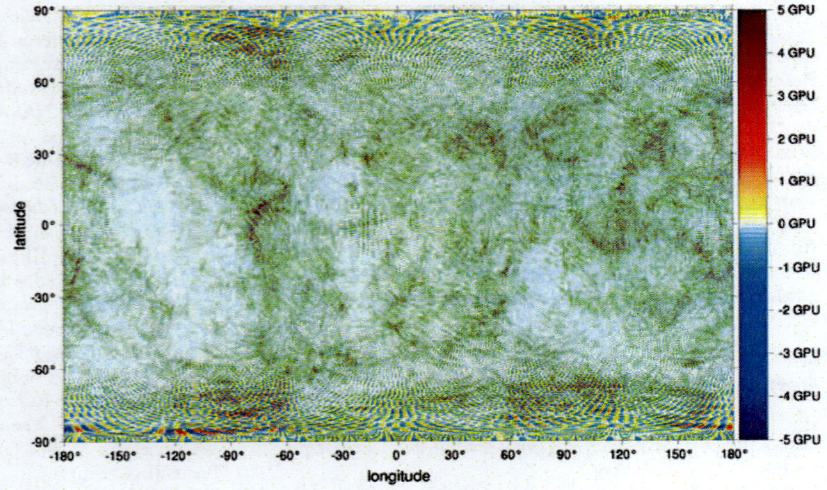

Fig. 28.8 Deviations of the 4th iteration $T_4^{(163842)}$ from T

The purpose of this experiment primarily was to check whether the applied formulas represent the entries of the matrix $[A^*]$ properly. In the experiment, therefore, Ω means but only the exterior of an ellipsoid of revolution with parameters given by the GRS80. As to Ω^*, here also the outer space of $\partial\Omega$ was taken, i.e. $\Omega^* = \Omega$. The points y_i were situated on an ellipsoid level at the depth of ca. $60\,km$. The gravity potential W was simulated by means of the EGM96 again since the number of parameters of this model is better balanced with the dimension of the approximation space $H_{163842}(\Omega^*)$ than in the case of the EGM08, which is much finer. Finally, $T = W - U$ has been used in quality of an "exact" solution.

In the first iteration step $T_0^{(n)} = 0$ was taken for an initial approximation, as in Sect. 6. By analogy $T_1^{(n)}$ was obtained in $H_{163842}(\Omega^*)$. The deviations of $T_1^{(163842)}$ from T are illustrated in Fig. 28.7 and globally may be characterized by the following values ~min.: $-7.4\,GPU$, max.: $6.4\,GPU$, rms: $2.1\,GPU$. Stress that these results actually offer a quality test showing the accuracy of the numerical representation of the matrix $[A^*]$. It is clear that the representation is not quite accurate.

Nevertheless several *subsequent iteration steps* that followed our scheme given by (28.29)–(28.33), though with an approximate $[A^*]$ (in the sense as above), led quickly to a considerable improvement. Deviations of $T_4^{(163842)}$ from T, as illustrated in Fig. 28.8, are substantially smaller. Globally they may be characterized by the following values $\sim$ min.: $-7.1\,GPU$, max.: $6.6\,GPU$, rms: $0.32\,GPU$.

The experiment has been also repeated, even for $\partial\Omega$ given by the ETOPO5 boundary (and $\partial\Omega^*$ being the ellipsoid of revolution as above). The results were very close to those already obtained (for $\partial\Omega$ being an ellipsoid of revolution) and clearly indicate convergence of the procedure. We take them as a strong stimulus and a reliable springboard for further investigations, analyses and numerical tests in solving the discussed topic. Our intentions are also supported by Sansò and Sacerdote (2008).

Acknowledgements The presentation of the paper at the VII Hotine-Marussi Symposium, Rome, Italy, July 6–10, 2009, was sponsored by the Ministry of Education, Youth and Sports of the Czech Republic through Projects No. LC506. The computations were performed in CINECA (Bologna) within the Project HPC-EUROPA++ (RII3-CT-2003–506079), supported by the European Community – Research Infrastructure Action under the FP6 "Structuring the European Research Area" Programme. All this support as well as the possibility to discuss the topic with Prof. F. Sansò, Milano, is gratefully acknowledged. Thanks go also to anonymous reviewers for valuable comments.

References

Bers L, John F, Schechter M (1964) Partial differential equations. Wiley, New York

Holota P (1991) On the iteration solution of the geodetic boundary-value problem and some model refinements. *Travaux de l'A.I.G.*, Tome 29, Paris, 1992, pp 260–289

Holota P (1997) Coerciveness of the linear gravimetric boundary-value problem and a geometrical interpretation. J Geodes 71:640–651

Holota P (1999) Variational methods in geoid determination and function bases. Phys Chem Earth Solid Earth Geodes 24(1):3–14

Holota P (2000) Direct method in physical geodesy. In: Schwarz KP (ed) Geodesy Beyond 2000, IAG Symposia, vol 121, Springer, New York, pp 163–170

Holota P (2001a) Variational methods in the recovery of the gravity field – Galerkin's matrix for an ellipsoidal domain. In: Sideris MG (ed) Gravity, Geoid and Geodynamics 2000, IAG Symposia, vol 123, Springer, New York, pp 277–283

Holota P (2001b) Variational methods in the representation of the gravitational potential. Cahiers du Centre Européen de Géodynamique et de Séismologie, vol 20, Luxembourg, 2003, pp 3–11

Holota P (2004) Some topics related to the solution of boundary-value problems in geodesy. In: Sansò F (ed) V Hotine-Marussi symposium on mathematical geodesy, IAG Symposia, vol 127, Springer, New York, pp 189–200

Holota P (2005) Successive approximations in the solution of weakly formulated geodetic boundary-value problem. In: Sansò F (ed) A window on the future of geodesy, IAG Symposia, vol 128, Springer, New York, pp 452–458

Holota P, Nesvadba O (2008). Model refinements and numerical solutions of weakly formulated boundary-value problems in physical geodesy. In: Xu P, Liu J, Dermanis A (eds) VI Hotine-Marussi symposium on theoretical and computational geodesy, IAG Symposia, vol 132, Springer, pp 320–326

Krarup T (1969) A contribution to the mathematical foundation of physical geodesy. Danish Geodetic Institute, Publ. No. 44, Copenhagen

Lemoine FG, Kenyon SC, Factor JK, Trimmer RG, Pavlis NK, Chinn DS, Cox CM, Klosko SM, Luthcke SB, Torrence MH, Wang YM, Williamson RG, Pavlis EC, Rapp RH, Olson TR (1998) The development of the joint NASA GSFC and National Imagery and Mapping Agency (NIMA) geopotential model EGM96. NASA/TP-1998–206861, NASA, GSFC, Greenbelt, Maryland

Moritz H (2000) Geodetic Reference System 1980. The Geodesist's Handbook. J Geodes 74:128–133

Nečas J (1967) Les méthodes directes en théorie des équations elliptiques. Academia, Prague

Nesvadba O, Holota P, Klees R (2007) A direct method and its numerical interpretation in the determination of the Earth's gravity field from terrestrial data. In: Tregoning P, Rizos C (eds) Dynamic Planet, IAG Symposia, vol 130, Chap. 54, Springer, New York, pp 370–376

Neyman YuM (1979) A variational method of physical geodesy. Nedra, Moscow (in Russian)

Rektorys K (1974) Variaicn´ý metody v inizenýrských problémech a v problémech matematické fyziky STNL, Praha. (In Czech)

Rektorys K (1977) Variational Methods. Reidel, Dordrecht-Boston

Sansò F (1986) Statistical methods in physical geodesy. In: Sünkel H (ed) Mathematical and numerical techniques in physical geodesy. Lecture notes in earth sciences, vol 7, Springer, New York, pp 49–155

Sansò F, Sacerdote F (2011) Least Squares, Galerkin and BVPs applied to the determination of global gravity filed models. In: Mertikas SP (ed) Gravity, Geoid and Earth Observations, IAG Symposia, vol 135, Springer, New York, pp 511–517

Tscherning CC (1975) Application of collocation. Determination of a local approximation to the anomalous potential of the Earth using "Exact" astro-gravimetric collocation. In: Brosowski B, Martensen E (eds) Methoden und Verfahren der Mathematischen Physik, Band 14, Bibliographisches Institut AG, Mannheim-Wien-Zürich, pp 83–110

On the Comparison of Radial Base Functions and Single Layer Density Representations in Local Gravity Field Modelling from Simulated Satellite Observations

29

M. Weigelt, W. Keller, and M. Antoni

Abstract

The recovery of local (time-variable) gravity features from satellite-to-satellite tracking missions is one of the current challenges in Geodesy. Often, a global spherical harmonic analysis is used and the area of interest is selected later on. However, this approach has deficiencies since leakage and incomplete recovery of signal are common side effects. In order to make better use of the signal content, a gravity recovery using localizing base functions can be employed. In this paper, two different techniques are compared in a case study using simulated potential observations at satellite level – namely position-optimized radial base functions and a single layer representation using a piecewise continuous density. The first one is the more common approach. Several variants exist which mainly differ in the choice of the position of the base function and the regularization method. Here, the position of each base is subject to an adjustment process. On the other hand, the chosen radial base functions are developed as a series of Legendre functions which still have a global support although they decay rapidly. The more rigorous approach is to use base functions with a strictly finite support. One possible choice is a single layer representation whereas the density is discretized by basic shapes like triangles, rectangles, or higher order elements. Each type of shape has its own number of nodes. The higher the number of nodes of a particular element, the more complicated becomes the solution strategy but at the same time the regularity of the solution increases. Here, triangles are used for the comparison. As a result, the radial base functions in the employed variant allow a modeling with a minimum number of parameters but do not achieve the same level of approximation as the discretized single layer representation. The latter do so at the cost of a higher number of parameters and regularization. This case study offers an interesting comparison of a near localizing with a strictly localizing base

M. Weigelt (✉) · W. Keller · M. Antoni
Geodätisches Institut der Universität Stuttgart,
D-70174 Stuttgart, Geschwister-Scholl-Str. 24D, Germany
e-mail: weigelt@gis.uni-stuttgart.de

N. Sneeuw et al. (eds.), *VII Hotine-Marussi Symposium on Mathematical Geodesy*, International Association of Geodesy Symposia 137, DOI 10.1007/978-3-642-22078-4_29, © Springer-Verlag Berlin Heidelberg 2012

function. However, results can currently not be generalized as other variants of the radial base functions might perform better. Also, the extension to a GRACE-type observable is desirable.

Keywords

Regional gravity field recovery • Radial base function • Single layer representation

1 Introduction

This research aims at the recovery of local gravity signal from satellite data as provided by the CHAllenging Minisatellite Payload (CHAMP) or the Gravity Recovery And Climate Experiment (GRACE). The common procedure is to derive spherical harmonic coefficients, which have a global support and need an infinite amount of coefficients in order to represent a space-limited function. As the development is truncated at a maximum degree, a spatially restricted signal cannot be fully represented by a spherical harmonic series, e.g. Barthelmes (1986) and Simons et al. (2006). Figure 29.1 shows an example of residual signal which has not been recovered by the spherical harmonic analysis using satellite-only data. For an arc crossing the Himalayan mountains in August 2003, the K-band range rate observation has been compared to numerically integrated velocity differences using GGM02S (Tapley et al. 2005) up to degree and order 110 which is the recommended maximum degree. The procedure has been repeated with GGM02C, complete to degree 150, which contains besides the GRACE observations also terrestrial observations. Since the GGM02S residuals are obviously correlated with the topography, gravity signal is still left in the K-band range rate observations. Note that the global support is not the only but one of the reasons for the discrepancies.

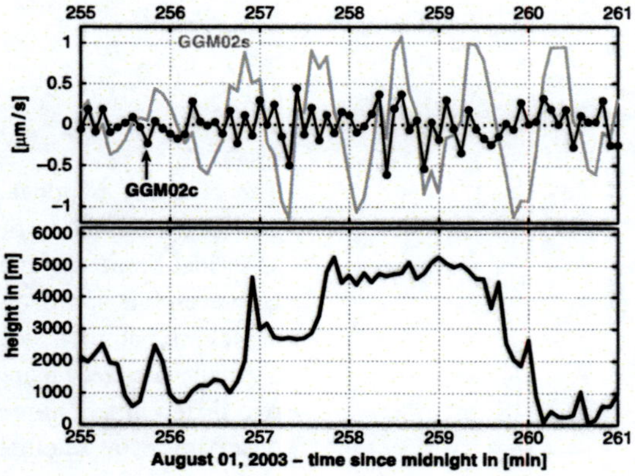

Fig. 29.1 *Top*: K-band observation vs. integrated velocity differences using GGM02S (*gray*) and GGM02C (*black*) for an arc crossing the Himalayan mountains in August 2003. *Bottom*: Topographic profile along the arc

As an alternative to the spherical harmonic solution, localizing base functions can be used. Here, the approximation quality of so-called position-optimized radial base functions (RBF) and a single layer representation using triangular elements is tested in a case study. It is shown that position-optimized radial base functions enable a reasonable approximation with a minimum number of base functions, whereas the single layer representation achieves better approximations at the cost of a higher number of parameters and regularization.

Section 2 outlines the processing scheme of the position-optimized radial base functions. Section 3 introduces the single layer representation and a simulation study is presented in Sect. 4, in which the two different approaches are applied to recover two simulated gravity signals from potential observations along a satellite orbit.

2 Position-Optimized Radial Base Functions

Radial base functions have been applied to satellite-to-satellite tracking data and terrestrial observations before. Eicker (2008) placed point masses on a regular grid and estimated local gravity solutions from CHAMP and GRACE. The approach remains linear as only a scale factor needs to be estimated, cf. (29.1), but regularization is necessary in order to counteract overparametrization. Barthelmes (1986) established the basis for an optimization of the position of point masses, which enables a minimization of their number at the cost of solving a non-linear problem. Antoni et al. (2007) considered the center of each base as well as the shape factor as unknown which are adjusted together with the scale factor. In order to solve the non-linear problem, an iterative procedure is established which is referred to as position-optimized radial base functions in the following. Wittwer (2009) combines both strategies by starting on a coarse grid which is subsequently refined by adding additional base functions at the location of the largest residuals. The depth of the base function is then subject to an adjustment using the closest observations. A data-adaptive strategy is introduced by Klees et al. (2008) which selects the optimal depth of the spherical radial base functions by applying a generalized cross validation.

Here, the procedure of Antoni et al. (2007) is applied. The radial base functions form a set of localizing functions which are isotropic, i.e. symmetric to the center point of the base function. The basic equation is given as

$$V = \frac{GM}{R} \sum_{b=1}^{B} \eta_b \sum_{n=1}^{N} \left(\frac{R}{r}\right)^{n+1} \sigma_b(n)\, P_n(\cos \psi_b), \tag{29.1}$$

where GM is the geocentric gravitational constant, R the mean Earth radius, B the number of base functions in use, N the maximum degree of development, η_b the scale factor and σ_b the shape parameter, which is degree dependent and defines the shape of the local base function. The Legendre polynomials P_n have as argument the cosine of the spherical distance ψ_b between the position of each base function and any point of interest (λ, ϑ, r), where λ is the longitude, ϑ the co-latitude and r the radius. The base functions are centered at the position $(\lambda_b, \vartheta_b, r_b)$.

Figure 29.2 shows the algorithm in order to estimate the unknown parameters $(\eta_b, \sigma_b, \lambda_b, \vartheta_b)$ of the radial base functions. The radius r_b is fixed to the mean Earth radius. The pre-processed signal may or may not be reduced using long-wavelength data, e.g. from a global model, which might be necessary in order to calculate local gravity solutions. Initial shape parameters are selected and initial positions need to be derived from the data. Since in the linear combination of radial base functions each coefficient can be positive or negative, each base function can produce a local maximum or minimum in the data. Therefore, local extrema are chosen as the initial guess for the positions of the base functions. They are used for a first (linear) adjustment in order to get initial values for the scale factors. All initial values are then improved by using a nonlinear least-squares solver, which is based on the Levenberg–Marquardt algorithm (Marquardt 1963). The function to be minimized is the sum of the squares of the simulated potential observations and the potential values produced by the radial base functions. Boundary conditions ensure that the parameters remain in predefined intervals. The procedure is repeatedly applied till the algorithm does not add new base functions. More details about the procedure can be found in Antoni et al. (2007) and Weigelt et al. (2008).

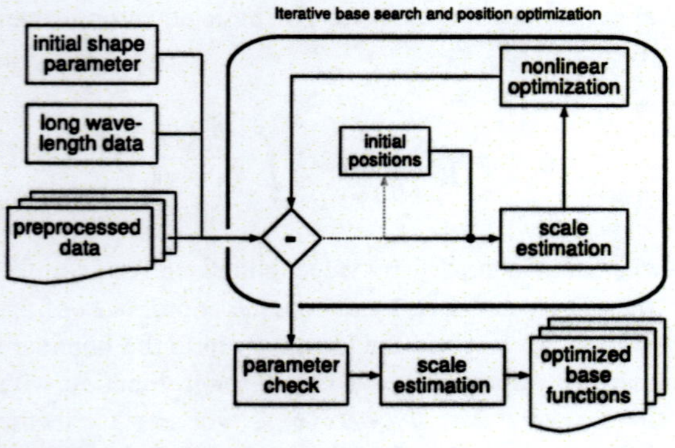

Fig. 29.2 Workflow for the iterative search and optimization of the radial base function parameters

3 Single Layer Representation with a Finite Number of Surface Elements

An alternative approach is to consider strictly space-limited functions. The potential is modeled by a single layer, which results in the necessity to evaluate a surface integral. The surface is decomposed into a finite number N of surface elements Ω_i. The unknown parameters are the surface mass density σ which are approximated by piecewise polynomials defined on rectangular or triangular elements. For the simulation study here, triangular elements are used and the single layer density is approximated by a linear polynomial on each triangular element. The basic equation is then given as

$$V(\mathbf{x}) = \frac{G}{4\pi \cdot R} \sum_{i=1}^{N} \sum_{k=1}^{3} \sigma_{i,k} \int_{\Omega_i} \frac{\Phi_i(\mathbf{y})}{\|\mathbf{x} - \mathbf{y}\|} d\mathbf{y}, \quad (29.2)$$

where $\sigma_{i,k}$ is the density value in the k-th vertex of the triangle Ω_i and $\mathbf{x}$ and $\mathbf{y}$ denote the coordinates of the point of interest and the location within the boundary element, respectively. Φ_i is the shape function with the property that $\Phi_i(\mathbf{y}_j) = \delta_{i,j}$, whereas $\mathbf{y}_j$ is the j-th vertex of Ω_i. The integral in (29.2) becomes a double integral over the triangle, which is planar in the latitude–longitude domain. Triangles offer a high flexibility in the placement of the vertices and thus in the discretization of the single layer density but the choice is far from being trivial. One possible choice is to search, by analogy to the procedure of the radial base functions, for extrema in small subset of the data, e.g. 1° square elements, and place the vertices there.

The idea of using a single layer representation with a finite number of (rectangular) boundary elements in satellite geodesy is not new. Rowlands et al. (2005) and Lemoine et al. (2007) use the so-called MASCON-approach to derive monthly time-variations of the gravity signal in selected basins from GRACE data. The surface mass densities are derived by numerically combining the partial derivatives of the K-band range observations towards the spherical harmonic coefficients with the partial derivatives of the spherical harmonics with respect to the surface mass densities. However, the single layer representation is expanded in spherical harmonics although it is possible to use elements with a distinct spatial support directly. The primary selection criteria are the order of the elements, the analytical and numerical evaluation of the integral in (29.2) and the necessity to apply regularization.

4 Simulation Study

The approximation quality of the two approaches is compared by applying the algorithms to potential observations with two different properties. In the first case (A, Fig. 29.3, left), the signal is formed by a single point mass at a depth of 125 km resulting in an isotropic signal structure. A more complex signal is created by placing 4,225 point masses at a depth of 120–130 km, shown in Fig. 29.3, right.

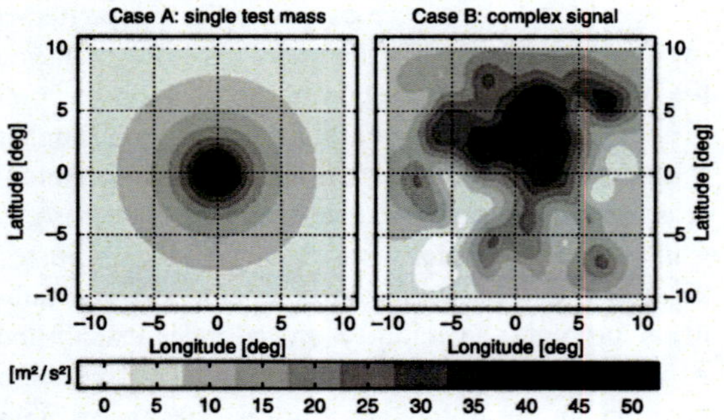

Fig. 29.3 Simulated gravity signal at the Earth's surface: in Case A, a single point mass is used in order to create a simple signal structure; in Case B, 4,225 point masses form a complex signal

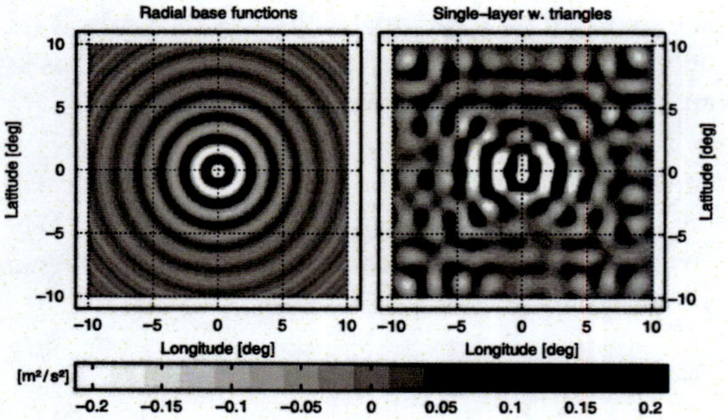

Fig. 29.4 Case A: difference between recovered signal and reference: radial base functions on the *left*, single layer representation using *triangles* on the *right*

For every source field, noise-free potential observations are generated along a Keplerian orbit with a height of 385 km for a period of 30 days and a 5 s sampling resulting in 3,204 data points in the area of interest. These are analyzed using the two algorithms and the results are validated in the spatial domain against a reference solution generated from the source field directly.

Figure 29.4 shows the results for the case of a single point mass. The single layer solutions inhibits slightly more artifacts but looking at the statistics in Table 29.1 reveals that both approaches achieve similar approximations. The important point is that the radial base functions achieve this quality with only 13 base functions and without regularization whereas in case of the single layer representation 416 base functions

Table 29.1 Statistics of the comparison for the two cases: the signal is given in absolute values in the denoted units; the difference Δ is given as relative quantity with respect to the signal

	Mean (m^2/s^2)	RMS (m^2/s^2)	Max (m^2/s^2)	Min (m^2/s^2)	Corr
Case A:					
Signal	11.4858	13.6770	63.7761	5.1237	–
Δ^{RBF}	0.0018 %	0.40 %	0.6 %	23.3 %	99.99 %
$\Delta^{Triangle}$	0.0016 %	0.80 %	2.1 %	36.3 %	99.98 %
Case B:					
Signal	18.5060	20.7034	1.6596	44.0252	–
Δ^{RBF}	0.5522 %	17.98 %	56.2 %	1156.3 %	92.02 %
$\Delta^{Triangle}$	0.0050 %	1.06 %	2.1 %	52.8 %	99.97 %

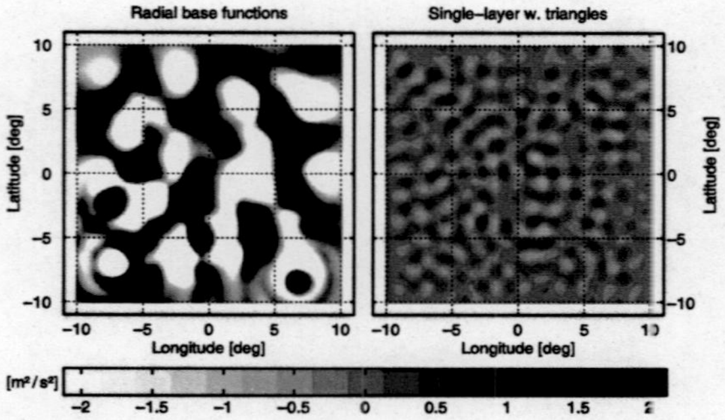

Fig. 29.5 Case B: difference between recovered signal and reference: radial base functions on the *left*, single layer representation using *triangles* on the *right*

(triangles) are used and regularization in form of a truncated singular value decomposition needs to be applied.

In case of the complex signal structure (Case B, Fig. 29.5) the single-layer representation outperforms the radial base functions simply due to the higher number of parameters but again at the cost of regularization. The algorithm of the radial base functions placed only 20 base functions and achieved a relative RMS-error of less than 20%. It also shows large maximum and minimum differences which are explainable by the low number of base functions. Obviously and with this low number, distinct features are easier to model than complex signals. Note that the radial base functions are able to achieve better approximation levels by introducing more base functions but regularization will become necessary, as well. On the other hand, the single layer representation seems to be a very flexible tool for the modeling of any type of structure in the data. It needs to be investigated if overparametrization can be minimized (or even avoided) by choosing a proper discretization of the single layer density.

Conclusion

The results shown here are currently very preliminary as the test scenario uses noise-free data, is restricted to potential observations and only two variants of the approaches are investigated. Nevertheless, it can already be concluded that radial base functions are able to achieve very good approximations with a minimum number of base functions, e.g. only 20 in case of the complex signal structure. For better level of approximations, either more base functions, e.g. on a grid, or other methods like the single layer representation need to be applied which will quickly lead to overparametrization. For a more complete picture of the performance of the two approaches, the tests need to be extended to other type of observables, e.g. the K-Band range rate of GRACE and or terrestrial observations, and to different variants of piecewise polynomial single layer representations as well as of radial base function representations.

References

Antoni M, Keller W, Weigelt M (2007) Representation of regional gravity fields by radial base functions. In: Sideris M (ed) Observing our changing Earth. International association of geodesy symposia, vol 133. Springer, Heidelberg, pp 293–299

Barthelmes F (1986) Untersuchungen zur Approximation des äusseren Gravitationsfeldes der Erde durch Punktmassen mit optimierten Positionen. Ph.D. Thesis, Zentralinstitut für Physik der Erde, Potsdam

Eicker A (2008) Gravity field refinement by radial base functions from in-situ satellite data. Ph.D. Thesis, Rheinische Friedrich-Wilhelms-Universität zu Bonn

Klees R, Tenzer R, Prutkin I, Wittwer T (2008) A data-driven approach to local gravity field modelling using spherical radial basis functions. J Geod 82:457–471. doi:10.1007/s00190-007-0196-3

Lemoine F, Luthcke S, Rowlands D, Chinn D, Klosko S, Cox C (2007) The use of mascons to resolve time-variable gravity from GRACE. In: Tregoning P, Rizos C (eds) Dynamic planet: monitoring and understanding a dynamic planet with geodetic and oceanographic tools. International association of geodesy symposia, vol 130. Springer, Heidelberg, pp 231–236. doi:10.1007/978-3-540-49350-1_35

Marquardt D (1963) An algorithm for least-squares estimation of nonlinear parameters. J Appl Math 11:431–441

Rowlands D, Luthcke S, Klosko S, Lemoine F, Chinn D, McCarthy J, Cox C, Anderson O (2005) Resolving mass flux at high spatial and temporal resolution using GRACE inter-satellite measurements. Geophys Res Lett 32(L04310):1–4. doi:10.1029/2004GL021908

Simons F, Dahlen F, Wieczorek M (2006) Spatiospectral concentration on a sphere. SIAM Rev 48(3):504–536. doi:10.1137/S0036144504445765

Tapley B, Ries J, Bettadpur S, Chambers D, Cheng M, Condi F, Gunter B, Kang Z, Nagel P, Pastor R, Pekker T, Poole S, Wang F (2005) GGM02 – an improved Earth gravity field model from GRACE. J Geod 79:467–478. doi:10.1007/s00190-005-0480-z

Weigelt M, Antoni M, Keller W (2008) Regional gravity field recovery from GRACE using position optimized radial base functions. In: Stelios M (ed) Gravity, geoid and earth observation. International association of geodesy symposia, vol 135. Springer, Heidelberg, pp. 139–146

Wittwer T (2009) Regional gravity field modelling with radial basis functions. Publications on Geodesy, vol 72. NCG Nederlandse Commissie voor Geodesie, Delft

Finite Elements Solutions of Boundary Value Problems Relevant to Geodesy

30

Z. Fašková, R. Čunderlík, and K. Mikula

Abstract

The paper is aimed at a solution to the boundary value problem (BVP) with the Dirichlet and the Neumann boundary conditions by the finite element method (FEM). The computational domain for global gravity field modeling is 3D space above the Earth bounded by the Earth's surface and upper spherical boundary. For local gravity field modeling on continental scale we choose only part of the Earth's surface and create four additional side boundaries. On the Earth's surface, the gravity disturbances generated from DNSC08 altimetry-derived data or EGM2008 geopotential model are considered. The disturbing potential on the upper spherical and side boundaries is generated from satellite model ITG-Grace. The derivation of FEM for this problem including the main discretization ideas is presented. Global quasigeoidal experiments and local refinements are performed. Later, solutions gained with linear and quadratic elements are compared and the influence of Dirichlet BC on the side boundaries on the local solution is studied. All numerical results are tested with potential generated from EGM2008 geopotential model.

Keywords

Boundary value problem with the Dirichlet and the Neumann boundary conditions • Finite element method • Global and local gravity field modeling

1 Introduction

Study of gravity field modeling is one of the most important theoretical and practical tasks of recent geodesy. Nowadays, the common approaches to the global gravity field modeling are using spherical harmonics and precise local modeling by FFT-based methods, see Sideris (1986), and geodetic collocation, see Tscherning (1978). From numerical methods, boundary element method (BEM) has been used by various groups, see Klees (1992), Klees et al. (2001) and Čunderlík et al. (2008). In case of the finite element method (FEM), the pioneering work has been done by Meissl (1981) and Shaofeng and Dingbo (1991). In spite of above-mentioned approaches, where the solution is sought on Earth's surface or its approximation, we present numerical solution computed in 3D computational domain above the Earth. We formulate BVP consisting of

Z. Fašková (✉) · R. Čunderlík · K. Mikula
Faculty of Civil Engineering,
Slovak University of Technology, 813 68 Bratislava,
Radlinského 11, Bratislava, Slovakia
e-mail: faskova@math.sk; cunderli@svf.stuba.sk;
mikula@math.sk

N. Sneeuw et al. (eds.), *VII Hotine-Marussi Symposium on Mathematical Geodesy*, International Association of Geodesy Symposia 137, DOI 10.1007/978-3-642-22078-4_30, © Springer-Verlag Berlin Heidelberg 2012

Laplace equation outside the Earth accompanied by the Neumann as well as the Dirichlet BC. FEM leads to the solution of sparse symmetric linear systems which give the disturbing potential solution in every discrete node of the 3D computational domain.

2 Formulation of the BVP with the Dirichlet and the Neumann BC

To formulate our BVP, we consider the linearized fixed gravimetric BVP (cf. Koch and Pope 1972; Holota 1997, 2005; Čunderlík et al. 2008).

$$-\Delta T(\mathbf{x}) = 0, \quad \mathbf{x} \in R^3 - \Omega, \tag{30.1}$$

$$\langle \nabla T(\mathbf{x}), \bar{s}(\mathbf{x}) \rangle = -\delta g(\mathbf{x}), \quad \mathbf{x} \in \partial\Omega, \tag{30.2}$$

$$T(x) \to 0 \quad as \quad |x| \to \infty, \tag{30.3}$$

where $T(\mathbf{x})$ is the disturbing potential defined as a difference between the real and normal gravity potential at any point $\mathbf{x}$ under the assumptions discussed in Holota (1997), $\delta g(\mathbf{x})$ is the gravity disturbance and $\bar{s}(\mathbf{x}) = \nabla U(\mathbf{x}) / |\nabla U(\mathbf{x})|$. It is very important.

In spite of the BVP (30.1)–(30.3) dealing with the infinite domain, in our approach we construct an artificial boundary $\Gamma_2 \subset \partial\Omega$ away from the approximate Earth surface Fig. 30.1. Moreover in continental experiments we restrict our computations only to a partial domain Ω depicted in Fig. 30.1 as well. The surface gravity disturbances in (30.2) represents the oblique derivative BC (neglecting the deflection of the vertical). In order to get the Neumann BC we project the oblique derivative BC into the boundary Γ_1, i.e., $\frac{\partial T(\mathbf{x})}{\partial n_{\Gamma_1}}$ is approximatelly equal to $-\delta g(\mathbf{x}) \cdot \cos\mu(\mathbf{x})$, where $\mu(\mathbf{x})$ is the angle $\angle(\bar{s}(\mathbf{x}), n_{\bar{\Gamma}1}(\mathbf{x}))$, see Čunderlík et al. (2008). On the upper spherical boundary $\Gamma_2 = \{x; |x| = R\}$ and side boundaries $\Gamma_{3,\ldots,6} \subset \partial\Omega$ the Dirichlet BC is prescribed.

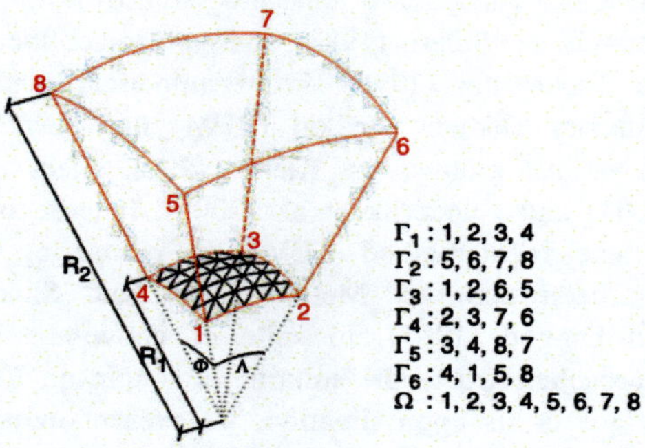

Fig. 30.1 Illustration of the computational domain

Then our BVP is defined as follows:

$$-\Delta T(\mathbf{x}) = 0, \quad \mathbf{x} \in \Omega, \tag{30.4}$$

$$\frac{\partial T(\mathbf{x})}{\partial n_{\Gamma_1}} = -\delta g^*(\mathbf{x}) = -\delta g(\mathbf{x}) \cdot \cos\mu(\mathbf{x}) \tag{30.5}$$

$$T(\mathbf{x}) = T_{SAT}(\mathbf{x}), \quad \mathbf{x} \in \Gamma_i, i = 2, \ldots, 6, \tag{30.6}$$

where T_{SAT} represents the disturbing potential generated from a satellite geopotential model and $\mu(\mathbf{x})$ is angle $\angle(\bar{s}(\mathbf{x}), n_{\bar{\Gamma}1}(\mathbf{x}))$, $\mathbf{x} \in \Gamma_1$. It is worth to notice that the Neumann problem (30.4)–(30.5) is in certain sense an approximation of the geodetic BVP because most of the Earth is flat (e.g. ocean) and there the normal to the ellipsoid is close to the normal of the surface; and also because many actual solution methods for BVP operate first a reduction to the ellipsoid.

3 Solution of the BVP by FEM

To derive the variational formulation of (30.4)–(30.6), we define the Sobolev space of test functions V that is the space of functions from $W_2^{(1)}(\Omega)$ which are equal to 0 on $\Gamma_i, i = 2, \ldots, 6$, in the sense of traces. We multiply the differential equation (30.4) by $v \in V$ and get

$$\int_{\Omega} \nabla T \cdot \nabla v \, dxdydz - \int_{\partial\Omega} \nabla T \cdot n \, v \, d\sigma = 0, \quad \forall v \in V. \tag{30.7}$$

Let the extension of Dirichlet BC given by T_{SAT} into the domain Ω be in $W_2^{(1)}(\Omega)$ and let $\delta g^* \in L^2(\Gamma_1)$. Then we define the weak formulation of our BVP (30.4)–(30.6) as follows: we look for a function T, such that $T - T_{SAT} \in V$ and

$$\int_{\Omega} \nabla T \cdot \nabla v \, dxdydz + \int_{\Gamma_1} \delta g^* \, v \, d\sigma = 0, \quad \forall v \in V. \tag{30.8}$$

Due to Brenner and Scott (2002) or Rektorys (1974), the solution of this problem always exists and is unique.

Table 30.1 Errors for the example of potential generated by unit sphere gained by comparing numerical solutions and exact solutions on subsequently refined grids

Exp.	A		B		C	
E	$\|\mathbf{u}-u\|_{L_2(\Omega)}$	EOC	$\|\mathbf{u}-u\|_{L_2(\Omega)}$	EOC	$\|\mathbf{u}-u\|_{L_2(\Omega)}$	EOC
2^3	0.006750	–	0.000378	–	0.005511	–
4^3	0.001053	2.68	0.000023	4.01	0.001018	2.43
8^3	0.000187	2.49	0.000001	3.40	0.000180	2.42
16^3	0.000036	2.37	–	–	0.000040	2.29

Since the FEM assumes discretization of the domain by subdomains called the finite elements, let choose finite dimensional subspace V_h that is corresponding to the finite element grid. In order to complete the discretization, we must select a basis of V_h, i.e., we choose the piecewise linear function $v_i \in V$ that is uniquely determined by value 1 at x_i and zero at every x_j, $i \neq j$. If we write $T^n(x, y, z) = \sum_{j=1}^{n} t_j v_j(x, y, z)$, plug it into the weak formulation (30.8) and consider test function $w = v_i$ we get

$$\sum_{j=1}^{n} t_j \phi(v_i, v_j) = q_i \quad i = 1, \ldots, n, \qquad (30.9)$$

where $q_i = -\int_{\Gamma_1} \delta g^* v_i \, d\sigma$. Then let the column vectors $(t_1, \ldots, t_n)$ and $(q_1, \ldots, q_n)$ be denoted by $\mathbf{t}$ and $\mathbf{q}$ and let $K = [K_{ij}]$ be matrix whose entries are $K_{ij} = \phi(v_i, v_j)$. We may rephrase (30.9) as

$$K\mathbf{t} = \mathbf{q}, \qquad (30.10)$$

which represents the linear system of equations for unknowns $\mathbf{t}$. The matrix K is called the stiffness matrix. More details can be found in Fašková et al. (2009).

4 Numerical Experiments

In the following section we present various numerical experiments by the FEM software – ANSYS.

The first experiments are theoretical only to illustrate the order of accuracy of FEM. We suppose potential $u(\Lambda, \Phi, R)$ generated by a homogeneous sphere with radius $R = 1\,[m]$ and we solve this problem in a space between $R = 1\,[m]$ and $R = 2\,[m]$. Since we know the exact solution, $u(\Lambda, \Phi, R) = 1/R$, we can compute the Dirichlet and Neumann BC. In all following experiments, the Neumann BC on the bottom boundary is applied. In Experiment A, on upper spherical and on side boundaries the Dirichlet BC is considered and linear elements are used. In experiment B, the same BC as in experiment A are considered, only quadratic elements are used. Finally, in the experiment C on the upper spherical boundary the Neumann and on the side boundaries the Dirichlet BC are considered. One can see that FEM with linear elements is second order accurate in L_2-norm and more than third order accurate with quadratic elements, Table 30.1.

The following numerical experiments deal with the global gravity field modelling. We have performed computations with very coarse grid – 3D 4-nodes elements $3° \times 3°$ based. As the input BC we use the DNSC08 gravity anomaly dataset, see Andersen et al. (2008), and we transform them into the surface gravity disturbances using EGM2008. As the Dirichlet BC on $\Gamma_i, i = 2, \ldots, 6$ we use the disturbing potential generated from the ITG-GRACE03S satellite geopotential model, see Mayer–Gúrr (2007), up to degree 180. We consider the space above the sphere of radius $R_1 = 6371\,[km]$, where the Neumann BC (30.5) is given, up to the sphere with radius $R_2 = 6871\,[km]$, where the Dirichlet BC (30.6) is considered. Results with the meshed domain are depicted in Fig. 30.2.

Since the further successive refinement yields to large memory requirements, we restrict our quasigeoidal modelling to area of Europe. The computational domain is again the space between two spheres, $R_1 = 6371\,[km]$ and $R_2 = 6871\,[km]$, spherical coordinates can be found in Table 30.2. Now the domains are meshed with 8-nodes linear elements and in one case with 20-nodes quadratic elements. Statistics of residuals between the local quasigeoidal solutions and EGM2008 can be found in Table 30.3. It is evident from experiment 1 and 2 that finer grid

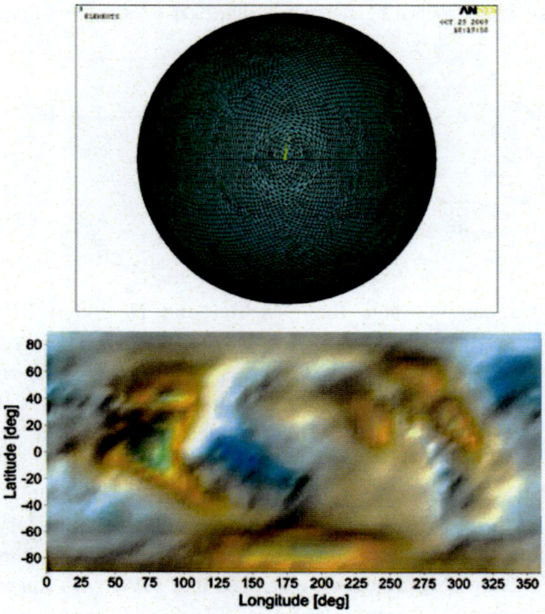

Fig. 30.2 Meshed computational domain and disturbing potential solution

Table 30.2 Geometry of computational domains and mesh statistics

Experiment	1	2	3	4
Sp. latitude	20, 70	20, 70	5, 25	5, 25
Sp. longitude	0, 50	0, 50	35, 55	35, 55
Resol. (deg)	0.2×0.2	0.08 × 0.08	0.04 × 0.04	0.08 × 0.08
Type of el.	linear	linear	linear	quadratic
No. of nodes	693,011	6,270,016	5,271,021	5,218,541
No. of elem.	625,000	5,859,375	5,000,000	1,250,000

Table 30.3 Statistics of residuals between the local quasigeoidal solutions and EGM2008

Experiment	1	2	3	4
No. of nodes on Γ_1	63,001	391,876	251,001	18,801
Min. residuum	−2.056	−1.252	−0.619	−0.479
Mean residuum	−0.027	0.058	−0.024	−0.023
Max. residuum	1.496	0.793	0.458	0.385
St. deviation	0.175	0.078	0.071	0.054

brings improving of results. One can see that using of quadratic elements in comparison with linear element also improve the solution, in spite of the fact that number of nodes in linear element's case is larger, experiment 3, 4 in Tables 30.2 and 30.3.

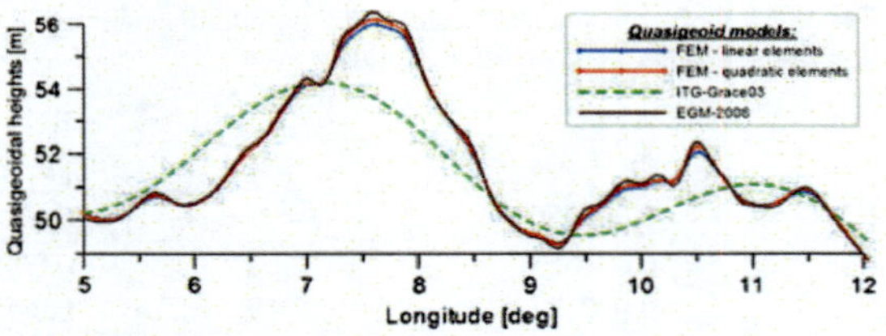

Fig. 30.3 Vertical profiles across the computational domain, $N46.00$

Figure 30.3 depicts the profiles across the FEM solution with linear and quadratic elements, EGM2008 and ITG-GRACE03S to illustrate an influence of the prescribed Dirichlet BC on the FEM solution. It is evident from the profiles and from the second plot in Fig. 30.4 too that a striping effect as well as eventual bias of the satellite geopotential model affects the FEM solution only in very close zones to the side boundaries. They have practically no impact on the central zones. One can see again that using quadratic elements improve the final solution especially in mountainous area, Fig. 30.3, third plot of Fig. 30.4.

Conclusion

The goal of this paper was to present the finite element method which looks for the numerical solution in 3D domains above the Earth's surface. We formulated the BVP in the 3D domain and we considered the Neumann as well as Dirichlet BCs on different parts of its boundary. On the Earth's surface we use the gravity disturbances generated from the DNSC08 altimetry-derived data or the EGM2008 geopotential model. For local quasigeoidal modeling we created additional four side boundaries where we considered, together with artificial upper boundary, the Dirichlet BC generated from ITG-GRACE03S satellite model. Our solutions were compared with EGM2008. We showed that for local quasigeoidal modeling is not necessary to integrate over the whole Earth's surface only over the domain above the area of interest. We also presented that the Dirichlet BC from satellite geopotential models fix the FEM solution but their eventual bias and striping effect do not influence the FEM solution in the central zones. It is also worth noting that using of quadratic elements give better results in mountainous areas though number of nodes is smaller than in the same case with linear elements. We also showed

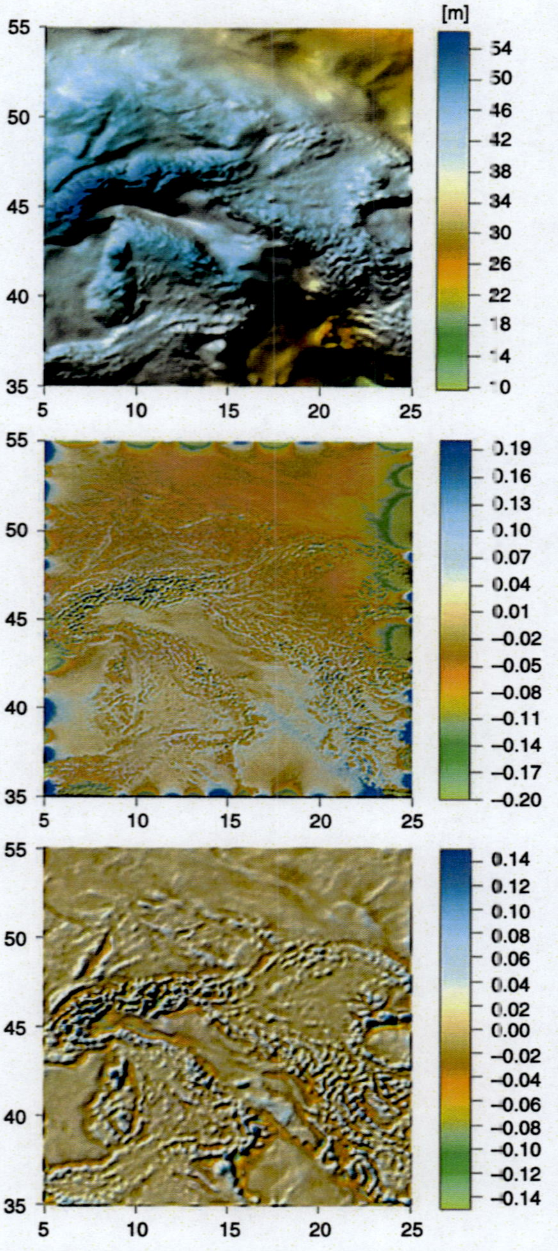

Fig. 30.4 Final quasigeoidal solution in middle part of Europe, residuals between FEM solution and EGM2008 and residuals between FEM solutions with quadratic and linear elements

that the further refinement of the discretization is straightforward and brings improving of the results. At the end we want to mention that the airborne gravimetry data can be applied as the Neumann BC on the upper boundary. All these results are promising for further investigation.

Acknowledgement Authors gratefully thank to the financial support given by grant VEGA 1/0269/09, APVV-LPP-216-06 and APVV-0351-07.

References

Andersen OB, Knudsen P, Berry P (2008) The DNSC08 ocean wide altimetry derived gravity field. Presented at EGU-2008, Vienna, Austria.

ANSYS www.ansys.com

Brenner SC, Scott LR (2002) The Mathematical Theory of Finite Element Methods 2nd edn. Springer, New York

Čunderlík R, Mikula K, Mojzeš M (2008) Numerical solution of the linearized fixed gravimetric boundary-value problem. J Geodes 82(1):15–29

Fašková Z, Čunderlík R, Mikula K (2009) Finite element method for solving geodetic boundary value problems. J Geodes doi: 10.1007/s00190-009-0349-7

Holota P (1997) Coerciveness of the linear gravimetric boundary-value problem and a geometrical interpretation J Geodes 71(10):640–651

Holota P (2005) Neumann's boundary-value problem in studies on Earth gravity field: weak solution 50 years of Research Institute of Geodesy, Topography and Cartography, vol 50(34), Prague, pp 49–69

Klees R (1992) Loesung des fixen geodaetischen Randwertprolems mit Hilfe der Randelementmethode DGK. Reihe C., Nr. 382, Muenchen

Klees R, Van Gelderen M, Lage C, Schwab C (2001) Fast numerical solution of the linearized Molodensky prolem. J Geodesy 75:349–362

Koch KR, Pope AJ (1972) Uniqueness and existence for the godetic boundary value problem using the known surface of the earth. Bull Géod 46:467–476

Mayer-Gürr T (2007) ITG-Grace03s: The latest GRACE gravity field solution computed in Bonn, presentation at GSTM+SPP, Potsdam

Meissl P (1981) The Use of Finite Elements in Physical Geodesy Report 313, Geodetic Science and Surveying, The Ohio State University

Pavlis NK, Holmes SA, Kenyon SC, Factor JK (2008) An Earth Gravitational Model to Degree 2160: EGM2008 General Assembly of EGU, Vienna, Austria

Rektorys K (1974) Variační metody v inženýrských problémech a v problémech matematické fyziky STNL, Praha. (In Czech)

Shaofeng B, Dingbo C (1991) The finite element method for the geodetic boundary value problem. Manuscripta Geodetica 16:353–359

Sideris MG, Schwarz KP (1986) Solving Molodensky's series by fast Fourier transform techniques Bull Geod 60:51–63

Tscherning CC (1978) Collocation and least squares methods as a tool for handling gravity field dependent data obtained through space research techniques. In: Hieber S, Guyenne TD (eds) On space oceanography, navigation and geodynamics European workshop. European Space Agency, pp 141–149

On Combination of Heterogeneous Gravitational Observables for Earth's Gravity Field Modelling

31

Pavel Novák

Abstract

The Earth's gravitational field is described in geodesy by the geopotential, a scalar function of position and time. Although it is not directly observable, its functionals such as first- and second-order directional derivatives can be measured by ground, airborne or spaceborne sensors. In geodesy, these observables are usually used for recovery of the geopotential at some known simple reference surface. Since no observation technique providing gravitational data is fully ideal, ground, airborne and spaceborne data collected with different accuracies, spectral contents, temporal and spatial distributions must be combined. An observation model for recovery of the geopotential is based on the Abel–Poisson equation modified to various gravitational observables. Integral kernels weight spatially contributions of particular observables as functions of their position. Models for different observables are combined exploring stochastic and design characteristics of actual observations.

Keywords

Geopotential • Gravity • Data combination • Abel–Poisson integral • Earth's gravitational field

1 Introduction

Traditional gravitational data collected in geodesy (ground gravity, deflections of the vertical, orbit perturbations, sea surface topography) have recently been extended for airborne and spaceborne gravitational and gradiometric data. Different combination methods have been proposed. One can combine observations, products derived from observations (global and local gravitational models) or products with observations (global gravitational models and local gravity data). The data combination method is discussed in this text.

Heterogeneous data can be used for derivation of a local, regional or global gravitational model. All types of recently available gravitational observables can be inverted into the geopotential by using a surface integral equation that represents a particular solution of the Dirichlet boundary-value problem (Kellogg 1929). Modifying this integral, a mathematical model for various data can be formulated with the geopotential derived in terms of its perturbations from the reference potential at the surface of the reference ellipsoid that generates this potential (Vaníček and Krakiwsky 1986).

P. Novák (✉)
University of West Bohemia, CZ-306 14 Pilsen, Univerzitní 22, Plzeň, Pilsen, Czech Republic
e-mail: panovak@kma.zcu.cz

N. Sneeuw et al. (eds.), *VII Hotine-Marussi Symposium on Mathematical Geodesy*, International Association of Geodesy Symposia 137, DOI 10.1007/978-3-642-22078-4_31, © Springer-Verlag Berlin Heidelberg 2012

The model represents a generalized version of Green's integral equations used in geodesy (Stokes, Hotine, Vening-Meinezs) that represent special forms of this integral equation. These integrals solve for the disturbing potential or its functionals from data distributed at some simple reference surface. For the model discussed in this article, input gravitational data are distributed on and outside the Earth's surface, locally or globally. The model allows for combined processing of heterogeneous gravitational data.

2 Field Equations

In order to proceed with mathematical formulations, three closed surfaces dividing the 3-D Euclidian space Ω are defined: Earth's surface (hypsometry and sea level) $\partial\Omega_t$, the geoid $\partial\Omega_g$ and the reference ellipsoid $\partial\Omega_e$. Assuming geocentric Cartesian coordinates, the (static) geopotential V satisfies at any point $\mathbf{x}$ with zero mass density the *Laplace differential equation*

$$\nabla^2 V(\mathbf{x}) = 0 \tag{31.1}$$

Moreover, the geopotential is also regular at infinity

$$V(\mathbf{x}) = \mathcal{O}\left(|\mathbf{x}|^{-1} \right) \tag{31.2}$$

with the Landau symbol $\mathcal{O}$ describing its inverse distance attenuation.

The *Dirichlet boundary-value problem* allows for evaluation of the geopotential in the mass-free space outside the Lipschitz boundary $\partial\Omega$ (smooth, closed and simply connected surface) on which the geopotential is continuously known

$$V(\mathbf{y}) = f(\mathbf{y}), \quad \mathbf{y} \in \partial\Omega(\mathbf{c}) \tag{31.3}$$

The algebraic vector $\mathbf{c}$ includes parameters of the boundary surface; for example, major and minor semi-axes a, b for the reference ellipsoid $\partial\Omega_e$, numerical coefficients in a harmonic series representation of the geoid $\partial\Omega_g$ or the topographical surface $\partial\Omega_t$. The boundary $\partial\Omega$ must completely contain all gravitating masses in order to satisfy the Laplace differential equation outside the boundary. For the point $\mathbf{x}$ outside the boundary $\partial\Omega$ with an infinitesimal measure $d\sigma$, this solution is the *Abel–Poisson surface integral*, e.g., (Kellogg 1929),

$$V(\mathbf{x}) = \frac{1}{|\partial\Omega|} \iint_{\partial\Omega} f(\mathbf{y})\, \mathcal{K}(\mathbf{x},\mathbf{y})\, d\sigma(\mathbf{y}) \tag{31.4}$$

The integration is performed over the solid geocentric angle $\Theta = \langle 0, \pi \rangle \times \langle 0, 2\pi \rangle$; the infinitesimal measure $d\sigma$ and the unitless integration kernel $\mathcal{K}$ will be functions of the curvilinear coordinates $\{\xi_1, \xi_2\}$ defined in the boundary and corresponding surface parameters $\mathbf{c}$ since $\mathbf{y} = \mathbf{y}(\mathbf{c}, \xi_1, \xi_2)$. The Dirichlet problem always has a unique solution for sufficiently smooth $\partial\Omega$ and Hölder-continuous f, e.g., (Krantz 1999). The integral kernel $\mathcal{K}$ (Green's function) can be derived for the reference ellipsoid $\partial\Omega_e(a, b, \varphi, \lambda)$ adopted as the close approximation of the geoid. In this case curvilinear coordinates $\{\varphi, \lambda\}$ represent the ellipsoidal latitude and longitude.

Since there are gravitating masses outside the reference ellipsoid (topography, atmosphere), the Laplace equation for the geopotential is not satisfied everywhere outside the reference ellipsoid. The gravitational potential δV of external masses must be accounted for

$$V^h(\mathbf{x}) = V(\mathbf{x}) - \delta V(\mathbf{x}) : \nabla^2 V^h(\mathbf{x}) = 0 \tag{31.5}$$

The potential δV can be computed by forward modeling (MacMillan 1958). Due to our limited knowledge of their geometry and mass density distribution, it can be computed only approximately, thus, the Laplace differential equation cannot rigorously be satisfied.

Only small perturbations of the reduced geopotential V^h from the reference gravitational potential U are solved in geodesy

$$T^h(\mathbf{x}) = V^h(\mathbf{x}) - U(\mathbf{x}) : \nabla^2 T^h(\mathbf{x}) = 0 \tag{31.6}$$

taking the advantage of the geodetic reference system (Moritz 1984) that approximates the Earth and its gravitational field. Solving for small values of the reduced harmonic disturbing potential T^h has also numerical advantages.

Using the reduced disturbing potential T^h harmonic everywhere outside the reference ellipsoid, the formulation of the Abel–Poisson integral in (31.4) reads

$$T^h(\mathbf{x}) = \frac{1}{|\partial\Omega_e|} \iint_{\Theta} T^h(\mathbf{y}_e) \mathcal{K}(\mathbf{x}, \mathbf{y}_e) d\sigma(\mathbf{y}_e) \tag{31.7}$$

The geocentric vector $\mathbf{y}_e$ refers to the surface of the reference ellipsoid. The sought geopotential V at the reference ellipsoid reads

$$V(\mathbf{y}_e) = T^h(\mathbf{y}_e) + \delta V(\mathbf{y}_e) + U(\mathbf{y}_e) \tag{31.8}$$

that can be deduced directly from 31.5 and 31.6.

3 Observables and Observation Equations

In the following, we shall assume that first- and second-order directional derivatives of the geopotential V are measured. These gravitational observables represent entries of the *gravitational vector*

$$\mathbf{g}(\mathbf{x}) = \nabla V(\mathbf{x}) \tag{31.9}$$

and the *gradiometric (Marussi) tensor*

$$\mathbf{G}(\mathbf{x}) = \nabla \otimes \nabla V(\mathbf{x}) \tag{31.10}$$

The symbol '⊗' stands for the Kronecker product of two vectors. Not all quantities in (31.9) and (31.10) are measurable with the same accuracy. Their availability and accuracy depend on the used instrumentation and particular observation technique.

The symbol $\mathscr{D}$ is used herein to symbolize any of the measurable functionals, i.e., a *general gravitational observable* is denoted as $\mathscr{D}V$. Gravitational observables reduced for the gravitational effect of masses outside the reference ellipsoid are defined through (31.5) and (31.6) by applying the operator $\mathscr{D}$

$$\mathscr{D}_{\mathbf{x}} T^h(\mathbf{x}) = \mathscr{D}_{\mathbf{x}} V(\mathbf{x}) - \mathscr{D}_{\mathbf{x}} \delta V(\mathbf{x}) - \mathscr{D}_{\mathbf{x}} U(\mathbf{x}) \tag{31.11}$$

The direct effect $\mathscr{D}\delta V$, e.g., (Novák and Grafarend 2005), fulfills two functions: reduced observations (1) correspond to the potential harmonic everywhere outside the reference ellipsoid, and (2) are smoothed out. The former condition is required for the application of the integral solution, the latter one then simplifies its numerical evaluation. The effect of the reference field $\mathscr{D}U$ can be computed using the geodetic reference system and Somigliana–Pizzetti theory, e.g., (Torge 2001).

Applying in (31.7) the operator $\mathscr{D}$ at the observation point $\mathbf{x}$ yields the observation equation for the general reduced gravitational observable

$$\mathscr{D}_{\mathbf{x}} T^h(\mathbf{x}) = \frac{1}{|\partial\Omega_e|} \iint_{\Theta} T^h(\mathbf{y}_e) \mathscr{D}_{\mathbf{x}} \mathscr{K}(\mathbf{x}, \mathbf{y}_e) d\sigma(\mathbf{y}_e) \tag{31.12}$$

Since the integration domain Θ is compact and functions $\mathscr{K}$ and $\mathscr{D}\mathscr{K}$ are continuous in both variables, the swap of differentiation and integration is legal. On the right-hand side of (31.12) one solves for the reduced geopotential T^h at the reference ellipsoid $\partial\Omega_e$ that must be corrected for the *indirect effect* and the reference potential U. The indirect effect δV defined as the gravitational potential of the external masses can be very large and its isostatic compensation is usually applied. Forms of the operator $\mathscr{D}$ for various observables are in Table 31.1 (derivative D, scale coefficient h).

Table 31.1 Operator $\mathscr{D}$ for various observables

Observable	Operator $\mathscr{D}$
Scalar gravimetry	$h_u D_u$
Vector gravimetry	∇
3-D gradiometry	$\nabla \otimes \nabla$
1-D gradiometry	$\langle \nabla \otimes \nabla\, \Delta\mathbf{x} \mid \Delta\mathbf{x} \rangle$
Inter-satellite gravimetry	$\langle \Delta\mathbf{x} \mid \delta\nabla \rangle$

4 Alternative Formulation of the Problem

In the previous section, we assumed that the unknown parameter is represented by the gravitational potential. However, one could look for a relation between the measured quantities as described above and the shape of the equipotential surface of the Earth's gravitational field with a stipulated value of the geopotential, namely V_0. Can we link the disturbing potential T to geometric perturbations of the particular equipotential surface (geoid) relatively to the reference ellipsoid (by definition the potential V_0 at the geoid corresponds to the potential U at the reference ellipsoid)?

We can write the Abel–Poisson integral of (31.4) for the geoid $\partial\Omega_g$ in the form

$$V(\mathbf{x}) = \frac{1}{|\partial\Omega_g|} \iint_{\Theta} V(\mathbf{y}_g) \mathscr{K}(\mathbf{x}, \mathbf{y}_g) d\sigma(\mathbf{y}_g) \tag{31.13}$$

Recognizing that $V(\mathbf{y}_g) = V_0$, then

$$V(\mathbf{x}) = \frac{V_0}{|\partial\Omega_g|} \iint_{\Theta} \mathscr{K}(\mathbf{x}, \mathbf{y}_g)\, d\sigma(\mathbf{y}_g) \tag{31.14}$$

This equation links the potential V with the value of potential at the geoid V_0 and its geometry $\partial\Omega_g$ described by the geocentric vectors $\mathbf{y}_g$. Let us assume that the external geopotential is known as well as the value V_0; the unknown (sought) parameters are the vectors $\mathbf{y}_g$ describing the geometry of the geoid in the geocentric coordinate frame. In (31.13), the measure of the geoid, its infinitesimal form $d\sigma$ and the integration kernel $\mathscr{K}$ are unknown. Obviously, the model cannot be applied rigorously. The measure of the geoid is unknown but its approximate value would merely represent a scaling error in the solution.

Taking the average radius of the ellipsoid, deviations of the geoid from the ellipsoid are at the order of 10^{-6}. The geoid is usually solved in terms of its deviations $\Delta\mathbf{y}$ from the reference ellipsoid, i.e., $\mathbf{y}_g = \mathbf{y}_e + \Delta\mathbf{y}$. Then for the scaling constant $\kappa = V_0\,|\partial\Omega_e|^{-1}$

$$V(\mathbf{x}) = \kappa \iint_{\Theta} \mathscr{K}(\mathbf{x}, \mathbf{y}_e + \Delta\mathbf{y})\,d\sigma(\mathbf{y}_e + \Delta\mathbf{y}) \quad (31.15)$$

Assuming the truncated Taylor expansion

$$\mathscr{K}(\mathbf{x}, \mathbf{y}_e + \Delta\mathbf{y})\,d\sigma(\mathbf{y}_e + \Delta\mathbf{y}) \approx \mathscr{K}(\mathbf{x}, \mathbf{y}_e)\,d\sigma(\mathbf{y}_e) + \Delta\mathbf{y}^T\,\nabla_{\mathbf{y}_e}\,[\mathscr{K}(\mathbf{x}, \mathbf{y})\,d\sigma(\mathbf{y})] \quad (31.16)$$

and reducing both sides of (31.15) for the reference potential U yields

$$T(\mathbf{x}) = \kappa \iint_{\Theta} \Delta\mathbf{y}^T \nabla_{\mathbf{y}_e}\,[\mathscr{K}(\mathbf{x}, \mathbf{y})\,d\sigma(\mathbf{y})] \quad (31.17)$$

or symbolically

$$T(\mathbf{x}) = \Delta\mathbf{y}\,\mathscr{M}(\mathbf{x}, \mathbf{y}_e) \quad (31.18)$$

This equation represents an approximate linear relation between the difference T of the actual gravitational potential from the reference gravitational potential at any point $\mathbf{x}$ outside the geoid and the global 3-D geometric deviations $\Delta\mathbf{y}$ of the geoid from the given reference ellipsoid generating the reference gravitational potential.

5 Model Combination

Discretizing the domain of the integral equation, the observation model has a form of a standard linear model explicit in the vector of observations $\mathbf{l}$, i.e., $\mathbf{l} = \mathbf{A}\,\mathbf{b}$. The observation vector may combine various types of the gravitational observables $\mathscr{D}V$ reduced for the effect of the reference gravitational field $\mathscr{D}U$ and the direct effect $\mathscr{D}\delta V$. Observed and reduced data refer to the observation points $\mathbf{x}$ on or outside the Earth's surface, i.e., no prior continuation or re-sampling of data are required. The vector of the unknown parameters $\mathbf{b}$ represents surface area means of the reduced disturbing potential T^h at the surface of the reference ellipsoid. Entries of the design matrix $\mathbf{A}$ can be evaluated as the kernel values multiplied by scaling parameters and discrete measures into which the surface of the reference ellipsoid was discretized.

Samples of the kernel functions (spatial weights for different gravitational observables) for a particular location (latitude 0 arcdeg, eastern longitude 90 arcdeg, height 5,000 m) are shown in Figs. 31.1–31.3 drawn in GMT (Wessel and Smith 1991). Figures 31.1 and 31.2 show angular components of the vector $\nabla\mathscr{K}$

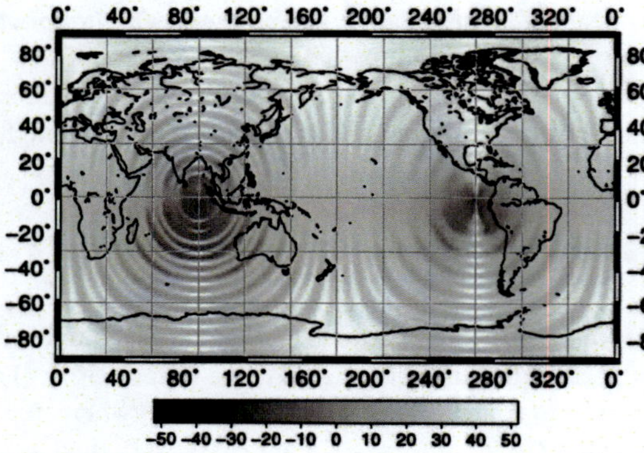

Fig. 31.1 Latitudinal derivative of the function $\mathscr{K}$ ($\mathrm{m}^{-1} \times 1\mathrm{e}5$)

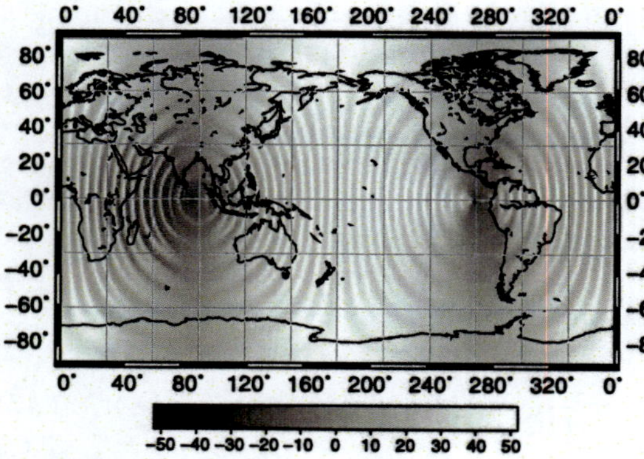

Fig. 31.2 Longitudinal derivative of the function $\mathscr{K}$ ($\mathrm{m}^{-1} \times 1\mathrm{e}5$)

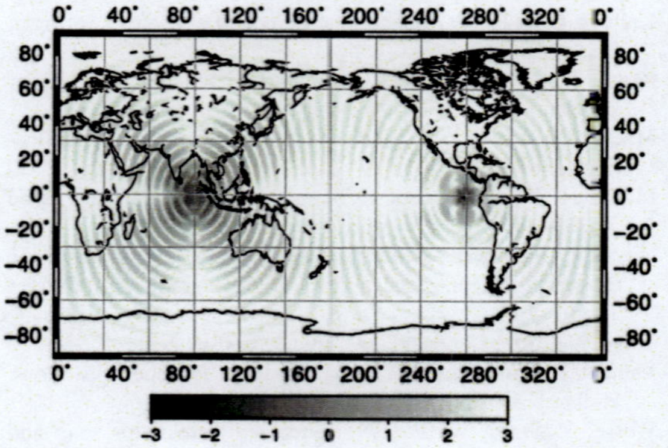

Fig. 31.3 Mixed lat/lon derivative of the function $\mathscr{K}$ ($m^{-2} \times 1e9$)

and Fig. 31.3 depicts one entry of $\nabla \otimes \nabla \mathscr{K}$, namely the mixed angular derivatives; all values are limited to degree 30. For this degree of expansion, their values in distant zones stretching to the anti-pole of the computation point are significant and cannot be neglected. Values in figures were scaled, see their captions.

Data characteristics affecting the application of the mathematical model are the observation noise, spectral content, spatial and temporal sampling. Concerning the signal-to-noise ratio, the most precise information is still coming from ground gravity surveys: point gravity values can routinely be measured with the accuracy up to $\pm 20\ \mu$Gal (Torge 2001).

Redundant ground gravity observations also allow for adjustment of gravimetric networks that may result in estimation of complete stochastic parameters of adjusted gravity. Ground gravity data is also affected by used instrumentation (drift) and uncertainties in 3-D positions of gravity stations.

Airborne data are limited geographically, although larger airborne surveys have been performed recently. The accuracy of airborne data is significantly worse compared to ground gravity data. Moreover, the spectral content of the airborne data is limited due to the high-frequency observation noise caused by airplane dynamics that is a priori removed from measured data. The routine accuracy of airborne data is at the level of ± 2 mGal for 5 km resolution (Forsberg et al. 2000).

Spaceborne data have global coverage except for the polar gaps. They suffer from the effect of changing geometry of the orbit in time: the so-called orbital resonance may affect their application. The spaceborne signal is also attenuated by the distance of the satellite from the gravitating masses and they are used only for recovery of a low-frequency component of the gravitational field. Most recent gradiometric data resolve the gravitational field at the level of ± 1 mGal for 100 km resolution (Drinkwater et al. 2003).

The combined model can be assembled as follows

$$\hat{\mathbf{b}} = \left(\mathbf{A}^T \mathbf{P}_l \mathbf{A}\right)^{-1} \mathbf{A}^T \mathbf{P}_l \mathbf{l},\ \mathbf{C}_{\hat{\mathbf{b}}} = \left(\mathbf{A}^T \mathbf{P}_l \mathbf{A}\right)^{-1} \quad (31.19)$$

While the weight matrix $\mathbf{P}_l = \mathbf{C}_l^{-1}$ describes stochastic properties of data (weight problem), the design matrix $\mathbf{A}$ reflects both their spatial distribution and spectral content (design problem). The weight matrix is based on estimated stochastic properties of various data types (data adjustment and validation). The design matrix is assembled by using band-limited spectral representation of kernel functions (data spectral properties).

Conclusion

Gravitational observables are considered as entries of the gravitational vector and/or the gradiometric tensor. The solution for the unknown geopotential in terms of discrete values at the international reference ellipsoid is formulated by using a single integral equation. No further approximation but discretization of its surface (integration domain) is required for the solution.

The solution of the model is based on the inverse integral inversion accompanied by one a priori and one a posteriori step. Both steps involve forward modeling and theory of the gravitational field of the biaxial ellipsoid. The reference ellipsoid serves as the boundary for the boundary-value problem as well as the reference for the external masses. The integral equations are used for formulation of linear relationships between available observables and geometric properties of the geoid defined in terms of its geometric deviations from the adopted reference ellipsoid.

Currently available gravitational data have limited spectral properties as well as various sampling and accuracy deficiencies. Their different spectral content can be accounted for by using respective spectral representations of the integration kernels since they allow for adjusting the model to spectral characteristics of particular gravitational data. Moreover, the model allows for the weighted data combination if stochastic properties of the gravitational data can a priori be estimated.

Acknowledgements The study was supported by the Czech Science Foundation (project 205/08/1103) and the Czech Ministry of Education, Youth and Sports (project MSM4977751301).

References

Drinkwater MR, Floberghagen R, Haagmans R, Muzi D, Popescu A (2003) GOCE: ESA's first Earth Explorer Core mission. Space Sci Series ISSI 18:419–432

Forsberg R, Olesen A, Bastos L, Gidskehaug A, Meyer U, Timmen L (2000) Airborne geoid determination. Earth Planets Space 52:863866

Kellogg OD (1929) Foundations of potential theory. Springer, Berlin

Krantz SG (1999) Handbook of complex variables. Birkhäuser, Boston

MacMillan WD (1958) Theory of the potential. Dover Publications, New York

Moritz H (1984) Geodetic Reference System 1980. Bull Géod 58:388–398

Novák P, Grafarend EW (2005) Ellipsoidal representation of the topographical potential and its vertical gradient. J Geodes 78:691–706

Torge W (2001) Geodesy. De Gruyter, Berlin

Vaníček P, Krakiwsky EJ (1986) Geodesy: The concepts. North Holland, Amsterdam

Wessel P, Smith WHF (1991) Free software helps map and display data. EOS Trans AGU 72:441

The Convergence Problem of Collocation Solutions in the Framework of the Stochastic Interpretation

32

Fernando Sansò and Giovanna Venuti

Abstract

The convergence of the collocation solution to the true gravity field is a problem defined long ago; some results were derived, in particular by T. Krarup, already in 1981. The problem is taken up again in the context of the stochastic interpretation of collocation theory and some new results are derived, showing that, when the potential T can be really continued down to a Bjerhammar sphere, reasonable convergence results hold true.

Keywords

collocation theory • convergence

1 Introduction

The convergence of the collocation solution $\widehat{T}$ to the true anomalous potential $T(P)$ is, according to Krarup (Krarup 1981), one of the most important problems of physical geodesy. Basically, the point is the following. We claim that the collocation solution, namely certain anomalous potential "predicted" from a discrete set of observations, is an "approximation" of the true potential. However, although this statement has an elementary meaning in the stochastic interpretation, where the variance of the prediction error is available, a careful analysis is needed when we want to give a precise interpretation to the word "approximation" from the functional point of view. In other words, when the observations become more and more numerous and dense close to the boundary, can we prove that our predicted potential $\widehat{T}$ approximates the true potential $T(P)$ in some functional sense, i.e. with respect to a suitable norm? The problem has been defined long ago (Sansò and Tscherning 1980), and a first satisfactory solution has been worked out by Krarup himself (Krarup 1991, 2006). In this formulation, a suitable set of observation points $\{P_i;\ i = 1, \ldots, N\}$ is given on the earth surface S (or on the telluroid), e.g. pointwise gravity disturbances (Moritz 1980), such that

$$\delta g_0(P_i) = \delta g(P_i) + \nu_i = -\frac{\partial T}{\partial h}(P_i) + \nu_i \quad (32.1)$$

with $\{\nu_i\}$ a sequence of white noise satisfying

$$E\{\nu_i\} = 0,\ \ \sigma^2_{\nu_i} \le \sigma_0^2. \quad (32.2)$$

Here, the choice of δg as observable is only to make the reasoning simpler; it could be substituted for different boundary values, for instance free air gravity anomalies.

F. Sansó · G. Venuti (✉)
DIIAR, Politecnico di Milano, P.zza Leonardo da Vinci 32, Milan, Italy
e-mail: fernando.sanso@polimi.it; giovanna.venuti@polimi.it

N. Sneeuw et al. (eds.), *VII Hotine-Marussi Symposium on Mathematical Geodesy*, International Association of Geodesy Symposia 137, DOI 10.1007/978-3-642-22078-4_32,

A deterministic collocation solution, according to Krarup (1969), is a function $\widehat{T} \in H_K$, belonging to the Hilbert space of functions harmonic down to a Bjerhammar sphere of radius R with reproducing kernel

$$K(P,Q) = \sum_{n=0}^{\infty} \sum_{m=-n}^{n} k_n S_{nm}(P) S_{nm}(Q) \quad (32.3)$$

where S_{nm} are the solid spherical harmonics, related to fully normalized surface spherical harmonics Y_{nm} as

$$S_{nm}(P) = \left(\frac{R}{r}\right)^{n+1} Y_{nm}(\sigma_P), \quad k_n \geq 0,$$

cf. (Sansò 1986), such that the hybrid norm functional

$$F(\widehat{T}) = \sum \frac{\left(\delta g_0(P_i) - [-\frac{\partial T}{\partial h}(P_i)]\right)^2}{\sigma_{\nu_i}^2} + \lambda \|\widehat{T}\|_{H_K}^2 \quad (32.4)$$

is minimized. The solution is

$$\widehat{T}_\lambda(P) = \sum_{ik} \left\{ -\frac{\partial}{\partial h_{Q_i}} K(P,Q_i) \right\} \cdot \left\{ -\frac{\partial}{\partial h_{Q_i}} \frac{\partial}{\partial h_{Q_k}} K(Q_i,Q_k) + \lambda \sigma_{\nu_i}^2 \delta_{ik} \right\}^{-1} \delta g_0(P_k). \quad (32.5)$$

The problem here is to prove whether and how $\widehat{T} \to T(P)$ when $N \to \infty$. Krarup's proof was that, by taking a suitable sequence $\lambda_N \to 0$, we can have $\widehat{T} \to T$. This happens in the sense of a "regular" Hilbert space H of functions harmonic outside the earth surface S, as it should be, cf. (Krarup 1981). The theorem is quite interesting although the question of the specific value of λ_N to be chosen for each N is indeed open and annoying.

In this paper, we shall look into the stochastic interpretation of collocation with the purpose of arriving at a different result of convergence. On one side, this result is less general, because we require the original function $T(P)$ to be harmonic outside the Bjerhammar sphere, what is mandatory for the stochastic interpretation; though, on the other side, it does not depend anymore on a parameter like λ.

2 A Preliminary Lemma

Lemma 1. *Let us assume $u(t)$ to be a centered random field on some set A, such that it is Lipschitz continuous in mean quadratic sense, cf. (Gikhman and Skorokhod 2004) and (Miranda 1970), i.e.*

$$0 \leq E\{[u(t) - u(t')]^2\} = 2\gamma(t,t') = C_0 - C(t,t') \leq \alpha |t - t'|; \quad (32.6)$$

assume that $\{t_i\}$ is a sequence which, for $N \to \infty$, becomes a set dense in A in a sense to be better specified later; assume that $u(t)$ is observed at t_i with noises ν_i satisfying (32.2)

$$Y_i = u(t_i) + \nu_i; \quad (32.7)$$

assume further that, for any $t \in A$ fixed,

$$\hat{u}(t) = \sum C(t,t_i)\{C(t_i,t_k) + \sigma_i^2 \delta_{ik}\}^{-1} Y_i \quad (32.8)$$

is the usual Wiener-Kolmogorov predictor and that

$$\mathcal{E}_N^2(t) = C(t,t) + - \sum C(t,t_i)\{C(t_i,t_k) + \sigma_i^2 \delta_{ik}\}^{-1} C(t,t_k) \quad (32.9)$$

is the corresponding prediction error; then,

$$\lim_{N \to \infty} \mathcal{E}_N^2(t) = 0 \quad (32.10)$$

Proof. Consider the linear estimator

$$\tilde{u}(t) = \frac{1}{N_{t,r}} \sum_{i=1}^{N} \chi(t,t_i,r) Y_i \quad (32.11)$$

where

$$\chi(t,t_i,r) = \begin{cases} 1 & |t - t_i| \leq r \\ 0 & \textit{otherwise,} \end{cases} \quad (32.12)$$

$$N_{t,r} = \sum_{i=1}^{N} \chi(t,t_i,r). \quad (32.13)$$

Since (32.8) is the best linear estimator of $u(t)$, we have indeed

$$\mathcal{E}_N^2(t) = E\{[u(t) - \hat{u}]^2\} \le \tilde{\mathcal{E}}_N^2(t) = E\{[u(t) - \tilde{u}]^2\}. \quad (32.14)$$

On the other hand, since $Y_i = u(t_i) + \nu_i$, we have

$$u(t) - \tilde{u} = \frac{1}{N_{t,r}} \sum_{i=1}^{N} [u(t) - u(t_i)]\chi(t, t_i, r) + \frac{1}{N_{t,r}} \sum_{i=1}^{N} \chi(t, t_i, r)\nu_i. \quad (32.15)$$

Therefore, the error propagation gives

$$\tilde{\mathcal{E}}_N^2(t) = \frac{1}{N_{t,r}^2} \sum_{ik} \chi(t, t_i, r)\chi(t, t_k, r) \cdot E\{[u(t) - u(t_i)][u(t) - u(t_k)]\} + \frac{1}{N_{t,r}^2} \sum_i \chi(t, t_i, r)\sigma_i^2. \quad (32.16)$$

With the help of the inequality $ab \le \frac{1}{2}a^2 + \frac{1}{2}b^2$, and recalling (32.2) and (32.6), from (32.16) we derive

$$\tilde{\mathcal{E}}_N^2(t) \le \frac{1}{N_{t,r}} \sum_i \chi(t, t_i, r)\gamma(t, t_i) + \frac{\sigma_0^2}{2N_{t,r}} \le \frac{\alpha}{2N_{t,r}} \sum_i \chi(t, t_i, r)|t - t_i| + \frac{\sigma_0^2}{N_{t,r}} \le \alpha r + \frac{\sigma_0^2}{N_{t,r}}. \quad (32.17)$$

With such a relation we are able now to specify the hypothesis on the sequence $\{t_i\}$; we assume that it is possible to find a sequence r_N such that

$$\lim_{N\to\infty} r_N = 0, \quad \lim_{N\to\infty} N_{t,r_N} = +\infty. \quad (32.18)$$

As an example, if $\{t_i\}$ tends to cover uniformly A, we shall have

$$N_{t,r_N} \ge \text{const}\, N r^2;$$

then, it is sufficient, for instance, to choose

$$r_N = \frac{1}{N^{\frac{1}{4}}},$$

to get both relations (32.18) verified. Therefore from (32.17) we have $\tilde{\mathcal{E}}_N^2(t) \to 0$ and, because of (32.14), (32.10) is proved. □

We conclude the paragraph noting that not only $\mathcal{E}_N^2(t) \to 0$ for each fixed t but, since according to (32.9)

$$\mathcal{E}_N^2(t) \le C(t, t), \quad (32.19)$$

it is enough to assume that $C(t, t)$ is bounded everywhere to have

$$\mathcal{E}_N^2(t) \le \text{const} \quad (32.20)$$

uniformly in N.

3 The Convergence of the Collocation Solution

We assume $T(P)$ to be a harmonic random function in the exterior Ω_B of a Bjerhammar sphere B, with zero mean and covariance

$$C(P, Q) = E\{T(P)T(Q)\} = \sum_{nm} \frac{c_n}{2n+1} S_{nm}(P)S_{nm}(Q). \quad (32.21)$$

The positive constants c_n are called full power degree variances and we assume that

$$\sum_{n=0}^{+\infty} c_n < C_0. \quad (32.22)$$

It is easy to see that, under (32.22), the function $C(P, Q)$ is continuous everywhere in Ω and harmonic in this set, so that

$$|C(P, Q)| \le C_0, \; \forall P, Q \in \Omega. \quad (32.23)$$

Since harmonic functions have derivatives of any order which are continuous in the domain of harmonicity, we can claim that similarly to (32.23)

$$|E\{\delta g(P)\delta g(Q)\}| = |C_{\delta g \delta g}(P, Q)| \le G_0, \quad \forall P, Q \in \Omega, \quad (32.24)$$

with G_0 a suitable constant. Now, let a regular, bounded surface S be given, totally embedded in Ω_B, and a sequence $\{P_i\}$ on S, at which $\delta g(P_i)$ is observed, in agreement with (32.1). The harmonicity of $C(P, Q)$ is certainly enough to guarantee that (32.6) is satisfied, on S and outside cf. (Sansò and Venuti 2005), so that Lemma 1 applies and we have

$$\lim_{N\to\infty} \mathscr{E}^2_{N,\delta g}(P) = 0 \quad \forall P \in S. \qquad (32.25)$$

In addition, according to the remark at the end of §2, we have

$$\mathscr{E}^2_{N,\delta g}(P) \le G_0, \qquad (32.26)$$

so that, by the dominated convergence theorem of Lebesgue (Riesz and Nagy 1965), we can state that

$$\lim_{N\to\infty} \int_S \mathscr{E}^2_{N,\delta g}(P)dS = \int_S \lim_{N\to\infty} \mathscr{E}^2_{N,\delta g}(P)dS = 0. \qquad (32.27)$$

On the other hand, by dint of Fubini's theorem (Riesz and Nagy 1965) the $\int_S$ and the average operation in (32.27), which is as well an integral operator, can be exchanged, so that we can claim

$$\lim_{N\to\infty} E\left\{\int_S [\delta g - \delta\hat{g}_N]^2 dS\right\}$$

$$= \lim_{N\to\infty} \int_S \mathscr{E}^2_{N,\delta g}(P)dS = 0. \qquad (32.28)$$

In this way, using a standard theorem of stochastic calculus (Gikhman and Skorokhod 2004), we come to prove Theorem 1.

Theorem 1. *Under the stated hypotheses on the random field $T(P)$ and on its covariance function (32.21), (32.22), on the surface S and on the sequence $\{P_i\}$, there is a subsequence $\{N_k\}$ such that*

$$\|\delta g(P) - \delta\hat{g}_{N_k}(P)\|_{L^2(S)} \to 0 \qquad (32.29)$$

almost surely, i.e., with probability $P = 1$.

It is enough now to recall a recent result given in (Sansò and Venuti 2008) to be able to establish the convergence of the collocation solution in the form of another theorem:

Theorem 2. *Under the assumption of Theorem 1 and further assuming that S is a regular star-shaped surface $S \equiv \{r = R(\sigma)\}$, such that its normal is never inclined more than 89° with respect to the radial unit vector $\underline{e}_r$, then, along the same sequence $\{N_k\}$ defined in Theorem 1, one has*

$$\|\nabla T(P) - \nabla\widehat{T}_{N_k}(P)\|_{L^2(S)} \to 0 \qquad (32.30)$$

almost surely.

4 Discussion of the Result

The fact that we can prove a result of convergence only with probability $P = 1$ should not bother the reader, since this is fully acceptable in the framework of a stochastic theory. The theorems presented here, for the case that on S we observe the gravity disturbance $\delta g(P)$, can be strongly generalized, as it will be done in a separate work. The strongest limitation of the present theory is indeed the assumption that $T(P)$ is from beginning harmonic down to some Bjerhammar sphere; on the other hand, without such an assumption the covariance structure (32.21), (32.22) looses its meaning and we would not be anymore in the realm of what we call least squares collocation theory.

References

Gikhman AV, Skorokhod AV (2004) The theory of stochastic processes. Springer, Berlin

Krarup T (1969) A contribution to the mathematical foundation of physical geodesy. Meddelelse N44 Geodaetisk Institut, Kobehavn

Krarup T (1981) A convergence problem in collocation theory. Boll Geod Sc Affini 40:225–240

Krarup T (1991) Approximation to the earth potential from discrete measurements. Proceedings of the Geodetic day in honour of A. Marussi. Atti Convegno Lincei pp 67–85

Krarup T (2006) Angelica returning, or the importance of a title. In: Borre K (ed) Mathematical foundations of geodesy. Springer, Berlin

Miranda C (1970) Partial differential equations of elliptic type. Springer, Berlin

Moritz H (1980) Advanced physical geodesy. Wichmann, Karlsruhe

Riesz F, Nagy Sz (1965) Leçons d'analyse functionelle. Gauthier-Villars, Paris

Sansò F, Tscherning CC (1980) Notes on convergence in collocation theory. Boll Geod Sc Affini 39:221–252

Sansò (1986) Statistical methods in physical geodesy. In: Sünkel (ed) Lecture notes in earth Sc. Springer, Berlin, pp 49–155

Sansò F, Venuti G (2005) Topological vector spaces of harmonic functions. J Geod 79:203–221

Sansò F, Venuti G (2008) On the explicit determination of stability constants for linearized geodetic BVP. J Geod 82: 909–916

On the Omission Errors Due to Limited Grid Size in Geoid Computations

33

Yan Ming Wang

Abstract

Based on the assumption that the ultra-high frequencies of the gravity field are produced by the topography variations, we compute the omission errors by using 3 arc-second elevation data from the Shuttle Radar Topography Mission (SRTM). It is shown that the maximum omission errors to the geoid are in the range of dm, cm and sub-cm level for grid sizes of 5″, 2″ and 1″ over the contiguous United States (CONUS), respectively. The results suggest that a 1 arc-minute grid size is sufficient for the 1-cm geoid, even for areas with very rough topography. The results also show that the omission errors to gravity are significant even for 1″ grid size, at which the smoothed-out gravity still reaches tens of mGals. The omission errors to gravity at a 5″ grid size peaks above 100 mGals, demonstrating the importance of correction of residual terrain to gravity observations in data gridding or block mean value computations.

The results are also compared with those based on Kaula's rule. While the omission errors based on Kaula's rule are ± 0.5 and ± 3.0 cm for 1″ and 5″ grid size, respectively, the RMS values of the omission error in this paper are ± 0.1 and ± 1.1 cm. The differences suggest Kaula's rule may overestimate the power of the gravity field at the ultra-high frequency band, which renders the convergence studies of the spherical harmonic series based on Kaula's rule questionable.

Keywords

Geoid • Omission error • RTM effect

1 Introduction

The geoid is a continuous equipotential surface containing all frequencies of the Earth's gravity field. Numerically, the geoid can only be computed at a limited grid size, so that high frequencies of the gravity field beyond the grid size are lost. The omission error is defined as the power of those lost high frequencies.

The omission error may be studied by using a series of spherical harmonics. Theoretically, the gravity field can be expressed as an infinite harmonic series. In practice, the series has to be limited to certain degree and order. For instance, the maximum degree of the EGM08 (Pavlis et al. 2008) is 2,160 which corresponds to a resolution of 5′. To reach 3″ resolution, the maximum degree of the spherical harmonic expansion

Y.M. Wang (✉)
National Geodetic Survey, Silver Spring, MD 20910, USA
e-mail: YAN.WANG@NOAA.GOV

N. Sneeuw et al. (eds.), *VII Hotine-Marussi Symposium on Mathematical Geodesy*, International Association of Geodesy Symposia 137, DOI 10.1007/978-3-642-22078-4_33,

should be 216,000, but that may become impractical for numerical computation. Since such a high degree and order of harmonic expansion is not available, Kaula's rule (Jekeli (2009)) may be used to characterize the gravity field at ultra-high frequencies. If Kaula's rule is valid at this bandwidth, a reasonable assessment of the power can be obtained in the statistical sense. Based on Kaula's rule, the omission errors are ± 3.0, and ± 0.5 cm for a spatial resolution 5″ and 1″, respectively.

Another way to study the omission error is to use numerical integration. Very dense coverage of gravity data in a local area is needed. However, very dense gravity coverage in an area, e.g., a gravity observation every 100 m in a 100 km × 100 km region, is rare. Nonetheless, 100 m or an even finer resolution digital elevation data are available regionally. In fact, the anomalous gravity field is produced by the topography and mass anomalies inside the Earth. It is known that the gravity field with wavelengths shorter than 100 km is mostly due to the topography. In this paper, we study the omission error of 5″ and finer grid sizes, so it is a plausible assessment that the power of a gravity field in a frequency band corresponding to 5″ to 3″ grid size is the only due to the topography variation. By removing the elevation at a certain resolution – the residual topography, traditionally called the residual terrain model (Forsberg (1984)) – may be the only main source of the ultra-high frequencies of the gravity field. Based on this assertion, we investigate the omission error by using the digital elevation of the SRTM in a 3″ (∼90 m) grid size. The motivation is simple: for the residual terrain is the main source of the ultra-high frequency of the gravity field, we can compute its effect on gravity and the geoid at different grid sizes. The following question will be answered: at what resolution does the gravity field have power less than 1 cm in geoid? In other words, what is the maximum grid size with the omission error to the geoid under 1 cm pointwise? In addition, what are the contributions of the residual terrain to gravity observed on the Earth's surface at different grid sizes?

2 Mathematical Formulation

The potential of the residual terrain at a point P can be computed by Newton's integral:

$$V_t(r_P, \mathbf{x}_P) = G \iint_\sigma \int_{r_C}^{r_S} \frac{\rho}{l} r^2 dr d\sigma \tag{33.1}$$

where G is Newton's gravitational constant; ρ is the mass density; r_P and $\mathbf{x}_P$ are the radial distance and the surface variables of computation point P, respectively; r_S is the radial distance of an arbitrary point on the Earth's surface S; r_C is the radial distance of an arbitrary point on a smoothed reference surface, and l is the distance between computation point P and current point:

$$l = \sqrt{r^2 - 2rr_P \cos\psi + r_P^2} \tag{33.2}$$

For a constant density of the residual terrain, the 3-D integral in (33.1) can be reduced into a surface integral (Martinec 1998; Sjöberg and Nahavandchi 1999; Heck (2003)):

$$V_t(r_P, \mathbf{x}_P) = G\rho \iint_\sigma k(r_S, r_C, \psi) d\sigma \tag{33.3}$$

with

$$k(r_S, r_C, \psi) = \frac{3r_P \cos\psi}{2}(l_{SP} - l_{CP}) \tag{33.4}$$

$$+ \frac{1}{2}(r_S l_{SP} - r_C l_{CP}) + \frac{1}{2}(-1 + 3 \times \cos^2\psi) r_P^2 \ln \frac{r_S - r_P \cos\psi + l_{SP}}{r_C - r_P \cos\psi + l_{CP}} \tag{33.5}$$

where l_{SP} and l_{CP} are the distances between point P and points on the Earth's surface S and the reference surface C:

$$l_{SP} = \sqrt{r_S^2 - 2r_S r_P \cos\psi + r_P^2} \tag{33.6}$$

$$l_{CP} = \sqrt{r_C^2 - 2r_C r_P \cos\psi + r_P^2} \tag{33.7}$$

The above equations are valid for points inside and outside of the topography. Since the contribution of the residual terrain to the geoid is in dm level in this study, the ellipsoidal effect is in sub-mm level and can be neglected. For the computation points located on the geoid, we can simply set $r_P = R$ in the kernel function.

The gravity of the residual terrain is the negative vertical derivatives of the potential V_t (ibid.):

$$g_t(r_P, \mathbf{x}_P) = -\frac{\partial V_t}{\partial r_P} = G\rho \int\int_{\sigma} k'(r_S, r_C, \psi) d\sigma \tag{33.8}$$

where

$$k' = -\frac{r_S r_P (1 - 6\cos^2\psi) + (3r_P^2 + r_S^2)\cos\psi}{l_{SP}} + \frac{r_C r_P (1 - 6\cos^2\psi) + (3r_P^2 + r_C^2)\cos\psi}{l_{CP}} - (-1 + 3\cos^2\psi) r_P \ln \frac{r_S - r_P\cos\psi + l_{SP}}{r_C - r_P\cos\psi + l_{CP}} \tag{33.9}$$

The reference surface is a smoothed Earth's surface. In an ideal case, this surface should only contain a lower band of frequencies of the topography. If the Earth's topography is expanded into an infinite spherical harmonic series, the reference surface should contain the spherical harmonics only to a certain degree and order. The residual terrain then contains all higher frequencies of the topography. There are some numerical difficulties to expand the topography into a spherical harmonic series to an ultra high degree and order, e.g., 100,000. A common practice to compute a reference surface is the use of the moving average. It is well known that the moving average is a low-pass filter, so the higher frequencies are effectively removed from the topography. The residual terrain contains only the frequencies smoothed out by the moving average.

In practice, the block-mean values are used. The block-mean values are the same as the moving averaged values, but with additional operation of data sampling. The omission error defined in this paper is the error due to use of the block mean values at a certain grid size.

3 Numerical Computations

We used the SRTM-DTED1 (3″) DEM for North and Central America and the Caribbean ($10° \leq \varphi \leq 60°$; $190° \leq \lambda \leq 308°$). This DEM comes in 2,384 $1° \times 1° 3'' \times 3''$ grids for $\varphi \leq 50°$ and $3'' \times 6''$ grids for $50° \leq \varphi \leq 60°$. It contains 13,992,031 gaps assigned an elevation value of –1, including large gaps of more than $1° \times 1°$ in Northern Florida and South Carolina and smaller continuous gaps in eastern Connecticut and Iowa. The majority of the rest of the gaps are concentrated along the Rocky Mountains, most notably in British Columbia and Mexico.

We also used the SRTM30/GTOPO30 global $30'' \times 30''$ DEM, which comes in 33 raster tiles – 27 of size $40° \times 50°$, and six $30° \times 60°$ tiles. The latter was mainly used to interpolate and fill the 3″ gaps in SRTM-DTED1.

The resulting tiles to the north of latitude 50° N were then interpolated into $3'' \times 3''$ grids, and all resulting tiles were mosaiced to produce the $3'' \times 3''$ SRTM window $10° \leq \varphi \leq 60°$; $190° \leq \lambda \leq 308°$.

To show the omission errors of various topographic features, we chose a latitude band at latitude 41.4042°. The profile of the elevation is drawn in the following figure.

Based on the roughness of the topography, we split the profile into three areas: rough area (Rocky Mountains, $236° \leq \lambda \leq 256°$), flat area (Middle West, $236° < \lambda \leq 280°$), and old mountains (Appalachian Mountains, $280° < \lambda \leq 290°$). The RMS values of the residual terrain for each area are shown in the following table.

Table 33.1 shows that RMS values of residual height are below 100 m; the RMS and extreme values increase with increasing grid size.

The contribution of residual terrains to geoid (omission error) is computed based on (33.3) and (33.5) on 3″ grid spacing (Fig. 33.1). The computation is undertaken in the spherical coordinate system and 1° integration area is chosen at which the computation errors are below 0.1 mm. The statistics of the omission errors are given in the following table (Table 33.2).

In the Rocky Mountains, the omission error of ±1.7 cm seems not significant, even at 5′ resolution. The error is reduced below ±1 cm at 2′ grid size, even

Table 33.1 Statistics of the residual height in meters (block-mean values removed). No. of samples: 63,306

Grid size	Rocky mount	Middle west	Appalachian	Total	Range
1′	45.8	7.5	36.9	32.4	−342/531
2′	65.5	9.0	45.4	44.8	−488/749
5′	121.7	13.7	58.6	79.3	−696/1,490

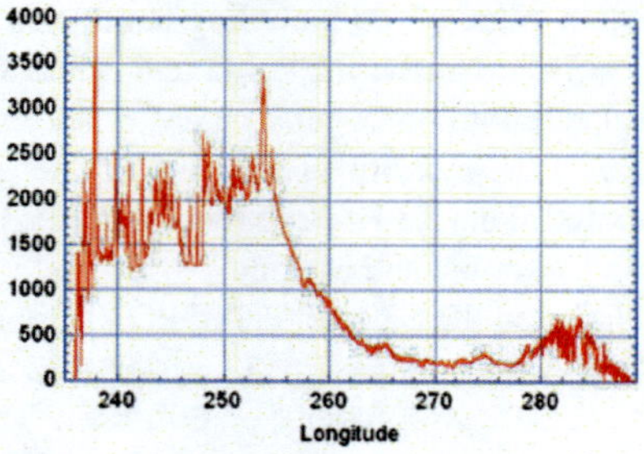

Fig. 33.1 Elevation profile along the longitude band in 3″ spacing

Table 33.2 RMS values and ranges of omission errors on geoid in cm. No. of samples: 63,306

Grid size	Rocky mount	Middle west	Appalachian	Total	Range
1′	±0.1	±0.0	±0.1	±0.1	−0.8/0.9
2′	±0.9	±0.1	±0.2	±0.5	−1.2/8.6
5′	±1.7	±0.3	±0.8	±1.1	−4.3/11.4

in the roughest mountainous area. At 1′ grid size, the RMS value of omission error is ±1 mm for the Rocky Mountains, ±0 mm for the Middle West, and ±1 mm for the Appalachian Mountains.

Pointwise, the omission error ranges from −4.3 cm to 11.4 cm in a 5″ grid size. The maximum error is not reduced significant in a 2′ grid size. However, the omission error is less than 1 cm pointwise in a 1″ grid size.

The omission errors to the geoid are plotted in the following figure (Fig. 33.2).

For a 5″ grid size, the omission error oscillates approximately ±5 cm in high mountain (the Rocky Mountains) regions, but swings around 0.5 cm in flat areas. It fluctuates around 3 cm in the Appalachian Mountains region. For a 2″ grid size, the omission error has meaningful values in the high mountains, other areas are generally insignificant. For a 1″ grid size, the omission error seems insignificant everywhere.

Based on Kaula's rule (e.g., Jekeli 2009), the omission error is ±0.5 cm, while the contribution of the residual terrain is ±0.1 cm for the 1″ grid size. Kaula's rule gives ±3.0 cm–3 times the contribution of the residual terrain for the 5″ grid size. The overestimation of by Kaula's rule at the high frequencies is also shown in Jekeli (2009).

It is also important to estimate the impact of limited grid sizes to the gravity. Equations (33.8) and (33.9) are evaluated along the longitude band on 3″ grid spacing. The computation is undertaken in the spherical coordinate system and 1° integration area is chosen

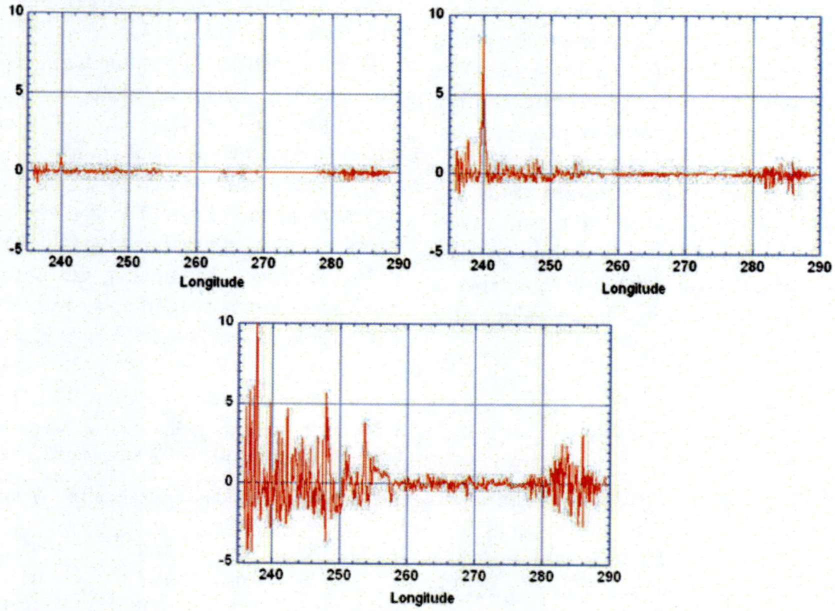

Fig. 33.2 Omission errors of the geoid at different grid sizes (from *top to bottom* 1″, 2″ and 5″)

Table 33.3 RMS values and ranges of omission error in gravity, units are in mGal. No. of samples: 63,306

Grid size	Rocky mount	Middle west	Appalachian	Total	Range
1′	±2.1	±0.1	±1.5	±1.4	−5.4/23.6
2′	±3.7	±0.2	±2.1	±2.4	−11.1/40.8
5′	±8.6	±0.4	±3.3	±5.5	−8.2/114.5

where the computation errors are below 0.1 mGal. The statistics are given in the following table.

Table 33.3 shows that the gravity of residual terrain is significant at every grid size. Even at 1″, the RMS value of gravity of residual terrain is ±1.4 mGal. The maximum gravity reaches 23.6 mGal in high mountains. For a 5″ grid size, the gravity of residual terrain peak is at 114.5 mGal in the Rocky Mountains, and its RMS value is as large as ±5.5 mGal. The results are somewhat unexpected, since we have seen the omission errors to the geoid is insignificant for most cases, and only have some meaningful magnitude in high mountain regions. The large contrast between the geoid and gravity is the evidence that the geoid is dominated by long wavelengths of the gravity field, while the gravity has large power in the high frequency band. To see the gravity changes with respect to location, the gravity of residual terrain (omission error to the gravity) is plotted in the following figure (Fig. 33.3).

As expected, all large gravity occurs in mountainous regions. In flat areas, the gravity of residual terrain is negligible.

4 Conclusions and Discussions

To estimate the omission errors to the geoid and gravity, we assume that the ultra- high frequencies of the gravity field are produced by the topographic variations. If this assumption is realistic, we can draw following conclusions:

1. The omission errors to the geoid can reach dm in high, rough mountains at the 5″ grid size. As expected, the roughness is the determining factor, for it represents the high frequencies of the topography. For the 1 cm geoid, the 5″ grid size is not suitable for mountainous regions.
2. The results show that omission error to the geoid is less than 1 cm pointwise everywhere in CONUS, if the 1′ grid size is used. This conclusion is based the assumption that high frequencies of the gravity field are produced by the residual terrain only.
3. The gravity due to the residual terrain is significant (larger than 1 mGal) even at a 1″ grid size. The results suggest that the RTM gravity may be useful

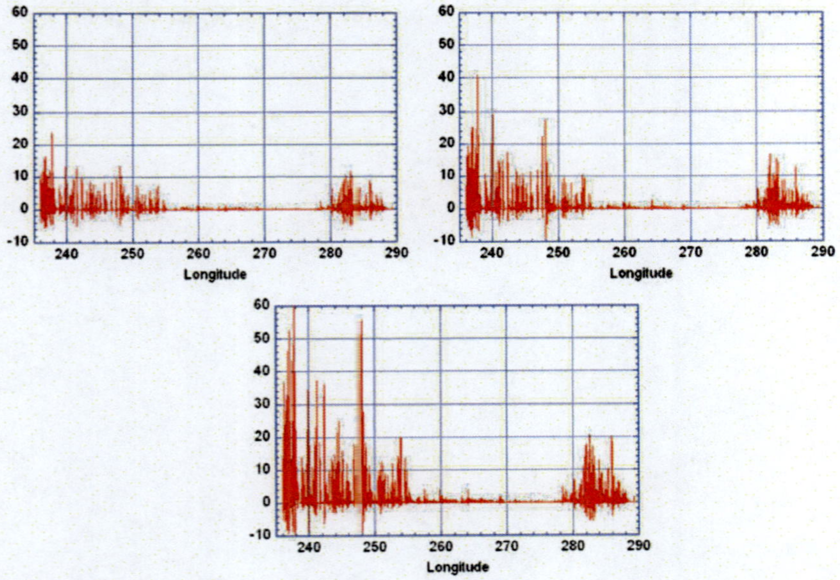

Fig. 33.3 Omission errors to the gravity (from top to bottom 1″, 2″ and 5″) in mGal

for removing the high frequencies from gravity observations before gridding or block mean value computations.

4. The omission errors to the geoid in this paper are smaller than those based on Kaula's rule. For a $1''$ grid size, the omission error is merely ± 0.1 cm verses ± 0.5 cm, and ± 1.1 cm verses ± 3.0 cm for a $5''$ grid size. The reason for the differences is that Kaula's rule overestimates the power of the gravity field at ultra-high frequencies, which renders the studies of convergence of the spherical harmonic series based on Kaula's rule questionable.

A constant density ($2.67\,\mathrm{g/cm^3}$) of the topographic masses is used in this paper. The actual density may vary 10% of the mean density adopted in this paper. From the results showed in the paper, the density variation is not significant enough to alter the above conclusions.

References

Forsberg R (1984) A study of terrain reductions, density anomalies and geophysical inversion methods in gravity field modeling. Dept. of Geodetic Science and Surveying. Report No. 355, Ohio State University, 1984

Jekeli C (2009) Omission error, data requirements, and the fractal dimension of the geoid. Proceedings of Hotine-Marussi Symposium 2009, in press

Heck B (2003) On Helmert's methods of condensation. J Geodes 77:155–170

Martinec Z (1998) Boundary-value problems for gravimetric determination of precise geoid. Lecture notes in Earth Sciences 73. Springer Berlin

Pavlis NK, Holmes SA, Kenyon SC, Factor JK (2008) An Earth gravitational model to degree 2,160: EGM2008. Presented at the 2008 General Assembly of the European Geosciences Union, Vienna, Austria, 13–18 April 2008

Sjöberg LE, Nahavandchi H (1999) On the indirect effect in the Stokes–Helmert method of geoid determination. J Geodes 73(2):87–93

An Oblique Derivative in the Direct BEM Formulation of the Fixed Gravimetric BVP

34

R. Čunderlík, K. Mikula, and R. Špir

Abstract

The fixed gravimetric boundary-value problem (FGBVP) represents an exterior oblique derivative problem for the Laplace equation. The boundary element method and the collocation with linear basis functions are used to get a numerical solution of FGBVP in which the oblique derivative is treated by its decomposition into normal and tangential components to the Earth's surface. The tangential components are expressed through the gradients of the linear basis functions.

This new numerical approach to the solution of FGBVP is applied to global gravity field modelling. Input surface gravity disturbances as the oblique derivative boundary conditions are generated from the DNSC08 gravity field model. The obtained numerical solution with the resolution of 0.1° is compared with the EGM2008 geopotential model at collocation points. A contribution of the tangential components to the solution is presented and discussed.

Keywords

Fixed gravimetric BVP • Oblique derivative problem • Direct BEM formulation • Collocation with linear basis functions

1 Introduction

The precise 3D positioning by GNSS has brought new opportunities for gravity field modelling. Today more attention is focused on the fixed gravimetric boundary-value problem (FGBVP). It represents an exterior oblique derivative problem for the Laplace equation, cf. e.g. Koch and Pope (1972), Bjerhammar and Svensson (1983) or Grafarend (1989). In order to solve such problem we use a direct formulation of the boundary element method (BEM) and the collocation with linear basis functions. In our previous approach we treated the oblique derivative problem by a projection of the oblique derivative onto the normal to the Earth's surface (Čunderlík et al. 2008). Such approach neglects a contribution of the tangential components. In this paper we follow a strategy based on a decomposition of the oblique derivative into normal and tangential components, cf. e.g. Balaš et al. (1989). Hence, we formulate the boundary integral equation (BIE) for the oblique derivative problem. Another approach based on an iterative treatment of the oblique derivative using the Galerkin method is described in (Nesvadba et al. 2007).

R. Čunderlík (✉) · K. Mikula · R. Špir
Department of Mathematics and Descriptive Geometry,
Faculty of Civil Engineering, Slovak University of Technology
Radlinkého 11, 813 68 Bratislava, Slovakia
e-mail: cunderli@svf.stuba.sk

N. Sneeuw et al. (eds.), *VII Hotine-Marussi Symposium on Mathematical Geodesy*, International Association of Geodesy Symposia 137, DOI 10.1007/978-3-642-22078-4_34, © Springer-Verlag Berlin Heidelberg 2012

Our objective in this paper is to derive the direct BEM formulation for FGBVP taking into account the oblique derivative and to implement it into the numerical scheme based on the collocation with linear basis functions. We focus on the evaluation of the tangential components that are expressed through the gradient of the linear basis functions. In the numerical experiments we apply our numerical approach to the global gravity field modelling. The obtained numerical solution with the resolution 0.1° is compared with our previous high-resolution results (Čunderlík and Mikula 2010). It allows us to estimate a contribution of the tangential components that were neglected before.

2 Direct BEM Formulation for FGBVP

The linearized FGBVP represents an exterior oblique derivative problem for the Laplace equation, cf. (Koch and Pope 1972), (Bjerhammar and Svensson 1983) or (Grafarend 1989)

$$\Delta T(x) = 0, \quad x \in R^3 - \Omega, \qquad (34.1)$$

$$\langle \nabla T(x),\, s(x)\rangle = -\delta g(x), \quad x \in \Gamma, \qquad (34.2)$$

$$T = O(|x|^{-1}) \; as \; x \to \infty. \qquad (34.3)$$

where T is the disturbing potential at any point x, δg is the surface gravity disturbance, the domain Ω represents the body of the Earth with its boundary Γ (the Earth's surface), $\langle , \rangle$ is the inner product of two vectors and

$$s(x) = -\nabla U(x)/|\nabla U(x)|, \quad x \in \Gamma, \qquad (34.4)$$

where U is the normal gravity potential. Equation (34.2) represents an oblique derivative boundary condition (BC) as the normal to the Earth's surface Γ does not coincide with the vector s defined by (34.4).

Let us apply the direct BEM formulation to (34.1)–(34.3). The main objective is to replace the Laplace equation (34.1) that governs the solution in the exterior domain R^3–Ω by an equation that reduces the solution to a problem on the boundary Γ (see e.g. (Brebbia et al. 1984) or (Schatz et al. 1990)). Such integral equations can be derived using Green's third identity or the method of weighted residual. It represents the boundary integral equation (BIE) (a detailed derivation is discussed in Čunderlík et al. 2008)

$$\frac{1}{2}T(x) + \int_\Gamma T(y)\frac{\partial G}{\partial n_\Gamma}(x,y)d_yS = \int_\Gamma \frac{\partial T}{\partial n_\Gamma}(y)G(x,y)\, d_yS, \quad x \in \Gamma, \qquad (34.5)$$

where dS is the area element, n_Γ is the unit outward normal vector to the boundary Γ (the Earth's surface) and the kernel function G is the fundamental solution of the Laplace equation,

$$G(x,y) = (4\pi|x-y|)^{-1}, \quad x,y \in R^3. \qquad (34.6)$$

The term $\partial T/\partial n_\Gamma$ in BIE (34.5) represents the normal derivative, while FGBVP includes the oblique derivative BC in (34.2). In order to derive BIE for the oblique derivative problem we follow the idea described in (Balaš et al. 1989). At first we decompose the vector ∇T into the normal and tangential components

$$\nabla T = \langle \nabla T, n_\Gamma\rangle\, n_\Gamma + \langle \nabla T, \tau\rangle\, \tau + \langle \nabla T, \rho\rangle\, \rho, \quad (34.7)$$

where τ and ρ are the unit tangential vectors. The unit vectors n_Γ, τ and ρ represent a local orthonormal triad. Then the oblique derivative term in (34.2) can be written in the form

$$\langle \nabla T, s\rangle = \langle\langle \nabla T, n_\Gamma\rangle\, n_\Gamma, s\rangle + \langle\langle \nabla T, \tau\rangle\, \tau, s\rangle + \langle\langle \nabla T, \rho\rangle\, \rho, s\rangle\,. \qquad (34.8)$$

After expressing $\langle \nabla T, n_\Gamma\rangle$ from (34.8) and inserting into (34.5), we get BIE for our oblique derivative problem

$$\frac{1}{2}T(x) + \int_\Gamma T(y)\frac{\partial G}{\partial n_\Gamma}(x,y)d_yS + \int_\Gamma \frac{\langle\langle \nabla T, \tau\rangle\, \tau, s\rangle}{\langle \mathbf{n}_\Gamma, \mathbf{s}\rangle}(y)G(x,y)d_yS + \int_\Gamma \frac{\langle\langle \nabla T, \rho\rangle\, \rho, \mathbf{s}\rangle}{\langle \mathbf{n}_\Gamma, \mathbf{s}\rangle}(y)G(x,y)d_yS = \int_\Gamma \frac{\langle \nabla T, \mathbf{s}\rangle}{\langle \mathbf{n}_\Gamma, \mathbf{s}\rangle}(y)G(x,y)d_yS, \quad x \in \Gamma. \qquad (34.9)$$

Here a contribution of the tangential components is expressed through the gradients of the unknown disturbing potential. The term $\langle \nabla T, s \rangle$ on the right-hand side of BIE (34.9) represents the oblique derivative BC defined in (34.2) and thus can be replaced by negative values of the input surface gravity disturbances. Then BIE (34.9) represents the direct BEM formulation for FGBVP defined by (34.1)–(34.3).

2.1 Collocation with Linear Basis Function and Tangential Components

As a numerical scheme for solving BIE (34.9) we use the collocation with linear basis functions. It means we consider a piecewise linear representation of the boundary functions T and δg on planar triangles that approximate the Earth's surface. Due to the limited extent of this paper, for more details we refer to (Čunderlík et al. 2008). Here we only focus on an evaluation of tangential components that are neglected in the aforementioned reference.

Considering BIE (34.9), a contribution of the tangential components is expressed through the gradients of the unknown disturbing potential. In case of the used collocation they can be expressed through the gradients of the linear basis functions

$$\nabla T(x) \approx \sum_{k=1}^{3} T_k \nabla \psi_k(x), \quad x \in \Delta\Gamma_j, \tag{34.10}$$

where T_k are unknown values of the disturbing potential at collocation points that represent vertices of the triangle $\Delta\Gamma_j$ and ψ_k are the linear basis functions at these points. Since the gradient of the linear basis function $\nabla\psi_j$ is constant on the whole triangle $\Delta\Gamma_j$, it can be expressed in the following way using the Green theorem

$$\begin{aligned}\nabla\psi_j &= \frac{1}{m(\Delta\Gamma_j)} \int_{\Delta\Gamma_j} \nabla\psi_j \, d\Delta\Gamma \\ &= \frac{1}{m(\Delta\Gamma_j)} \int_{\partial\Gamma_j} \psi_j \bar{\eta} \, d\partial\Gamma, \end{aligned} \tag{34.11}$$

where $m(\Delta\Gamma_j)$ is the area of the triangle $\Delta\Gamma_j$ and $\bar{\eta}$ is the normal vector to its sides $\partial\Gamma$. Considering a fact that the jth linear basis function equals to 1 at the jth collocation point and to 0 at others vertices of the triangle $\Delta\Gamma_j$, i.e. at mth and kth collocation points, (34.11) can be simplify into the form

$$\nabla\psi_j = \frac{1}{2m(\Delta\Gamma_j)} \left[l_{jm}\bar{\eta}_{jm} + l_{jk}\bar{\eta}_{jk} \right], \tag{34.12}$$

where l_{jm} and l_{jk} are the lengths of the sides of the triangle intersecting at the jth collocation point and $\bar{\eta}_{jm}$ and $\bar{\eta}_{jk}$ are the normal vectors to these sides.

In such a way we get the discrete form of BIE (34.9) that subsequently yields a linear system of equations. For more details how to compute the integral operators see (Čunderlík et al. 2008).

3 Numerical Experiments

Numerical experiments deal with the global gravity field modelling. At first we approximate the Earth's surface by a triangulated surface. Vertices of this triangulation represent the collocation points. Their horizontal positions are generated by the algorithm developed in (Čunderlík et al. 2002). A mesh size of the triangular elements is 0.1° in latitude. It represents 4,860,002 collocation points regularly distributed over the whole Earth's surface. Vertical positions of the collocation points are interpolated from the DNSC08 mean sea surface (Andersen et al. 2008) at oceans, and SRTM30PLUS-V5.0 global topography model (Becker et al. 2009) on lands.

The input surface gravity disturbances as the oblique derivative BC are generated from the DNSC08 gravity field model (Andersen et al. 2008). In order to reduce large memory requirements we eliminate far zones contributions using the ITG-GRACE03S satellite geopotential model (Mayer-Gürr 2007) up to 180° and the iterative procedure described in (Čunderlík and Mikula 2010).

Final large-scale computations were accomplished on parallel computers with 128 GB of the distributed memory using standard MPI (Message Passing Interface) subroutines for the code parallelization (Aoyama and Nakano 1999). As a linear solver we use the nonstationary iterative methods BiConjugate Gradient Stabilized (BiCGSTAB) method (Barrett et al. 1994), which is suitable for dense and nonsymetric matrices.

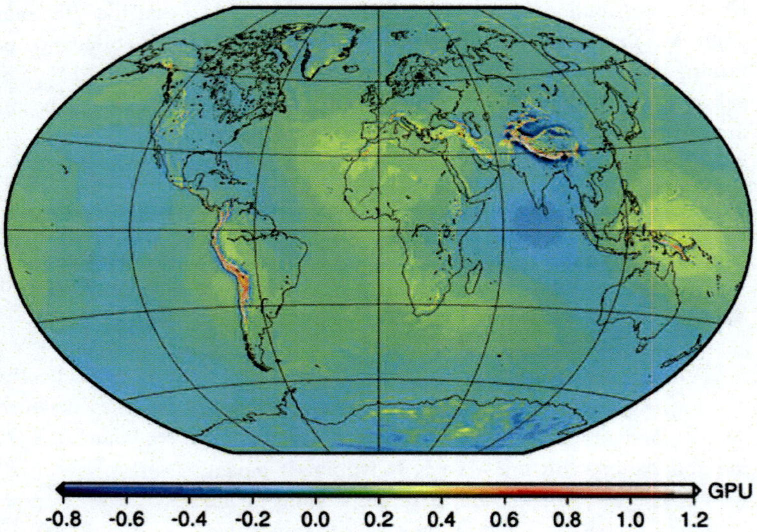

Fig. 34.1 Residuals between the numerical solution by BEM and EGM2008 evaluated at collocation points ($1\text{GPU} = 10\,\text{m}^2\,\text{s}^{-2}$)

The obtained numerical solution with the resolution of 0.1° is compared with the EGM2008 geopotential model (Pavlis et al. 2008) up to degree 2,160. Residuals at the collocation points are depicted in Fig. 34.1. Basic statistical characteristics of such residuals are presented in Table 34.1. A comparison with our previous high-resolution results where the oblique derivatives are projected into the normal derivatives (Čunderlík and Mikula 2010) shows a contribution of the tangential components (Fig. 34.2, Table 34.2).

Table 34.1 Statistical characteristic of residuals between the numerical solution by BEM and EGM2008 at collocation points

AREA	TOTAL	OCEANS	LANDS
Mean	0.062GPU	0.056GPU	0.075GPU
Std.	0.106GPU	0.078GPU	0.154GPU
Min	−1.468GPU	−0.372GPU	−1.468GPU
Max	4.393GPU	0.397GPU	4.393GPU

Table 34.2 Contribution of the tangential components of the oblique derivative ($1\text{GPU} = 10\,\text{m}^2\text{s}^{-2}$)

Area	Total	Oceans	Lands
Mean	0.065GPU	0.057GPU	0.086GPU
Std.	0.035GPU	0.024GPU	0.049GPU
Min	−0.091GPU	0.024GPU	−0.091GPU
Max	0.717GPU	0.313GPU	0.717GPU

4 Discussion and Conclusions

A decomposition of the oblique derivative into the normal and tangential components allows us to formulate the direct BEM formulation for FGBVP. The derived integral relation in (34.9) represents BIE for the oblique derivative problem. An evaluation of the tangential components through the gradients of the linear basis functions in case of the collocation is an efficient approach. The presented numerical experiments show that a contribution of the tangential components can not be neglected, especially in the mountainous areas of the high deflections of verticals (Fig. 34.2). It is evident from Table 34.2 and Fig. 34.2 that their contribution is almost everywhere positive and globally increases the numerical solution. The maximum values up to 0.7GPU (∼0.7 m) are in Himalayas and Andes. Their contribution influences surroundings in far distances (Fig. 34.2).

The obtained gravity field model represents the numerical solution of the linearized FGBVP with the resolution of 0.1°. Its comparison with EGM2008 (spherical harmonics up to degree 2,160) shows better agreement at oceans (Fig. 34.1) where the standard deviation of residuals at collocation points is 0.078 GPU (Table 34.1). Here negative residuals in area of the global minimum (south of India) and positive residuals in area of the global maximum

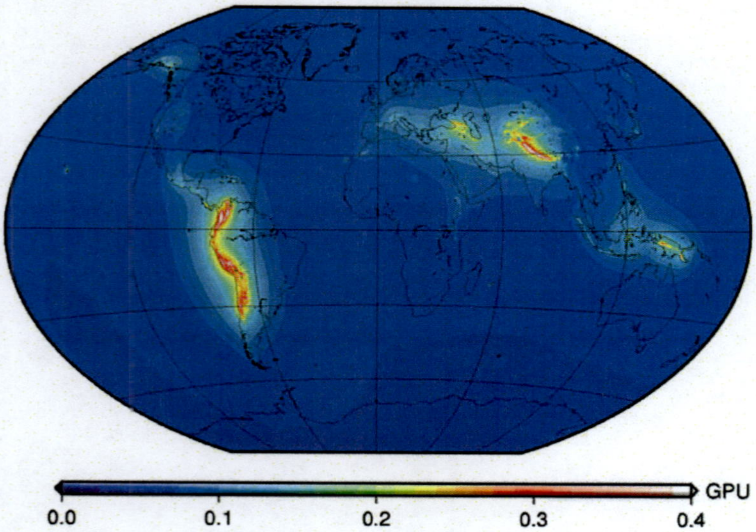

Fig. 34.2 Contribution of the tangential components of the oblique derivative (1GPU $= 10\,\mathrm{m^2\,s^{-2}}$)

(Oceania) ranging up to ± 0.3 GPU indicates a small discrepancy. The highest residuals up to 4 GPU are in Himalayas and Andes that significantly influence the standard deviations of residuals on lands (0.15 GPU). Such high values and an evident correlation with the topography motivate for a further investigation of our approach.

Acknowledgements Authors gratefully thank to the financial support given by grants: VEGA 1/0269/09, APVV-LPP-0216–06 and APVV-0351–07.

References

Andersen OB, Knudsen P, Berry P (2008) The DNSC08 ocean wide altimetry derived gravity field. Presented at EGU-2008, Vienna, Austria, April 2008

Aoyama Y, Nakano J (1999) RS/6000 SP: Practical MPI programming. IBM, Poughkeepsie

Balaš J, Sládek J, Sládek V (1989) Stress analysis by boundary element methods. Elsevier, Amsterdam

Barrett R, Berry M, Chan TF, Demmel J, Donato J, Dongarra J, Eijkhout V, Pozo R, Romine C, Van der Vorst H (1994) Templates for the solution of linear systems: Building blocks for iterative methods. http://www.netlib.org/templates/Templates.html

Becker JJ, Sandwell DT, Smith WHF, Braud J, Binder B, Depner J, Fabre D, Factor J, Ingalls S, Kim S-H, Ladner R, Marks K, Nelson S, Pharaoh A, Sharman G, Trimmer R, vonRosenburg J, Wallace G, Weatherall P (2009) Global bathymetry and elevation data at 30 arc seconds resolution: SRTM30_PLUS, revised for Marine Geodesy

Bjerhammar A, Svensson L (1983) On the geodetic boundary-value problem for a fixed boundary surface – satellite approach. Bull Géodes 57:382–393

Brebbia CA, Telles JCF, Wrobel LC (1984) Boundary element techniques, theory and applications in engineering. Springer, New York

Čunderlík R, Mikula K, Mojzeš M (2002) 3D BEM application to Neumann geodetic BVP using the collocation with linear basis functions. In: Proceedings of ALGORITMY 2002, conference on scientific computing, Podbanské: 268–275

Čunderlík R, Mikula K, Mojzeš M (2008) Numerical solution of the linearized fixed gravimetric boundary-value problem. J Geodes 82:15–29

Čunderlík R, Mikula K (2010) Direct BEM for high-resolution gravity field modeling. Stud Geophys Geodes, Vol. 54, No. 2, pp. 219–238

Grafarend EW (1989) The geoid and the gravimetric boundary-value problem. Rep 18 Dept Geod, The Royal Institute of Technology, Stockholm

Koch KR, Pope AJ (1972) Uniqueness and existence for the geodetic boundary value problem using the known surface of the earth. Bull Géod 46:467–476

Mayer-Gürr T (2007) ITG-Grace03s: The latest GRACE gravity field solution computed in Bonn. Presentation at GSTM+SPP, Potsdam, 15–17 Oct 2007

Nesvadba O, Holota P, Klees R (2007) A direct method and its numerical interpretation in the determination of the Earth's gravity field from terrestrial data. In: Tregoning P, Rizos C (eds) Dynamic Planet. IAG Symposia, vol 130, Springer, New York, pp 370–376

Pavlis NK, Holmes SA, Kenyon SC, Factor JK (2008) An Earth gravitational model to degree 2,160: EGM2008, presented at the 2008 General Assembly of EGU, Vienna, Austria, April 13–18, 2008

Schatz AH, Thomée V, Wendland WL (1990) Mathematical theory of finite and boundary element methods. Birkhäuser, Basel

Generalizing the Harmonic Reduction Procedure in Residual Topographic Modeling 35

Ove Christian Omang, Carl Christian Tscherning, and Rene Forsberg

Abstract

In gravity field modeling measurements are usually located on or above the terrain. However, when using the residual topographic modeling (RTM) method, measurements may end up inside the masses after adding the mean topography. These values do not correspond to values evaluated using a harmonic function. A so-called harmonic correction has been applied to gravity anomalies to solve this problem. However, for height anomalies no correction has been applied. To generalize the correction to e.g. height anomalies we interprete that the vertical gravity gradient inside the masses multiplied by height equals the correction. In principle the procedure is applicable to all gravity field functionals. We have tested this generalization of the procedure which consist in determining equivalent quantities in points Q on the mean surface if this surface is *in free air*. The procedure has as data the reduced values in P inside the masses but considered as being located at the mean surface. Numerical tests with height anomaly data from New Mexico and Norway as control data show that for gravity anomalies the general procedure is better than using the original harmonic correction procedure.

Keywords

Residual topographic modeling • Generalized harmonic reduction • Harmonic correction

1 Introduction

When using RTM for gravity field modeling (Forsberg and Tscherning 1981; Forsberg 1984), measurements may be located inside the masses added using a mean topography. They do therefore not anymore correspond to values evaluated using a harmonic function. For gravity anomaly data a so-called harmonic correction, $-4\pi G\rho \cdot \Delta h$ (Forsberg 1984), is applied, while no correction has been applied on for example height anomalies. G is the gravitational constant, ρ the mass density and Δh the difference between the altitude of

O.C.D. Omang (✉)
Geodetic Institute, Norwegian Mapping Authority,
3507 Hønefoss, Norway
e-mail: OVE.OMANG@STATKART.NO

C.C. Tscherning
Niels Bohr Institute, University of Copenhagen, Copenhagen, Denmark

R. Forsberg
Danish National Space Center, Denmark

N. Sneeuw et al. (eds.), *VII Hotine-Marussi Symposium on Mathematical Geodesy*, International Association of Geodesy Symposia 137, DOI 10.1007/978-3-642-22078-4_35, © Springer-Verlag Berlin Heidelberg 2012

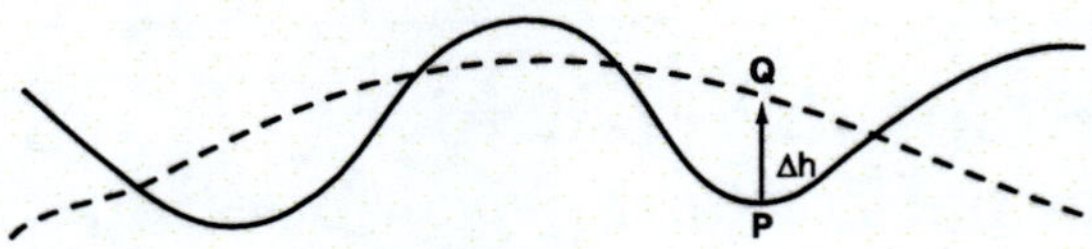

Fig. 35.1 The *dashed curve* shows the mean surface, the *solid curve* the actual topography

the point (Q) on the mean surface and the observation point (P), see Fig. 35.1. However, the correction is equal to the vertical gravity gradient inside the masses multiplied by Δh, if the horizontal gravity gradients are equal to zero. This interpretation makes it possible to generalize the correction to e.g. height anomalies.

We have implemented and tested an alternative procedure which is applicable to all types of gravity field observables. The basic idea is to continue the observations to the reference surface used in RTM, an idea which was already discussed in Elhabiby et al. (2009). However in order to continue the observation the vertical gradient has to be known. In fact the harmonic correction is such a vertical gravity gradient, if the horizontal gradients are considered equal to zero. This is due to Poisson's equation, Torge (2001), eq. (3.31). From this we may compute the harmonic correction for height-anomalies, ζ_c, using that the difference between the anomalous potential (T) in P and Q is equal to the integral of the gravity disturbance inside the masses from P to Q. Hence

$$\zeta_c = \frac{-4\pi G\rho\Delta h^2}{\gamma} \tag{35.1}$$

where γ is normal gravity.

Further development of this procedure requires that we are able to compute the gravity gradients inside the (homogeneous) mass, however, in this paper we have assumed that the horizontal gravity gradients are equal to zero.

We have tested the procedure by predicting height anomalies from gravity anomalies in the New Mexico White Sands test area and in an area in Western Norway with very deep fjords (Dahl and Forsberg 1999).

2 Generalized Harmonic Correction

The gravity potential of the Earth, W, may be split in two parts, $W = U + T$. Here U is an Earth Gravity Model (EGM) which includes the centrifugal potential, so that T becomes a harmonic function outside the masses. The EGM includes the contribution from the topography from wavelengths corresponding nearly to the maximal degree and order of the EGM, i.e. 360 for EGM96 (Lemoine et al. 1998) and 2,160 for EGM08 (Pavlis et al. 2008). Consequently point measurements (and sometimes airborne measurements) may be located inside the masses we have artificially placed outside the reference ellipsoid. For gravity anomalies, a so-called harmonic correction was introduced by Forsberg (1984) in order to solve this problem. We will here use the alternative procedure described above which for gravity anomalies turns out to give nearly the same result (when height anomalies are predicted) as when the harmonic correction is applied.

The basic equation is, with P an arbitrary point outside the *real* masses, T the anomalous potential, with subscript *rtm* for the contribution from the residual topography and *res* the remaining part:

$$T(P) = T_{rtm}(P) + T_{res}(P) \tag{35.2}$$

If the point P is inside the *artificial* masses, the generalized method is to use the observation at Q, see Fig. 35.1. For T and all other gravity functionals this means:

$$T(Q) = T_{rtm}(P) + T_{res}(Q) \tag{35.3}$$

where a harmonic correction had been applied in P. The harmonic correction has the function that it moves/continues the quantity from P to Q inside the masses, disregarding horizontal gradients. In order to determine an estimate of $T_{res}(Q)$ we use a Taylor expansion:

$$\tilde{T}_{res}(Q) = T_{res}(P) + \frac{\partial T}{\partial h}|_P \cdot \Delta h \tag{35.4}$$

Where we compute the vertical derivative using Least Squares Collocation (LSC) implemented in the GRAVSOFT (Forsberg and Tscherning 2008) program GEOCOL (Tscherning 1974).

The above equations may be written down for different functionals, L, such as the gravity disturbance δg, the gravity anomaly Δg or the height anomaly ζ. Equation 35.4 rewritten for gravity anomalies

$$\Delta\tilde{g}_{res}(Q) = \Delta g_{fa}(P) - \Delta g_{egm}(P) - \Delta g_{rtm}(P) - T_{zz}\cdot\Delta h, \quad (35.5)$$

and for height anomaly

$$\tilde{\zeta}_{res}(Q) = \zeta(P) - \zeta_{egm}(P) - \zeta_{rtm}(P) - \frac{\partial T}{\partial r}\cdot\Delta h, \quad (35.6)$$

where subscript fa is free-air anomaly, egm is the gravity/height anomaly signal from the EGM96, rtm is the terrain effect including the harmonic correction, T_{zz} is the vertical gravity gradient, $T_{zz}\cdot\Delta h$ and $\frac{\partial T}{\partial r}\cdot\Delta h$ is the first term of the Taylor expansion when moving from P to Q.

Note that data which are outside the mean surface are not at all changed in this procedure.

3 Computational Procedure

The GRAVSOFT program TC, used to compute the RTM effects, has been modified so that it optionally will produce a file where the height data-column in the output data-record contains the mean-surface height of the point Q if the point is inside the masses. This delivers a file with $L(T_{res}(P)) \approx L(T_{res}(Q))$ values associated with the point Q.

The values are then subtracted from the data (from which the effect of a global model already has been removed) and we obtain residual values. The effect of the new harmonic correction of height anomalies is for the Norwegian data seen as a further smoothing of the residual height anomalies from 0.44 to 0.34 m, see Table 35.4. With these data the vertical derivative may be computed using LSC (Moritz 1980).

4 Numerical Tests

We have selected our two test areas, New Mexico and Sognefjord, based on availability of control data, gravity data, and roughness of the topography.

4.1 New Mexico

The first test area is located in New Mexico, USA, covering an area from the Mexico border (El Paso) in south to Albuquerque in north, and stretching about 3.0° east–west. The New Mexico test area is located in an area starting at 1,000 m above sea level and has mountain areas ranging up to 3,350 m.

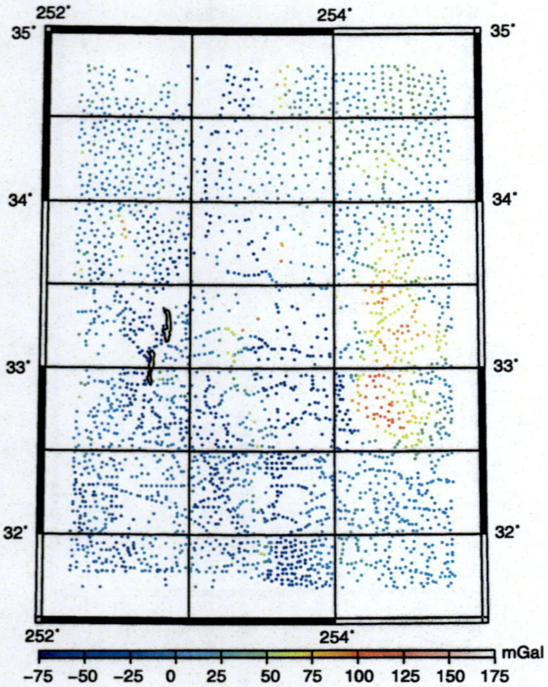

Fig. 35.2 Distribution of the 2,920 free-air gravity anomalies in the New Mexico test area

Within the geographical area 31.7° to 34.8° N, 252.2° to 254.8° E a total of 2,920 free-air gravity anomaly measurements were selected, see Fig. 35.2 for distribution and Table 35.1 for statistics. 1,979 points were inside the masses and their values were changed from line 3 to 4 in Table 35.1, while 941 points were unaffected.

As control data we have selected 20 height anomaly points (GPS/leveling) in New Mexico. Since the US height system uses orthometric height the GPS/leveling points were transformed to height anomaly data using the quasigeoid minus geoid

Table 35.1 Statistics of 2,920 gravity points in New Mexico test area. All values in mGal

	Mean	Std. dev	Min	Max
Δg_{fa}	9.182	30.405	−58.700	162.500
$-\Delta g_{egm}$	−2.932	21.283	−74.792	126.430
$-\Delta g_{rtm}$	0.282	13.153	−41.020	45.739
$-T_{zz}\cdot\Delta h$	0.308	13.035	−40.697	45.739

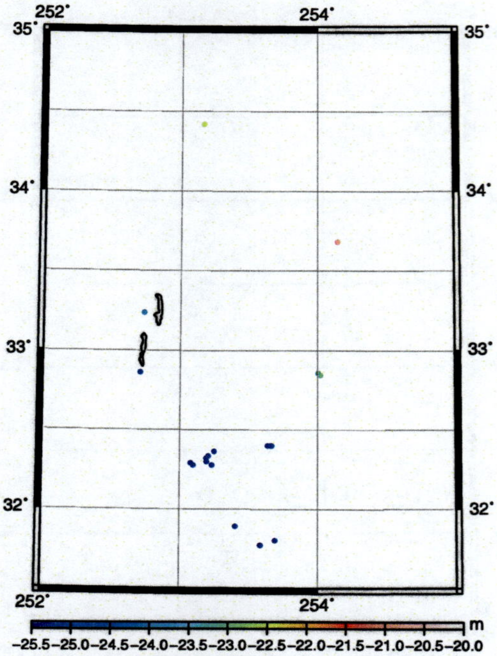

Fig. 35.3 Distribution of the control data in the New Mexico test area

Table 35.2 Statistics of the 20 control points in New Mexico test area. All values in meter

	Mean	Std. dev	Min	Max
ζ_{gps}	−24.268	1.083	−25.059	−20.917
$-\zeta_{egm}$	0.040	0.159	−0.330	0.305
$-\zeta_{hc}$	0.167	0.135	−0.107	0.396
$-\zeta^{Q}_{\frac{\partial T}{\partial r}\cdot\Delta h}$	0.166	0.122	−0.075	0.367
ζ_{gps}-ζ_{egm}-$\zeta_{tc_{wohc}}$	−0.897	0.159	−1.268	−0.632

separation formula (see e.g. Dahl and Forsberg (1999)). The data distribution is illustrated in Fig. 35.3 and statistics are given in Table 35.2. They are distributed in height from 1,130 to 1,662 m above sea level. Eighteen GPS/leveling points were inside the masses and their values where affected by the new procedure, while two points were unaffected.

4.2 Sognefjord

The Sognefjord test area was mainly selected due to its topography. The topography ranges from sea level to 2468.73 m.

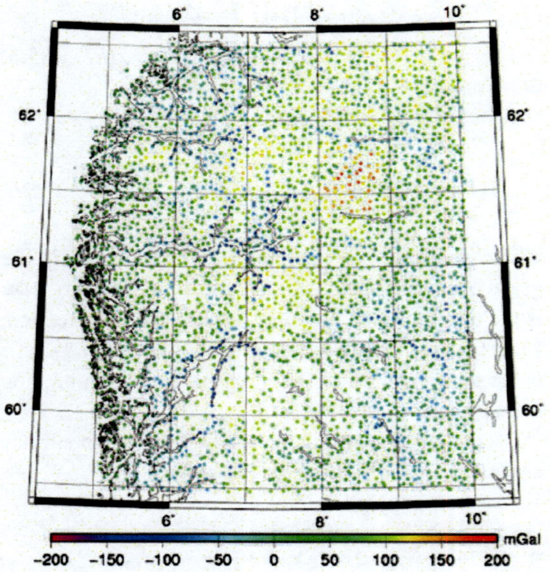

Fig. 35.4 Distribution of the free-air gravity anomalies in the Sognefjord test area

Table 35.3 Statistics of 2,877 gravity points. All values in mGal

	Mean	Std. dev	Min	Max
Δg_{fa}	29.926	53.978	−113.750	205.580
$-\Delta g_{egm}$	−1.176	46.547	−169.219	131.595
$-\Delta g_{rtm}$	1.070	12.060	−36.120	45.000
$-T_{zz}\cdot\Delta h$	1.147	11.823	−35.075	42.250

In the Sognefjord test area we have selected land only gravity data within the region 59.5° to 62.5° N, 4.5° to 10.0° E. A total of 2,877 free-air gravity anomaly measurements were selected, see Fig. 35.4 for distribution and Table 35.3 for statistics. 1,369 points were inside the masses and their values were changed from line 3 to 4 in Table 35.3, while 1,508 points were unaffected, by the generalized procedure.

As control data we have selected 131 height anomaly data (GPS/leveling) around the Sognefjord. The data distribution is illustrated in Fig. 35.5 and statistics are given in Table 35.4. They are distributed in height from 1 to 1424.7 m above sea level. 116 GPS/leveling points were inside the masses and their values were altered by the generalized procedure, while 15 points were unaffected.

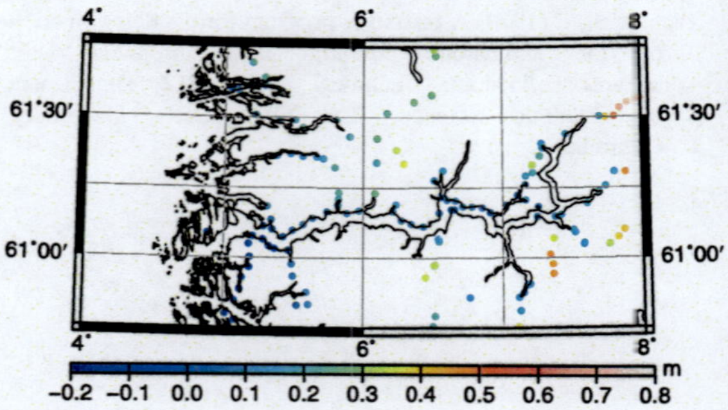

Fig. 35.5 Illustrates the difference between height anomalies estimated from gravity data and control data in Sognefjord. It refers point Q, see line 3 in Table 35.5. This figure also illustrates the distribution of the GPS/leveling data in the Sognefjord test area

Table 35.4 Statistics of the 131 control points in Sognefjord test area. All values in meter

	Mean	Std. dev	Min	Max
ζ_{gps}	46.270	0.416	45.372	47.604
$-\zeta_{egm}$	0.841	0.445	0.019	1.841
$-\zeta_{hc}$	0.796	0.347	0.163	1.370
$-\zeta^{Q}_{\frac{\partial T}{\partial r}\cdot\Delta h}$	0.736	0.296	0.200	1.254
ζ_{gps}-ζ_{egm}-$\zeta_{tc_{wohc}}$	0.857	0.387	0.155	1.441

5 Results and Conclusions

The generalized method gives an improved smoothing of the gravity and height anomaly data compared to the original harmonic correction method. In Tables 35.1 and 35.3 the gravity anomalies in line 4 are slightly smoother than line 3, which represents the original method. The original harmonic correction of the control data is given in line 5 of Tables 35.2 and 35.4 while the other lines indicates the new procedure. As the tables indicate the new method give smoother data than the original harmonic correction method.

A significant improvement is obtained when including harmonic correction for the height anomalies, compare line 1 and 2 in Table 35.5. An improvement of fit of 2.6 cm and 0.4 cm in the Sognefjord and New Mexico area, respectively.

By moving the point from P to Q in the Sognefjord test area improves the fit even further to 15.7 cm, see line 3 in Table 35.5 and Fig. 35.5, while in the New Mexico test area the result is slightly worse than using harmonic correction in point P, but better than not using. This may be due to the relatively low number of control points and their poor distribution or that we already are at the error level of the data set.

The numerical tests with height anomaly data from New Mexico, USA and Sognefjord, Norway as control data shows that for gravity anomalies the general procedure gives better to similar results compared to using the original harmonic correction procedure.

Table 35.5 Statistics of height anomalies estimated from gravity data compared with GPS/leveling control points. Subscript *hc* is harmonic correction, while *nohc* is no harmonic correction. All values in meter

Δg	ζ_{GPS}	Mean	Std. dev	Max	Min
Sognefjord					
P_{hc}	P_{nohc}	0.287	0.197	0.702	−0.140
P_{hc}	P_{hc}	0.226	0.171	0.714	−0.120
Q	Q	0.175	0.157	0.712	−0.124
New Mexico					
P_{hc}	P_{nohc}	−0.891	0.052	−0.780	−0.974
P_{hc}	P_{hc}	0.172	0.048	0.276	0.111
Q	Q	0.171	0.050	0.282	0.109

Acknowledgements We would like to thank the reviewers for their comments and suggestions on this manuscript.

References

Dahl O, Forsberg R (1999) Different ways to handle topography in practical geoid determination. Phys Chem Earth (A) 24(1):41–46

Elhabiby M, Sampietro D, Sanso F, Sideris M (2009) Bvp, global models and residual terrain correction. In: IAG symp, vol 113. Springer, Berlin, pp 211–217

Forsberg R (1984) A study of terrain reductions, density anomalies and geophysical inversion methods in gravity field modelling. Technical Report 355, Dept of Geodetic Science and Surveying, Ohio State University, Columbus

Forsberg R, Tscherning, CC (1981) The use of height data in gravity field approximation by collocation. J Geophys Res 86:7843–7854

Forsberg R, Tscherning CC (2008) An overview manual for the GRAVSOFT geodetic gravity field modelling programs, 2nd edn. DTU Space and University of Copenhagen

Lemoine F, Kenyon S, Factor J, Trimmer R, Pavlis N, Chinn D, Cox C, Klosko S, Luthcke S, Torrence M, Wang Y, Williamson R, Pavlis E, Rapp R, Olson T (1998) The development of the joint nasa gsfc and the national imagery and mapping agency (nima) geopotential model egm96. Technical Report NASA/TP-1998-206861, Goddard Space Flight Center, Greenbelt

Moritz H (1980) Advanced physical geodesy. Abacus, Turnbridge Wells Kent, Kent

Pavlis N, Holmes S, Kenyon S, Factor J (2008) An earth gravitational model to degree 2160: Egm2008. Presented at the 2008 General Assembly of the European Geosciences Union, Vienna
Torge W (2001) Geodesy, 3 edn. De Gruyter, Berlin
Tscherning C (1974) A fortran iv program for the determination of the anomalous potential using stepwise least squares collocation. Technical Report 212, Department of Geodetic Science, The Ohio State University, Columbus

Error Propagation in Geodetic Networks Studied by FEMLAB

36

Kai Borre

Abstract

Geodetic networks can be described by discrete models. The observations may be height differences, distances, and directions. Geodesists always make more observations than necessary and estimate the solution by using the principle of least squares. Contemporary networks often contain several thousand points. This leads to so large matrix problems that one starts thinking of using continous network models. They result in one or more differential equations with corresponding boundary conditions.

The Green's function works like the covariance matrix in the discrete case. If we can find the Green's function we also can study error propagation through large networks. Exactly this idea is exploited for error propagation studies in large geodetic networks.

To solve the boundary value problems we have used the FEMLAB software. It is a powerful tool for this type of problems. The M-file was created by Daniel Bertilsson. Modifying the code is so simple that a student can do it. We demonstrate some results obtained this way.

1 Introduction

Plane geodetic networks can be described by graphs the nodes of which are points with known or unknown coordinates. The observations describe the edges of the graph and at the same time serve as a means for estimating the coordinates of the unknown nodes. We always make more observations than necessary and then use the method of least squares for at least two reasons, (1) a unique solution of the problem, and (2) estimation of the *a posteriori* covariance matrix of the parameters.

The mathematical abstract of the method is the following: The least-squares problem of a single triangle is set up. Next, the original observations are transformed into *pseudo-observations*. In leveling they represent the *slope of terrain* in the direction of the coordinate axes and the closing error of this triangle. In the case of distance and azimuth networks, they are linear functions of the elements of the *metric tensor*. Henceforth we compute the weighted square sum of residuals. The kernel of this symmetric form is analogous to what in elasticity theory is called the *stiffness matrix*.

Now we look at the network consisting of such elementary triangles as a whole. Here the crucial point appears, namely, how to prolong the discrete scalar

K. Borre (✉)
Danish GPS Center, Aalborg University, Fredrik Bajers Vej 7C, DK-9220 Aalborg, Denmark
e-mail: borre@gps.aau.dk

N. Sneeuw et al. (eds.), *VII Hotine-Marussi Symposium on Mathematical Geodesy*, International Association of Geodesy Symposia 137, DOI 10.1007/978-3-642-22078-4_36,

and vector fields to continuous functions. We solve this problem by following simple principles from the method of finite elements. The weighted square sum of the residuals can now be extended to an approximating Riemann–Stieltjes sum over the entire network. The search for a minimum of this sum leads to a variational problem, the solution of which is a boundary value problem of the Neumann type.

This boundary value problem is the continuous analogue to the normal equations. The inverse of the partial differential operator (including the boundary conditions) is an integral operator, the kernel of which is Green's function of the operator concerned. Green's function acts in an analogous way with the usual covariance matrix of the original discrete least-squares problem.

So the method yields Green's function that is an approximate substitute for an n-dimensional covariance matrix. Instead of trying to grasp an n-dimensional covariance matrix, we are furnished with a single covariance function that depends on certain characteristic parameters for the network. Such a simplification of course has its costs. We lose information about local phenomena of the network, but we obtain a problem that can be studied and solved analytically.

2 Continuous Model

This section jumps from matrix equations to differential equations. The following text follows ideas from Strang (2007).

Instead of a finite number of heights along a straight leveling line, the physical system covers an interval like $0 \le x \le L$. The pseudo-observation $e = du/dx$, viz. slope of terrain, and weights $c(x)$ vary from one point to the next. Therefore height differences between individual nodes must be replaced by derivatives.

To take this step we copy the main idea of calculus. The slope of a curve comes from the average slope over increasingly small intervals: $\big(u(x+\Delta x) - u(x)\big)/\Delta x$ approaches the slope du/dx. Height differences are replaced by the slope of terrain.

The equilibrium of our small piece requires the difference in weighted errors at its two ends to balance the external error:

$$\Big(c\frac{du}{dx}\Big)_{x+\Delta x} - \Big(c\frac{du}{dx}\Big)_x = \frac{d}{dx}(cb).$$

The observed minus the computed observational value is denoted by b. Dividing by Δx, we approach the differential ***equation of equilibrium***

$$-\frac{d}{dx}\Big(c\frac{du}{dx}\Big) = -\frac{d}{dx}(cb). \qquad (36.1)$$

This is the matrix equation $A^{\mathrm{T}}CAx = A^{\mathrm{T}}Cb$ in the continuum limit.

You can already see the analogies between the discrete case and the continuous case. We list them in Table 36.1 and then discuss them individually.

Clearly the basic framework is not changed. The simplest step comes from C. Multiplication by a diagonal matrix, which is $y_i = c_i e_i$ in the discrete case, becomes multiplication by a function, $w(x) = c(x)e(x)$. Similarly A and its transpose reflect the geometry

$$A = \frac{d}{dx} \quad \text{and} \quad A^{\mathrm{T}} = -\frac{d}{dx}. \qquad (36.2)$$

Differential equations come with boundary conditions. We can specify u to a certain value, especially zero. Such a condition is called a ***Dirichlet boundary condition*** or a fixed boundary condition. Alternatively, we can specify the derivative of u to a certain value, again especially zero. Then we speak of a ***Neumann boundary condition*** or a free boundary condition. We must specify boundary conditions at both ends. In this paper we only consider the Neumann condition.

Table 36.1 Analogies between the discrete and continuous realm

Discrete	Continuous
Nodal unknown x	Displacement u
Error $e = b - Ax$	Pseudo-observation $e = du/dx$
Weighted error $y = Ce$	Weight × pseudo-observation $w = ce$
Node law $A^{\mathrm{T}}y = 0$	Equilibrium $-dw/dx = b$
Incidence matrix A	Differential operator d/dx
Weight matrix C	Multiplication by $c(x)$
Transposed matrix A^{T}	Transposed operator $-d/dx$
Normal equation $A^{\mathrm{T}}CAx = A^{\mathrm{T}}Cb$	Differential equation $-\frac{d}{dx}\big(c\frac{du}{dx}\big) = -\frac{d}{dx}(cb)$
Fixed node	Displacement boundary condition $u(0) = 0$
Error-free condition	Weighted error boundary condition $w(L) = 0$

So far we focused on leveling networks that have one unknown per node. However, there are other network types that include two or more unknowns per node. In networks where we measure distances and azimuths we always have the basic unknowns (u, v) that describe the ***displacements of the node***.

Due to inadequate modeling of the atmospheric refraction index used for reduction of the electrooptic distance measurements, we are facing situations where all distance measurements taken at a certain node have a change of scale α in common. So we add a small correction term to the observation taken; subsequently we call them *relative* observations.

The most common way of providing direction information is to perform direction measurements by means of a theodolite. The horizontal reading circle of the instrument has a gradation of 360° or more commonly 400 grades. However the direction of the zero mark makes an arbitrary angle with the north direction in the individual setup of the theodolite. There are simple methods for estimating an approximate value of the azimuth of the zero direction. This approximate value needs a small angular correction β (radians). The parameter β is common to all directions measured at a specific node.

We summarize the possible four observation types:

	Absolute	Relative
Distance	1	2
Direction	3	4

In our opinion it is a matter of taste or philosophy which model to use. The most often used observation types are 1 and 4. However, there are specific national traditions. In the Netherlands types 2 and 4 are preferred.

Given the observation type a quadratic form of weighted square sum can be formed. The minimum of this form leads to the equation of equilibrium. The weight is denoted c_i where i can be any number from 1 to 4.

In order to obtain explicit solutions of our problem we restrict the graph to consists of equilateral triangles and the weights to be constants and common to all edges. This leads to what is called a *homogeneous* and *isotropic* network.

We may have contributions from all observation types. For reasons of simplification we introduce the parameters A, B, a, and b:

$$\begin{aligned} A &= 3(c_1 + c_2) + (c_3 + c_4) \\ B &= (c_1 + c_2) + 3(c_3 + c_4) \\ a &= 4c_2 \\ b &= 4c_4. \end{aligned} \tag{36.3}$$

The factors $1, 3, 4$ originate from the triangular mesh.

3 Geodetic Network Boundary Value Problem

Adding contributions from all observation types, we arrive at the following set of partial differential equations $\mathcal{O}^*\mathcal{S}\mathcal{O}w = F$ for the ***homogeneous, isotropic*** case:

$$-\left[\begin{array}{cc|cc} A\partial_{xx} + B\partial_{yy} & (A-B)\partial_{xy} & a\partial_x & -b\partial_y \\ (A-B)\partial_{xy} & B\partial_{xx} + A\partial_{yy} & a\partial_y & b\partial_x \\ \hline a\partial_x & a\partial_y & a(\frac{l^2}{3}\Delta + 2) & 0 \\ -b\partial_y & b\partial_x & 0 & b(\frac{l^2}{3}\Delta + 2) \end{array}\right] \times \left[\begin{array}{c} u \\ v \\ \hline \alpha \\ \beta \end{array}\right] = \left[\begin{array}{c} F_u \\ F_v \\ \hline F_\alpha \\ F_\beta \end{array}\right]. \tag{36.4}$$

The forces $F_u, F_v, F_\alpha, F_\beta$ depend on the observations and may be found by adding the vectors $\mathcal{O}^*\mathcal{S}g$ for all four observation types.

Next we derive the boundary conditions for our actual problem. Without giving details we quote the result

$$\begin{bmatrix} An_1\partial_x+Bn_2\partial_y & \frac{A-B}{2}(n_1\partial_y+n_2\partial_x) & an_1 & -bn_2 \\ \frac{A-B}{2}(n_1\partial_y+n_2\partial_x) & Bn_1\partial_x+An_2\partial_y & an_2 & bn_1 \\ an_1 & an_2 & \frac{l^2}{3}a(n_1\partial_x+n_2\partial_y) & 0 \\ -bn_2 & bn_1 & 0 & \frac{l^2}{3}b(n_1\partial_x+n_2\partial_y) \end{bmatrix} \begin{bmatrix} u \\ v \\ \alpha \\ \beta \end{bmatrix} = \begin{bmatrix} f_u \\ f_v \\ f_\alpha \\ f_\beta \end{bmatrix}. \tag{36.5}$$

There are good reasons for expecting that the boundary forces $f_u, f_v, f_\alpha, f_\beta$ vanish if the boundary is shifted a half side length away from the physical boundary of the network. However, we have not been able to establish a rigorous proof for this conjecture.

We now introduce complex variables and let z and ζ denote points inside the given region Ω and z_0 denotes a point on the boundary ω. The general problem is the following:

$$\begin{aligned} \Delta u(z) &= f(z) && \text{in } \Omega \\ \frac{\partial u(z_0)}{\partial \boldsymbol{n}_{z_0}} &= g(z_0) && \text{on } \omega. \end{aligned} \tag{36.6}$$

For our particular problem we try to construct an approximate Neumann function. To this end we write instead of $f(z)$ the delta function $\delta(z,\zeta)$. However, this does not lead to an admissible u; we have to subtract the average value of the delta function, which is simply $1/|\Omega|$. Thus we must try to solve

$$\begin{aligned} -\Delta N(z,\zeta) &= \delta(z,\zeta) - \frac{1}{|\Omega|} && \text{in } \Omega \\ \frac{\partial N(z_0,\zeta)}{\partial \boldsymbol{n}_{z_0}} &= 0 && \text{on } \omega. \end{aligned} \tag{36.7}$$

Hence we let the *geodetic PDEs* have right side $\delta - 1/|\Omega|$ and zero boundary values. This corresponds to propagation of *random errors*. Alternatively, if we define the functions $F_u, F_v, F_\alpha, F_\beta$ different from zero, we are studying the propagation of *systematic errors*.

Finally we have to specify the shape of the boundary ω of the region Ω in which we want to solve the problem, and assign weights c_1, c_2, c_3, and c_4. One or more of these weights can be zero, but at least one must be positive. For given weights c_i, we can immediately compute A, B, a, and b according to (36.3).

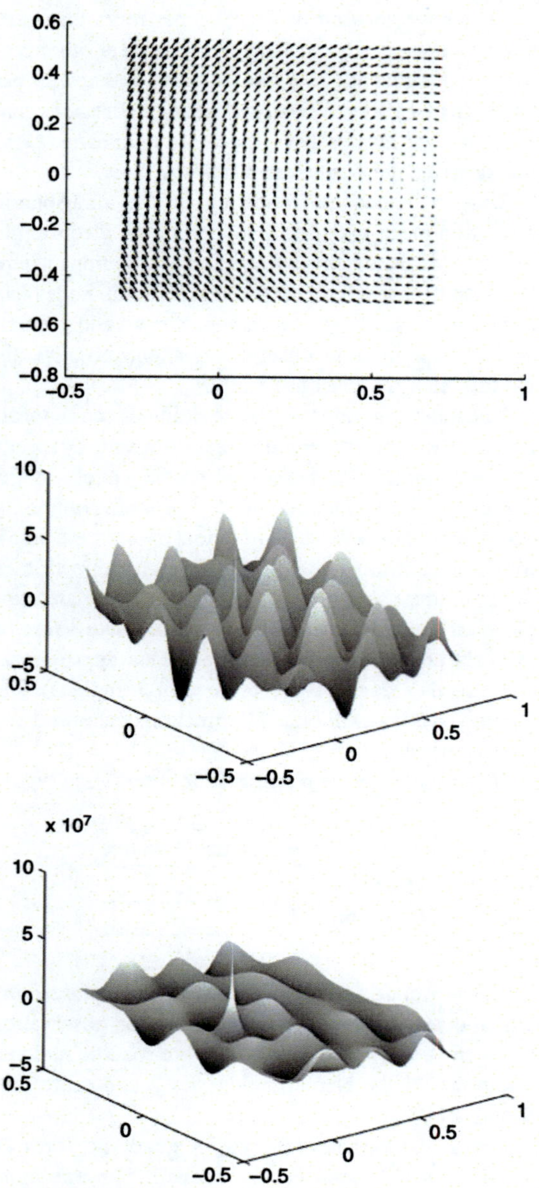

Fig. 36.1 Solution of the geodetic boundary value problem for a square region Ω: Displacement field $\boldsymbol{w} = (u, v)$ (*top*) and the scalar α (*middle*) and β fields (*bottom*) for weights $c_2 = 10$ and $c_1 = c_3 = c_4 = 0.000\,001$

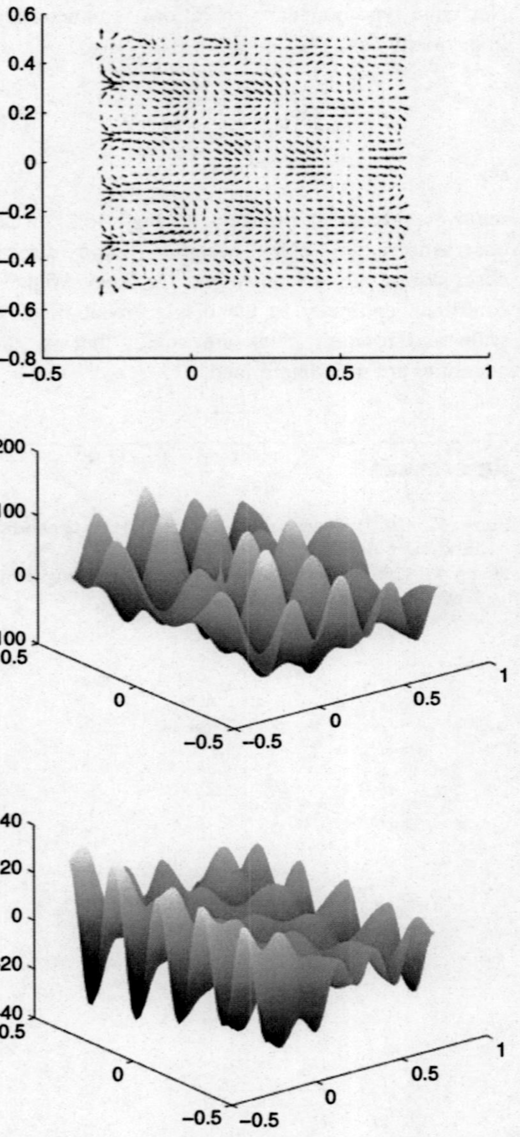

Fig. 36.2 Solution of the geodetic boundary value problem for a square region Ω: Displacement field $w = (u, v)$ (*top*) and the scalar α (*middle*) and β fields (*bottom*) for weights $c_1 = c_2 = 1$ and $c_3 = c_4 = 10$

4 MATLAB Code

It is a great pleasure to present to all readers an efficient *M*-file that solves our boundary value problem. It is written by Daniel Bertilsson and we are grateful that he so generously has made it available to us. It is called geonet. The computational power comes from the FEMLAB software that is distributed by Comsol.

```
% GEONET
% This script solves the system of PDEs for geodetic networks (4.23)
% with homogeneous boundary conditions (4.29).
% The components of the right side of (4.23) are all taken to be
% delta - 1/(area of region).
% The right side of (4.29) is 0.
% The solution components (u, v) are plotted as vector field arrows.
% The components alpha and beta are plotted as graphs, also colored
% according to function value (figures 2 and 3, respectively).
%
% The script uses functions from the finite element software
% FEMLAB that is distributed by COMSOL.

% The following parameters define the problem:
c1 = 1e-6;
c2 = 10;
c3 = 1e-6;
c4 = 1e-6;
l = 0.1;
% The solution will have oscillations of wavelength comparable to l.
% Thus, if you take a smaller l, you need to refine the mesh in order
% to resolve the oscillations. This is done by changing the value hmax
% below to a smaller value.
geo = 'rectangle'; % Geometry type. Can be 'ellipse' or 'rectangle'
cx = 0.2; cy = 0;  % Coordinates of the center of the ellipse/rectangle
lx = 1; ly = 1;    % Width and height of the ellipse/rectangle

%%%%%%%%%%%%%%%%%%%%%%%%%%%%%%%%%%%%%

% Define variables
clear fem;
fem.dim = 'u' 'v' 'alpha' 'beta' ;
fem.variables = 'A' 3*(c1+c2)+c3+c4 ...
         'B' c1+c2+3*(c3+c4) ...
         'a' 4*c2 'b' 4*c4 'K' l^2/3 ;
% Define geometry
switch geo
 case 'rectangle'
  fem.geom = rect2(cx-lx/2, cx+lx/2, cy-ly/2, cy+ly/2);
  area = lx*ly;
 case 'ellipse'
  fem.geom = ellip2(cx, cy, lx/2, ly/2);
  area = pi*lx*ly/4;
end
fem.geom = geomcsg( fem.geom, , point2(0,0) );
% Create mesh
hmax = 0.02;          % Maximum side for the triangles in the mesh
hmax`orig = 0.0001;   % Local hmax around the origin
lr = get(fem.geom, 'lr');
nr`orig = find(isfinite(lr1));  % Point number in the mesh for the origin
fem.mesh = meshinit(fem, 'hmax', hmax, [nr`orig; hmax`orig]);
% Natural boundary conditions are default, so we don't need to spec-
ify them.
% Define coefficients in the PDE
fem.equ.c = 'A' 'B'...
         0 '(A-B)/2'; '(A-B)/2' 0 'B' 'A'...
         0            0     'K*a'...
         0            0   0   'K*b' ;
fem.equ.al = 0 ...
         0    0...
         'a' 0 0 'a' 0...
         0 '-b' 'b' 0 0 0 ;
fem.equ.a = 0 0 '-2*a' '-2*b' ;
f = -1/area;
fem.equ.f = f f f f ;
% The delta functions on the right-hand side are specified directly in the
% right-hand side vector L occurring in the FEM discretization
% of the problem.
np = size(fem.mesh.p, 2);        % Number of points in mesh
```

```
fem.mat.L = zeros(4 * np, 1);
fem.mat.L(nr`orig) = 1;          % The delta function gives a contribution at
fem.mat.L(nr`orig + np) = 1;     % the origin, for all four components
fem.mat.L(nr`orig + 2 * np) = 1;
fem.mat.L(nr`orig + 3 * np) = 1;
% Solve problem
fem.sol = femlin(fem);
% Plot the four components of the solution
figure
postplot(fem, 'arrowdata', 'u','v', 'arrowxspacing', 30, 'arrowyspacing', 30);
figure
postplot(fem, 'tridata', 'alpha', 'triz', 'alpha', 'trimap', 'jet(4096)');
figure
postplot(fem, 'tridata', 'beta', 'triz', 'beta', 'trimap', 'jet(4096)');
```

Figures 36.1 and 36.2 show solutions of the boundary value problem given by (36.4) and (36.5). We use various weights c_i and a square region Ω.

We add some comments on the figures. The solutions oscillate with a wavelength comparable to the meshwidth l. Note that the α and β fields have large variances when they are poorly determined. Figure 36.1 demonstrates a network with relative distance measurements. The rotation of the network is poorly determined since a distance network with Neumann type boundary conditions is undetermined in its orientation. The condition

$$\iint_{\Omega} (\partial_y u - \partial_x v)\, d\Omega = 0 \tag{36.8}$$

removes this indeterminacy. In Fig. 36.2 distance observations are given weight 1 and direction observations are given weight 10. They "fight" to contribute optimally to the displacement field. The scale and rotation fields are small; that is, these quantities are well determined.

References

Borre K (2001) Plane networks and their applications. Birkhäuser, Boston

Strang G (2007) Computational science and engineering. Wellesley-Cambridge Press, Wellesley

Smoothing the Gradiometric Observations Using Different Topographic–Isostatic Models: A Regional Case Study

37

J. Janák, F. Wild-Pfeiffer, and B. Heck

Abstract

In terrestrial and airborne gravity field determination the formulae for the gravitational potential and its first order derivatives have been used, while the second order derivatives are related to the analysis of upcoming satellite gravity gradiometry missions of GOCE type. Especially there, the reduction of topographic and isostatic effects is important to produce a smooth gravity field suitable for downward continuation.

In this paper various isostatic models, namely the models of Airy–Heiskanen (A–H) and Pratt–Hayford (P–H), the combination of the Airy–Heiskanen model (land area) and the Pratt–Hayford model (ocean area), the first (H1) condensation model of Helmert as well as a crust density model are analyzed for a GOCE-like satellite orbit in two selected regions: Japan and central part of Europe.

The different topographic–isostatic effects are compared with respect to the degree of smoothing of the measured satellite gradiometric data. The results of this paper can serve as a base for further investigations of the suitability of particular reduction models for downward continuation.

Keywords

Satellite gravity gradiometry • isostatic models (Airy–Heiskanen • Pratt–Hayford) • first condensation model of Helmert • crust density model.

1 Introduction

As GOCE (*G*ravity Field and Steady-State *O*cean *C*irculation *E*xplorer) has been successfully launched in March 17, 2009, the preparation of effective utilization of GOCE data for all variety of scientific purposes is essential. While the majority of applications are focused on global approach, this paper attempts to emphasis some possibilities of regional applications.

J. Janák (✉)
Department of Theoretical Geodesy, Slovak University of Technology, Radlinského 11, 813 68 Bratislava, Slovakia
e-mail: juraj.janak@stuba.sk

F. Wild-Pfeiffer
Institute of Navigation, University of Stuttgart, Breitscheidstraße 2, 70174 Stuttgart, Germany

B. Heck
Geodetic Institute, Karlsruhe Institute of Technology (KIT), Englerstraße 7, 76131 Karlsruhe, Germany

N. Sneeuw et al. (eds.), *VII Hotine-Marussi Symposium on Mathematical Geodesy*, International Association of Geodesy Symposia 137, DOI 10.1007/978-3-642-22078-4_37,

One important aspect is the determination of the effect of topographic and isostatic masses on the particular components of the gravity gradient tensor. If this effect is significant, it will be possible to smooth the GOCE data before processing and ensure e.g. a better stability for downward continuation of the data (see Janák and Wild-Pfeiffer 2010).

2 Test Regions

Eastern Asia as the first test region (Region A) is surrounded by parallels 15° and 56° and by meridians 116° and 157° (the area is 41° × 41°). Approximately 56% of the surface of Region A is covered by ocean. A relief of the Earth's surface in Region A is shown in Fig. 37.1.

Europe as the second test region (Region B) is surrounded by parallels 30° and 71° and by meridians −10° and 31° (the area is 41° × 41°). With respect to Region A, Region B is shifted towards the pole and it is predominantly covered by continent. A relief of the Earth's surface in Region B is plotted in Fig. 37.2. The statistics of elevation for both regions is shown in Table 37.1.

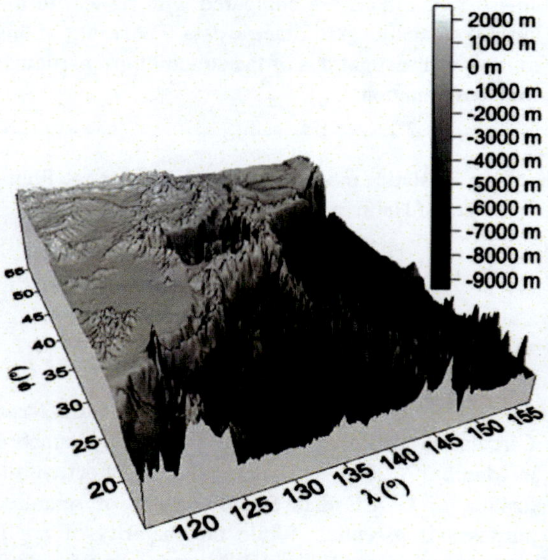

Fig. 37.1 Region A represented by ETOPO2v2c with the resolution of 10′

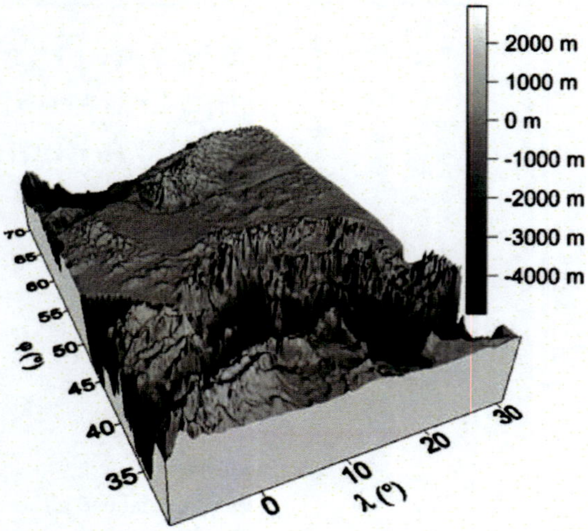

Fig. 37.2 Region B represented by ETOPO2v2c with the resolution of 10′

Table 37.1 Statistical parameters of 10′ mean elevations generated from ETOPO2v2c model

Parameter	Region A	Region B
Min. (m)	−9,371	−4,978
Max. (m)	2,627	2,934
Mean (m)	−2,334	−346
Range (m)	11,998	7,912
Std. (m)	2,987	1,235
Dry land (%)	44	56
Ocean (%)	56	44

3 Modelling of the Topographic–Isostatic Effects

The topographic–isostatic effects for both regions have been computed in altitude of 260 km (approximate GOCE flight level) using global numerical integration. The integration domain is subdivided into three parts: the vicinity of the computation point, the external domain and the poles, see Figs. 37.3 and 37.4.

As the integration is performed in spherical geographical coordinates, the integration elements are bounded by meridians and parallels while meridians are convergent towards the poles. Therefore the polar regions are integrated separately according to the integration scheme presented in Fig. 37.4.

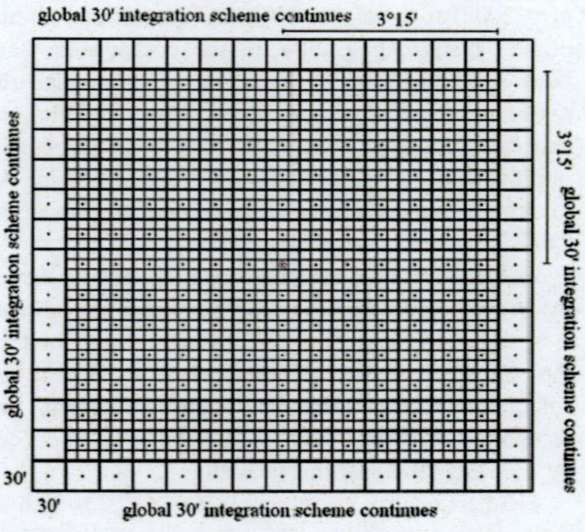

Fig. 37.3 Subdivision of the spherical integration domain in the vicinity of the computation point vertically projected onto a reference sphere

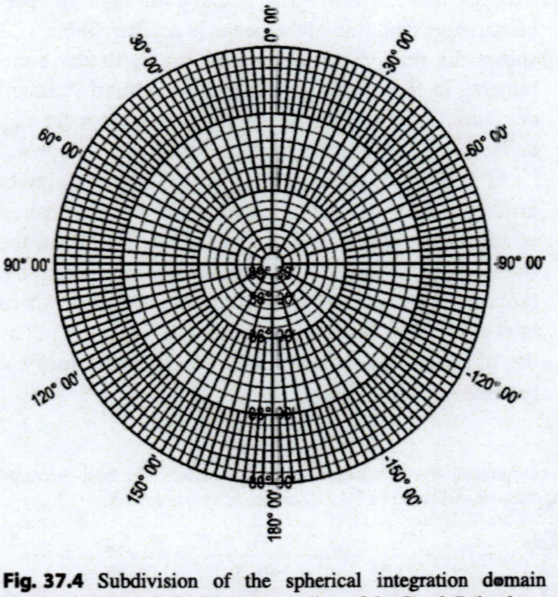

Fig. 37.4 Subdivision of the spherical integration domain around the North Pole (The surrounding of the South Pole obeys the same scheme)

The aim of the numerical experiment is the quantification of the degree of smoothing of the disturbing gravity tensor components in GOCE flight level after applying a topographic–isostatic model.

Modelling the topographic–isostatic effects according to Airy–Heiskanen (A–H), Pratt–Hayford (P–H), the combined model, the first Helmert condensation

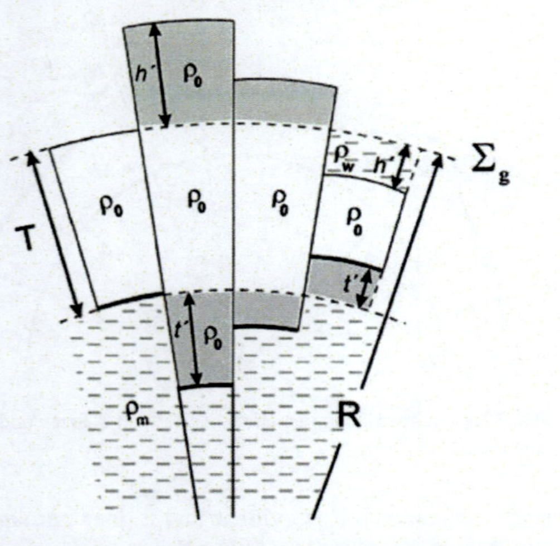

Fig. 37.5 Schematic explanation of the Airy–Heiskanen model

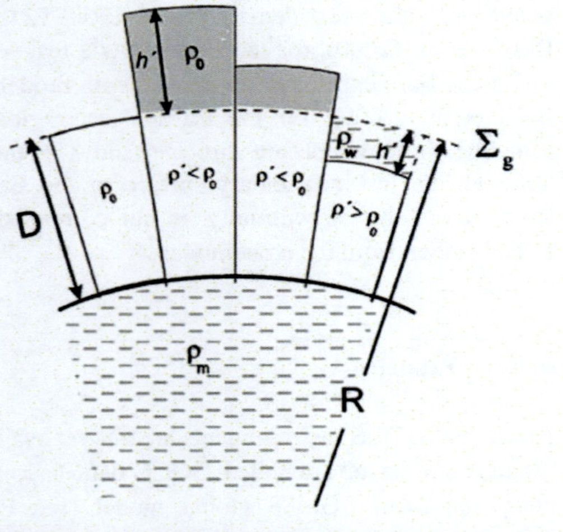

Fig. 37.6 Schematic explanation of the Pratt–Hayford model

model (H1) and a crust model, the following parameters have been used: radius of the reference sphere $R = 6{,}371{,}009\,\mathrm{m}$, standard column height $T = 25\,\mathrm{km}$, mean ocean water density $\rho_w = 1{,}030\,\mathrm{kg\,m^{-3}}$, mean Earth crust density $\rho_0 = 2{,}670\,\mathrm{kg\,m^{-3}}$ and density of the upper part of the mantle $\rho_m = 3{,}270\,\mathrm{kg\,m^{-3}}$, standard column depth $D = 100\,\mathrm{km}$ and depth of the condensation layer $d = 21\,\mathrm{km}$, see Figs. 37.5–37.7 and Wild-Pfeiffer and Heck (2008).

The combined model consists of a mixture of the A–H model over the continents and the P–H model over the oceans. The second condensation model of Helmert has not been considered because from

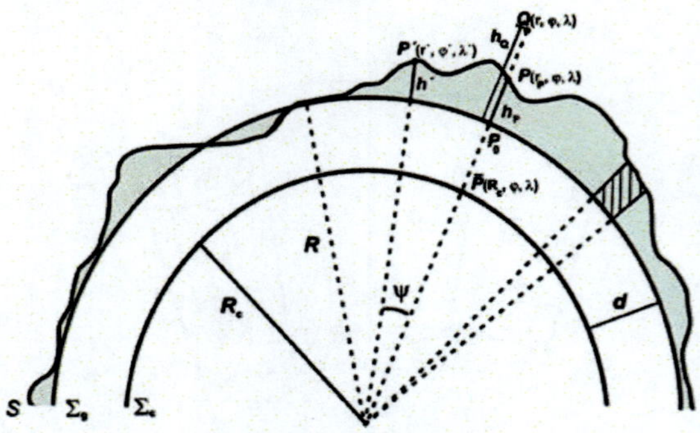

Fig. 37.7 Schematic explanation of Helmert's first condensation model

previous research it is evident that it does not smooth the disturbing gravity tensor components see Janák and Wild-Pfeiffer (2010).

The last topographic–isostatic model has been computed using the crust density model CRUST2.0, see Bassin et al. (2000), for the Eastern Asia test region and using the new European Moho depth model, see Grad et al. (2009), for the Europe test region. In principle, these models are similar to the A–H model, although the lower boundary surface of the Earth's crust, the Moho discontinuity, is not computed but directly taken from the crust model.

4 Results

As the real GOCE measurements are not yet available for users, a set of simulated GOCE data have been computed from EGM08 global model (see Pavlis et al. 2008) up to degree 360. The degree of smoothing of the disturbing gravity tensor components has been quantified by comparison of the standard deviation and range before and after applying a particular topographic–isostatic reduction. The relative improvements for particular components in Region A in terms of range and standard deviation are shown in Table 37.2.

In Fig. 37.8, one particular component $T_{\varphi\varphi}$ using the A–H topographic–isostatic compensation model is shown. It is apparent that the application of the topographic–isostatic reduction has a significant smoothing effect. Based on Table 37.2 it can be recognized that the range of undulation is reduced by 23% after applying this reduction.

The best relative smoothing in Table 37.2 is 34.6% and belongs to the range of T_{rr} component. The effects based on the CRUST2.0 model do not provide any smoothing.

The relief in the European test region is less broken than in the Eastern Asia test region and the percental representation of the ocean is smaller. Table 37.3 shows the relative improvements for particular components in Region B in terms of range and standard deviation. In Fig. 37.9 one particular component $T_{\varphi\varphi}$ using the P–H topographic–isostatic model is shown.

In Fig. 37.9, analogously to Fig. 37.8, the gravity gradients are significantly smoothed by application of the topographic–isostatic reduction. Based on the information of Table 37.3 it can be recognized that the range of undulation is reduced by almost 38%, which is also the best relative smoothing in Table 37.3. The use of the new European Moho depth model does not provide any smoothing.

Table 37.2 Percental smoothing of the disturbing gravity tensor components over the Eastern Asia test region. The best percental smoothing in terms of range and standard deviation for every component is indicated in *bold*. *Dash* means no smoothing

Model	Crit.	$T_{\varphi\varphi}$	$T_{\varphi\lambda}$	$T_{\varphi r}$	$T_{\lambda\lambda}$	$T_{\lambda r}$	T_{rr}
A–H	Range	**23.2**	24.3	18.4	21.7	**26.3**	25.3
	Std.	19.1	15.8	18.7	**20.2**	12.6	18.8
P–H	Range	20.7	31.1	18.6	**28.3**	25.6	33.7
	Std.	9.8	15.8	10.8	18.0	11.4	14.8
A–H/P–H	Range	20.8	**31.1**	18.5	27.8	25.6	**34.6**
	Std.	11.4	**17.0**	12.6	18.1	12.3	15.5
H1	Range	18.4	22.6	**18.7**	21.7	23.0	24.5
	Std.	**19.1**	16.6	**19.0**	19.5	**12.9**	**19.1**
CRUST 2.0	Range	–	–	–	–	–	–
	Std.	–	–	–	–	–	–

Table 37.3 Percental smoothing of the disturbing gravity tensor components over the Europe test region. The best percental smoothing in terms of range and standard deviation for every component is indicated in *bold*. *Dash* means no smoothing

Model	Crit.	$T_{\varphi\varphi}$	$T_{\varphi\lambda}$	$T_{\varphi r}$	$T_{\lambda\lambda}$	$T_{\lambda r}$	T_{rr}
A–H	Range	21.9	8.3	19.7	3.8	7.6	13.8
	Std.	19.0	**12.6**	16.6	**4.6**	**8.5**	**12.3**
P–H	Range	**37.6**	17.6	**33.2**	5.3	2.1	**24.4**
	Std.	**23.5**	11.0	**19.2**	-	2.3	11.7
A–H/P–H	Range	34.1	**19.2**	31.9	**8.4**	4.8	24.4
	Std.	22.0	10.6	18.3	-	3.3	11.4
H1	Range	20.0	9.6	19.1	5.8	**8.9**	14.0
	Std.	17.7	11.8	15.6	4.5	8.2	11.6
New Moho	Range	–	–	–	–	–	–
	Std.	–	–	–	–	–	–

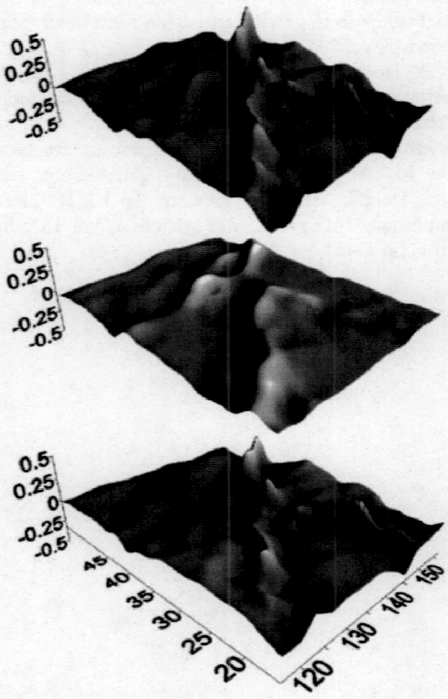

Fig. 37.8 *Top*: disturbing gravity tensor component $T_{\varphi\varphi}$ generated from EGM08 global model in Region A; *Middle*: topographic–isostatic effect computed from the Airy–Heiskanen model; *Bottom*: $T_{\varphi\varphi}$ component corrected for the topographic–isostatic effect. Units in *z*-axis are Eötvös units (EU)

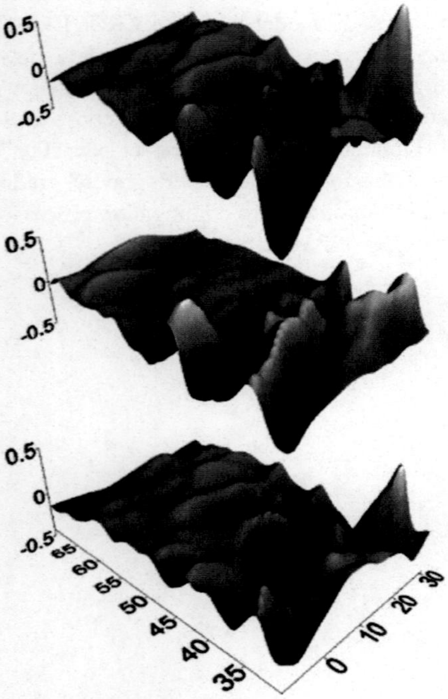

Fig. 37.9 *Top*: disturbing gravity tensor component $T_{\varphi\varphi}$ generated from EGM08 global model in Region B; *Middle*: topographic–isostatic effect computed from the Pratt–Hayford model; *Bottom*: $T_{\varphi\varphi}$ component corrected for topographic–isostatic effect. Units in *z*-axis are Eötvös units (EU)

Conclusion

Using the topographic–isostatic reductions of Airy–Heiskanen, Pratt–Hayford, the combined model or the first Helmert condensation model, it is possible to smooth the simulated gradiometric data based on EGM08 model. The relative degree of smoothing represented by comparison of standard deviations is about 20%, while the relative degree of smoothing represented by comparison of ranges can reach more than 30%. Although the percentage of smoothing is significant, a large part of the disturbing gravity gradient signal still remains uncompensated. The reason may be the insufficiently modelled density and the uncompensated signal from deeper mass structures. The choice of the best topographic–isostatic model depends on the region of interest and also on the particular component of gravity tensor.

The topographic–isostatic models based on existing Moho depth models (CRUST2.0 or the new European Moho depth model) provide very large topographic–isostatic effects and their application on the disturbing gravity tensor does not improve the smoothness of the data.

The comparison of the Moho depth model based on the A–H model with the CRUST2.0 and the new European Moho depth model shows that the A–H model generates a much shallower and smoother Moho boundary. As a consequence, the A–H topographic–isostatic effect is smaller.

The behaviour of smoothed gravity gradiometry data in the downward continuation process will be the object of further research.

Acknowledgements The research presented in this paper has been partially supported by the Slovak national projects VEGA 1/0775/08 and APVV 0351–07.

References

Bassin C, Laske G, Masters G (2000) The current limits of resolution for surface wave tomography in North America. EOS Trans Am Geophys Union 81:F897

Grad M, Tiira T, ESC Working Group (2009) The Moho depth map of the European Plate. Geophys J Int 176: 279–292

Janák J, Wild-Pfeiffer F (2010) Comparison of various topographic–isostatic effects in terms of smoothing gradiometric observations. In Mertikas S (ed) Proc. of IAG international symposium gravity, geoid and earth observation. IAG Symposia, vol 135. Springer, Berlin, pp 377–382

Pavlis NK, Holmes SA, Kenyon SC, Factor JK (2008) An earth gravitational model to degree 2160 EGM2008. Geophysical Research Abstracts EGU 2008

Wild-Pfeiffer F, Heck B (2008) Topographic and isostatic reductions for use in satellite gravity gradiometry. In: Xu P, Liu J, Dermanis A (eds) Proc. of the VI. Hotine-Marussi symposium. IAG proc., IAG symposia, vol 132. Springer, Berlin, pp 49–55

New Expressions of the Gravitational Potential and Its Derivatives for the Prism

38

Maria Grazia D'Urso

Abstract

We present novel expressions for the gravitational potential and its first derivative induced by a prism, having a constant mass density, at an observation point coincident with a prism vertex. They are obtained as a special case of more general formulas which can be derived for an arbitrary homogeneous polyhedron. Remarkably, the expressions presented in the paper entail a reduced computational burden with respect to alternative ones reported in the specialized literature.

Keywords

Newtonian gravitational field • First-order gradient • Right rectangular parallelepiped • Prism

1 Introduction

The computation of the gravitational effects due to given mass distributions is a basic issue in applied geophysics and geodesy (Tsoulis 1999). Thus, analytical formulas for the gravity field of the most common model bodies, e.g. a prism, a cylinder, a cone, etc. have been contributed in the past (Kellogg 1929; MacMillan 1930).

In particular, the prism plays an important role mostly in studies on local gravity field modelling when the so called flat-Earth approximation is sufficient (Nagy 1966; Banerjee and DasGupta 1977; Nagy et al. 2000).

However, analytical expressions for the prism exhibit singularities when the computation of the potential is carried out at corners (Tsoulis 2000); more generally such a situation does occur also for polyhedral bodies (Petrović 1996).

Aim of this paper is to briefly illustrate the basic ideas of a novel approach for expressing, exclusively by line integrals, the gravitational potential and its derivatives for an arbitrary homogeneous polyhedron. The formulas, obtained by a suitable application of the Gauss theorem, highlight from the very beginning the potential sources of singularities and clearly indicate how to correctly take them into account. In this way there is no need to derive ad-hoc formulas (Tsoulis 2000) or include a-posteriori corrections as shown, e.g., in (Tsoulis and Petrović 2001).

Specialization of the derived expressions to the case of a prism and to an observation point P coincident with a prism vertex yields final formulas which require a reduced number of algebraic operations with respect to others formulas reported in the literature (Tsoulis 2000). Due to space limitations, the case of P lying along an edge or on a face of the prism,

M.G. D'Urso (✉)
Di.M.S.A.T. - Università di Cassino, 03043 Cassino (FR), via G. Di Biasio 43, Cassino, Italy
e-mail: durso@unicas.it

N. Sneeuw et al. (eds.), *VII Hotine-Marussi Symposium on Mathematical Geodesy*, International Association of Geodesy Symposia 137, DOI 10.1007/978-3-642-22078-4_38,

as well as the computation of the second-order gradient of the potential, will be reported in a forthcoming paper.

2 Gravitational Potential of a Prism

Let us consider a right rectangular parallelepiped (prism) Ω having a distribution of mass of constant density δ. The gravitational potential U induced at P is defined by the well-known Newton integral

$$U(P) = G\delta \int_{\Omega} \frac{1}{(\mathbf{r}\cdot\mathbf{r})^{1/2}} dV \qquad (38.1)$$

where G is the gravitational constant and $\mathbf{r}$ the vector connecting an arbitrary point of Ω with P.

We are interested to compute the potential U, as well as its first derivative, irrispective of the position of P; however, due to space limitations, we shall confine ourselves to the case in which P coincides with one of the vertices of the prism.

Accordingly, we shall consider a three-dimensional (3D) cartesian reference frame x, y, z having origin at P and axes parallel to the edges of the prism; we further introduce a 2D frame ξ_1 and η_1 in the plane of the first face, ABCD, of the prism. The lenghts of the edges parallel to x, y, z will be denoted by a, b and c respectively, see e.g. Fig. 38.1.

Our goal is to compute the potential (38.1) by expressing the volume integral as a sum of quantities pertaining to the vertices of each face of the prism. The result is obtained by first transforming the 3D integral into the sum of 2D integrals extended to each face of

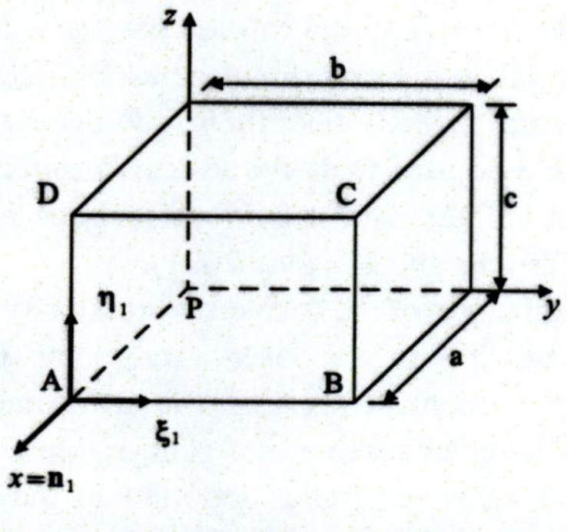

Fig. 38.1 The prism and the 2D reference frame for the first face

the prism and, then, express each of the 2D integrals as a sum of 1D integrals extended to each edge of the generic face.

However, in the application of Gauss theorem, special attention has to be paid to possible singularities of the scalar field appearing in (38.1). Actually, the potential diverges if $P \in \Omega$, i.e. if $\mathbf{r} = \mathbf{o}$. To formally handle this case we invoke the results (Tang 2006)

$$\operatorname{div} \frac{\mathbf{r}}{(\mathbf{r}\cdot\mathbf{r})^{3/2}} = \operatorname{div}[\mathbf{h}(\mathbf{r})] = 0 \quad \text{if } \mathbf{r} \neq \mathbf{o} \qquad (38.2)$$

where *div* stands for the divergence operator and

$$\int_{\Omega} \varphi(\mathbf{r}) \operatorname{div}[\mathbf{h}(\mathbf{r})] dV = \begin{cases} 0 & \text{if } \mathbf{o} \notin \Omega \\ m(\mathbf{o})\varphi(\mathbf{o}) & \text{if } \mathbf{o} \in \Omega \end{cases} \qquad (38.3)$$

m representing the measure, expressed in radians, of the solid angle of the intersection between Ω and a spherical neighbourhood of the singularity point $\mathbf{o}$. Stated equivalently, the divergence of the field $\mathbf{h}(\mathbf{r})$ represents the Dirac delta function Δ at the origin.

Invoking the differential identity (Tang 2006)

$$\operatorname{div}(\varphi\mathbf{u}) = \operatorname{grad}\varphi \cdot \mathbf{u} + \varphi \operatorname{div}\mathbf{u}, \qquad (38.4)$$

where *grad* denotes the gradient operator and $\mathbf{u}$ is an arbitrary C^1 vector field, it turns out to be

$$\begin{aligned} \int_{\Omega} \operatorname{div} \frac{\mathbf{r}}{(\mathbf{r}\cdot\mathbf{r})^{1/2}} dV &= \int_{\Omega} \operatorname{div}[(\mathbf{r}\cdot\mathbf{r})\mathbf{h}(\mathbf{r})]\, dV \\ &= \int_{\Omega} \operatorname{grad}(\mathbf{r}\cdot\mathbf{r}) \cdot \mathbf{h}(\mathbf{r}) dV + \int_{\Omega} (\mathbf{r}\cdot\mathbf{r}) \operatorname{div}[\mathbf{h}(\mathbf{r})] dV \\ &= 2\int_{\Omega} \frac{dV}{(\mathbf{r}\cdot\mathbf{r})^{1/2}} + \int_{\Omega} (\mathbf{r}\cdot\mathbf{r})\Delta(\mathbf{o}) dV \end{aligned} \qquad (38.5)$$

where we have exploited the result $\operatorname{grad}(\mathbf{r}\cdot\mathbf{r}) = 2\mathbf{r}$.

Due to the peculiar properties of the Dirac delta function the last integral on the right-hand side is always zero, whether or not $\mathbf{o}$ belongs to Ω. Actually, in the former case $\varphi(\mathbf{r}) = \mathbf{r}\cdot\mathbf{r}$ vanishes at the origin while, in the latter one, the divergence of the field is zero on account of (38.2).

Thus, the potential U can be expressed, whatever is the position of P with respect to the domain Ω, by means of the formula

$$U(P) = G\delta \int_{\Omega} \frac{1}{(\mathbf{r}\cdot\mathbf{r})^{1/2}} dV = \frac{G\delta}{2} \int_{\Omega} \operatorname{div} \frac{\mathbf{r}}{(\mathbf{r}\cdot\mathbf{r})^{1/2}} dV \tag{38.6}$$

which represents, upon application of Gauss theorem, the basic tool for the actual computation of the gravitational potential.

In fact, denoting by $\mathrm{Fr}(\Omega)$ the boundary of Ω, that is the collection of its six faces F_i, and by $\mathbf{n}$ the unit vector, pointing outwards, at a generic point of $\mathrm{Fr}(\Omega)$ one has

$$U(P) = \frac{G\delta}{2} \int_{Fr(\Omega)} \frac{\mathbf{r}\cdot\mathbf{n}}{\sqrt{\mathbf{r}\cdot\mathbf{r}}} dA = \frac{G\delta}{2} \sum_{i=1}^{6} \int_{F_i} \frac{\mathbf{r}_i\cdot\mathbf{n}_i}{\sqrt{\mathbf{r}_i\cdot\mathbf{r}_i}} dA_i \tag{38.7}$$

To express the previous 2D integrals in terms of line integrals, we need a further application of Gauss theorem, now in the plane of each face. To this end we consider the orthogonal projection of the point P on the i-th face, say it P_i, and assume this point as the origin of a local 2D reference frame, see e.g. Fig. 38.1.

Denoting by $\boldsymbol{\rho}_i = (\xi_i, \eta_i)$ the position vector of each point of the i-th face with respect to P_i and observing that the product $d_i = \mathbf{r}_i \cdot \mathbf{n}_i$ is constant over each face, the previous expression can be equivalently written as

$$\begin{aligned} U(P) &= \frac{G\delta}{2} \sum_{i=1}^{6} d_i \int_{F_i} \frac{dA_i}{(\mathbf{r}_i\cdot\mathbf{r}_i)^{1/2}} \\ &= \frac{G\delta}{2} \sum_{i=1}^{3} d_i \int_{F_i} \frac{dA_i}{(\boldsymbol{\rho}_i\cdot\boldsymbol{\rho}_i + d_i^2)^{1/2}} \end{aligned} \tag{38.8}$$

where the sum has been limited to the faces which do not contain P so that $d_i \neq 0$.

From now on, 2D vector fields will be denoted by Greek bold symbols to distinguish them from the 3D ones. Furthermore, to shorten the formulas reported in the sequel, we shall adopt the more compact notation

$$f(\boldsymbol{\rho}, k) = (\boldsymbol{\rho}\cdot\boldsymbol{\rho} + k^2)^{1/2}; \quad \mathbf{g}(\boldsymbol{\rho}) = \boldsymbol{\rho}/(\boldsymbol{\rho}\cdot\boldsymbol{\rho}) \tag{38.9}$$

where k is an arbitrary scalar.

The last integral in (38.8) can be further simplified by applying once more the differential identity (38.4) to a generic 2D field $\boldsymbol{\rho}$ as follows

$$\begin{aligned} \operatorname{div}[f(\boldsymbol{\rho},k)\,\mathbf{g}(\boldsymbol{\rho})] &= grad[f(\boldsymbol{\rho},k)]\cdot\mathbf{g}(\boldsymbol{\rho}) \\ &\quad + f(\boldsymbol{\rho},k)\, div[\mathbf{g}(\boldsymbol{\rho})] \end{aligned} \tag{38.10}$$

Being $\operatorname{div}[\mathbf{g}(\boldsymbol{\rho})] = 0$ when $\boldsymbol{\rho} \neq \mathbf{o}$, the vector field $\mathbf{g}$ plays in the 2D case the same role as $\mathbf{r}/(\mathbf{r}\cdot\mathbf{r})^{3/2}$ in the 3D one. Observing further that $\operatorname{grad} f = \boldsymbol{\rho}/f(\boldsymbol{\rho},k)$ and denoting by F an arbitrary 2D domain, one has

$$\int_F \frac{dA}{f(\boldsymbol{\rho},k)} = \int_F \left[\operatorname{div}\left[\frac{f(\boldsymbol{\rho},k)\boldsymbol{\rho}}{\boldsymbol{\rho}\cdot\boldsymbol{\rho}}\right] - f(\boldsymbol{\rho},k)\operatorname{div}[\mathbf{g}(\boldsymbol{\rho})]\right] dA \tag{38.11}$$

Since $\operatorname{div}[\boldsymbol{\rho}/(\boldsymbol{\rho}\cdot\boldsymbol{\rho})]$ represents the 2D Dirac delta, the last integral above becomes

$$\int_F (\boldsymbol{\rho}\cdot\boldsymbol{\rho} + k^2)^{1/2} \operatorname{div}[\mathbf{g}(\boldsymbol{\rho})] dA = k\alpha \tag{38.12}$$

where α is the angular measure, expressed in radians, of the intersection between the boundary of F and a circular neighbourhood of the point $\boldsymbol{\rho} = \mathbf{o}$.

The previous formulas can be usefully employed for computing the last integral in (38.8). Actually, making use of Gauss theorem and denoting in turn by F_1, F_2 and F_3 the faces of the prism orthogonal to the axes x, y and z and not containing P, it turns out to be

$$U(P) = \frac{G\delta}{2} \sum_{i=1}^{3} d_i \left\{ \int_{Fr(F_i)} \frac{f(\boldsymbol{\rho}_i, d_i)(\boldsymbol{\rho}_i\cdot\boldsymbol{\nu}_j)}{\boldsymbol{\rho}_i\cdot\boldsymbol{\rho}_i} ds_i - d_i\frac{\pi}{2} \right\} \tag{38.13}$$

where $d_1 = a$, $d_2 = b$, $d_3 = c$ and $\boldsymbol{\nu}_j$ is the 2D normal to the j-th side of the face F_i contained in the plane of the face.

Being the product $\boldsymbol{\rho}_i \cdot \boldsymbol{\nu}_j$ constant over the j-th side of $Fr(F_i)$ the previous formula can also be written as

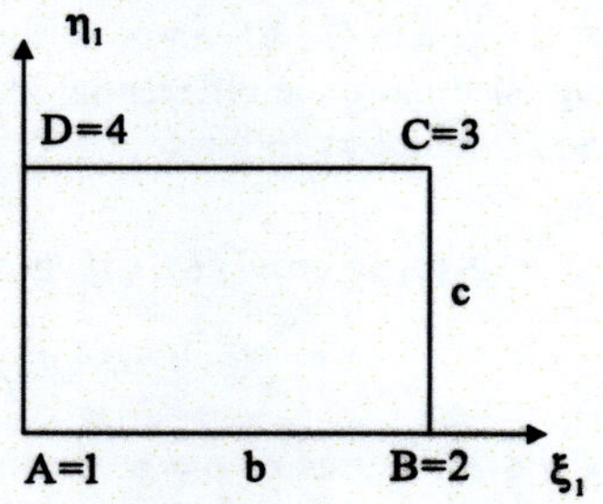

Fig. 38.2 2D reference frame for the first face of the prism

$$U(P) = \frac{G\delta}{2}\sum_{i=1}^{3} d_i \left\{ \sum_{j=1}^{4} \beta_{ij} \int_{l_{ij}} \frac{f(\boldsymbol{\rho}_i, d_i)}{\boldsymbol{\rho}_i \cdot \boldsymbol{\rho}_i} ds_{ij} - d_i \frac{\pi}{2} \right\}$$
$$= \frac{G\delta}{2}\sum_{i=1}^{3} d_i \left\{ I_i - d_i \frac{\pi}{2} \right\} \qquad (38.14)$$

where $\beta_{ij} = \boldsymbol{\rho}_i \cdot \boldsymbol{\nu}_j$ and s_{ij} is the abscissa along edge j, denoted by l_{ij}, of the face F_i.

To be specific and take the algebraic manipulations to the lesser extent, we shall detail only the calculation of I_1 since the other two integrals can be obtained from the latter straightforwardly. Numbering the vertices of face 1 as in Fig. 38.2, one gets

$$I_1 = \sum_{j=2}^{3} \beta_{1j} \int_{l_{ij}} \frac{[\boldsymbol{\rho}_1(s_{1j}) \cdot \boldsymbol{\rho}_1(s_{1j}) + a^2]^{1/2}}{\boldsymbol{\rho}_1(s_{1j}) \cdot \boldsymbol{\rho}_1(s_{1j})} ds_{1j} \qquad (38.15)$$

since β_{11} and β_{14} are both zero.

Setting $\lambda_{1j} = s_{1j}/l_{1j}$, where $\lambda_{1j} \in [0, 1]$, it turns out to be:

$$\boldsymbol{\rho}_1(s_{1j}) = \boldsymbol{\rho}_j + \lambda_j(\boldsymbol{\rho}_{j+1} - \boldsymbol{\rho}_j) \qquad (38.16)$$

where, to avoid prolification of symbols, the suffix $(\cdot)_1$ used to denote the first face has been omitted on the right-hand side. Thus, $\boldsymbol{\rho}_j$ and $\boldsymbol{\rho}_{j+1}$ have been used to denote the position vectors of the end vertices of the j-th edge of the face 1 in place of the more correct symbols $\boldsymbol{\rho}_{1j}$ and $\boldsymbol{\rho}_{1,j+1}$. The same holds true for λ_j.

According to (38.16) one has:

$$\boldsymbol{\rho}_1(s_{1j}) \cdot \boldsymbol{\rho}_1(s_{1j}) = \boldsymbol{\rho}_j \cdot \boldsymbol{\rho}_j + 2\lambda_j \boldsymbol{\rho}_j \cdot (\boldsymbol{\rho}_{j+1} - \boldsymbol{\rho}_j)$$
$$+ \lambda_j^2 (\boldsymbol{\rho}_{j+1} - \boldsymbol{\rho}_j) \cdot (\boldsymbol{\rho}_{j+1} - \boldsymbol{\rho}_j) \qquad (38.17)$$

so that

$$\boldsymbol{\rho}_1(s_{12}) \cdot \boldsymbol{\rho}_1(s_{12}) = b^2 + c^2 \lambda_2^2 \qquad (38.18)$$

and

$$\boldsymbol{\rho}_1(s_{13}) \cdot \boldsymbol{\rho}_1(s_{13}) = c^2 + b^2 (\lambda_3 - 1)^2 \qquad (38.19)$$

for edges 2 and 3 respectively. Being also $\beta_{12} = b$ and $\beta_{13} = c$ it turns out to be

$$I_1 = bc \left[\int_0^1 \frac{\sqrt{a^2 + b^2 + c^2\lambda_2^2}}{b^2 + c^2\lambda_2^2} d\lambda_2 + \int_0^1 \frac{\sqrt{a^2 + c^2 + b^2(\lambda_3 - 1)^2}}{c^2 + b^2(\lambda_3 - 1)^2} d\lambda_3 \right] \qquad (38.20)$$

Being further

$$\int \frac{\sqrt{e + gx}}{f + gx} = \sqrt{\frac{e-f}{fg}} \arctan \sqrt{\frac{e-f}{f}} \sqrt{\frac{g}{e + gx^2}} x + \frac{\ln\left[2\left(\sqrt{g}x + \sqrt{e + gx^2}\right)\right]}{\sqrt{g}} \qquad (38.21)$$

where x can stand both for λ and $\lambda - 1$, one finally has

$$I_1 = A_1 + A_2 + L_1 + L_2 \qquad (38.22)$$

where it has been set

$$A_1 = a \arctan \frac{ac}{b\sqrt{a^2 + b^2 + c^2}} \qquad (38.23)$$

$$A_2 = a \arctan \frac{ab}{c\sqrt{a^2 + b^2 + c^2}} \qquad (38.24)$$

$$L_1 = b \ln \frac{\sqrt{a^2 + b^2 + c^2} + c}{\sqrt{a^2 + b^2}} \qquad (38.25)$$

$$L_2 = c \ln \frac{\sqrt{a^2 + c^2}}{\sqrt{a^2 + b^2 + c^2} - b} \qquad (38.26)$$

The previous expressions do not suffer from any singularity except for the trivial case $b = 0$ or $c = 0$.

By adopting the same procedure detailed above, the integrals I_2 and I_3 can be obtained as well. However, due to the simmetry of the problem at hand, I_2 can be obtained by setting in the expression of I_1 the ordered triple $\{b, c, a\}$ in place of $\{a, b, c\}$. The same substitution can be carried out to derive I_3 starting from the expression of I_2.

Upon algebraic manipulations by means of the standard addition formula of the logarithms, the resulting expression for $U(P)$ reads

$$U(P) = \frac{G\delta}{2}\left\{\sum_{i=1}^{3}(U_{iatn} + U_{ilog}) - \frac{\pi}{2}h\right\} \tag{38.27}$$

where $h = a^2 + b^2 + c^2$,

$$U_{1atn} = (a^2 + b^2)\arctan\frac{ab}{c\sqrt{a^2 + b^2 + c^2}} \tag{38.28}$$

and

$$U_{1log} = ab\ln\frac{\sqrt{a^2 + b^2 + c^2} + c}{\sqrt{a^2 + b^2 + c^2} - c} \tag{38.29}$$

Setting in the expressions of U_{1atn} and U_{1log} the ordered triple $\{b, c, a\}$ in place of $\{a, b, c\}$, the additional terms U_{2atn} and U_{2log} can be respectively obtained. A further permutation of the triple $\{b, c, a\}$ to $\{c, a, b\}$ needs to be invoked to get U_{3atn} and U_{3log} in turn from U_{2atn} and U_{2log}.

It is worth noting that formula (38.27) exhibits the same computational complexity of formula (32) in (Tsoulis 2000).

3 First-Order Derivative of the Potential of a Prism

The gravitation vector, i.e. the gradient of the potential at P, is obtained by considering the following identity

$$\begin{aligned}\operatorname{grad} U(P) &= G\delta\int_{\Omega}\operatorname{grad}\frac{1}{(\mathbf{r}\cdot\mathbf{r})^{1/2}}\,dV\\ &= G\delta\int_{Fr(\Omega)}\frac{\mathbf{n}}{(\mathbf{r}\cdot\mathbf{r})^{1/2}}\,dA\end{aligned} \tag{38.30}$$

where use has been made of Gauss theorem. Being $\mathbf{n}$ constant on each face one has, on account of (38.8)

$$\begin{aligned}\operatorname{grad} U(P) &= G\delta\sum_{i=1}^{6}\mathbf{n}_i\int_{F_i}\frac{dA_i}{(\mathbf{r}_i\cdot\mathbf{r}_i)^{1/2}}\\ &= G\delta\sum_{i=1}^{6}\mathbf{n}_i\int_{F_i}\frac{dA_i}{(\boldsymbol{\rho}_i\cdot\boldsymbol{\rho}_i + d_i^2)^{1/2}}\end{aligned} \tag{38.31}$$

where $\mathbf{n}_i$ is the unit normal vector to the i-th face. Due to the particular shape of the prism only two faces do actually contribute to each component of the gradient. For instance, only faces 1 and 4, orthogonal to the x axis, contribute to the first component of the gradient $[\operatorname{grad} U(P)]_x$ since $\mathbf{n}_1 = \{1, 0, 0\}$ and $\mathbf{n}_4 = \{-1, 0, 0\}$.

Accordingly, formula (38.31) specializes to

$$[\operatorname{grad} U(P)]_x = G\delta\left[\int_{F_1}\frac{dA_1}{(\boldsymbol{\rho}_1\cdot\boldsymbol{\rho}_1 + a^2)^{1/2}} - \int_{F_4}\frac{dA_4}{(\boldsymbol{\rho}_4\cdot\boldsymbol{\rho}_4)^{1/2}}\right] \tag{38.32}$$

The two integrals above can be computed by means of formula (38.11) as specialized in the braces appearing in formula (38.13); specifically, the first integral becomes

$$\int_{F_1}\frac{dA_1}{(\boldsymbol{\rho}_1\cdot\boldsymbol{\rho}_1 + a^2)^{1/2}} = I_1 - a\frac{\pi}{2} \tag{38.33}$$

where I_1 is provided by (38.22). The second integral in (38.32), concerning the face F_4 passing through the origin, is obtained setting $a = 0$ in the right-hand side of (38.33). Thus, the final expression is

$$\begin{aligned}[\operatorname{grad} U(P)]_x = G\delta\Bigg\{&A_1 + A_2 + L_1 + L_2 - a\frac{\pi}{2}\\ &- b\ln\frac{\sqrt{b^2 + c^2} + c}{b}\\ &- c\ln\frac{c}{\sqrt{b^2 + c^2} - b}\Bigg\}\end{aligned} \tag{38.34}$$

which is always non-singular apart from the trivial cases $b = 0$ and $c = 0$. Notice that the numerical evaluation of I_1 can be usefully employed for the computation both of $U(P)$ and $[\operatorname{grad} U(P)]_x$.

The standard addition formula of the logarithms and the generalized one for the inverse tangent function proved in (D'Urso and Russo 2002) yield the more compact expression

$$[\text{grad}\, U(P)]_x = G\delta \left\{ a \arctan \frac{-bc}{a\sqrt{a^2+b^2+c^2}} + b \ln \frac{b(\sqrt{a^2+b^2+c^2}+c)}{\sqrt{a^2+b^2}(\sqrt{b^2+c^2}+c)} + c \ln \frac{\sqrt{a^2+c^2}(\sqrt{b^2+c^2}-b)}{c(\sqrt{a^2+b^2+c^2}-b)} \right\} \tag{38.35}$$

which requires less numerical computations with respect to formula (29) in (Tsoulis 2000), yet providing the same result.

Expressions for $[\text{grad}\, U(P)]_y$ can be obtained by replacing the ordered triple $\{a, b, c\}$ with $\{b, c, a\}$ in the formula above. The expression of $[\text{grad}\, U(P)]_z$ can be obtained from $[\text{grad}\, U(P)]_y$ in a similar way.

Conclusions

A general procedure for computing the potential and its first-order gradient for a prism has been illustrated. The proposed approach addresses in a unified framework any source of singularity, indipendently from the position of the observation point P with respect to the prism, without the necessity of ad-hoc formulations or a-posteriori corrections. In the special case of P coincident with a prism corner, the resulting formulas turn out to be computationally more effective than the analogous ones reported in (Tsoulis 2000).

References

Banerjee B, DasGupta SP (1977) Gravitational attraction of a rectangular parallelepiped. Geophys 42:1053–1055

D'Urso MG, Russo P (2002) A new algorithm for point-in polygon tests. Surv Rev 36–284:410–422

Kellogg OD (1929) Foundations of potential theory. Springer, Berlin

MacMillan WD (1930) Theoretical mechanics, vol. 2: the theory of the potential. Mc-Graw-Hill, New York

Nagy D (1966) The gravitational attraction of a right rectangular prism. Geophys 31:362–371

Nagy D, Papp G, Benedek J (2000) The gravitational potential and its derivatives for the prism. J Geodesy 74:553–560

Petrović S (1996) Determination of the potential of homogeneous polyhedral bodies using line integrals. J Geodesy 71:44–52

Tang KT (2006) Mathematical methods for engineers and scientists. Springer, Berlin

Tsoulis D (1999) Analytical and numerical methods in gravity field modelling of ideal and real masses. Deutsche Geodätische Kommission, Reihe C, Heft Nr. 510, München

Tsoulis D (2000) A note on the gravitational field of the right rectangular prism. Bollettino di Geodesia e Scienze Affini, LIX-1:21–35

Tsoulis D, Petrović S (2001) On the singularities of the gravity field of a homogeneous polyhedral body. Geophys 66:535–539

Part VI

Satellite Gravity Theory

Comparison of Kinematic Orbit Analysis Methods for Gravity Field Recovery

39

T. Reubelt, N. Sneeuw, and E.W. Grafarend

Abstract

Gravity recovery from kinematic orbits is possible at three levels: (1) coordinates (integral of Fredholm type, i.e. boundary value problem for short arcs), (2) velocity (integrals of motion, e.g. energy balance approach) and (3) accelerations, which are directly connected to the force function by the equation of motion (e.g. acceleration approach). For CHAMP, these three approaches have been applied successfully. With the advent of e.g. GOCE and SWARM, and given the uncertainty of a GRACE follow-on mission, the kinematic orbit analysis methods have gained further relevance. In this paper, the aforementioned approaches are compared by means of simulations. Important issues as the influence of the correlation of kinematic orbit errors and data weighting are investigated. A big advantage of the acceleration approach is its simplicity and speed due to missing integration. As it amplifies noise, the numerical differentiation in approaches 2 and especially 3 are regarded as a bottleneck. However, the noise of kinematic orbits is highly correlated, which reduces the effect of noise amplification, such that approaches 2 and 3 are not affected. Simulations based on white and correlated orbit noise show that both the acceleration approach and the boundary value problem lead to promising results. However, an advantage of the acceleration approach is that the non-diagonal elements can be neglected in data weighting (for a conventional kinematic orbit sampling interval of 30 s) without significant loss of accuracy, which reduces the computational effort significantly.

Keywords

CHAMP • GRACE • GOCE • SWARM • Kinematic orbits • Orbit analysis • Numerical differentiation • Acceleration approach • Gravity field recovery

1 Introduction

With the launch of the CHAMP satellite the gravity field recovery entered a new era: for the first time, the global, homogeneous and complete coverage of the Earth with measurements of the same accuracy, sensitivity, resolution and homogeneity is

T. Reubelt (✉) · N. Sneeuw · E.W. Grafarend
Institute of Geodesy, University of Stuttgart,
Geschwister-Scholl-Strasse 24D, 70174 Stuttgart, Germany
e-mail: reubelt@gis.uni-stuttgart.de;sneeuw@gis.uni-stuttgart.de;grafarend@gis.uni-stuttgart.de

N. Sneeuw et al. (eds.), *VII Hotine-Marussi Symposium on Mathematical Geodesy*, International Association of Geodesy Symposia 137, DOI 10.1007/978-3-642-22078-4_39,

realized. This is possible due to a low orbit height (270–450 km), near-polar orbits and especially sophisticated measurement principles of the new low Earth orbiter (LEO) missions CHAMP, GRACE and GOCE. With CHAMP, the detection of the long-wavelengths of the static gravitational field up to SH (spherical harmonic) degree $L_{\max} = 80$ is enabled by means of the high-low satellite-to-satellite tracking (hl-SST) and the application of orbit analysis methods. In CHAMP's successors GRACE and GOCE the hl-SST is helpful to improve the low frequencies in the gravity field recovery. In future the ESA magnetic field mission SWARM can be used for the purpose of gravity field recovery by means of orbit analysis.

In contrast to former satellites of the pre-CHAMP era, the measurement techniques of the new satellites capture the free fall motion of test masses with a dense and homogeneous coverage. This enables a more or less in-situ gravity field recovery (Ilk et al. 2008). In case of orbit analysis the in-situ principle becomes feasible by the determination of precise kinematic orbits: the free fall of a test mass with respect to the gravity field is observed directly and densely in three dimensions. This in-situ character allows short-arc or even point-wise analysis. The advantage becomes clear, if the classical approach (Reigber 1989) for orbit analysis is regarded. Its drawbacks stemming from the analysis of long arcs, namely large computational efforts and summation of errors in the force function, can be avoided.

Kinematic orbit analysis can be regarded as a two-step-procedure (Reubelt 2009), where (1) the kinematic orbits are computed from the pure geometric information of hl-SST and (2) the SH coefficients are estimated from the analysis of the kinematic orbits. Although kinematic orbit analysis methods have been regarded with scepticism at the beginning of CHAMP, they have been established now due to advanced analysis methods and especially due to significant improvements in kinematic orbit determination (accuracy of 1–3 cm, Švehla and Földváry 2006). This enables similar or better accuracy in gravity field recovery compared to the classical method (Reubelt 2009).

As summarized by Ilk et al. (2008) the analysis of kinematic orbits is possible at three levels (Fig. 39.1): (1) at the level of coordinates as integral of Fredholm type, (2) at the level of velocity by solving the integrals of motion and (3) at the level of accelerations, which are directly connected to the force function by the

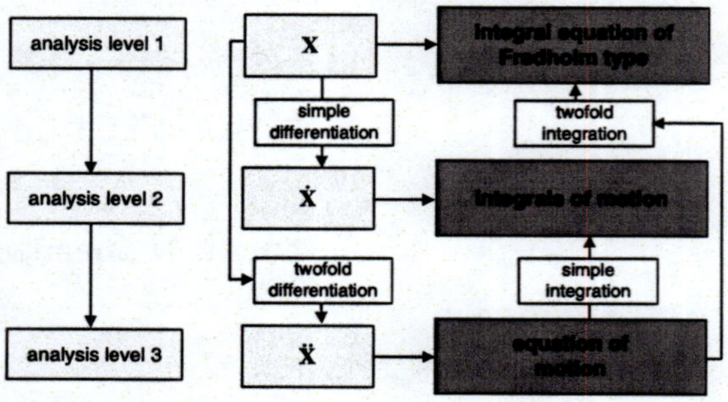

Fig. 39.1 The three analysis levels of kinematic orbit analysis (after Ilk et al. 2008)

equation of motion. The velocity and acceleration can be derived from the kinematic orbits by means of numerical differentiation. The integrals of motion and the integral equation of Fredholm type can be derived (besides some possible other manipulations) by simple or twofold integration of the equation of motion

$$\ddot{\mathbf{X}}(t) = \mathbf{f}\left(t, \mathbf{X}(t), \dot{\mathbf{X}}(t), \mathbf{p}\right) \tag{39.1}$$

where $\mathbf{X}(t)$ is the satellite's orbit, $\mathbf{f}$ the specific force acting on it, being a function of time t, position $\mathbf{X}$, velocity $\dot{\mathbf{X}}$ and the parameters $\mathbf{p}$ (of gravity field, surface forces, ...).

For CHAMP, approaches based on these three levels have been applied successfully in kinematic orbit analysis. The approaches mainly established are the boundary value problem (BVP) for short arcs (level 1), the energy balance approach (EBA) (level 2) and the acceleration approach (ACA) (level 3), which are considered and compared in this paper. It should be mentioned that further approaches can be formulated at the different analysis levels, especially for the integrals of motion (Ilk et al. 2008).

Since all three analysis levels (or approaches) are based on the same equation (Fig. 39.1) it can be argued that they should lead to identical or at least very similar results, if appropriate methods for integration or numerical differentiation are selected and correct error-propagation is applied. However, as former comparisons show (Ditmar and van Eck van der Sluijs 2004; Mayer-Gürr et al. 2005a), this is not the case. Although a similar accuracy for BVP and ACA was obtained for simulations, a decreased accuracy (factor 1.5–2) was found for the EBA. A weakness of these former comparisons is, that these were mainly considered for white orbit noise. As newer investigations of kinematic orbits show, kinematic orbit noise is highly correlated (Švehla and Földváry 2006; Reubelt 2009), which has significant effects on the accuracy of gravity field estimation (Reubelt 2009). Another aspect so far not considered in detail is the influence of data weighting. Data weighting of huge, correlated data sets is a demanding task due to (1) a large computational/memory effort and (2) a limited knowledge of the true measurement noise. Thus it has to be considered if data-weighting has an impact at all. In this paper, a comparison of the three mentioned approaches is made considering both uncorrelated and correlated orbit errors as well as the influence of data-weighting.

2 Orbit Analysis Methods

2.1 Boundary Value Problem for Short Arcs

A realisation of analysis level 1 is the solution of (39.1) as a boundary value problem for short arcs, as it was applied successfully by Mayer-Gürr et al. (2005b) in CHAMP-data analysis. By a twofold integration the equation of motion can be formulated as a BVP, which results in the integral equation of Fredholm type

$$\mathbf{X}(t) = (\mathbf{X}_n - \mathbf{X}_0)\frac{t - t_0}{t_n - t_0} + \mathbf{X}_0 - \frac{1}{t_n - t_0} \times \int_{t_0}^{t_n} K(t, t')\mathbf{f}\left(t', \mathbf{X}(t), \dot{\mathbf{X}}(t), \mathbf{p}\right) \mathrm{d}t' \quad (39.2)$$

where the satellite's position $\mathbf{X}$ at time t is described via the arc's boundary values ($\mathbf{X}_0 = \mathbf{X}(t_0)$, $\mathbf{X}_n = \mathbf{X}(t_n)$) and an integral of the specific force function $\mathbf{f}$ which is composed of tidal and surface forces and the gravity field parameters $\mathbf{p}_{grav}$ ($K(t, t')$ is the integral kernel). The integral can be solved numerically, e.g. by shifting polynomials with short length (8, 10 points), and a linear relation between the kinematic orbit points and the searched-for SH coefficients can be established. Typical orbit lengths used for integration are 30–90 minutes in order to avoid accumulation of errors in the force function. The BVP can be regarded as the most direct method since the kinematic orbit is used directly as observation. The minimum set of estimates are the SH coefficients and the boundary values.

2.2 Energy Balance Method

The EBA is based on the idea of energy conservation and can be derived from the equation of motion by multiplication with $\dot{\mathbf{X}}$ and integration over t. The EBA can be formulated in the space-fixed system as

$$E_{kin}(t) - V(\mathbf{X}(t), \mathbf{p}_{grav}) - R(t) - \int_{t_0}^{t} \mathbf{f}_{dist}(t) \cdot \dot{\mathbf{X}}(t)\mathrm{d}t = E_0 \quad (39.3)$$

where $E_{kin} = |\dot{\mathbf{X}}(t)|^2/2$ is the kinetic energy per unit mass, V is the gravitational potential, $R(t) = \omega(\mathbf{X}\dot{\mathbf{Y}} - \mathbf{Y}\dot{\mathbf{X}})$ is the rotational potential and E_0 is the energy constant (the "Hamiltonian"). The loss/gain of energy due to tides and surface forces is considered by means of the integral over the generated disturbing specific forces $\mathbf{f}_{dist}$. In order to apply the EBA, the energy constant and the energy drift caused by accelerometer offsets have to be determined in a pre-step. While the former can be estimated by comparison to existing SH-models, the latter can be determined also by crossover-analysis. The SH coefficients can be estimated from the "pseudo-observation" potential V, which can be determined from the kinetic energy E_{kin} by (39.3). The needed velocity $\dot{\mathbf{X}}$ is derived from the kinematic orbit $\mathbf{X}(t)$ by means of numerical differentiation. The EBA was applied successfully by several researchers for CHAMP (e.g. Weigelt 2007).

2.3 Acceleration Approach

By application of the acceleration approach the equation of motion (1) can be directly evaluated. Detailed descriptions and investigations can be found in Reubelt et al. 2003) and Reubelt (2009). Here the ACA is outlined briefly: At first, the satellite's acceleration vector $\ddot{\mathbf{X}}$ is determined by means of numerical differentiation from the kinematic orbit $\mathbf{X}$ in the space fixed system. After reduction of disturbing accelerations caused by tides and surface forces the acceleration $\ddot{\mathbf{X}}_{grav}$ caused by the terrestrial gravitational field is obtained. This can be balanced now by the equation of motion with a SH model of the gravitational attraction vector $\boldsymbol{\Gamma}_{grav} = \mathrm{grad}\, V(\mathbf{X}(t), \mathbf{p}_{grav})$, and so the SH coefficients can be determined. This means that the searched-for SH coefficients are estimated from the "pseudo-observations" $\ddot{\mathbf{X}}_{grav}$. The ACA offers in contrast to the BVP and the EBA a simple method with a short computation time (no integration) and an easy implementation. As a drawback, the numerical differentiation is regarded which amplifies noise. Nevertheless, promising results have been obtained from CHAMP by application of the ACA with a similar accuracy as from other approaches (Reubelt 2009; Ditmar et al. 2006).

3 Numerical Differentiation

As a major disadvantage of the EBA and especially the ACA the numerical differentiation is regarded, which has the undesirable property of amplifying noise with ω_k^2 ($\omega_k \hat{=}$ angular frequency). However the effect of numerical differentiation depends on the type of noise in the original data. In contrast to white noise, which is the same on every frequency ω_k, most of the noise is conserved in the low frequencies and decreases at high frequencies ω_k for a correlated noise scenario. If numerical differentiation is applied, the total noise of the correlated noise scenario remains smaller than for white noise scenario. This is demonstrated in Fig. 39.2 for the comparison between white kinematic orbit noise (standard deviation $\sigma_X = 2\,\text{cm}$, correlation $\rho_{i,j} = 0$ between two orbit points (X_i,X_j), sampling time $\Delta t = 30\,\text{s}$) and correlated kinematic orbit noise ($\sigma_X = 2\,\text{cm}$, $\rho_{i,j} = 0.93^{|i-j|}$, $\Delta t = 30\,\text{s}$). While the absolute noise level of the kinematic orbit for both cases is the same, the noise level of the determined accelerations differs significantly. The applied correlations lead to accelerations, whose noise level is decreased by a factor 5 compared to the white noise case ($1.5 \times 10^{-5}\,\text{m/s}^2$ vs. $8 \times 10^{-5}\text{m/s}^2$). The numerical differentiation method applied (9-point-Gregory-Newton-interpolation, e.g. Reubelt et al. 2003) can be expressed in terms of coordinate differences (baselines) $\Delta \mathbf{X}_i = \mathbf{X}_{i+1} - \mathbf{X}_i$. This also explains the low noise increase by numerical differentiation for correlated errors. Since the behaviour of correlated kinematic orbit noise is much smoother, the baseline errors (0.75 cm vs. 2.8 cm) and thus the acceleration errors are much lower.

This means, that numerical differentiation must not be considered as a drawback for the ACA and EBA if the kinematic orbit noise is (highly) correlated. Indeed, it is correlated as investigations of the variance-covariance-matrices (VCM) of kinematic orbits show. Here, the correlation length is about 25 min, and the correlation of adjacent orbit point errors (sampling time $\Delta t = 30\,\text{s}$) is about 0.93 (Švehla and Földváry 2006). Moreover, as comparisons between kinematic and reduced-dynamic orbits show, the differences between them for coordinates/baselines/accelerations are 2 cm/ 0.65 cm/$1.2 \times 10^{-5}\,\text{m/s}^2$ (Reubelt et al. 2006), which confirms the correlation scenario of Fig. 39.2. Here, the reduced-dynamic orbit serves as smooth reference for evaluation.

In total this means, that numerical differentiation may not reduce the accuracy of the ACA and EBA if the orbit noise is highly correlated.

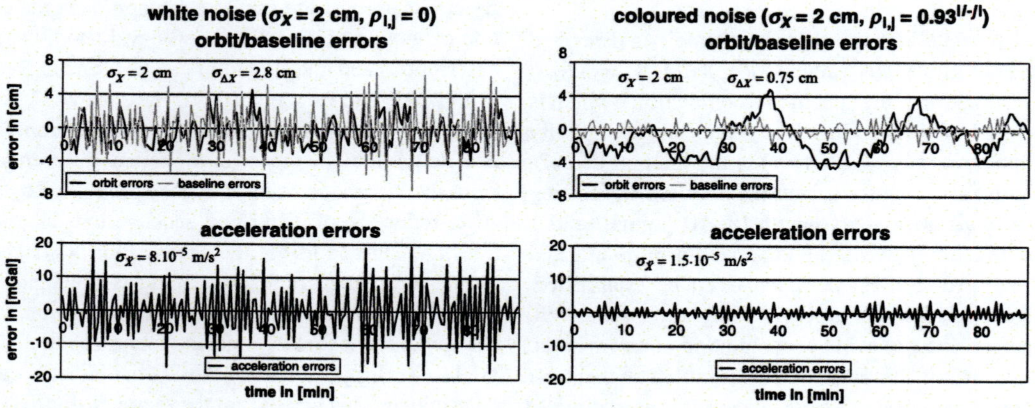

Fig. 39.2 Noise amplification by numerical differentiation for white (*left*) and highly correlated (*right*) orbit errors

4 Stochastic Properties and Weighting

Data weighting in satellite gravimetry is a difficult task. On the one hand, the real stochastic properties of the exploited data sets are not known, which means that an adequate data-weighting is not guaranteed. On the other hand, the data sets are normally correlated, which means that their VCM's are fully or at least largely occupied. For the huge data sets collected by satellites, the handling (storage, inversion, ...) of such huge VCM's becomes very demanding or even impossible. Within the scope of the new missions methods for handling large correlated data sets have been developed (e.g. Ditmar et al. 2007). But their application still is very time-consuming and is mainly possible for simplified or generalized cases. Thus it has to be considered if the non-diagonal elements (the covariances) in the VCM have a big impact in data-weighting. If the gain of accuracy is marginal, they might be neglected for the advantage of better computational performance.

Figure 39.3 shows the covariance-functions for the pseudo-observations of the three investigated approaches for white and correlated orbit errors (and a sampling time $\Delta t = 30$ s). Since the kinematic orbit is used directly as observation in the BVA, this means an uncorrelated VCM (a unit matrix) for white orbit noise and a highly correlated VCM with a large correlation length for correlated orbit errors. Therefore (1) the correlations in data-weighting should not be neglected in the BVP (except for white orbit noise) and (2) the correlations of orbit errors should be known exactly for an adequate data-weighting due to the direct dependence.

In contrast, the covariance functions of the kinetic energy and the accelerations (pseudo-observations of the EBA and ACA) (1) are always correlated, but with a small correlation length (< 2 data points) and (2) their shape is almost independent from the correlation of the orbit errors. This means, that due to (2) the correlations of the orbit errors do not have to be known exactly for an adequate data-weighting and due to (1) the question arises, if the correlations in the weight matrix for the EBA and ACA have a big impact at all.

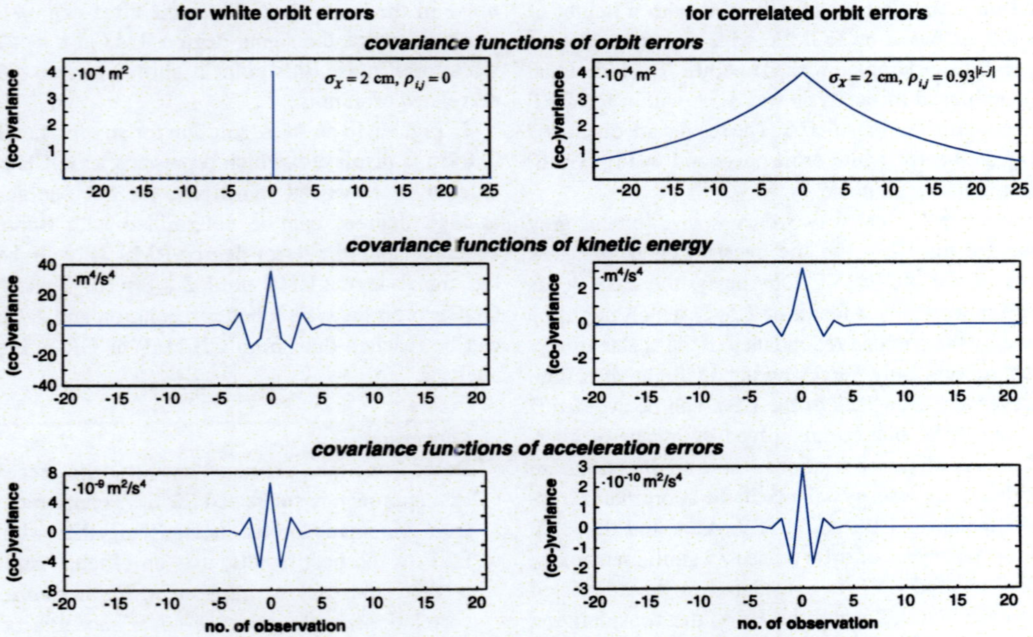

Fig. 39.3 Covariance functions of kinematic orbit errors, kinetic energy and acceleration errors for the two cases: white orbit noise (*left*), correlated orbit noise (*right*); sampling time of kinematic orbit $\Delta t = 30$ s

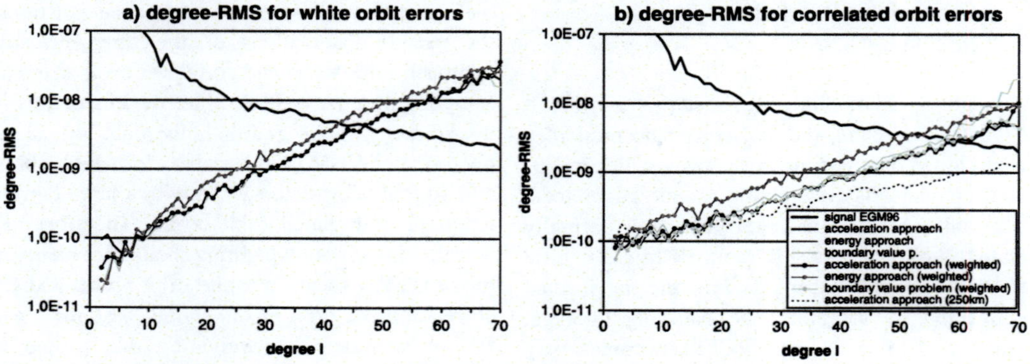

Fig. 39.4 Errors of the different kinematic orbit analysis methods in terms of degree-RMS-curves; (**a**) for white orbit noise, (**b**) for correlated orbit noise

This seems to be an advantage of the ACA and EBA in contrast to the BVP.

5 Results

The different approaches were tested for both cases, white and correlated orbit noise, and weighted and unweighted gravity field solutions have been estimated by means of least squares adjustment. The results are based on a simulated 1-month orbit with a height of $h = 400$ km based on EGM96 with maximum degree $L_{\max} = 70$ and sampled with $\Delta t = 30$ s. The individual arc length used for analysis was 1.5 h within the BVP and 3 h within the EBA/ACA. The results are displayed in Fig. 39.4a) for white orbit noise and in Fig. 39.4b) for correlated orbit noise.

There are basically two curves visible: the grey curves for the EBA and the overlaid dark/light grey curves of the ACA/BVP. Obviously, the EBA leads to results worse by a factor of 1.5–2, which might be caused by the reduced redundancy of the scalar energy which reflects only forces acting in flight direction. However this drawback of the EBA can be avoided if the integrals of motion are solved by means of alternative energy balance relations where all directions are considered, e.g. energy balance in the coordinate directions (Löcher and Ilk 2007). The ACA and the BVP lead for both types of orbit noise to a similar accuracy, if the weighted BVP is considered in the correlated noise case. For ACA and the EBA, the consideration of the correlations in data weighting (the weighted solutions) has almost no influence on the accuracy compared to the unweighted solutions, as concluded in Sect. 4. Only for the white noise case, data weighting has a positive effect for the very low degrees ($l < 7$) for the ACA. In contrast, data weighting within the BVP is important for correlated orbit errors (Fig. 39.4b), as deduced in Sect. 4.

In Sect. 3 it was argued that the approaches based on numerical differentiation (ACA, EBA) might benefit from correlated orbit errors. However, as shown in Fig. 39.4, also the BVP benefits from correlated orbit noise in the same way. While the error-degree-RMS curve intersects the signal-degree-RMS at $l \approx 47$ for white orbit noise, this point is shifted to $l \approx 58$ for correlated orbit noise.

In Fig. 39.4b an ACA-solution for an orbit height of 250 km is displayed, which represents a GOCE orbit. Due to an enhanced sensitivity, the SH coefficients at high degrees can be determined with improved accuracy and the error-degree-RMS is well below the signal-degree-RMS until $L_{\max} = 70$. Thus, from GOCE orbit analysis a better resolution and accuracy can be reached than from CHAMP or GRACE orbit analysis.

Conclusion

Based on the previous results, the acceleration approach is suggested for the gravity recovery from kinematic orbits. It leads together with the BVP to the best results, it is an efficient and fast method and easy to implement. Furthermore, the correlations of the variance-covariance matrix can be neglected in data weighting without significant loss of accuracy for a conventional kinematic orbit sampling of $\Delta t = 30$ s. This means that (1) the large computational effort involved with data weighting can be avoided and (2) the stochastic properties (i.e. the correlations) of the observation noise do not have to be known for the data inversion. The method is not only proposed for CHAMP, but also for combination with the other measurement types in GRACE and GOCE as well as for other GPS-tracked LEOs, e.g. SWARM.

References

Ditmar P, van Eck van der Sluijs A (2004) A technique for modelling the Earth's gravity field on the basis of satellite accelerations. J Geodesy 78:12–33

Ditmar P, Kuznetsov V, van Eck van der Sluijs A, Schrama E, Klees R (2006) 'DEOS_CHAMP-01C_70': a model of the Earth's gravity field computed from accelerations of the CHAMP satellite. J Geodesy 79:586–601

Ditmar P, Klees R, Liu X (2007) Frequency-dependent data weighting in global gravity field modelling from satellite data contaminated by non-stationary noise. J Geodesy 81(1): 81–96

Ilk KH, Löcher A, Mayer-Gürr T (2008) Do we need new gravity field recovery techniques for the new gravity field satellites? In: Xu P, Liu J, Dermanis A (eds) VI Hotine-Marussi symposium on theoretical and computational geodesy, IAG Symposium Wuhan, China 29 May–2 June, 2006, IAG Symposia, vol 132, pp 3–9

Löcher A, Ilk KH (2007) A validation procedure for satellite orbits and force function models based on a new balance equation approach. In: Tregoning P, Rizos C (eds) Dynamic planet – monitoring and understanding a dynamic planet with geodetic and oceanographic tools, IAG Symposium, Cairns, Australia, 22–26 August, 2005, IAG Symposia, vol 130, pp 280–287

Mayer-Gürr T, Feuchtinger M, Kusche J (2005a) A comparison of various procedures for global gravity field recovery from CHAMP orbits. In: Reigber C, Lühr H, Schwintzer P, Wickert J (eds) Earth observation with CHAMP-results from three years in orbit. Springer, New York, pp 151–156

Mayer-Gürr T, Ilk KH, Eicker A, Feuchtinger M (2005b) ITG-CHAMP01: a CHAMP gravity field model from short kinematic arcs over a one-year observation period. J Geodesy 78(7–8):462–480

Reigber C (1989) Gravity field recovery from satellite tracking data. In: Sanso F, Rummel R (eds) Theory of satellite geodesy and gravity field determination, Lecture Notes in Earth Sciences, vol 25. Springer, New York, pp 197–234

Reubelt T, Austen G, Grafarend EW (2003) Harmonic analysis of the earth's gravitational field by means of semi-continuous ephemerides of a low earth orbiting GPS-tracked satellite. Case study: CHAMP. J Geodesy 77(5–6): 257–278

Reubelt T, Götzelmann M, Grafarend E (2006) Harmonic analysis of the earth's gravitational field from kinematic CHAMP orbits based on numerically derived satellite accelerations. In: Flury J, Rummel R, Reigber C, Rothacher M, Boedecker G, Schreiber U (eds) Observation of the earth system from space. Springer, New York, pp 27–42

Reubelt T (2009) Harmonische Gravitationsfeldanalyse aus GPS-vermessenen kinematischen Bahnen niedrig fliegender Satelliten vom Typ CHAMP, GRACE und GOCE mit einem hoch auflösenden Beschleunigungsansatz. DGK Reihe C, Nr. 632, München.

Švehla D, Földváry L (2006) From kinematic orbit determination to derivation of satellite velocity and gravity field. In: Flury J, Rummel R, Reigber C, Rothacher M, Boedecker G, Schreiber U (eds) Observation of the earth system from space. Springer, New York, pp 177–192

Weigelt M (2007) Global and local gravity field recovery from satellite-to-satellite tracking, UCGE Report 20248. Ph.D. Thesis, Department of Geomatics Engineering, University of Calgary, Canada

Inclination Functions: Orthogonality and Other Properties

40

Nico Sneeuw

Abstract

The two complementary properties of orthogonality and completeness are well known for spherical harmonics. The addition theorem is an expression of the latter. Since inclination functions are related to spherical harmonics it can be expected that orthogonality and completeness properties exist for them as well. The Wagner-Gooding identities are identified as versions of the addition theorem for inclination functions. Orthogonality of inclination functions is derived here. Due to their complementarity, it is advised to use these properties together when testing algorithms for the numerical calculation of, e.g., Legendre and inclination functions.

Keywords

Addition theorem • Completeness • Inclination functions • Orthogonality

1 Introduction

The properties of orthogonality and completeness for given function systems are complementary. For instance orthogonality of spherical harmonics is a pre-requisite for spectral analysis. Completeness on the other hand, as expressed by the addition theorem, guarantees that any square-integrable function on the sphere can be synthesized by the given set of base functions.

Due to their complementary nature both properties could and should be used together when validating algorithms for function computation. For spherical harmonics, and more particularly for Legendre-functions the two properties are briefly recapitulated from literature in Sect. 2.

The main objective of this contribution is to consider orthogonality and completeness for inclination functions $F_{lmk}(I)$. It is emphasized here that neither the algorithms for $F_{lmk}(I)$- or $P_{lm}(\cos\theta)$-calculation, nor their validation itself is object of this manuscript. Surely, since inclination functions are derived from spherical harmonics, both properties should exist. Indeed, the so-called Wagner-Gooding identities are identified as addition theorems in disguise in Sect. 3. Orthogonality of inclination functions, has not been discussed in literature. Therefore, the orthogonality property of $F_{lmk}(I)$ will be derived in Sect. 4. This allows the complementary validation of algorithms for inclination function computation.

N. Sneeuw (✉)
Institute of Geodesy, University of Stuttgart, Stuttgart, Germany
e-mail: sneeuw@gis.uni-stuttgart.de

N. Sneeuw et al. (eds.), *VII Hotine-Marussi Symposium on Mathematical Geodesy*, International Association of Geodesy Symposia 137, DOI 10.1007/978-3-642-22078-4_40,

2 Spherical Harmonics $Y_{lm}(\sigma)$

Starting point of the following considerations are the two complementary properties of fully normalized, complex-valued spherical harmonics $Y_{lm}(\sigma)$:

Orthogonality:

$$\frac{1}{4\pi}\iint_{\sigma} Y_{lm}(\sigma)Y^{*}_{l'm'}(\sigma)\mathrm{d}\sigma = \delta_{ll'}\delta_{mm'}. \quad (40.1)$$

Addition theorem:

$$\frac{1}{2l+1}\sum_{m=-l}^{l} Y_{lm}(\sigma)Y^{*}_{lm}(\sigma') = P_l(\cos\psi), \quad (40.2)$$

in which $Y_{lm}(\sigma) = Y_{lm}(\theta,\lambda) = P_{lm}(\cos\theta)\exp(im\lambda)$ with θ and λ the spherical co-latitude and latitude, respectively. Moreover, $Y^{*}_{lm}(\sigma)$ refers to the complex conjugated spherical harmonic, i.e. $P_{lm}(\cos\theta)\exp(-im\lambda)$. As a further remark on notation: it is assumed throughout this work that all functions P_{lm} and F_{lmk} are fully normalized. The more conventional overbar is not used. One exception to this notation is made for the Legendre polynomial P_l in (40.2), and equations that derive from it, which is a non-normalized function in this context.

The link between the addition theorem and completeness of the function system is realized by the infinite sum over the degree l, leading to the Dirac function:

$$\sum_{l=0}^{\infty}\sum_{m=-l}^{l} Y_{lm}(\sigma)Y^{*}_{lm}(\sigma') = \delta(\psi).$$

2.1 Testing P_{lm}-Algorithms 1: Orthogonality

The property (40.1) is reduced now to an orthogonality of associated Legendre functions:

$$\frac{1}{2}\int_{0}^{\pi} P_{lm}(\cos\theta)P_{l'm}(\cos\theta)\sin\theta\mathrm{d}\theta = 2(2-\delta_{m,0})\delta_{ll'}. \quad (40.3)$$

Through Gauss-Legendre quadrature, e.g. Stroud and Secrest (1966), we obtain the discretized version

$$\sum_{i=1}^{N} P_{lm}(\cos\theta_i)P_{l'm}(\cos\theta_i)w_i = 2(2-\delta_{m,0})\delta_{ll'}. \quad (40.4)$$

If the right hand side is integrated in the quadrature weights w_i, the equivalent matrix version reads:

$$P^{\mathsf{T}}WP = I, \quad (40.5)$$

in which the matrix P is defined for constant order m and variable degree l and co-latitude θ:

$$P = \begin{pmatrix} P_{mm}(\cos\theta_1) & P_{m+1,m}(\cos\theta_1) & \cdots & P_{Lm}(\cos\theta_1) \\ P_{mm}(\cos\theta_2) & P_{m+1,m}(\cos\theta_2) & \cdots & P_{Lm}(\cos\theta_2) \\ \vdots & \vdots & \ddots & \vdots \\ P_{mm}(\cos\theta_N) & P_{m+1,m}(\cos\theta_N) & \cdots & P_{Lm}(\cos\theta_N) \end{pmatrix}.$$

As a demonstration, Fig. 40.1 displays the result of applying (40.4) for validating a typical two-point recursion for P_{lm}-with constant order m and increasing degree l.

$$P_{lm} = W_{lm}\left[\cos\theta P_{l-1,m} - W^{-1}_{l-1,m}P_{l-2,m}\right] \quad (40.6)$$

with $W_{lm} = \sqrt{\frac{(2l+1)(2l-1)}{(l+m)(l-m)}}$. For brevity, the starting values have been omitted here.

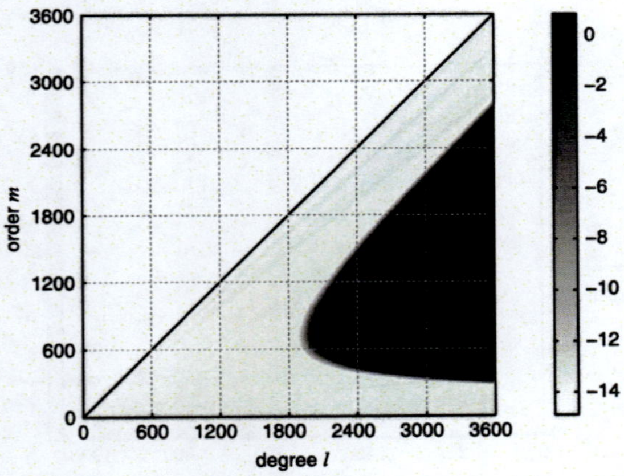

Fig. 40.1 Testing Legendre function algorithms through orthogonality. Displayed is $\log(\varepsilon_{lm})$ with $\varepsilon_{lm} = \mathrm{diag}(P^{\mathsf{T}}WP) - 1$ up till maximum degree $L = 3{,}600$

It shows that this particular recursion breaks down for double-precision arithmetic beyond degree $l = 2000$, although (near-)zonals and (near-)sectorials behave well up till the tested maximum degree $L = 3600$. Again, it is not the purpose of this paper to discuss the algorithms themselves, e.g. the need for scaling in this particular recursion.

2.2 Testing P_{lm}-Algorithms 2: Addition Theorem

Also the addition theorem (40.2) is reduced now to its variant for associated Legendre functions by letting $\lambda = \lambda'$:

$$\frac{1}{2l+1}\sum_{m=0}^{l} P_{lm}(\cos\theta)P_{lm}(\cos\theta') = P_l(\cos\psi), \tag{40.7}$$

again with the caveat that the right hand side denotes an unnormalized Legendre polynomial. Particularly for $\theta = \theta'$ and, hence, $\psi = 0$ a useful version arises:

$$\frac{1}{2l+1}\sum_{m=0}^{l} P_{lm}^2(\cos\theta) = P_l(1) = 1, \tag{40.8}$$

which is visualized in Fig. 40.2.

But also for $\theta \neq \theta'$ a useful test of (40.7) is demonstrated in Fig. 40.3.

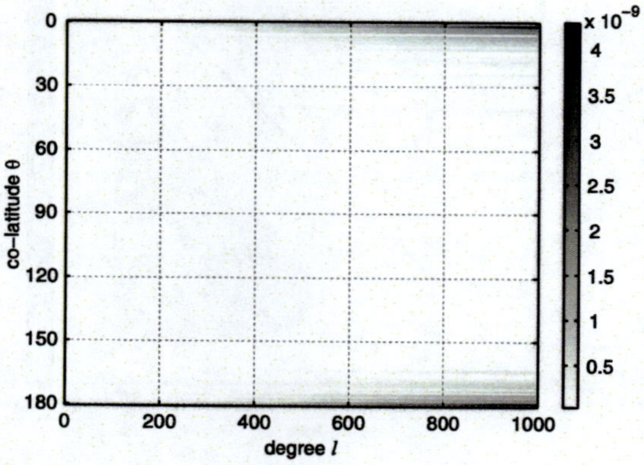

Fig. 40.2 Numerical test using the addition theorem of Legendre functions for $\theta = \theta'$. Displayed is the discrepancy in (40.8)

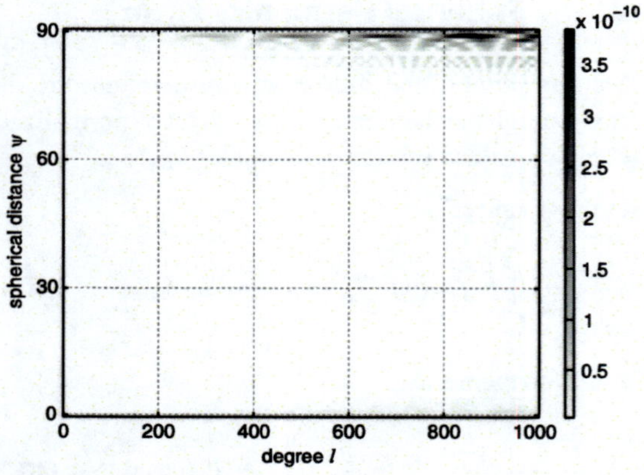

Fig. 40.3 Numerical test for $\theta = 90°$ and $\theta' \in [0, 90°]$. Displayed is the discrepancy in (40.7)

3 Inclination Functions $F_{lmk}(I)$

Inclination functions $F_{lmk}(I)$ arise naturally when a spherical harmonic $Y_{lm}(\sigma)$ is expressed in a new coordinate system, which has been rotated through three Euler angles. The spherical harmonic is transformed through representation coefficients $D_{lmk}(\alpha, \beta, \gamma)$, e.g. Edmonds (1957):

$$Y_{lm}(\sigma) = \sum_{k=-l}^{l} D_{lmk}(\alpha, \beta, \gamma) Y_{lk}(\sigma'), \tag{40.9}$$

$$\text{with:}\quad D_{lmk}(\alpha, \beta, \gamma) = e^{im\alpha} d_{lmk}(\beta) e^{ik\gamma}. \tag{40.10}$$

When rotating into the orbital coordinate system, in which the orbital plane is the new equator $\theta' = \frac{\pi}{2}$ and the new x-axis continuously points toward the satellite, we have:

$$Y_{lm}(u, I, \Lambda) = \sum_{k=-l}^{l} i^{k-m} e^{i(ku+m\Lambda)} d_{lmk}(I) Y_{lk}(0, 0), \tag{40.11}$$

in which the orbital elements u, I and Λ are defined (see also Fig. 40.4):

$$u = \omega + \nu = \text{argument of latitude}$$

$$\Lambda = \Omega - \text{GAST} = \text{longitude of ascending node}$$

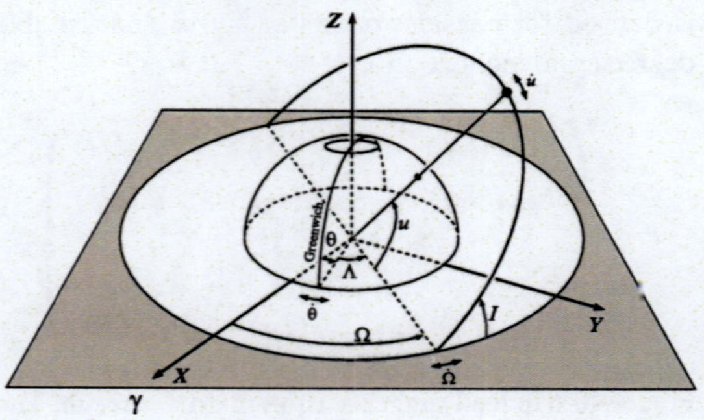

Fig. 40.4 Orbit configuration. (Note: θ means GAST here)

Sneeuw (1992) defines complex-valued normalized inclination functions $F_{lmk}(I)$ as:

$$F_{lmk}(I) = i^{k-m} d_{lmk}(I) P_{lk}(0). \tag{40.12}$$

Except for normalization and for being complex-valued, the notation here deviates from the conventional Kaula inclination functions $F_{lmp}(I)$, which uses the non-negative index p. In contrast $F_{lmk}(I)$ uses $k = l - 2p$ which may be negative and runs in steps of two, due to the term $P_{lk}(0)$ which is zero for $l - k$ odd. For further details of rotating spherical harmonics and of the definition of inclination functions it is referred to Sneeuw (1992).

3.1 Testing F_{lmk}-Algorithms 1: Addition Theorem

Two invariances of inclination functions are given in Gooding and Wagner (2008). The restricted relation

$$\sum_{m=0}^{l} F_{lmk}^2(I) + F_{lm,-k}^2(I) = (1 + \delta_{k,0}) P_{lk}^2(0), \tag{40.13}$$

also known as Wagner's conjecture, Wagner (1983), was proved by Sneeuw (1992) by:

- Using the symmetry $F_{l,m,-k} = (-1)^k F_{l,-m,k}$.
- Employing the definition (40.12).
- Considering that the coefficients $d_{lmk}(I)$ are an orthonormal matrix representation.

Thus, we obtain $\sum_{m=-l}^{l} F_{lmk}^2(I) = P_{lk}^2(0)$, indeed. The general invariance

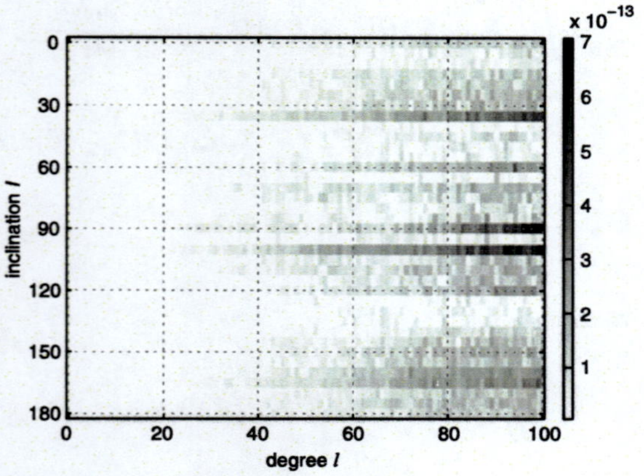

Fig. 40.5 Wagner-Gooding invariance: residual of (40.14)

$$\sum_{k=-l}^{l} \sum_{m=0}^{l} F_{lmk}^2(I) = 2l + 1 \tag{40.14}$$

can then be derived by further summation over k. It basically reflects the addition theorem (40.8). Sneeuw (1991) used the restricted invariance successfully for testing inclination function algorithms. The use of the general invariance for testing inclination functions is demonstrated in Fig. 40.5, where the residuals of the invariance are given for inclinations in the domain [0°; 180°] and for degrees l up to 100.

4 Orthogonality of $F_{lmk}(I)$

Orthogonality of inclination functions has, to the author's knowledge, so far not been discussed in literature. The property of orthogonality does not refer here to the fact that the d_{lmk}-symbols in (40.10) and (40.12) are an orthonormal matrix representation. Instead, the aim here is a relation of the type:

$$\int_I F_{lmk}(I) F_{l'm'k'}(I) \overset{?}{=} \delta_{\ldots},$$

in which the right hand side would be an expression with Kronecker δ-symbols.

Starting point for finding such a relation is the orthogonality of SO(3) representation symbols $D_{lmk}(\Omega) = D_{lmk}(\alpha, \beta, \gamma)$, e.g. Edmonds (1957):

$$\frac{1}{8\pi^2}\iiint\limits_{\Omega} D_{lmk}(\Omega)D^*_{l'm'k'}(\Omega)\mathrm{d}\Omega = \frac{\delta_{ll'}\delta_{mm'}\delta_{kk'}}{2l+1}$$

$$\Rightarrow \frac{1}{8\pi^2}\iiint\limits_{\Omega} \mathrm{e}^{i(m-m')\alpha}\mathrm{e}^{i(k-k')\gamma} d_{lmk}(\beta)d^*_{l'm'k'}(\beta)\mathrm{d}\Omega$$

$$= \frac{1}{2l+1}\delta_{ll'}\delta_{mm'}\delta_{kk'},$$

with

$$\iiint\limits_{\Omega} \ldots \mathrm{d}\Omega = \int\limits_0^{2\pi}\int\limits_0^{\pi}\int\limits_0^{2\pi} \ldots \sin\beta \mathrm{d}\alpha \mathrm{d}\beta \mathrm{d}\gamma.$$

The orthogonality of the trigonometric functions leads to $\delta_{mm'}\delta_{kk'}$, implying that we can at most hope to find an orthogonality of inclination functions of equal orders $m = m'$ and $k = k'$. We first extract the orthogonality of representation coefficients d_{lmk}:

$$\frac{1}{2}\int\limits_0^{\pi} d_{lmk}(\beta)d_{l'mk}(\beta)\sin\beta \mathrm{d}\beta = \frac{(-1)^{k-m}\delta_{ll'}}{2l+1}. \tag{40.15}$$

Applying definition (40.12) yields the sought for relation:

$$\frac{1}{2}\int\limits_0^{\pi} F_{lmk}(I)F_{l'mk}(I)\sin I \mathrm{d}I = \frac{1}{2l+1}P^2_{lk}(0)\delta_{ll'}. \tag{40.16}$$

4.1 Testing F_{lmk}-Algorithms 2: Orthogonality

The relation (40.16) can now be used to test algorithms for inclination function computation. Or, vice versa, we can now numerically test the orthogonality itself. In analogy to the discretization (40.4) we also apply Gauss-Legendre quadrature to numerically evaluate (40.16):

$$\sum_{i=1}^{N} F_{lmk}(I_i)F_{l'mk}(I_i)w_i = \frac{1}{2l+1}P^2_{lk}(0)\delta_{ll'}.$$

the right hand side can be accomodated into the Gauss-Legendre weights, such that we arrive at the matrix equivalent $F^{\mathsf{T}}WF = I$, in which the matrix F is defined for constant orders m and k and variable degree l and inclination I:

$$F = \begin{pmatrix} F_{mmk}(I_1) & F_{m+2,m,k}(I_1) & \cdots & F_{Lmk}(I_1) \\ F_{mmk}(I_2) & F_{m+2,m,k}(I_2) & \cdots & F_{Lmk}(I_2) \\ \vdots & \vdots & \ddots & \vdots \\ F_{mmk}(I_N) & F_{m+2,m,k}(I_N) & \cdots & F_{Lmk}(I_N) \end{pmatrix}.$$

Figure 40.6 indeed suggests a unit matrix as result. The off-diagonal entries are *numerically zero*. The question whether this is – up to rounding errors – a unit matrix is answered by Fig. 40.7, which show the diagonal

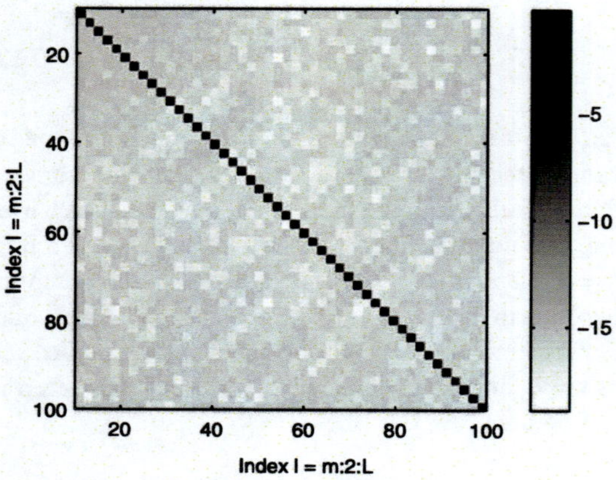

Fig. 40.6 $\log(F^{\mathsf{T}}WF)$ with $m = k = 10$ and l, l' from 10 to 100

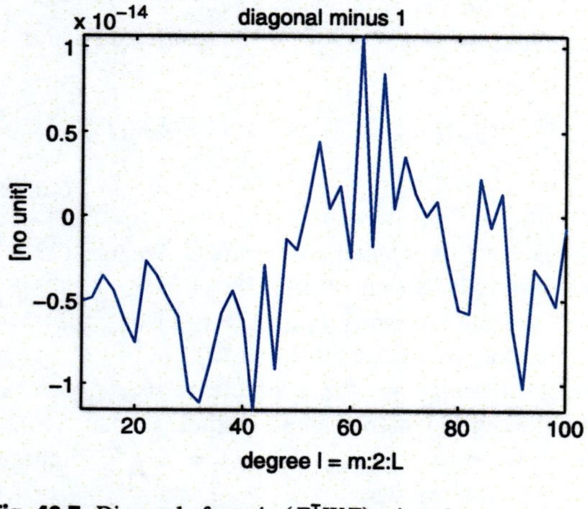

Fig. 40.7 Diagonal of matrix $(F^{\mathsf{T}}WF)$ minus 1

entries minus 1. Both figures successfully demonstrate the validity of the orthogonality relation (40.16).

Conclusion

The properties of completeness and orthogonality – two fundamental properties of function systems – are suggested here for testing algorithms for calculating such function systems. They are complementary properties and, hence, should be used in tandem. One property may have more diagnostic capability in the spectral domain (l, m, k), whereas the other may identify strengths and weaknesses in the spatial domain (θ, I).

This principle is followed here for Legendre functions $P_{lm}(\cos\theta)$ and for inclination functions $F_{lmk}(I)$. The completeness property for Legendre functions is represented by the addition theorem. For inclination functions the Wagner-Gooding invariances play such a role. The orthogonality property for inclination functions, not known in literature so far, was derived here.

Acknowledgements The numerical work of Ms Yang Xue in preparing this manuscript is highly appreciated. The discussions over the past years with R.H. Gooding on the intricate properties of inclination functions have been an inspiration.

References

Edmonds AR (1957) Angular momentum in quantum mechanics. Princeton University Press, Princeton

Gooding RH, Wagner CA (2008) On the inclination functions and a rapid stable procedure for their evaluation together with derivatives. Celest Mech Dyn Astr 101:247–272, DOI 10.1007/s10569-008-9145-6

Sneeuw NJ (1991) Inclination functions: Group theoretical background and a recursive algorithm. Technical report 91.2, TU Delft, Faculty of Geodetic Engineering

Sneeuw NJ (1992) Representation coefficients and their use in satellite geodesy. Manuscr Geod 17:117–123

Stroud A, Secrest D (1966) Gaussian quadrature formulas. Prentice Hall, Englewood Cliffs, NJ

Wagner CA (1983) Direct determination of gravitational harmonics from low-low GRAVSAT data. J Geophys Res 88(B12):10,309–10,321

Properties and Applications of EOF-Based Filtering of GRACE Solutions

41

Siavash Iran Pour and Nico Sneeuw

Abstract

Several filter strategies have been proposed in the past to deal with noise in the GRACE gravity field solutions, particularly with the North-South stripes. These strategies have led to several families of filters, roughly characterized by the properties stochastic vs. deterministic and isotropic vs. anisotropic. One type of filter that has been shown to be very effective is based on Empirical Orthogonal Functions (EOF) analysis in combination with white-noise testing of the resulting time series in the spectral domain (Wouters and Schrama: Geophys. Res. Lett. 34, 2007 doi: 10.1029/2007GL032098). In this paper, we recast the EOF procedure into a filter equation, i.e. the filter transfer is described explicitly. This allows us to emphasize the characteristics of the EOF-based filter. Moreover, our formulation provides an easy means to propagate the GRACE fields into degree variances and spatial covariance functions.

Keywords

EOF filtering • GRACE • Gravity field

1 Introduction

The twin "Gravity Recovery and Climate Experiment" (GRACE) satellites observe the time varying gravity which is the sum of all mass variations in the Earth System. However, the main problem in the GRACE solutions of Stokes coefficients is the increasing noise at higher degrees in the spherical harmonics (SH) coefficients (Swenson and Wahr 2006).The noises appear as unphysical North-South striping patterns in the spatial domain maps (for example in the Equivalent Water Height (EWH) maps) which have a strong correlation between the even and odd degree coefficients. The stripes can be largely suppressed by weighting the SH coefficients by a Gaussian smoothing function, but since the errors have a non-isotropic character, while the function is isotropic, a large smoothing radius is required for removing the stripes which by itself causes a significant loss of information in the GRACE solutions (Chen et al. 2007).

One approach to remove those stripes is using Empirical Orthogonal Functions (EOF). In the EOF approach, the data matrix is decomposed into three matrices: $D = U\Sigma V^{\mathsf{T}}$, where U and V are orthonormal

S.I. Pour (✉) · N. Sneeuw
Institute of Geodesy, University of Stuttgart, 70174 Stuttgart, Geschwister-Scholl-Str. 24D, Germany
e-mail: siavash@gis.uni-stuttgart.de; sneeuw@gis.uni-stuttgart.de

N. Sneeuw et al. (eds.), *VII Hotine-Marussi Symposium on Mathematical Geodesy*, International Association of Geodesy Symposia 137, DOI 10.1007/978-3-642-22078-4_41,

matrices containing the eigenvectors of $D^{\mathrm{T}}D$ and DD^{T}, respectively. Σ is a diagonal matrix which has the singular values corresponding to the eigenvectors (Preisendorfer 1988). The individual modes of the decomposed data matrix (the columns of the eigenvector matrices) are tested for the white noise. The cumulative power spectra of each mode is compared to the cumulative power spectrum of ideal white noise through the Kolmogorov-Smirnov Test (KS test) and the data matrix is reconstructed by using only the eigenvectors and the singular values of the non-white noise modes.

In this contribution, we present the EOF-KS test filtering approach in the spectral domain where EOF filtering is done on the SH coefficients. The filter transfer is described explicitly which allows us to emphasize the EOF-based filter characteristics.

2 Mathematical Model

In our approach, we substitute the EOF decomposition plus KS-test filtering by its equivalent formula $D' = UL\Sigma V^{\mathrm{T}}$, where L is a diagonal matrix with zero diagonal elements when the modes pass the KS test (noises) and one when they do not (signals). However, since $U^{\mathrm{T}}U = I$ (identity matrix), the formula can be rewritten as:

$$D' = ULU^{\mathrm{T}}U\Sigma V^{\mathrm{T}} \tag{41.1}$$

Therefore, we can define the filtered data matrix D' as:

$$D' = ULU^{\mathrm{T}}D = F_u D \tag{41.2}$$

where we call F_u "Left Filter Operator" which:

$$F_u = ULU^{\mathrm{T}} \tag{41.3}$$

In a similar way, we can define "Right Filter Operator" $F_v = VLV^{\mathrm{T}}$ which can be used for our data filtering purpose as: $D' = DF_v$. Through this model, the filtered degree RMS can be also calculated by error propagation law as:

$$Q_{D'} = F_u Q_D F_u^{\mathrm{T}} \tag{41.4}$$

3 EOF Analysis in the Spectral Domain

3.1 Analysis

In the spectral analysis approach, the data are the SH coefficients estimated up to a degree and order 120 in the period of 10/2002 to 05/2008. First, the missing months are interpolated, that is of course not necessary for the EOF analysis but for the construction of PSD in the white noise recognition step which requires continues data sets, then the monthly data are smoothed by different Gaussian smoothing filter radii and sorted in an order-wise way to remove the correlation between the spherical harmonics as a function of degree (Swenson and Wahr 2006). This means that we keep the order of each data matrix fixed. Then, the EOF analysis is separately done on C_{lm} and S_{lm} coefficients matrices as the data matrices. As an example, the following matrix shows the time series of the C_{lm} coefficients for the specific order of m.

$$C_m = \begin{pmatrix} C_{mm}(t_1) & C_{mm}(t_2) & \cdots & C_{mm}(t_n) \\ C_{m+1,m}(t_1) & C_{m+1,m}(t_2) & \cdots & C_{m+1,m}(t_n) \\ \vdots & \vdots & \ddots & \vdots \\ C_{lm}(t_1) & C_{lm}(t_2) & \cdots & C_{lm}(t_n) \end{pmatrix}$$

3.2 Testing

The analysis provides the modes which are tested for the white noise through the KS test for significance level of 5% (Wouters and Schrama 2007). Figure 41.1 shows those modes which are kept for the reconstruction step for the different orders C_{lm} and S_{lm}, also the signal percentage of each order retained after EOF-KS test filtering.

3.3 Filtering

The filter operator F_u is built based on our testing approach explained before. As examples, the operators for two different orders (for C_{lm}) $m = 0$ and $m = 10$

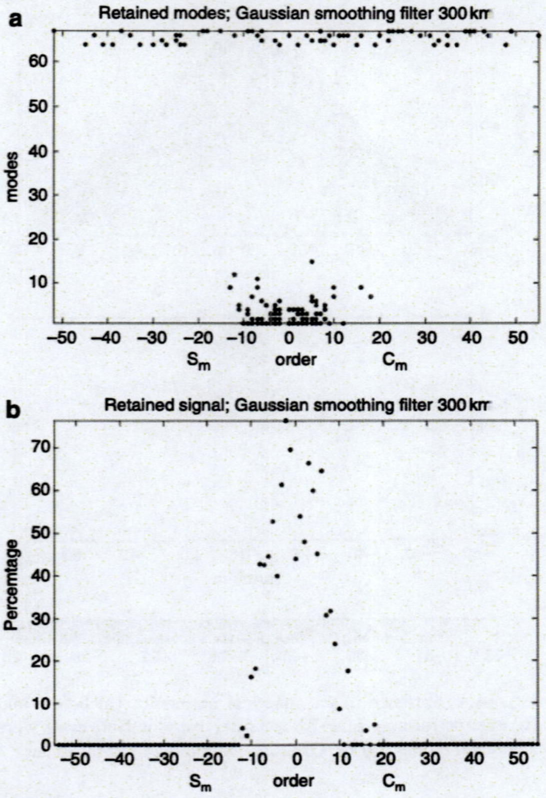

Fig. 41.1 The C_{lm} and S_{lm} modes (a) and the energy (b) pass the KS test after performing 300 km Gaussian filter, $\alpha = 5\%$

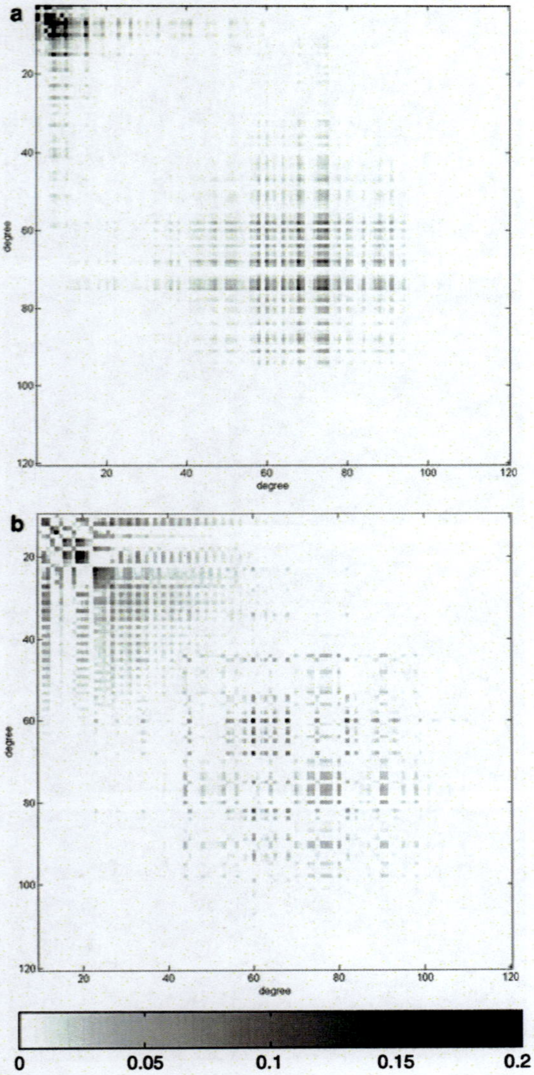

Fig. 41.2 The EOF filter operator F_u for two different orders (for C_{lm}) $m = 0$ (a) and $m = 10$ (b) after performing the Gaussian smoothing filter 300 km

for the 300 km Gaussian smoothing filter are shown in Fig. 41.2. The figures show more significant values for the lower degrees, however some patterns can be seen in the middle.

3.4 Reconstruction

After the reconstruction, the order-wise matrices are again transferred to the monthly data matrices. Then, the reconstructed data are compared to each other for the different smoothing radii. Figure 41.3 shows the EWH anomaly for May 2008 for two different Gaussian smoothing radii, with and without EOF and white noise analysis.

For each Stokes coefficient (K_{lm}) the variance is calculated by the following formula:

$$Var[K_{lm}] = \left[\frac{1}{N} \sum_{i=1}^{N} (K_{lm}(t_i))^2 \right]^{\frac{1}{2}} \tag{41.5}$$

where N is the number of months in the study.

The variance in each spherical harmonic before and after performing EOF-KS test filter operator for the studied period of this contribution are presented in Fig. 41.4.

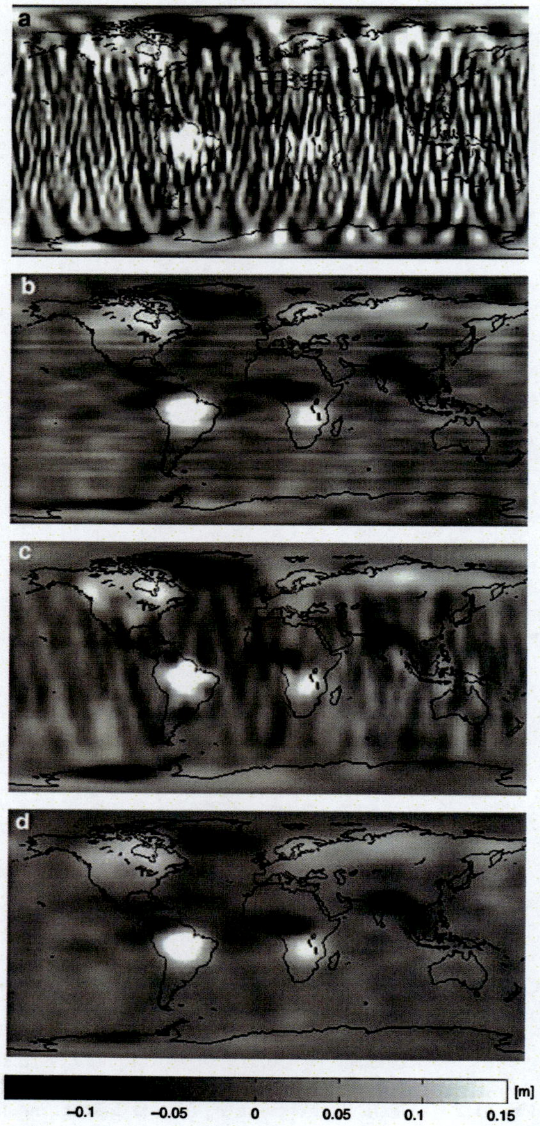

Fig. 41.3 EWH anomaly maps for May 2008, smoothed by (**a**) only Gaussian filter 300 km radius, (**b**) Gaussian filter 300 km radius in addition to EOF-KS test filter, (**c**) only Gaussian filter 500 km radius, (**d**) Gaussian filter 500 km radius in addition to EOF-KS test filter

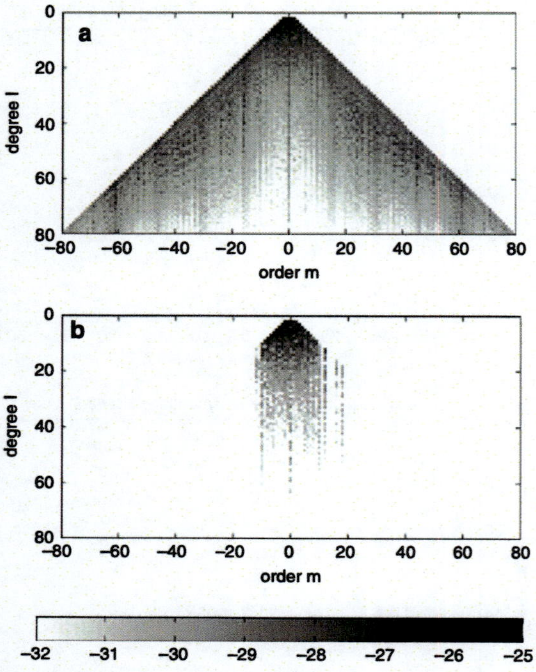

Fig. 41.4 Variance in the spherical harmonics (**a**) before and (**b**) after performing EOF-KS test filter operator. Both cases were smoothed by the 300 km Gaussian smoothing filter in advance

4 Discussion

The reconstructed EWH maps provided by the spectral domain analysis of the GRACE data show a strong capability of EOF analysis for de-striping the patterns which is due to the ability of performing EOF analysis for individual orders (Iran Pour et al. 2009). Also examples in the spectral domain show that the EOF analysis with white noise test but with smaller smoothing Gaussian filters result in maps comparable to those which are just smoothed by the Gaussian smoothing filters but with larger radii. The KS test is used as a tool for white noise recognition for the modes provided by EOF analysis which are performed through the filtering operators. It can be seen (Fig. 41.1 as an example) that some of the last modes of some orders pass the test (recognized as signals). This can be explained by the non-white noise property of the modes, although it is important to mention that those last modes have very small singular values and therefore their contributions to the data reconstruction are also very small. However, the test is claimed for the white noise recognition, while the modes could represent other types of noises which were not studied in this work.

It can also be seen that the EOF analysis with the white noise test smooths the patterns which may diminish the effects of some geophysical phenomena. This means that more investigations are needed for the testing strategies and the noise recognition methods in the future. Also, the properties of spatial covariance functions provided by the mathematical model of this contribution will be the subject to the further research.

References

Chen JL, Wilson CR, Famiglietti JS, Rodell M (2007) Attenuation effect on seasonal basin-scale water storage changes from GRACE time-variable gravity J Geod 81:237–245, doi: 10.1007/s00190-006-0104-2

Iran Pour S, Bentel K, Sneeuw N (2009) EOF-based filtering of GRACE gravity field solutions: A comparison between spectral and spatial approaches Geophysical Research Abstracts, Vol. 11, EGU2009-2853-1, Vienna, Austria 2009

Preisendorfer RW (1988) Principal component analysis in meteorology and oceanography, Elsevier, Amsterdam

Swenson S, Wahr J (2006) Post-processing removal of correlated errors in GRACE data Geophys Res Lett 33, doi: 10.1029/2005GL025285

Wouters B, Schrama EJO (2007) Improved accuracy of GRACE gravity solutions through empirical orthogonal function filtering of spherical harmonics Geophys Res Lett 34, doi: 10.1029/2007GL032098

Numerical Investigation of Different Gravity Models in Orbit Propagation of Two Short CHAMP and GRACE-A Arcs

42

D. Tsoulis and T.D. Papanikolaou

Abstract

Dynamic orbit determination is based on the solution of the equation of motion and the accurate modeling of all forces acting on the satellite. The present contribution aims at investigating the role of a geopotential model which is used to evaluate the Earth's gravitational component in the procedure of dynamic orbit determination. For this reason we focus on the results obtained from the use of different available Earth gravity models in orbit propagation of current Low Earth Orbiters (LEOs). The obtained short arcs are compared with the corresponding Rapid Science Orbit (RSO) datasets. For the contribution of the dynamic component the gravity models JGM-3, EGM96, EIGEN-CHAMP03S, EIGEN-GRACE02S, EIGEN-CG03C, EIGEN-GL04C, EIGEN-GL04S1, EGM2008 and EIGEN-5C are used, while for the RSO information data from the CHAMP (CHAllenging Mini-satellite Payload) and GRACE (Gravity Recovery And Climate Experiment) satellites are applied. Orbit propagation is based on numerical integration of the equation of motion according to the 7th order Runge–Kutta–Nyström method. Differences between the two orbit types attempt to indicate the band-limited behaviour of the Earth's gravity model.

Keywords

Orbit propagation • Earth's gravity models • Runge–Kutta numerical integration

1 Introduction

Dynamic orbit determination is a fundamental tool in satellite orbit analysis. The accuracy of the dynamic approach depends strongly on the modeling errors of the forces acting on the satellite. Among all forces acting on a satellite orbiting the Earth, the major contribution is given by the Earth's static gravity field.

The mathematical relation between gravity field and satellite motion has been the topic of extensive investigations by many authors. First satellite gravity models were determined using almost exclusively Kaula's linear perturbation theory (Kaula 1966) and satellite observations of that time i.e. Doppler and optical observations, while during the last decade numerous new gravity models emerged from the analysis of CHAMP and GRACE data applying the concept of GPS high-low Satellite-to-Satellite Tracking (hl-SST).

D. Tsoulis (✉) · T.D. Papanikolaou
Department of Geodesy and Surveying, Aristotle University of Thessaloniki, University Box 440, 54124, Thessaloniki, Greece
e-mail: tsoulis@auth.gr

N. Sneeuw et al. (eds.), *VII Hotine-Marussi Symposium on Mathematical Geodesy*, International Association of Geodesy Symposia 137, DOI 10.1007/978-3-642-22078-4_42, © Springer-Verlag Berlin Heidelberg 2012

One of the available tools in estimating the accuracy of a gravity model is through satellite orbit fits (Foerste et al. 2008; Tapley et al. 2005). Orbit analysis provides an initial measure regarding the gravity model's evaluation. The present contribution aims at quantifying numerically the dependency between a geopotential model and the corresponding orbit. As we focus on the role of the used geopotential model, all other gravitational and non-gravitational effects are ignored assuming that this does not affect in a major sense the comparison between the orbit differences. We apply orbit propagation to short arcs based only on a known Earth's gravity model and using as LEOs the CHAMP and GRACE-A satellites. The extrapolated orbits are compared with the corresponding Rapid Science Orbits, which in the case of CHAMP are characterized by an accuracy level of 12 cm (Michalak et al. 2003). Although orbits of better quality are available RSO's accuracy is considered satisfactory for the present investigation. Orbit comparison indicates the deviation of each extrapolated orbit arc as a function of the used gravity model. Here, we focus on the relative differences between the extrapolated orbits depending on a variable spherical harmonic coefficients range. The RSO is used in this sense as a common reference orbit.

2 Theoretical Background

The basic mathematical description of a satellite orbit is given by Newton's second law

$$m\ddot{\mathbf{r}} = \mathbf{F}, \tag{42.1}$$

where $\ddot{\mathbf{r}}$ denotes the acceleration vector, m the satellite's mass and $\mathbf{F}$ the sum of all forces acting on the satellite. The Earth's gravitational attraction expresses the major contributor in $\mathbf{F}$, its magnitude exceeding that of all the other forces by several orders of magnitude (Beutler 2005). Equation (42.1) describes the motion of a satellite's center of mass in an inertial reference system while Earth gravity models refer to a certain geocentric terrestrial reference system.

Equation (42.1) is a second-order differential equation and its solution requires analytical or numerical integration methods. Here, we apply an efficient numerical method that belongs to the family of Runge–Kutta–Nyström methods. This class of algorithms is especially designed for the direct integration of second order differential equations and for that reason is preferable for orbit computations. There are also other favourite methods in LEOs analysis, e.g., multistep and collocation methods (Montenbruck and Gill 2000; Beutler 2005). Here a 7th order Runge–Kutta–Nyström method, RKN7(6)-8, has been applied as proposed by Dormand and Prince (1978).

Differences between orbits may refer either to the geocentric reference frame or the orbital frame. The choice of the latter is more suitable for the study of modern methodologies e.g. satellite altimetry, SST tracking or satellite gradiometry. The transformation between the inertial reference frame and the orbital frame is required to express finally the differences in the radial, transverse (along-track) and normal (cross-track) components. The required mathematical scheme can be found in Rosborough (1986), while a more rigorous transformation is presented in Casotto (1993) which has been applied here.

3 Data Processing

In order to initialize the numerical integration, initial values of position and velocity components are required at a single epoch. These values are taken from the RSO data at an arbitrary instance. During the numerical integration the integration interval was set to the constant value of 30 s.

The length of the computed orbit arcs is chosen to be 93 min which is approximately close to the revolution period of CHAMP and GRACE satellites.

Earth orientation parameters (EOP) that describe the Earth's rotation are available by the Earth Orientation Center. EOP are used to compute the transformation matrix between the terrestrial reference frame and the inertial reference frame. The EOP used here were determined according to the IERS Conventions 2003 and derived from the combined solution C04. The interpolation of EOP at the distinct computation times was performed through polynomial (Lagrange) interpolation. Due to the limited interest in high accuracy requirements during this procedure, diurnal and semi-diurnal tidal variations that affect the EOP data were not included in the computations.

The RSO is a dynamic orbit and the used gravity model during the adjustment procedure is GRIM5-C1 (70 × 70) updated with 2 months of CHAMP data (Michalak et al. 2003). The RSO data were provided by the ISDC (Information System and Data Center) of the GeoForschungsZentrum (GFZ) Potsdam. They refer to ITRF96 and are linked to the IERS Conventions 1996 while they are sampled at an interval of 30 s. The discrepancy between the different IERS conventions has a geometric effect on the computed orbit differences. However the magnitude of this disagreement proves to be smaller than the accuracy of the present orbit propagation. In order to quantify this effect we performed computations of a single orbit arc by using different solutions of EOP for the same date, as these were available from IERS through 2007. The comparison of these computations led to differences which varied between 0.1 mm to a few millimeters. Thus, the deviations caused by the aforementioned discrepancy are assumed to be negligible in the frame of the present analysis.

The approach that is described above has been applied to CHAMP and GRACE-A satellites for 8.10.2007 from 00:00 to 01:33 which at that time were at altitudes of 341 km and 467 km, respectively. The incorporation of the different gravity models has been carried out cumulatively, i.e. the computation of the gravitational part for each degree has been performed using all previous coefficients starting from degree 2 with a degree interval equal to 1 for the range $2 < n < 150$ and an interval of 50 for the higher degrees.

Orbit comparison is represented by the root mean squares (RMS) of the differences between the computed GRACE-A and CHAMP arcs and the corresponding RSO data respectively. The RMS variations are provided in the Figs. 42.1–42.4 for the three orbital components as well as in the Tables 42.1 and 42.2 for the radial component. The performed calculations reveal (a) a degree bandwidth (up to $n = 70$) with large differences varying between tens of meters and a few decimeters, (b) a threshold around $n = 120$ for GRACE-A and $n = 150$ for CHAMP above which the orbit differences appear unaltered until the n_{max} of the corresponding model and (c) minimum magnitudes of the differences in the order of few decimeters for specific degree values in the bandwidth $30 < n < 45$.

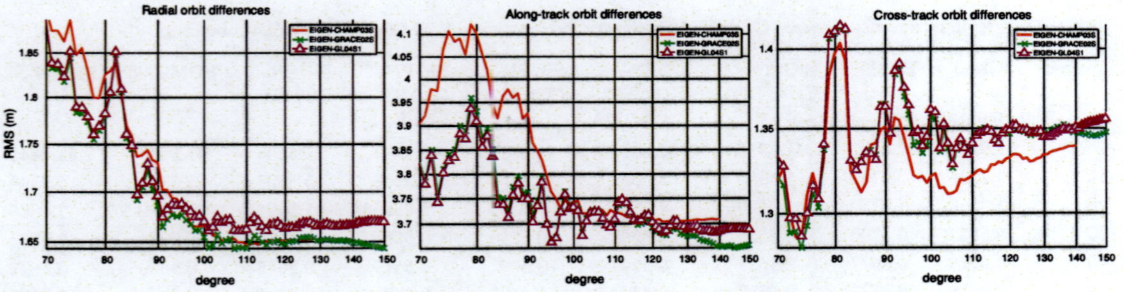

Fig. 42.1 RMS differences between CHAMP propagated orbit arcs and RSO data for satellite-only gravity models

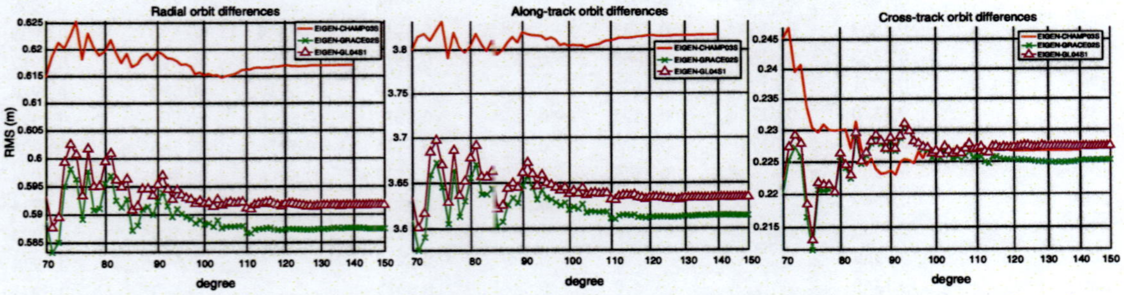

Fig. 42.2 RMS differences between GRACE-A propagated orbit arcs and RSO data for satellite-only gravity models

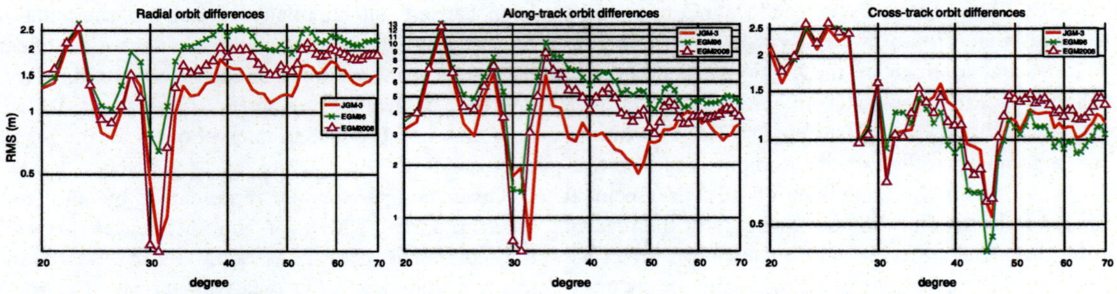

Fig. 42.3 RMS differences between CHAMP propagated orbit arcs and RSO data for combined gravity models

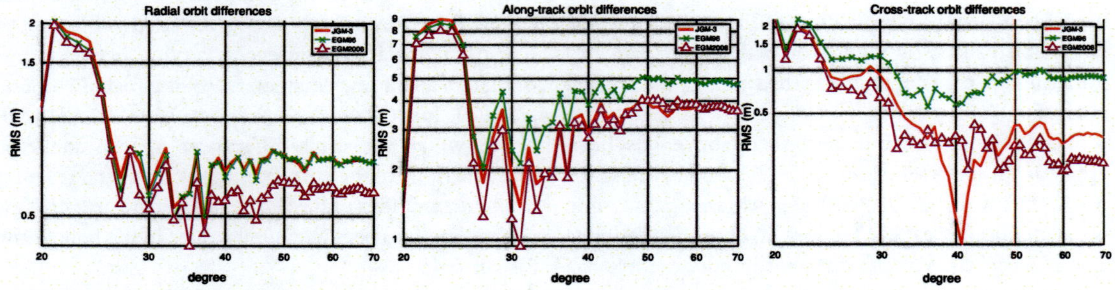

Fig. 42.4 RMS differences between GRACE-A propagated orbit arcs and RSO data for combined gravity models

Table 42.1 RMS differences between GRACE-A propagated orbit arcs and RSO orbit for the radial component

Degree	JGM-3	EGM96	EIGEN-CHAMP03S	EIGEN-GRACE02S	EIGEN-CG03C	EIGEN-GL04C	EIGEN-GL04S1	EGM2008	EIGEN-5C
	Radial RMS differences (m)								
2	41.9266	41.8882	41.8935	41.8825	41.8814	41.8834	41.8834	41.8808	41.8845
10	5.2528	5.3169	5.3308	5.3369	5.3363	5.3353	5.3353	5.3360	5.3349
20	1.0960	1.0752	1.0580	1.0587	1.0589	1.0586	1.0586	1.0580	1.0586
30	0.5882	0.5796	0.5295	0.5288	0.5280	0.5284	0.5284	0.5271	0.5283
40	0.6929	0.6637	0.5628	0.5597	0.5602	0.5619	0.5619	0.5600	0.5622
50	0.7818	0.7732	0.6587	0.6392	0.6402	0.6421	0.6421	0.6410	0.6427
60	0.7622	0.7570	0.6395	0.6122	0.6148	0.6167	0.6167	0.6156	0.6172
70	0.7410	0.7400	0.6150	0.5890	0.5915	0.5933	0.5933	0.5923	0.5938
80		0.7403	0.6198	0.5956	0.5981	0.5995	0.5995	0.5983	0.6000
90		0.7404	0.6196	0.5924	0.5934	0.5954	0.5954	0.5944	0.5959
100		0.7367	0.6152	0.5883	0.5895	0.5919	0.5919	0.5907	0.5925
110		0.7363	0.6162	0.5868	0.5833	0.5911	0.5911	0.5900	0.5917
120		0.7371	0.6170	0.5874	0.5895	0.5919	0.5918	0.5907	0.5926
130		0.7370	0.6170	0.5875	0.5894	0.5917	0.5916	0.5906	0.5925
140		0.7369	0.6171	0.5876	0.5893	0.5916	0.5916	0.5905	0.5924
150		0.7369		0.5875	0.5894	0.5917	0.5916	0.5905	0.5924
200		0.7369			0.5894	0.5917		0.5905	0.5924
250		0.7369			0.5894	0.5917		0.5905	0.5924
300		0.7369			0.5894	0.5917		0.5905	0.5924
360		0.7369			0.5894	0.5917		0.5905	0.5924

Table 42.2 RMS differences between CHAMP propagated orbit arcs and RSO orbit for the radial component

Degree	JGM-3	EGM96	EIGEN-CHAMP03S	EIGEN-GRACE02S	EIGEN-CG03C	EIGEN-GL04C	EIGEN-GL04S1	EGM2008	EIGEN-5C
	Radial RMS differences (m)								
2	114.0916	114.099	114.0967	114.1034	114.1049	114.103	114.103	114.1064	114.1018
10	8.3415	8.3593	8.3131	8.3151	8.3154	8.315	8.315	8.3155	8.3147
20	1.3111	1.3478	1.5463	1.5446	1.542	1.5425	1.5425	1.5408	1.5432
30	0.4729	0.7857	0.2269	0.2264	0.2286	0.2277	0.2277	0.2285	0.2277
40	1.6808	2.3628	1.8511	1.8569	1.8624	1.8606	1.8606	1.8591	1.8612
50	1.2333	2.0211	1.6374	1.6140	1.6211	1.6174	1.6174	1.6155	1.6181
60	1.7386	2.3328	2.0096	1.9870	1.9952	1.9925	1.9925	1.9896	1.9931
70	1.5093	2.2336	1.9040	1.8842	1.8936	1.8889	1.8889	1.8866	1.8889
80		2.1268	1.8259	1.7809	1.7886	1.7829	1.7829	1.7847	1.7836
90		2.0392	1.7305	1.6863	1.6983	1.6961	1.6961	1.6940	1.6963
100		2.0068	1.6663	1.6549	1.6703	1.6704	1.6704	1.6648	1.6701
110		2.0021	1.6475	1.6438	1.6664	1.6626	1.6627	1.6561	1.6612
120		2.0057	1.6513	1.6481	1.6702	1.6652	1.6636	1.6573	1.6635
130		2.0092	1.6531	1.6513	1.6732	1.6679	1.6658	1.6603	1.6661
140		2.0090	1.6522	1.6492	1.6730	1.6675	1.6670	1.6594	1.6657
150		2.0117		1.6440	1.6753	1.6698	1.6701	1.6622	1.6684
200		2.0106			1.6742	1.6687		1.6611	1.6671
250		2.0106			1.6742	1.6688		1.6611	1.6672
300		2.0106			1.6742	1.6688		1.6611	1.6672
360		2.0106			1.6742	1.6688		1.6611	1.6672

4 Concluding Remarks

The main feature of the presented computational scheme is that it permits the evaluation of a cumulative contribution of an available gravity model in the process of numerical integration of a real LEO orbit. This results to an independent assessment tool for the performance of a given gravity model expressed at all three orbital components (radial, cross-track, along-track) at satellite altitude. The incorporation of different models led to significant numerical discrepancies for the same orbital arc. Furthermore, a major difference between CHAMP and GRACE-A could be pointed out in terms of the observed threshold, above which the orbit differences remained unaltered. It is clear, however, that significant extensions of the present approach have to be performed in order to enable a more detailed or even band-limited analysis of the different gravity models. Due to the simplification of the current forces model the obtained results represent a rather pessimistic measure for the quality of the different geopotential models. Among the issues to be tackled in the frame of our ongoing and future work we should mention (a) the examination of more and longer orbit arcs of different satellites scattered over larger epochs including seasonal extremes of characteristic non-gravitational effects such as solar radiation and (b) the extension of the present elementary model towards a real force model, by adding non-gravitational forces and third-body perturbations, which would enhance the interpretation goals of the present approach.

Acknowledgements The authors acknowledge the ISDC data center for providing the RSO data files (http://isdc.gfz-potsdam.de/), Earth Orientation Center for providing EOP data (http://hpiers.obspm.fr/eop-pc/) and the International Centre for Global Earth Models (ICGEM) for providing the various Earth's gravity models. Financial support by the European Space Agency under Project ESTEC/22316/09/NL/CBI is gratefully acknowledged. The first author acknowledges further a Research Fellowship by the Alexander von Humboldt Foundation and his host Professor Nico Sneeuw at the Geodetic Institute, University of Stuttgart.

References

Beutler G (2005) Methods of celestial mechanics – Vol.I, Vol.II. Springer, Berlin

Casotto S (1993) Position and velocity perturbations in the orbital frame in terms of classical elements perturbations. Cel Mech Dynam Astron 55:209–221

Dormand JR, Prince PJ (1978) New Runge-Kutta algorithms for numerical simulation in dynamical astronomy. Cel Mech Dynam Astron 18:223–232

Foerste C et al (2008) The GeoForschungsZentrum Potsdam/Groupe de Recherche de Gèodésie Spatiale satellite-only and combined gravity field models: EIGEN-GL04S1 and EIGEN-GL04C. J Geodesy 82:331–346

Kaula WM (1966) Theory of satellite geodesy. Dover Publications, New York

Michalak G, Baustert G, König R, Reigber C (2003) CHAMP rapid science orbit determination - status and future prospects. In: Reigber C, Lühr H, Schwintzer P (eds) First CHAMP mission results for gravity, magnetic and atmospheric studies. Springer, Berlin, pp 98–103

Montenbruck O, Gill E (2000) Satellite orbits - models, methods and applications. Springer, Berlin

Rosborough GW (1986) Satellite orbit perturbations due to the geopotential, Rep.CSR-86–1, Center for Space Research, University of Texas at Austin

Tapley B et al (2005) GGM02 - an improved Earth gravity field model from GRACE. J Geodesy 79:467–478

Different Representations of the Time Variable Gravity Field to Reduce the Aliasing Problem in GRACE Data Analysis

43

Torsten Mayer-Gürr, Enrico Kurtenbach, and Annette Eicker

Abstract

The projected accuracy of the GRACE satellite mission has not been reached yet. One reason among others is the inaccurate modelling of the temporal variations in the analysis procedure by monthly or weekly mean fields, which can be shown in a simple simulation scenario. Two approaches to improve the temporal modeling are presented here: on the one hand the representation in terms of continuous temporal basis functions and on the other hand the increase of the temporal resolution to daily gravity field solutions by the use of the Kalman filter approach.

Keywords

GRACE • Time-variable gravity field • Aliasing • Kalman filter

1 Introduction

For more than 7 years the twin satellite mission GRACE (Tapley et al. 2004) has measured the Earth's gravity field and its temporal variations with unprecedented accuracy. But the projected accuracy (baseline accuracy) has not been reached yet, which is on the one hand due to errors and inaccuracies in the background models used to reduce high-frequency temporal variations within the standard GRACE gravity field recovery process. On the other hand, inaccurate modelling of the temporal variations by monthly mean fields also leads to the typical GRACE error striping pattern, which will further be investigated in this article. The influence of an inaccurate representation in the time domain will be shown in Sect. 2. Hereupon two alternative approaches to improve the temporal GRACE solutions will be presented: on the one hand the representation by continuous temporal basis functions and on the other hand an increase of the temporal resolution up to daily gravity field solutions.

T. Mayer-Gürr (✉) · E. Kurtenbach · A. Eicker
Institute of Theoretical Geodesy and Satellite Geodesy, Graz University of Technology Steyrergasse 30/III, 8010 Graz, Austria
e-mail: mayer-guerr@tugraz.at; kurtenbach@geod.uni-bonn.de; eicker@geod.uni-bonn.de

2 Simulation

To investigate the effect of temporal aliasing we simulated a GRACE scenario. ITG-Grace03s serves as pseudo real gravity field up to degree $n = 60$. In this simulation it is assumed that high frequency variations generated by mass transports in the atmosphere and the ocean are exactly known. Therefore, these variations can perfectly be removed from the data by models and are not included in the observations anymore. The simulated observations are corrupted by white noise

N. Sneeuw et al. (eds.), *VII Hotine-Marussi Symposium on Mathematical Geodesy*, International Association of Geodesy Symposia 137, DOI 10.1007/978-3-642-22078-4_43,

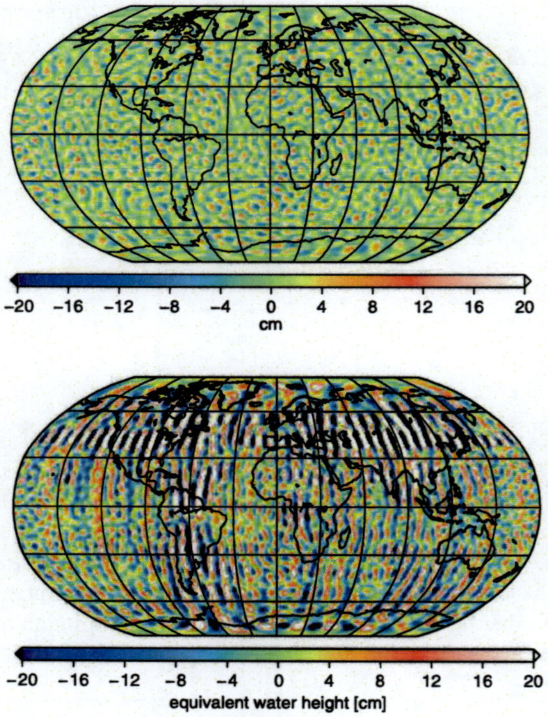

Fig. 43.1 Errors in a simulated GRACE gravity field solution in terms of equivalent water heights up to degree $n = 60$. *Top*: recovery of a static gravity field, *Bottom*: recovery of a field inlcluding annual signal

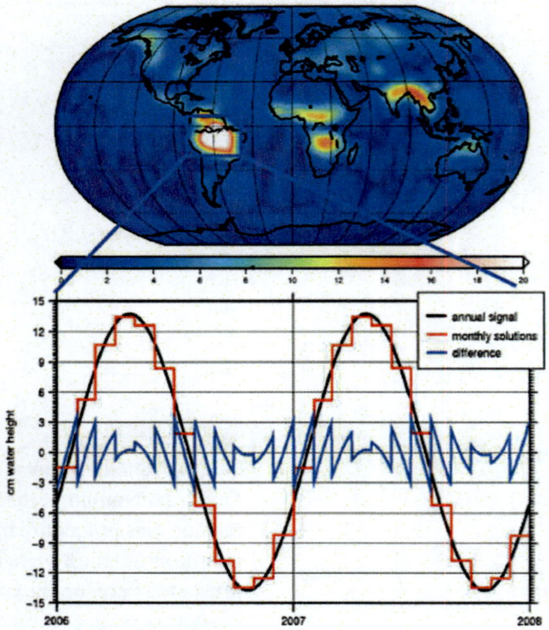

Fig. 43.2 Simulated annual signal. *Top*: amplitudes in terms of equivalent water heights. *Bottom*: time variations in the amazon basin

with a standard deviation of $1\,\mu\text{m/s}$ for the range-rates, 2 cm for the kinematic orbits, and $3 \cdot 10^{-10}\,\text{m/s}^2$ for the accelerometer observations. In a first step a gravity field is estimated from 1 month of data. The same method is used as for the gravity field solution ITG-Grace03, see (Mayer-Gürr et al. 2007) and (Mayer-Gürr 2006). For this setting the acccuracy expected before GRACE launch (baseline accuracy) can be recovered and the noise in the solution is very small. Figure 43.1 (top) shows the difference between the recovered solution and the pseudo real field in terms of equivalent water heights. In the next step the same simulation is repeated, the only difference is that some time variable gravity field signal was added to the observations. To keep the simulation as simple as possible only an annual signal is considered. This signal is derived from the time variable part of the ITG-Grace03 solution and is smoothed spatially to avoid small features that cannot be recovered by the GRACE mission. Figure 43.2 (top) shows the spatial distribution of the amplitudes in terms of equivalent water heights. The errors of the recovered solution are shown in Fig. 43.1. The errors are significantly larger compared to the recovery of the static solution and show the striping pattern as it is well known from the real GRACE solutions. We would like to point out that only a signal changing slowly in time (i.e. the introduced annual variations) causes these stripes and no high frequency signal is included in the simulation. To explain this surprising result the time variations in the Amazon basin are investigated (see Fig. 43.2, bottom). The simulated input signal is shown in black. The estimated GRACE solutions in terms of monthly mean values are displayed in red. The difference between the true signal and the recovered solution (blue line) can reach more than 3 cm of equivalent water height in this region. The annual signal contains a large amount of information which cannot be explained by monthly mean solutions and therefore results in aliasing errors appearing in the striping pattern. We conclude from the result of this simulation scenario that a parameterization of the time variable gravity field by monthly means is not an adequate choice.

3 Representation by Splines in Time Domain

The gravity field representation by a mean value over a certain time span (e.g. monthly or weekly) can be expressed as a linear combination of piecewise constant functions,

$$V(\lambda,\theta,r,t) = \sum_i V_i(\lambda,\theta,r)\Psi_i(t), \qquad (43.1)$$

where Ψ_i is equal to one only in the specific time span:

$$\Psi_i(t) = \begin{cases} 1 & \text{if } t_i \leq t < t_{i+1} \\ 0 & \text{otherwise.} \end{cases} \qquad (43.2)$$

The position-dependency $V_i(\lambda,\theta,r)$ can be expressed by a series of spherical harmonics,

$$V_i(\lambda,\theta,r) = \frac{GM}{R}\sum_{n=0}^{\infty}\left(\frac{R}{r}\right)^{n+1}\sum_{m=-n}^{n} c_{nm}Y_{nm}(\lambda,\theta), \qquad (43.3)$$

where λ, θ, r are the spherical coordinates; GM is the Earth's gravitational constant; R is the semi-major axis of a reference ellipsoid; c_{nm} are the spherical harmonic coefficients; and Y_{nm} are the fully normalized spherical harmonics. Alternatively the potential can be represented by an expansion in terms of space-localizing basis functions, see for example (Eicker 2008).

However, a representation in terms of mean values is not a suitable choice as has been shown in Sect. 2. Most of the geophysical processes causing mass changes have a continuous character so the choice of basis function $\Psi_i(t)$ with a smooth transition seems to be more natural. In case of ITG-Grace03s (Mayer-Gürr et al. 2007), the temporal variations are parameterized by quadratic splines,

$$\Psi_i(t) = \begin{cases} \frac{1}{2}\tau_{i-1}^2 & \text{if } t_{i-1} \leq t < t_i, \\ -\tau_{i-1}^2 + \tau_{i-1} + \frac{1}{2} & \text{if } t_i \leq t < t_{i+1}, \\ \frac{1}{2}\tau_{i-1}^2 - \tau_{i-1} + \frac{1}{2} & \text{if } t_{i+1} \leq t < t_{i+2}, \\ 0 & \text{otherwise,} \end{cases} \qquad (43.4)$$

where $\tau = (t - t_i)/(t_{i+1} - t_i)$ is the normalized time. These quadratic splines are illustrated in Fig. 43.3.

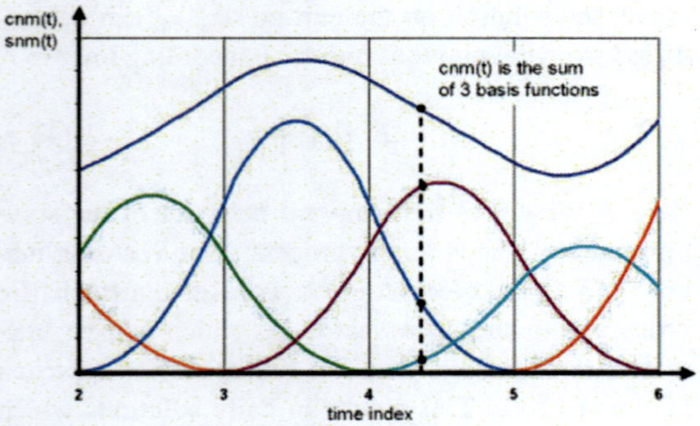

Fig. 43.3 Schematic illustration of quadratic splines as continuous temporal basis functions

The drawback of this approach is the fact that only geophysical processes with a smooth behaviour in time can well be approximated. For example the effect of the Sumatra-Andaman earthquake does not belong to this type of processes.

4 The Kalman Filter Approach

Instead of using smooth functions to approximate continous processes one could decrease the analysed time span to recover fast time variations as detailed as possible (both smooth and discontinous processes). The existing solutions are for example 10-day fields (Lemoine et al. 2007) or even weekly fields (Flechtner et al. 2009). In the approach presented here we carry the idea to extremes and try to derive daily solutions.

The processing of GRACE measurements (K-band range data, GPS and accelerometer observations) for one day can be formulated in a Gauss Markov Model according to

$$l_t = A_t x_t + v_t, \qquad (43.5)$$

where l_t is the vector containing the GRACE measurements, x_t are the unknown gravity field parameters (in this case spherical harmonic coefficients), A_t is the design matrix and v_t contains the measurement noise for each day t.

Solving the normal equation for each day individually provides dissatisfactory results due to insufficient data coverage. In order to nevertheless reach such a high temporal resolution, another piece of information is needed to stabilize the solution. Assuming that the gravity field parameters cannot change in an arbitrary range, the solution on the current day x_t can be predicted from the previous one x_{t-1} according to

$$x_t = B_t x_{t-1} + w_t. \tag{43.6}$$

Here B_t describes the temporal behavior of the solutions and w_t stands for the process noise vector. Equations (43.5) and (43.6) together constitute a stochastic process with the state vector x_t, which can be estimated, for example, within a Kalman filter approach (Kalman 1960). This results in daily solutions which are not completely independent but for each day only the areas covered by GRACE observations are updated.

As a first approximation B_t can be chosen as the identity matrix when expecting that the gravity field coefficients remain nearly constant from one day to the next. Whereas strictly complying with this assumption would lead to a static field, the introduction of the process noise vector allows variations within a certain range. The stochastic behaviour of this zero-mean noise vector, characterized by its covariance matrix $Q_t = Cov(w_t, w_t)$, is accounted for in the Kalman filter approach and can be derived empirically by the analysis of geophysical models.

It has to be pointed out that in the GRACE estimation process only this empirical covariance matrix Q_t is taken into account, but not the model results themselves. Thus the estimated gravity field solution will not be biased towards the applied geophysical model. Inaccuracies in the model are therefore less critical.

As in the GRACE analysis process short-term mass variations such as atmospheric mass redistributions or ocean tides are reduced in a de-aliasing step (Flechtner et al. 2009), the residual signal is mainly dominated by variations in continental hydrology (and, in specific regions, ice mass variations or long-term processes such as postglacial rebound and seismic activities), see for example (Güntner 2008).

Following the method described in (Kurtenbach et al. 2009) the empirical auto-covariance function of the WaterGAP Global Hydrology Model (WGHM), given as as daily equivalent water heights (EQWH) over the continents on a 0.5° × 0.5° grid (for details see Döll et al. 2003 and Hunger and Döll 2008), is derived.

The resulting auto-covariance function is a two-dimensional function depending on time and on the spatial distance. In Fig. 43.4 the temporal correlations are displayed for different spatial distances (indicated

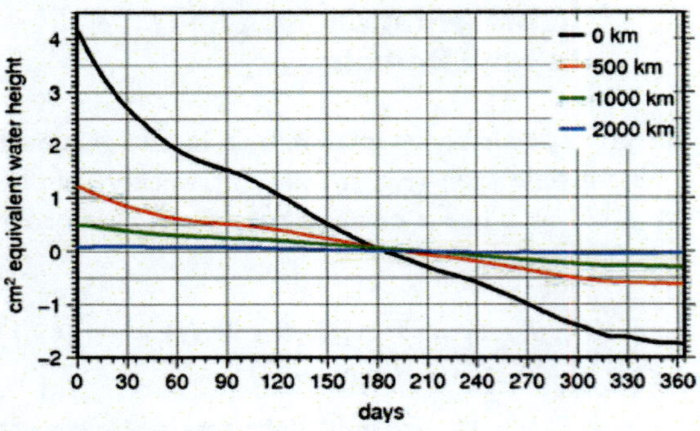

Fig. 43.4 Empirical covariance function of the residual stochastic part of the signal (annual and semi-annual signal reduced)

by different colors) on the surface of the sphere. The deterministic part representing the annual and semi-annual period was reduced and only the residual stochastic part is shown. The covariance matrix of the process noise can now be derived from the empirical covariance function. Since for daily snapshots the variations of the Earth's gravity field from one day to the next are essential, the covariance function is evaluated at the time step $\Delta t = 1$ day.

To show the applicability of the Kalman filter approach 2 years of GRACE L1B data were processed, i.e. the years 2006 and 2007. The process is initialized with the static solution ITG-Grace03 and the analysis procedure equals the one applied in the calculation of the ITG-Grace03s time series, (Mayer-Gürr et al. 2007). Results of the Kalman solution are presented in Fig. 43.5, where the spatial distribution of the temporal variances of the daily solutions are displayed for the time span 2006–2007 (top). They are compared to the corresponding variances of the GFZ-RL04 weekly (middle) and monthly (bottom) gravity field models. The weekly models are provided up to degree and order 30 and the monthly models up to degree and order 120. For further details on theses two solutions, see Flechtner et al. (2009). Because of the implicit filtering taking place during the Kalman filter estimation, the unfiltered GFZ solutions had to be filtered as well. Therefore, to both GFZ-RL04 solutions a non-isotropic decorrelation filter (DDK), see Kusche (2007), was applied. Even though a direct comparison of the results is difficult due to the different filtering techniques, all three illustrations depict very similar spatial patterns. In the large river basins the

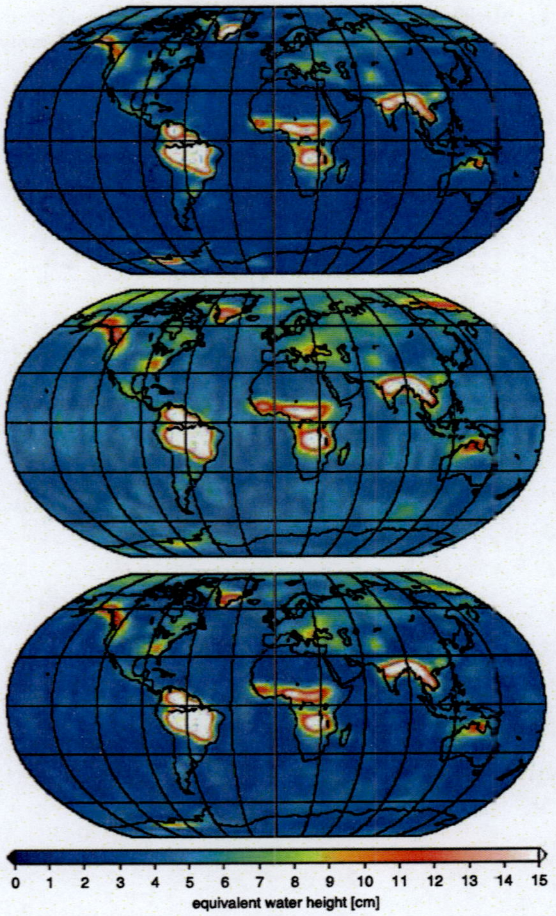

Fig. 43.5 Temporal variability in terms of RMS values of different GRACE solutions. *Top*: daily Kalman solutions. *Middle*: weekly GFZ-RL04 solutions. *Bottom*: monthly GFZ-RL04 solutions. Both GFZ solutions filtered with non-isotropic decorrelation filter (DDK)

daily solutions show a variability comparable to the weekly or monthly solutions. This indicates that by using the Kalman filter process the signal is not damped more strongly than in case of applying the DDK filter.

Conclusion

It could be shown that slowly changing temporal variations cannot be well approximated in the GRACE gravity field recovery process by the standard monthly mean fields. Inadequate representation leads to aliasing errors and the typical GRACE striping pattern. In this article two approaches to reduce this problem were presented. Because most of the geophysical processes observed by GRACE are continuous in time, it is obvious to use continuous temporal basis functions to model these variations. To capture also discontinuous processes, the temporal resolution was increased up to daily gravity field solutions. Solving the normal equation for each day individually provides unsatisfactory results due to insufficient data coverage. Instead, to reach such a high temporal resolution, another piece of information in terms of temporal correlations were introduced in the Kalman filter approach. The first approach has successfully been used in practise (Mayer-Gürr et al. 2007), the principal applicability of the Kalman filter to GRACE L1B data analysis was shown in (Kurtenbach et al. 2009).

These considerations will gain even more importance in case of future satellite missions featuring significantly higher measurement accuracy.

References

Döll PF, Kaspar F, Kaspar B (2003) A global hydrological model for deriving water availability indicators: model tuning and validation, J Hydrol 270:105–134

Eicker A (2008) Gravity Field Refinements by Radial Basis Functions from In-situ Satellite Data. Dissertation University of Bonn.

Flechtner F, Dahle Ch, Neumayer KH, Koenig R, Foerste Ch (2009) The release 04 CHAMP and GRACE EIGEN gravity field models. In: Flechtner F, Gruber T, Guentner A, Mandea M, Rothacher M, Wickert J (eds) Satellite geodesy and earth system science – observation of the earth from space. Springer, Berlin, in preparation

Güntner A (2008) Improvement of global hydrological models using GRACE data. Surv Geophys doi:10.1007/s10712-008-9038-y

Hunger M, Döll P (2008) Value of river discharge data for global – scale hydrological modeling. Hydrol Earth Syst Sci 12:841–861

Kalman RE (1960) A new approach to linear filtering and prediction problems. Trans ASME Series D J Basic Eng 82:35–45

Kurtenbach E, Mayer-Gürr T, Eicker A (2009) Deriving daily snapshots of the Earth's gravity field from GRACE L1B data using Kalman filtering. Geophys Res Lett 36:L17102, doi:10.1029/2009GL039564

Kusche J (2007) Approximate decorrelation and non-isotropic smoothing of time-variable GRACE-type gravity field models. J Geodes 81:733–749

Lemoine J-M, Bruinsma S, Loyer S, Biancale R, Marty J-C, Perosanz F, Balmino G (2007) Temporal gravity field models inferred from GRACE data. Adv Space Res 39:1620–1629, doi:10.1016/j.asr.2007.03.062

Mayer-Gürr T (2006) Gravitationsfeldbestimmung aus der Analyse kurzer Bahnboegen am Beispiel der Satellitenmissionen CHAMP und GRACE. Dissertation at the University of Bonn.

Mayer-Gürr T, Eicker A, Ilk KH (2007) ITG-Grace02s: A GRACE gravity field derived from short arcs of the satellites orbit. In: Proceedings of the 1st International Symposium of the International Gravity Field Service "Gravity Field of the Earth", Istanbul, 193–198

Tapley BD, Bettadpur S, Watkins M, Reigber Ch (2004) The gravity recovery and climate experiment: mission overview and early results. Geophys Res Lett 31:L09607

Part VII

Earth Oriented Space Techniques and their Benefit for Earth System Studies

44

Multi-Sensor Monitoring of Low-Degree Gravitational Changes

J.L. Chen and C.R. Wilson

Abstract

Earth gravity change is caused by mass redistribution within the Earth system, including air and water redistribution in the atmosphere, ocean, land, and cryosphere, and mass variation of the solid Earth (in the core, mantle, and crust). Gravity change can be quantified by geodetic measurements and numerical climate models. We estimate time series of low-degree gravitational variations, ΔC_{21}, ΔS_{21}, and ΔC_{20} using four different techniques, from the Gravity Recovery and Climate Experiment (GRACE), Earth Orientation Parameters (length of day and polar motion), advanced climate models (including atmospheric, oceanic, and hydrologic models), and satellite laser ranging. We compare these independent estimates at different time scales, and discuss major uncertainties for the various techniques. Independent estimates of ΔC_{21}, ΔS_{21}, and ΔC_{20} are important for validating the geodetic techniques and for improving understanding of large scale and low frequency mass redistribution within the Earth system.

Keywords

GRACE • Low degree gravity • Earth rotation • Climate models • SLR

1 Introduction

This study examines global aspects of space geodetic data and data-assimilating climate models by comparing observations and estimates of gravity field degree 2 spherical harmonic (SH) variations at time scales of a month and longer. Variations in all coefficients (except degree 2, order 2) are determined from three independent space geodetic observational data sets (GRACE gravity mission; Earth Orientation Parameters (EOP); Satellite laser ranging to multiple satellites (SLR)) and from climate model predictions of atmospheric mass redistribution, barotropic wind-driven ocean mass redistribution, and land surface hydrology (excluding polar regions). SLR has been an effective technique for measuring low degree gravitational changes, especially the lowest degree even zonal harmonic ΔC_{20} (also called –J2 in the literature) with time series extending over more than three decades (e.g. Yoder et al. 1983; Rubincam 1989; Cheng and Tapley 2004).

J.L. Chen (✉)
Center for Space Research, University of Texas, Austin, TX 78712, USA
e-mail: chen@csr.utexas.edu

C.R. Wilson
Department of Geological Sciences, Jackson School of Geosciences University of Texas, Austin, TX 78712, USA
e-mail: crwilson@mail.utexas.edu

N. Sneeuw et al. (eds.), *VII Hotine-Marussi Symposium on Mathematical Geodesy*, International Association of Geodesy Symposia 137, DOI 10.1007/978-3-642-22078-4_44, © Springer-Verlag Berlin Heidelberg 2012

Excitations of polar motion (X, Y) and length-of-day (LOD) due to surface mass load variations are proportional to changes in degree-2 SH (i.e., ΔC_{21}, ΔS_{21}, and ΔC_{20}) (e.g., Wahr 1982; Eubanks 1993; Gross et al. 2004a). Therefore, estimates of ΔC_{21}, ΔS_{21}, and ΔC_{20} from accurately measured Earth rotational changes are possible, provided that wind and ocean current contributions can be estimated using numerical ocean and atmospheric models. GRACE satellite gravimetry provides another means for measuring these low degree gravitational changes (Tapley et al. 2004).

This investigation builds upon earlier studies (Chen et al. 2004; Chen and Wilson 2003, 2008) by extending the period of observation, and employing improved estimates. Modern observations of mass redistribution from the GRACE mission, and estimates of climate-related sources from data assimilating models are only available for the last few years, since 2002 in the case of GRACE, so even a few additional years is significant. Despite their short duration, GRACE and other recent observations offer a chance to understand causes of low frequency EOP changes in the overlapping period of the last few years. In principle, this should allow the longer EOP record to serve as a proxy for climate change that extends many decades into the past.

The paper begins with a summary of data sources and data processing. Time series sampling is dictated by GRACE data to be at approximately monthly intervals. This requires filtering and decimation of other space geodetic and climate series, available more frequently. For example, the EOP series are daily, and climate model time series provide several samples each day. For the four estimates (GRACE, EOP, climate models, SLR), there are 3 time series each for SH degree 2, order 0, and order 1 (cosine and sine). These are compared at time scales of less than 1 year (intra-annual), 6 months and 1 year (seasonal) and at greater than 1 year (interannual).

2 Data Processing

2.1 EOP Estimates

Earth Orientation Parameters (EOP: length of day and polar motion) can be converted to standard excitation functions, which in turn are related to changes in degree 2 SH coefficients (Chen and Wilson 2003).

$$\Delta C_{21} = -(1+k_2') \cdot \sqrt{\frac{3}{5}} \cdot \frac{(C-A)}{1.098 R^2 M} \cdot \chi_1^{mass}$$
$$\Delta S_{21} = -(1+k_2') \cdot \sqrt{\frac{3}{5}} \cdot \frac{(C-A)}{1.098 R^2 M} \cdot \chi_2^{mass}$$
$$\Delta C_{20} = -(1+k_2') \cdot \frac{3}{2\sqrt{5}} \cdot \frac{C}{0.753 R^2 M} \cdot \chi_3^{mass} \tag{44.1}$$

Here R is Earth radius, M is Earth mass, (C, A) are polar and equatorial moments of inertia, and k_2' is load Love number of degree 2. χ_1, χ_2, and χ_3 are excitation functions of polar motion (X, Y) and LOD, respectively. These relations between EOP excitations and ΔC_{21}, ΔS_{21}, and ΔC_{20} require that angular momentum effects of winds and ocean currents are first removed from EOP variations. That is, observed excitation functions must be corrected by subtracting the portion not due to mass redistribution:

$$\chi_i^{mass} = \chi_i^{observed} - \chi_i^{wind} - \chi_i^{current} \qquad i = 1, 2, 3 \tag{44.2}$$

Wind and current effects must be estimated from global atmospheric and oceanic data assimilating numerical models. As the results below demonstrate, this can be done reasonably well, even for χ_3 for which about 90% of the variance is due to the winds (Gross et al. 2004b).

EOP time series are IERS C04 daily values, converted to $\chi_i^{observed}$. Then the wind contribution is removed from each with daily NCEP reanalysis winds from surface to 10 mbar, for 17 layers (Kalnay et al. 1996). After this, the series are smoothed by a 11-day moving average filter (with the weighting of the 1st and last days as 0.5 and others as 1) and decimated at 10-day intervals (the same as ocean model data), and ocean current effects (from horizontal velocities) are removed using an ECCO data assimilating 46 layers ocean model (run kf066b) (Fukumori et al. 1999, 2000). More information on computing atmospheric wind and ocean current excitations is available in Chen and Wilson (2008). Finally, the 10-day samples are interpolated into daily intervals, smoothed by a 31-day moving average filter, and then decimated to monthly values corresponding to GRACE sample times.

2.2 GRACE and Climate Model Estimates

Several GRACE gravity field solutions are available for study. Here we use Release 4 (RL04) series from the University of Texas Center for Space Research (CSR) and the German (Potsdam) GFZ series. The CSR series covers the period April 2002 to February 2009 and the GFZ series from August 2002 to February 2009. Solutions are approximately, but not exactly at monthly intervals.

There are three main contributions of the climate to changes in the gravity field: mass redistribution within the oceans; within the atmosphere; and due to terrestrial water storage variations. The first two of these are calculated from numerical models as part of the GRACE processing system at 3-hour intervals and removed from GRACE range-rate data in a step called "dealiasing" (Bettadpur 2007). We restore atmospheric and oceanic contributions using the GAC products provided by the GRACE project (Bettadpur 2007). After restoring atmospheric and oceanic contributions, GRC (CSR or GFZ) values for ΔC_{21}, ΔS_{21}, and ΔC_{20} should include all climate sources, and be comparable to other results, such as EOP and SLR estimates.

An independent estimate of terrestrial water storage effects is obtained from the NASA Global Land Data Assimilation System (GLDAS-NOAH), which includes soil moisture and snow contributions, but omits deeper storage change in groundwater, and excludes Greenland and Antarctica (Rodell et al. 2004). We impose mass conservation by distributing evenly over the oceans any GLDAS water surplus or deficit over land. The resulting Atmosphere-Ocean-Water time series, the sum of GLDAS and GRACE dealiasing fields is called AOW.

2.3 SLR Estimates

Satellite laser ranging to multiple satellites, including Starlett, Ajisai, Stella, and Lageos 1 and 2, provides sufficient information about Earth's gravitational field changes to obtain monthly samples in all degree 2 SH coefficients, and some higher degree coefficients. SLR time series of ΔC_{20} are available with GRACE Release 4 (RL04) products, as described in GRACE technical note TN-05 (Cheng and Ries 2008).

For ΔC_{21}, ΔS_{21} we use a series available only through mid-2007 (MK Cheng, personal communication), which is based on GRACE RL01 standards. RL01 lacked an ocean pole tide correction and used an outdated solid earth pole tide model. Therefore, these SLR ΔC_{21} and ΔS_{21} series require a pole tide correction, which for the oceans is the Desai model (Desai 2002) used in GRACE RL04, and for the solid earth pole tide, is estimated as the difference between solid earth pole tide standards used in GRACE RL04 and RL01.

3 Time Series Comparisons

There are four independent techniques or sources (GRC, EOP, AOW, and SLR), each providing time series of the three components ΔC_{21}, ΔS_{21}, and ΔC_{20}. After a visual assessment of the time series, comparisons will be made over three different ranges of periods, short (intra-annual, less than a year), seasonal, and long period (interannual, more than a year).

Figure 44.1a,b,c shows the four time series for each SH coefficient. Some general conclusions can be drawn from the figure. First, all four estimates show (for each SH coefficient) similar variances. This indicates no major errors contaminating them, and implies errors of comparable size among the four. A second common feature is evident seasonal variations of similar size. Third, several of the time series look reasonably well correlated, at least at periods exceeding several months. Fourth, there is a suggestion of similar long period variations in several, at periods exceeding about one year. These observations justify more detailed analysis, considering separately the three time scales.

3.1 Annual Variations

Annual and semi-annual sinusoids were fit by unweighted least squares to each SH time series. Table 44.1 gives amplitudes and phases. Seasonal frequency estimates are similar among all series for each SH coefficient.

3.2 Short Period Variations

A general observation is that though of similar variance, time series do not appear well correlated for much of the period. We remove seasonal (annual and semiannual) signals and variations at period longer

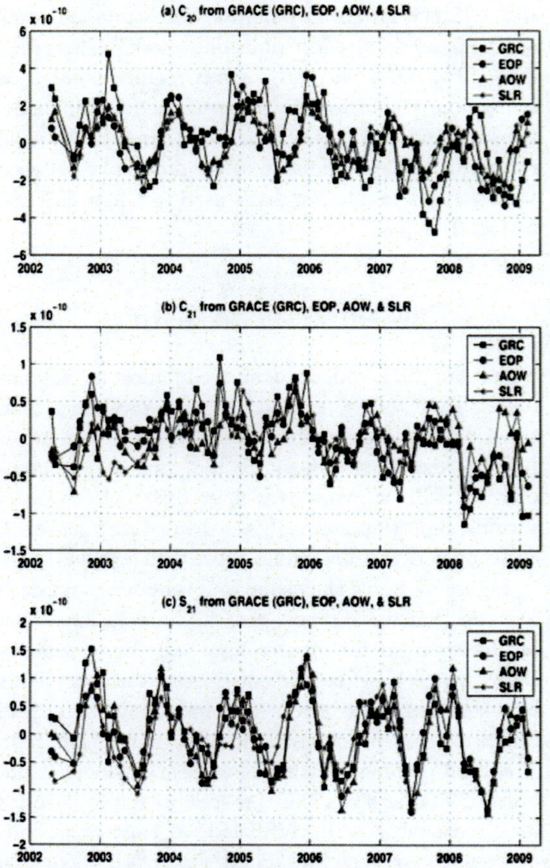

Fig. 44.1 Changes in dimensionless SH coefficients from the four independent techniques. Series have been resampled to match GRACE times as discussed in the text. Time series show similar variance and time scales of variation, including seasonal changes. SLR series for ΔC_{21} and ΔS_{21} both end in mid-2007, as discussed in the text. Figure: **(a)** *(top)* ΔC_{20}, **(b)** *(middle)* ΔC_{21}, **(c)** *(bottom)* ΔS_{21}

Table 44.1 Amplitude and phase of annual and semiannual ΔC_{20} variations estimated from GRACE (GRC) CSR and GFZ RL04 solutions, Earth rotation (EOP), geophysical modes (AOW), and SLR. The phase is defined as ϕ in $\sin(2\pi(t - t_0) + \phi)$, where t_0 refers to 0 hour on January 1

Gravity change	Annual amplitude phase ($\times10^{-10}$) (deg)	Semiannual amplitude phase ($\times10^{-10}$) (deg)
ΔC_{21} (GRC/CSR)	0.24 160	0.09 140
ΔC_{21} (GRC/GFZ)	0.18 153	0.06 150
ΔC_{21} (EOP)	0.22 154	0.07 138
ΔC_{21} (AOW)	0.24 118	0.06 201
ΔC_{21} (SLR)	0.19 139	0.11 168
ΔS_{21} (GRC/CSR)	0.67 109	0.15 230
ΔS_{21} (GRC/GFZ)	0.64 104	0.17 243
ΔS_{21} (EOP)	0.70 116	0.23 251
ΔS_{21} (AOW)	0.87 100	0.15 238
ΔS_{21} (SLR)	0.61 113	0.04 164
ΔC_{20} (GRC/CSR)	1.34 30	0.08 110
ΔC_{20} (GRC/GFZ)	1.56 62	0.65 114
ΔC_{20} (EOP)	1.34 58	0.75 53
ΔC_{20} (AOW)	1.37 47	0.11 169
ΔC_{20} (SLR)	1.46 43	0.25 181

than a year using unweighted least squares fit and show the intraseasonal time series of ΔC_{21} and ΔS_{21} in Fig. 44.2 and ΔC_{20} in Fig. 44.3. Figure 44.3a, reveals a number of larger "events" in the GRC time series not evident with others. These are likely due to noise. Figure 44.3b, omitting GRC and EOP series, shows remarkably high correlation between AOW and SLR time series, with a correlation coefficient of 0.93 (see Table 44.2 details). This suggests that instances of poor correlation evident in Fig. 44.3a are likely due to errors in GRC and EOP series. In the case of EOP, it was already noted that the wind accounts for about 90% of observed LOD variance and subtracting wind and ocean current effects (via climate models) leaves the residual ΔC_{20} very susceptible to contamination by errors of the wind fields. In the case of GRC, various error sources can be identified. One recognized source is imperfections in ocean tide models, which may alias to longer period changes in ΔC_{20}. For example, errors in the S_2 tide will appear as a 161 day alias, and in K_2 at 1,362 days (or 3.73 years) (Knudsen 2003), possibly being confused with other long term variability in the relatively short GRC series. Indeed, Fourier power spectra of GFZ and CSR ΔC_{20} series (Fig. 44.4) show peaks near these two periods, and the CSR peak is larger, suggesting a greater problem relative to GFZ. Figure 44.5 shows that after subtracting 161 and 1,362 day sinusoids, there remain large differences between the two GRC series. The two are almost out of phase near the end of 2007 (and in early 2008).

There is a significant increase in the correlation coefficients between AOW and SLR ΔC_{20} in the present study than that from Chen and Wilson (2008) (e.g., 0.93 vs. 0.75). This increase is likely partly due to the improved SLR data processing method using the GRACE RL04 standards, as compared to the RL01 standards adopted in the SLR ΔC_{20} data in

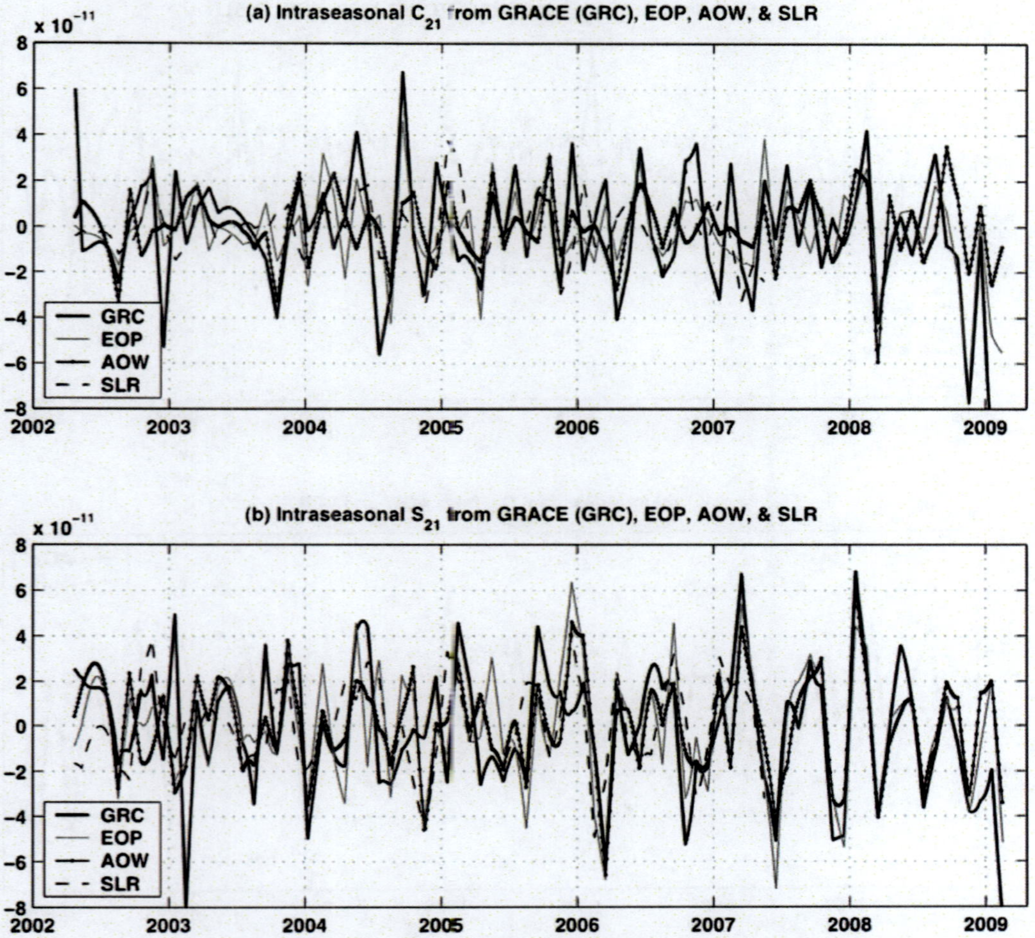

Fig. 44.2 Interseasonal variations in ΔC_{21} and ΔS_{21} components after annual, semi-annual, and variations at periods longer than 1 year are removed using unweighted least squares fit

Chen and Wilson (2008). Another likely cause to this improvement is the reduced end effect of the low-pass filter used in LOD data, due to extended length of time series (see Chen and Wilson 2008 for details), which results in better agreement of SLR ΔC_{20} with AOW estimates at the end of the time series (i.e., the last half year or so).

3.3 Interannual Variations

Interannual (long period) variations are of great interest because GRACE is able to observe a broad spectrum of SH coefficients, which should lead to understanding of the geographical sources of Earth system mass redistribution at interannual periods. As noted, this may allow EOP series (especially polar motion or ΔC_{21} and ΔS_{21}) to serve as proxies for climate change for a number of decades into the past. A preliminary examination, shown in Fig. 44.6 suggests this is feasible. Interannual variations in GFZ and CSR solutions show similar oscillations and trends, and agree reasonably well with EOP solutions for ΔC_{21} and ΔS_{21} including evidence of similar trends. It is notable that there is a lack of a trend, in the AOW series in Fig. 44.6. This is due to recognized limitations of land surface models (GLDAS), which will contribute almost all the variance at long periods

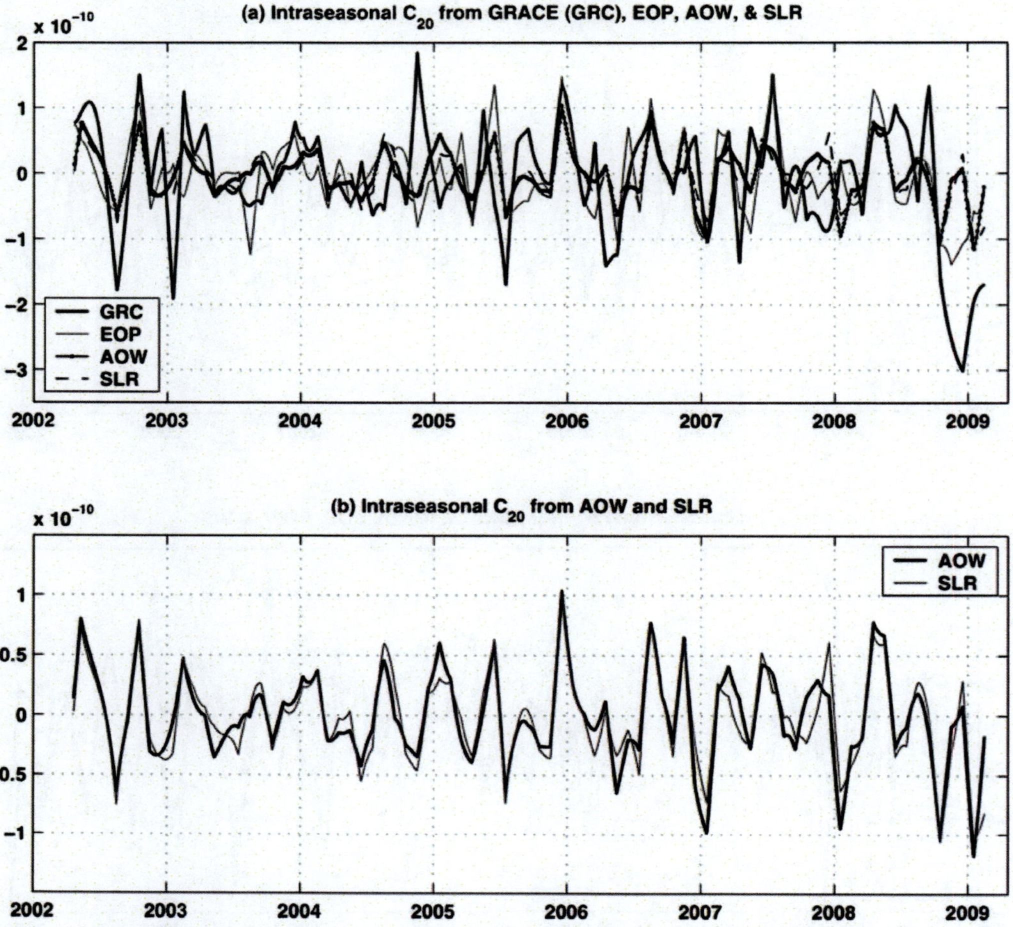

Fig. 44.3 Intraseasonal (short period) variations for all four series (**a**, *upper*) and SLR and AOW only (**b**, *lower*). Excellent agreement shown in the lower panel suggests various errors may contaminate EOP and GRC series. EOP estimates of ΔC_{20} are susceptible to errors in removing about 90% of the variance due to wind and current effects. Errors with GRACE estimates are discussed in more detail in the text

Table 44.2 Correlation coefficients at zero lag of intraseasonal ΔC_{21}, ΔS_{21}, and ΔC_{20} time series, among estimates from GRACE (GRC), Earth rotation (EOP), geophysical models (AOW), and SLR. The 99% significance level is about 0.32. This table is an update of Table 2 of Chen and Wilson (2008)

Correlation pair	ΔC_{21}	ΔS_{21}	ΔC_{20}
GRC/EOP	0.73	0.65	0.54
GRC/AOW	0.42	0.66	0.46
GRC/SLR	0.39	0.40	0.48
EOP/AOW	0.55	0.85	0.68
EOP/SLR	0.55	0.56	0.68
AOW/SLR	0.33	0.52	0.93

to AOW. (Atmospheric and oceanic mass redistribution are less likely to be important at these periods because accumulation of mass is physically less likely than in the various terrestrial water reservoirs). GLDAS and similar models have known limitations including omission of ground water storage and lack of polar ice sheet effects. Both sources are strong candidates for long period variability. In addition, trends in both GRC and EOP time series include a contribution from postglacial rebound, not part of the AOW series.

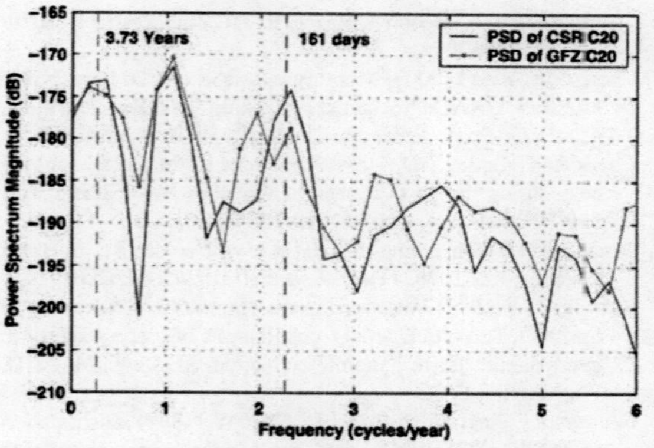

Fig. 44.4 Power spectra of the two GRACE time series indicating peaks in the vicinity of anticipated alias of S_2 and K_2 ocean tide model errors (with periods of 161 days and 3.73 years, respectively, marked by *dashed lines*)

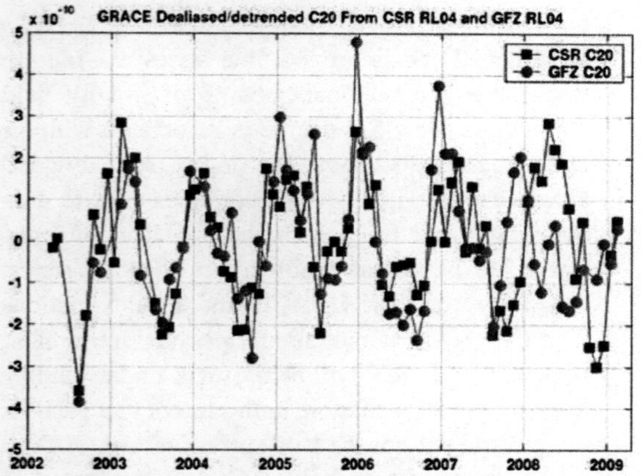

Fig. 44.5 The two GRACE time series after subtraction of best-fit sinusoids at alias periods of K_2 and S_2 tide errors. Trends are also removed using least squares fit

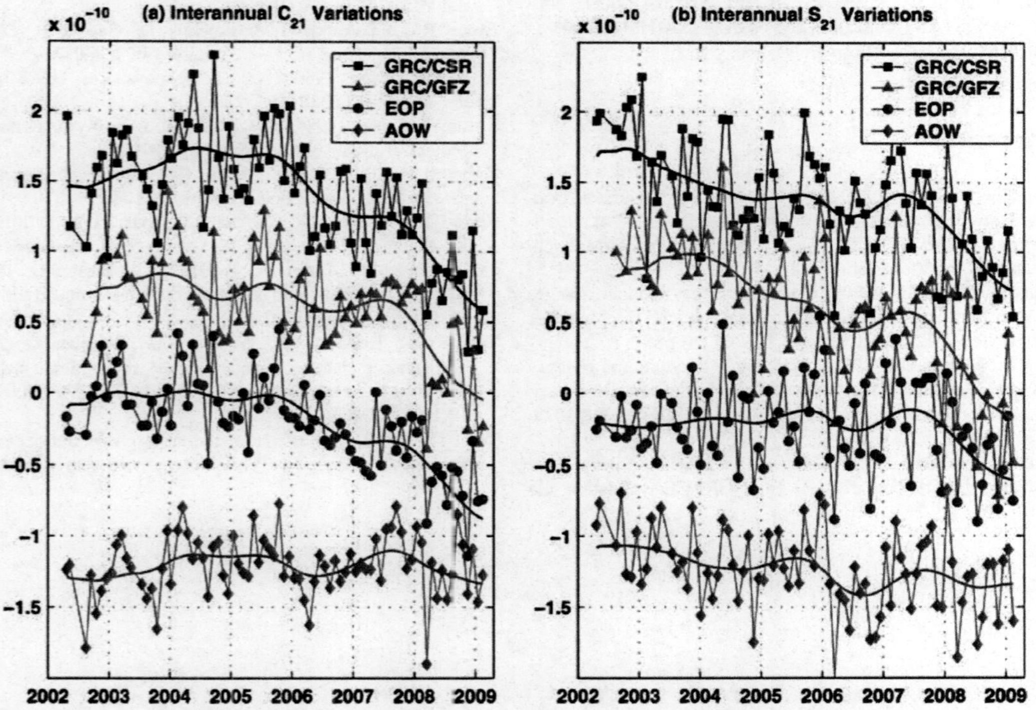

Fig. 44.6 Interannual variations in ΔC_{21} and ΔS_{21} plotted with non-seasonal residuals. AOW shows smaller trends and less variability, but the other series (GRC/CSR, GRC/GFZ and EOP) exhibit similar oscillations at periods of several years. While the time series are too short to quantify the correlation, it supports the possibility of using the GRC series to understand the mass variations evident in the EOP (polar motion) series. For clarity, we add offsets to the time series. SLR ΔC_{21} and ΔS_{21} time series cover a shorter time span and are excluded here

Conclusion

Lengthened and improved time series continue to verify that the dominant source of gravity field variations at low SH degree is associated with air and water redistribution. We find reasonably good agreement among all four techniques used to estimate variations in each of the 3 SH coefficients. There are persistent problems in GRACE series, especially for C_{20}. GRACE and EOP variations in ΔC_{21}, ΔS_{21} at interannual periods are similar, suggesting GRACE will be useful in understanding long period polar motion, both present and past. At interannual periods the land surface model GLDAS does not show trends in ΔC_{21}, ΔS_{21} due to a combination of model inadequacies, and unmodeled sources, such as polar ice and post glacial rebound.

Acknowledgments The authors would like to thank the two anonymous reviewers for their insightful comments, which led to improved presentation of the results. This research was supported by a NASA PECASE Award (NNG04G060G), NSF IPY Program (ANT-0632195), and NGA grants (HM1582-09-1-0001, HM1582-07-1-2032).

References

Bettadpur S (2007) CSR Level-2 Processing Standards Document for Product Release 04, GRACE 327–742, Revision 3.1, The GRACE Project, Center for Space Research, University of Texas at Austin

Chen JL, Wilson CR (2003) Low degree gravitational changes from Earth rotation and geophysical models. Geophys Res Lett 30(24):2257. doi:10.1029/2003GL018688

Chen JL, Wilson CR (2008) Low degree gravitational changes from GRACE, Earth rotation, geophysical models, and satellite laser ranging. J Geophys Res 113:B06402. 10.1029/2007JB005397

Chen JL, Wilson CR, Tapley BD, Ries J (2004) Low degree gravitational changes from GRACE: validation and interpretation. Geophys Res Lett 31(22):L22607. 10.1029/2004GL021670

Cheng MK, Ries J (2008) Monthly estimates of C20 from 5 SLR satellites, GRACE Technical Note #05, The GRACE Project, Center for Space Research, University of Texas at Austin

Cheng MK, Tapley BD (2004) Variations in the Earth's oblateness during the past 28 years. J Geophys Res (Solid Earth) 109(B18):9402–+. doi:10.1029/2004JB003028

Desai S (2002) Observing the pole tide with satellite altimetry. J Geophys Res 107(C11):3186. doi:10.1029/2001JC001224

Eubanks TM (1993) Variations in the orientation of the earth. In Smith D, Turcotte D (eds) Contributions of space geodesy to geodynamic: Earth dynamics, Geodyn. Ser., vol. 24, AGU, Washington, D.C., pp 1–54

Fukumori I, Raghunath R, Fu L, Chao Y (1999) Assimilation of TOPEX/POSEIDON data into a global ocean circulation model: How good are the results? J Geophys Res 104:25647–25665

Fukumori I, Lee T, Menemenlis D, Fu L-L, Cheng B, Tang B, Xing Z, Giering R (2000) A dual assimilation system for satellite altimetry, Joint TOPEX/POSEIDON and Jason-1 Science Working Team Meeting, Miami Beach, Fla., 15–17 November.

Gross RS, Blewitt G, Clarke PJ, Lavallée D (2004a). Degree-2 harmonics of the Earth's mass load estimated from GPS and Earth rotation data. Geophys Res Lett 31:L07601. doi:10.1029/2004GL019589

Gross RS, Fukumori I, Menemenlis D, Gegout P (2004b) Atmospheric and oceanic excitation of length-of-day variations during 1980–2000. J Geophys Res 109:B01406. doi:10.1029/2003JB002432

Kalnay EM et al (1996) The NCEP/NCAR 40-year reanalysis project. Bull Am Meterol Soc 77:437–471

Knudsen P (2003) Ocean tides in GRACE monthly averaged gravity fields. Space Science Reviews 108:261–270

Rodell M, Houser PR, Jambor U, Gottschalck J, Mitchell K, Meng C-J, Arsenault K, Cosgrove B, Radakovich J, Bosilovich M, Entin JK, Walker JP, Lohmann D, Toll D (2004) The global land data assimilation system. Bull Amer Meteor Soc 85(3):381–394

Tapley BD, Bettadpur S, Watkins MM, Reigber C (2004) The gravity recovery and climate experiment: mission overview and early results. Geophys Res Lett 31(9):L09607. 10.1029/2004GL019920

Wahr JM (1982) The effects of the atmosphere and oceans on the earth's wobble. 1. Theory. Geophys J Roy Astron Soc 70:349–372

Using Swarm for Gravity Field Recovery: First Simulation Results

45

Xinxing Wang and Reiner Rummel

Abstract

The CHAMP satellite, equipped with a GPS receiver and an accelerometer, has yielded an enormous increase in accuracy of global satellite-only gravity field models and it has proved the concept of high-low satellite to satellite tracking (SST) for gravity field recovery. Therefore, we are interested whether the Swarm mission, consisting of three CHAMP-like satellites, is suitable for gravity field determination as well. This article presents the results of two simulation studies for static and time-variable gravity field recovery. The latter may become especially relevant after the mission end of GRACE and CHAMP.

Swarm's potential of recovering the global gravity field is investigated by simplified simulations. GPS baselines between the three satellites are used as observations. Generally, the quality of static field recovery can be better than that of CHAMP with the help of the more accurate GPS baseline measurements. A 24-months simulation with temporal variations shows that Swarm may have the potential of recovering the long wavelength part of hydrology signal.

Keywords

Swarm • Gravity field recovery • GPS baseline measurements • temporal variation

1 Introduction

The Swarm mission is an approved satellite mission of ESA, which aims at the global determination of the outer, inner and lithospheric magnetic field of the Earth as well as the temporal evolution of the inner and outer field. Swarm is scheduled to be launched in 2011 and is expected to give new insights into the Earth's interior and climate. More details about the mission can be found e.g. in Olsen et al. (2007). The Swarm constellation consists of two low orbiting satellites (Swarm A and B) and a third one at higher altitude (Swarm C). The two lower satellites will fly initially in a circular orbit at 450 km altitude, which is similar to the orbit altitudes of CHAMP and GRACE, while the higher satellite will fly at 530 km. Each of these three satellites carries onboard a GPS receiver and an accelerometer. They make it possible to do precise continuous 3D positioning and to separate gravitational and non-gravitational forces. Thus it is comparable to three single CHAMP-like gravity

X. Wang (✉) · R. Rummel
Institute of Astronomical and Physical Geodesy, Technische Universität München, Munich, Germany
e-mail: xinxing.wang@bv.tum.de; rummel@bv.tum.de

N. Sneeuw et al. (eds.), *VII Hotine-Marussi Symposium on Mathematical Geodesy*, International Association of Geodesy Symposia 137, DOI 10.1007/978-3-642-22078-4_45, © Springer-Verlag Berlin Heidelberg 2012

solutions. Moreover, thanks to spaceborne differential GPS, one can determine the relative position vectors between these three satellites. These GPS baselines can be used as observations for gravity field determination and they are expected to bring an improvement against the single solutions, as the relative positions can be determined with an accuracy about one order of magnitude higher than the absolute positions of the individual satellites. The absolute GPS position accuracy for CHAMP is about 2–3 cm and the relative position of GRACE -based on GPS only- has been reported to be of mm-accuracy (Visser 2006). Based on these performances, we believe that the same accuracy of the relative positions can be achieved for the Swarm mission.

In this article we try to investigate the possibility of gravity field recovery from Swarm via simplified simulations with the energy balance approach. Both static and time-variable field recovery will be discussed. The energy balance approach and its utilization in gravity field recovery has been discussed and implemented in geodesy for years. Gerlach et al. (2003) and Han (2004) have used this approach for CHAMP and GRACE, respectively and have demonstrated its application in practise.

In next section we will give some general aspects about this approach, followed by a short discussion about the stability of the Swarm constellation. In Sect. 3, closed-loop simulations will be described and the results will be presented in Sect. 4. A brief conclusion will be drawn in Sect. 4.2.

2 General Aspects of Gravity Field Recovery from Swarm

2.1 Energy Balance Approach

The energy balance approach is based on the energy conservation law, which states that the sum of potential and kinetic energy is constant. In an Earth-fixed reference frame the relationship is:

$$T(x,t) = \frac{1}{2}\dot{\mathrm{x}}(t)^2 - Z(\mathrm{x},t) - U(\mathrm{x},t) - \int a \cdot d\mathrm{x}(t) + C \qquad (45.1)$$

where $T(x,t)$ is the disturbing potential, $\dot{x}(t)$ is the velocity vector of the satellite, $Z(x,t)$ is the centrifugal potential, $U(x,t)$ is the normal potential, C is a unknown constant (Jacobi constant) and $\int a \cdot dx(t)$ sums up effects from all non-conservative and time-variable forces, such as air drag, the tidal effect of Sun and Moon or time-variable gravity signals. In this article all non gravitational forces are assumed to be perfectly measured by the onboard accelerometer and are therefore not considered in the simulation. Time-variable gravity field is considered and treated differently for the cases of static and temporal gravity field recovery, see Sect. 3. All parameters in (45.1) are time series along the satellite orbit except the constant C. If C can be determined, the disturbing potential along the orbit can be calculated. Details of the calculation of C can be seen in Gerlach et al. (2003).

For the baseline measurement, the difference of disturbing potential between the satellites (e.g. Swarm A and B) can be formed as a function of the relative velocity vector $\Delta\dot{x}_{A,B}$:

$$\Delta T_{A,B} = T_B - T_A = \frac{1}{2}\Delta\dot{x}_{A,B}^2 + \dot{x}_A \cdot \Delta\dot{x}_{A,B} - \Delta Z_{A,B} - \Delta U_{A,B} - \Delta C_{A,B} \qquad (45.2)$$

This difference serves as pseudo observation to recover the potential coefficients of the gravity field.

2.2 Stability of the Swarm Constellation

The Swarm B satellite has a longitude shift of approximately 1.4° relative to Swarm A, which corresponds to roughly 160 km distance at the equator. In this respect the Swarm A-B situation is very similar to the GRACE constellation with two close satellites. The distance between GRACE A and B is about 220 km. The difference is that Swarm A-B offers cross-track information while GRACE A-B provides the along-track information. Because of the different initial orbit altitudes and the slightly different inclination of Swarm C and A/B ($I_C = 88°$, $I_A = I_B = 87.4°$), the orbit plane of C will drift away from that of A and B. Swarm C is designed in such a way that it starts with an almost identical orbit plane as Swarm A (B) and drifts away from A (B) with increasing local time difference in order to have measurements in different regions at the same time (Kotsiaros 2009). After 4.5 years the local time difference should be about 12 h.

In order to take advantage from the accuracy of the GPS baseline measurement, it is mandatory to have 5 or 6 GPS satellites in common view. As the Swarm configuration changes over time, this condition is not always fulfilled for the baseline A-C and B-C. One is able to derive usable A-C and the B-C baselines during one quarter of the mission (Gerlach and Visser 2006). In contrast, the constellation A-B can always be used as observation.

As we intend to use the Swarm baselines and to benefit from the lower altitude also during the end of the mission phase, it is important that this configuration can stay stable throughout the mission. As discussed above Swarm C will drift away from A and B and the baseline measurements between A-C and B-C will be difficult due to lack of sufficient common view with the GPS satellites. Thus, the focus of the stability study is on the Swarm A-B baseline. As is well known, the Earth's gravity field, in particular the flattening of the Earth, causes strong precessional motion superimposed by oscillations on the orbit. For the purpose of formation keeping, at least the relative drift between Swarm A and B should be eliminated (Sneeuw et al. 2006). For our case, with Kepler elements be given (Olsen et al. 2007), the formation will stay stable under a J2-perturbation if the following constraints can be fulfilled (Sneeuw et al. 2006)

$$\begin{cases} \Delta\dot{\Omega}_{AB} = \dot{\Omega}_B - \dot{\Omega}_A = 0 \\ \Delta\dot{\omega}_{AB} = \dot{\omega}_B - \dot{\omega}_A = 0 \\ \Delta\dot{M}_{AB} = \dot{M}_B - \dot{M}_A = 0 \end{cases} \tag{45.3}$$

where $\dot{\Omega}$, $\dot{\omega}$ and $\dot{M}$ are all functions of a, e and I. Equation (45.3) is fulfilled in the case of Swarm A-B because A and B are designed in such a way that they share the same orbit height, eccentricity and inclination throughout the mission. Under real conditions, active orbit control manoeuvres are planed to maintain the stability of the constellation (Olsen et al. 2007). Therefore we assume the baseline A-B is stable throughout the mission time and it is always possible to use this baseline as observation for gravity field recovery.

3 Simulation Description

In this section, the simulation processes will be described. Two simulations have been carried out, aiming at the recovery of the static field and the time-variable hydrology signal, denoted as simulation 1 and 2, respectively. For both simulations, the orbit height of Swarm A/B is set to 350 km and 500 km for C, which corresponds approximately to the constellation 3 years after launch to benefit from lower altitude. The OMCT-ECMWF data set in 6-hourly resolution is used for the combination of atmosphere and ocean signal (AO). For the hydrology signal we use the LaD model (H) in monthly solution from January 2003 for 24 months. The latter is also linearly interpolated into 6-hourly resolution within January 2003 for simulation 1 to be consistent with the AO signal. The other simulation parameters are listed in Tables 45.1 and 45.2. In simulation 1 the static EGM96 is superimposed with time series of AOH signal and it is tried to remove the latter via integration of the corresponding accelerations of the time variations along the orbit. The recovered field will then be compared with EGM96 to d/o 70.

In simulation 2 we try to recover the long wavelength part of the hydrology signal. For this purpose a 24-months simulation is done, in which the static EGM96 field is superimposed with the LaD model with monthly resolution. Here we assume all other time variations can be perfectly removed, thus the recovered monthly fields contain only the static and hydrology signal. To reduce computational effort, we only try to resolve to degree and order 30. Furthermore we assume that the hydrology signal is not changing within 1 month. Zenner (2006) has shown that this LaD model contains mainly annual and semi-annual signal, which justifies our assumption. Thus although this assumption is not perfectly true, its can be regarded as acceptable.

In both simulations the observations are contaminated with white noise with different noise levels, which are listed in Table 45.3. For absolute position and velocity we use 1 cm and 0.1 mm/s, for baseline position and velocity 1 mm and 0.01 mm/s, respectively.

Table 45.1 Simulation 1 conditions for static field recovery

Duration	Sampling	EGM96 d/o	Time variation
1 month (01.2003)	30 s	70	AOH (6 h) d/o 70

Table 45.2 Simulation 2 conditions for hydrology recovery

Duration	Sampling	EGM96 d/o	Time variation
24 months	30 s	30	H (monthly) d/o 30

Table 45.3 Noise level for simulations (in mm and mm/s)

	Abs. p	Abs. v	Rel. p	Rel. v
noise (σ)	10	0.1	1	0.01

For both simulations, the disturbing potential is calculated through the energy balance approach. In case of GPS baseline measurements, the difference in disturbing potential is computed and used as pseudo observation. The output potential coefficients will be computed through least squares spherical harmonic analysis. Here we apply the "brute-force" analysis and directly calculate the coefficients from the disturbing potential. A comparison between the input and output gives an error measure of the simulations.

4 Simulation Results

Let us now turn to the results of the simulations. Simulation 1 aims at the recovery of the static field and will be presented in 4.1. Simulation 2 investigates the possibility of recovery of the long wavelength part of hydrology and will be described in 4.2. A brief conclusion will be drawn in Sect. 4.2.

4.1 Results of Simulation with Static Field

In this part we present the results of the static gravity field recovery simulations. Gerlach and Visser (2006) has shown similar simulation results. Additionally to his simulation we superimpose AOH signal to EGM96 and then remove their contribution via integration of corresponding accelerations. The results are shown in the degree RMS values of Fig. 45.1. From the plot we can draw the following information: (1) among the single satellite solutions, A and B have the same quality and C is inferior due to the higher altitude; (2) baseline A-B suffers from poorly determined zonal coefficients due to its pure cross-track constellation, but its curve appears less steep towards higher degrees than the other curves. This indicates it may be helpful for a better determination of higher degree coefficients; (3) the baseline A-C offers a good result for the chosen month despite that only about 25% observation can be used to form this baseline. But as the orbit planes of A/B and C change with time, the geometry of the A-C constellation can become so bad that this baseline is not usable for gravity determination. As the orbits drift, the A-C baseline will be mostly usable in the Polar Regions. It therefore suffers from an inhomogeneous

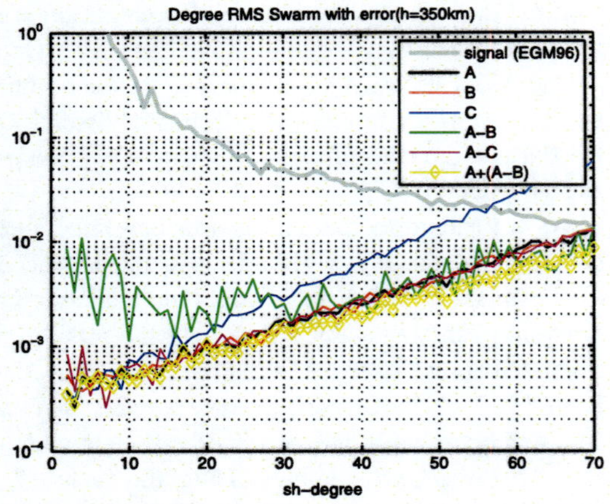

Fig. 45.1 Degree RMS plot for static field recovery

distribution of the observations over the Earth's surface; (4) The combined solution $A + (A - B)$ offers the best solution. This combination is an interesting choice: it combines the good zonal coefficients solutions from A with the better determined higher degree coefficients from A-B. In general, it is possible to determine the static field up to degree and order 70 with Swarm. With the baseline solution we can expect to go to even higher degrees. The results can be expected to be better than those of CHAMP (Gerlach and Visser 2006).

4.2 Results of Simulation with Time Variation

In this part the hydrology signal in monthly intervals is added to the static field. The output field is a static field composed of EGM96 and the input hydrology signal. If we subtract EGM96 from the output, we expect to get the recovered hydrology signal. The recovery result is first presented in a degree RMS plot in Fig. 45.2. Then we take the two coefficients C20 and C30 as examples to investigate the possibility to recover the long wavelength part of the signal.

In the degree RMS plot a GRACE curve is plotted for comparison, which is obtained by adding a $10^{-3}\,m^2/s^2$ potential noise to the potential difference between leading and following satellite (Han 2004). Here the leading satellite is Swarm A and the following satellite is about 200 km away on the same orbit. It is

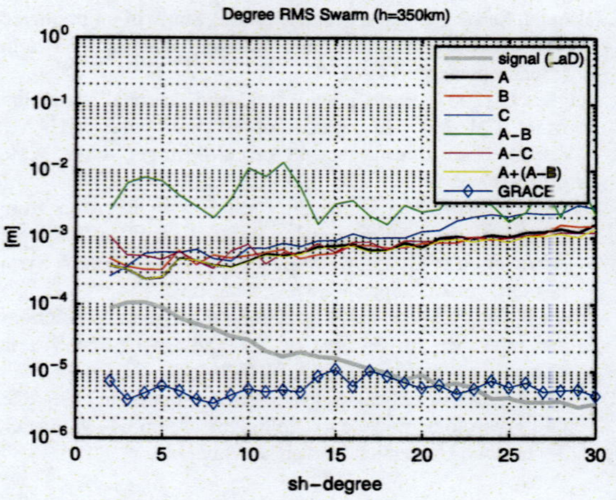

Fig. 45.2 Degree RMS plot for hydrology recovery

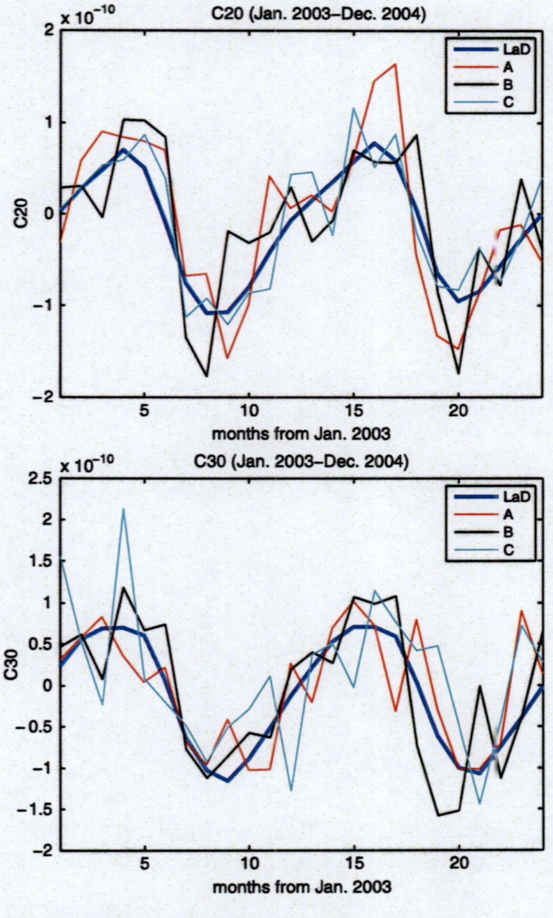

Fig. 45.3 C20 and C30 recovery from Swarm

clear that with a GRACE-like accuracy one would be able to determine the hydrology signal to about degree 25. In contrast, it seems to be impossible for Swarm to recover the signal. All the curves including baseline solutions show an error much higher than the hydrology signal itself. If we only focus on the very long wavelength part, e.g. C20 and C30 (Fig. 45.3), a recovery can be possible. All the single satellite solutions fit the signal curve well in the simulation period of 2 years. Both the amplitude and phase of the recovered coefficients seem to agree with the model. Baseline solutions are not shown here as their solutions are orders of magnitude worse than the single solutions and require further investigations. Compared to similar study for CHAMP (Reigber 2007), we have three independent solutions from Swarm. The simulation results seem to be promising.

Conclusion

The Swarm mission consists of three CHAMP-like satellites and it is possible to use Swarm for gravity field recovery. Besides the absolute GPS positions also the baselines between the satellites can be used as observations, which could improve the quality of recovery. Swarm can be used for static field recovery up to degree 70 and higher, with the help of the combination of single satellite and baseline solution. It might also have potential for the determination of long periodic time variations in its low spherical harmonics. While the behaviour of baseline observations to temporal variations requires further studies, one can at least conclude that three CHAMP-like solutions and thus more redundancy are feasible. In comparison to GRACE, the GPS baseline accuracy is by far not at the level of the K-band microwave link and accelerometers, and therefore the quality of gravity field recovery by Swarm is not comparable to that of GRACE in both static and time-variable field recovery. But considering that Swarm is by no means a gravity mission, these results are actually satisfying and should be considered. Swarm is expected to be launched in 2011 and it will likely be still useful for gravity field recovery to about 2016. In this sense Swarm can be regarded as a welcome complementary mission to the dedicated gravity missions in the near future.

References

Gerlach Ch, Sneeuw N, Visser P, Svehla D (2003) CHAMP gravity field recovery with the energy balance approach: first results. In: First CHAMP mission results for gravity, magnetic and atmospheric studies. Springer Verlag, New York, pp 134–139

Gerlach Ch, Visser P (2006) Swarm and gravity, possibilities and expectations for gravity field recovery. In: Proceedings of first Swarm international workshop, ESA WPP-261

Han S-C (2004) Efficient determination of global gravity field from satellite-to-satellite tracking. Celest Mech Dyn Astron 88:69–102

Kotsiaros S (2009) Development of algorithms and tools for data analysis, data visualization, and data validation for the Swarm satellite mission, ESA working paper 2355, pp 20–27

Olsen N, Sabaka T, Gaya-Pique L (2007) Study of an improved comprehensive magnetic field inversion analysis for swarm (Final report), pp 14–23

Reigber C (2007) Earth gravity field and seasonal variability from CHAMP. In: Earth observation with CHAMP results from three years in orbit. Springer, New York, pp 25–30

Sneeuw N, Sharifi MA, Keller W (2006) Gravity Recovery from formation flight missions. In: Proceedingsof International Association of Geodesy Symposia, VI Hotine-Marussi Symposium, vol 132. Springer, New York, pp 29–34

Visser P (2006) Space-borne gravimetry: progress, predictions and relevance for Swarm. In: Proceedings of first Swarm international workshop, ESA WPP-261

Zenner L (2006) Zeitliche Schwerefeldvariationen aus GRACE und Hydrologiemodellen (Diplomarbeit), pp 64–80. IAPG, Technische Universität München, Munich

Consistent Modeling of the Geodetic Precession in Earth Rotation

46

E. Gerlach, S. Klioner, and M. Soffel

Abstract

A highly precise model for the motion of a rigid Earth is indispensable to reveal the effects of non-rigidity in the rotation of the Earth from observations. To meet the accuracy goal of modern theories of Earth rotation of 1 microarcsecond (μas) it is clear, that for such a model also relativistic effects have to be taken into account. The largest of these effects is the so called geodetic precession.

In this paper we will describe this effect and the standard procedure to deal with it in modeling Earth rotation up to now. With our relativistic model of Earth rotation (Klioner et al. 2001) we are able to give a consistent post-Newtonian treatment of the rotational motion of a rigid Earth in the framework of General Relativity. Using this model we show that the currently applied standard treatment of geodetic precession is not correct. The inconsistency of the standard treatment leads to errors in all modern theories of Earth rotation with a magnitude of up to 200 μas for a time span of one century.

Keywords

General Relativity • relativistic Earth rotation • geodetic precession/nutation • rigidly rotating multipoles

1 Introduction

Geodetic precession/nutation is the largest relativistic effect in Earth rotation. This effect has been discovered already a few years after the formulation of General Relativity (de Sitter 1916) and very early it was recognized to be important for Earth rotation. It results mainly in a slow rotation of a geocentric locally inertial reference frame with respect to remote celestial objects roughly about the ecliptic normal. Due to its relatively large magnitude of about 1.9″ per century, which is 3×10^{-4} of the general precession, corresponding corrections are used in all standard theories of precession and nutation since the IAU 1980 theory (Seidelmann 1982).

The standard way to consider geodetic precession in Earth rotation theories up to now was the following: firstly, using purely Newtonian equations, one computed the orientation of the Earth in a geocentric, locally inertial reference frame. To obtain the solution with respect to the kinematically non-rotating Geocentric Celestial Reference System (GCRS) the

E. Gerlach (✉) · S. Klioner · M. Soffel
Lohrmann Observatory, Dresden University of Technology, Dresden, Germany
e-mail: Enrico.Gerlach@tu-dresden.de; Sergei.Klioner@tu-dresden.de; Michael.Soffel@tu-dresden.de

N. Sneeuw et al. (eds.), *VII Hotine-Marussi Symposium on Mathematical Geodesy*, International Association of Geodesy Symposia 137, DOI 10.1007/978-3-642-22078-4_46, © Springer-Verlag Berlin Heidelberg 2012

corrections for geodetic precession were then simply added, as described for example in Bretagnon et al. (1997, Sect. 8). These corrections can be calculated separately, since they are completely independent of the rotational state of the Earth, e. g. Brumberg et al. (1991).

The purpose of this work is to demonstrate that the standard way of applying the geodetic precession is not correct. After stating the problem in the following Section, we explain shortly our relativistic model of Earth rotation used for this study in Sect. 3. In Sect. 4 we describe how the corrections for the geodetic precession can be computed, while in Sect. 5 two different, but equivalent and correct ways to obtain a GCRS solution are given. In the last section of this paper we compare our solution to published ones and draw concluding remarks.

Throughout the paper we will use the following conventions:

- Lower case Latin indices take the values 1, 2, 3.
- Repeated indices imply the Einstein's summation irrespective of their positions, e. g. $x^i y^i = x^1 y^1 + x^2 y^2 + x^3 y^3$.
- ε_{abc} is the fully antisymmetric Levi-Civita symbol, defined as $\varepsilon_{abc} = (a-b)(b-c)(c-a)/2$.
- Vectors are set boldface and italic: $\boldsymbol{X} = X^i$, while matrices are set boldface and upright: $\mathbf{P} = P^{ij}$
- The choice to use index or vector notation for a specific formula is done with regard to readability and clarity.

2 The GCRS and Geodetic Precession

The Geocentric Celestial Reference System is officially adopted by the IAU to be used to describe physical phenomena in the vicinity of the Earth and, in particular, the rotational motion of the Earth. The GCRS is connected with the Barycentric Celestial Reference System (BCRS) by a generalized version of the Lorentz transformation. This transformation was chosen in such a way that the GCRS spatial coordinates $\boldsymbol{X}_K$ are kinematically non-rotating with respect to the BCRS coordinates, i. e. no additional spatial rotation of the coordinates is involved in the transformation from one system to the other (Soffel et al. 2003).

Since the origin of the GCRS coincides with the geocenter and the Earth is moving in the gravitational field of the Solar system a local inertial frame with spatial coordinates $\boldsymbol{X}_D$ slowly rotates in the GCRS:

$$X^i_D = R^{ij}(T)\, X^j_K. \tag{46.1}$$

Here R^{ij} is an orthogonal matrix, the time variable T is the Geocentric Coordinate Time TCG. Due to this rotation the equations of motion in the GCRS contain a Coriolis force. The locally inertial analogon of the GCRS is called dynamically non-rotating. This rotation between the kinematically non-rotating GCRS and its dynamically non-rotating counterpart is called geodetic precession. The angular velocity of geodetic precession $\boldsymbol{\Omega}_{\mathrm{GP}}$ is given by

$$\boldsymbol{\Omega}_{\mathrm{GP}} \approx \frac{1}{c^2}\sum_A \frac{GM_A}{r^3_{EA}}\left[\left(2\boldsymbol{v}_A - \frac{3}{2}\boldsymbol{v}_E\right)\times \boldsymbol{r}_{EA}\right], \tag{46.2}$$

where c is the speed of light in the vacuum, G the gravitational constant, $\boldsymbol{v}_A$ the BCRS velocity of the body A with mass M_A, $\boldsymbol{v}_E$ is the velocity of the geocenter, $\boldsymbol{r}_{EA}$ the vector from body A to the geocenter and r_{EA} its Euclidean norm. The angular velocity $\boldsymbol{\Omega}_{\mathrm{GP}}$ corresponds to the orthogonal matrix R^{ij} so that the respective kinematical Euler equations read

$$\Omega^a_{\mathrm{GP}} = \frac{1}{2}\varepsilon_{abc}\, R^{db}(T)\,\frac{d}{dT}R^{dc}(T). \tag{46.3}$$

This equation can be easily verified by direct substitution of the matrix elements.

3 Model of Earth Rotation

A complete and profound discussion of our relativistic model of Earth rotation can be found in Klioner et al. (2001, 2010). For the purposes of this work we neglect all other relativistic effects except for the geodetic precession. In particular, we neglect relativistic torques, relativistic time scales, and relativistic scaling of various parameters. Then, in dynamically non-rotating coordinates $\boldsymbol{X}_D$ the equations of rotational motion of the Earth can be written as

$$\frac{d}{dT}\boldsymbol{S}_D = \boldsymbol{L}_D, \tag{46.4}$$

$$\boldsymbol{S}_D = \boldsymbol{S}_D(\mathbf{P}_D; \mathscr{A}, \mathscr{B}, \mathscr{C}), \tag{46.5}$$

$$\boldsymbol{L}_D = \boldsymbol{L}_D(\mathbf{P}_D; C_{lm}, S_{lm}; \boldsymbol{x}_{AD}). \tag{46.6}$$

Here $\boldsymbol{L}_D$ is the torque and $\boldsymbol{S}_D$ the angular momentum in the dynamical non-rotating frame and $\mathbf{P}_D$ is a time-dependent orthogonal matrix transforming the coordinates $\boldsymbol{X}_D$ to a terrestrial reference system $\boldsymbol{Y}$, where the gravitational field of the Earth is constant:

$$Y^a = P_D^{ab}(T)\, X_D^b. \qquad (46.7)$$

$\mathscr{A}, \mathscr{B}, \mathscr{C}$ are the principle moments of inertia of the Earth, C_{lm}, S_{lm} are the coefficients of the gravitational field of the Earth in $\boldsymbol{Y}$ and $\boldsymbol{x}_{AD}$ are the BCRS coordinates $\boldsymbol{x}_A$ of body A (Sun, Moon, etc.) rotated by the geodetic precession:

$$x_{AD}^i = R^{ij}(T)\, x_A^j. \qquad (46.8)$$

The matrix $\mathbf{P}_D$ can be parametrized by Euler angles φ, ψ and ω in the usual way (Bretagnon et al. 1997). Thus, these three angles as functions of time T represent a solution of (46.4)–(46.6). Note that the only difference between a purely Newtonian solution of Earth rotation and (46.4)–(46.6) is that the torque should be computed by using rotated positions $\boldsymbol{x}_{AD}$ of external bodies and not the normal BCRS positions $\boldsymbol{x}_A$. This reflects the fact that the coordinates $\boldsymbol{X}_D$ rotate with respect to the BCRS.

The corresponding equations in the kinematically non-rotating GCRS take the form

$$\frac{d}{dT}\boldsymbol{S}_K = \boldsymbol{L}_K + \boldsymbol{\Omega}_{\rm GP} \times \boldsymbol{S}_K, \qquad (46.9)$$

$$\boldsymbol{S}_K = \boldsymbol{S}_K(\mathbf{P}_K; \mathscr{A}, \mathscr{B}, \mathscr{C}; \boldsymbol{\Omega}_{\rm GP}), \qquad (46.10)$$

$$\boldsymbol{L}_K = \boldsymbol{L}_K(\mathbf{P}_K; C_{lm}, S_{lm}; \boldsymbol{x}_A). \qquad (46.11)$$

The torque $\boldsymbol{L}_K$ is defined by the same functional form as $\boldsymbol{L}_D$, but the coordinates of external bodies are taken directly in the BCRS. Equation (46.9) contains an additional Coriolis torque proportional to $\boldsymbol{\Omega}_{\rm GP}$. Besides this, the angular momentum $\boldsymbol{S}_K$ in the GCRS explicitly depends on the geodetic precession $\boldsymbol{\Omega}_{\rm GP}$. For details of these equations see Klioner et al. (2010) and references therein.

From (46.1) and (46.7) it is clear, that the solutions of these two sets of equations are related by

$$P_K^{ab} = P_D^{ac}\, R^{cb}. \qquad (46.12)$$

4 Computing the Geodetic Precession

To determine the effect of geodetic precession one has to compute the matrix $\mathbf{R}$. This can be done by a numerical integration of (46.2)–(46.3). It should be remarked that care has to be taken how to represent this matrix properly. To avoid the discontinuities that can arise when using the common Euler angles to describe an arbitrary rotation, we decided to use quaternions to represent this matrix. With matrix $\mathbf{R}$, solution $\mathbf{P}_D$ and (46.12) we can calculate the differences $\delta\varphi$, $\delta\psi$ and $\delta\omega$ between the Euler angles φ, ψ and ω corresponding to matrix $\mathbf{P}_K$ and those corresponding to $\mathbf{P}_D$.

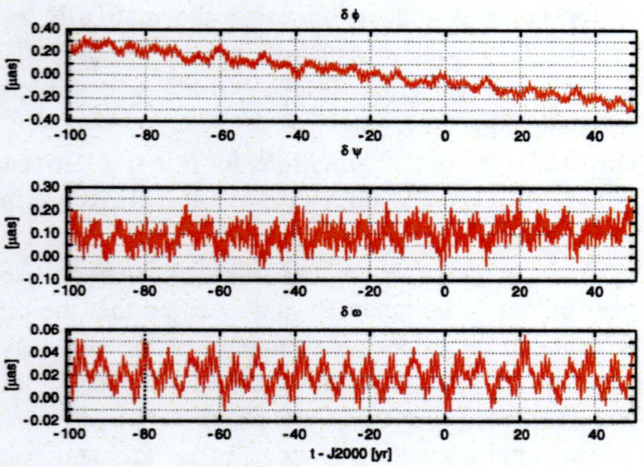

Fig. 46.1 Differences (in μas) for the analytical solution for geodetic precession derived by Bretagnon et al. (1998) and our numerical solution. The sign error in the SMART solution (Bretagnon et al. 1998) for angle φ is corrected here

In the process of computing and verifying the results we have found and corrected a sign error for the correction induced by the geodetic precession for angle φ in Bretagnon et al. (1998). Taking this sign error into account the differences between the analytical solution for $\delta\varphi$, $\delta\psi$ and $\delta\omega$ derived by Brumberg et al. (1991) and Bretagnon et al. (1998) and our numerical solution are below 1 μas. They are shown in Fig. 46.1. The remaining differences are explained by the limited accuracy of the analytical treatment of this effect by the other authors compared to our numerical result.

5 Computing the GCRS Solution

According to the equations given in Sect. 3 there are two ways to compute the matrix $\mathbf{P}_K$ corresponding to the solution of the rotational motion of the Earth with respect to the kinematically non-rotating GCRS:

1. One can numerically integrate (46.4)–(46.6) and obtain the solution $\mathbf{P}_D$ with respect to dynamically non-rotating coordinates $\boldsymbol{X}_D$. Then one can correct

for geodetic precession using the matrix $\mathbf{R}$ and (46.12) to rotate the solution into the GCRS.

2. One can integrate (46.9)–(46.11) and directly obtain $\mathbf{P}_K$.

Obviously, the initial conditions for (46.4)–(46.6) and (46.9)–(46.11) are again related by (46.12) taken at the initial epoch.

We have implemented both of the above mentioned possibilities to compute $\mathbf{P}_K$ and verified that the differences in φ, ψ and ω computed in the two ways represent only numerical noise at the level of 0.001 μas and less after 100 years of integration.

The implementation is done in an efficient way. The numerical integration of the matrix $\mathbf{R}$ runs for example simultaneously with the numerical integration for $\mathbf{P}_D$. The relative running times between a purely Newtonian integration and both ways described above are given in Table 46.1. Further details on our numerical code and its capabilities can be found for example in Klioner et al. (2008).

A purely Newtonian model differs from (46.4)–(46.6) only by the positions of the solar system bodies used to compute the torque on the Earth: the Newtonian model uses the BCRS ephemeris directly, while for (46.4)–(46.6) one has to rotate this ephemeris according to (46.8). *It is this rotation that has never been considered before in any theory of Earth rotation, which represents the main source of inconsistency in the standard way of taking the geodetic precession into account.* The effect of the rotation of the ephemeris on the Euler angles φ, ψ and ω is shown in Fig. 46.2. One finds that the error due to this inconsistency amounts to 200 μas after 100 years of integration.

A summary of the interrelations between the correct solutions for the Earth rotation in dynamically and kinematically non-rotating coordinates as well as Newtonian and "kinematically non-rotating" solution derived in the standard, inconsistent way is given schematically in Fig. 46.3.

Table 46.1 Relative CPU times for various integrations

	Relative CPU times
Newtonian case	1.00
Integration of $\mathbf{P}_D$ with rotated ephemeris	1.21
Direct integration of $\mathbf{P}_K$	1.08

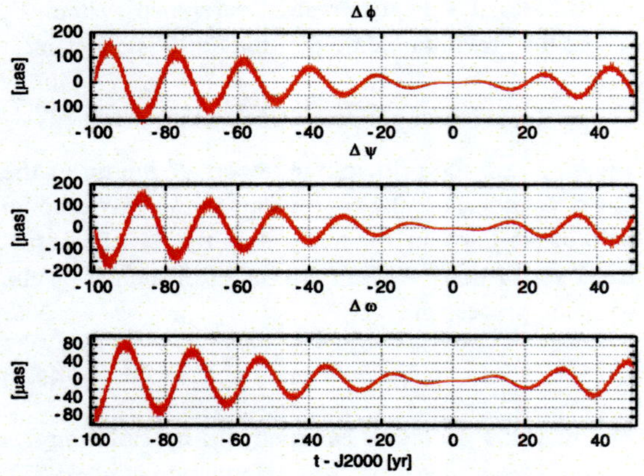

Fig. 46.2 Differences (in μas) for the Euler angles between a purely Newtonian solution and the correct solution in dynamically non-rotating coordinates. The latter is obtained by using ephemeris data rotated according to (46.8)

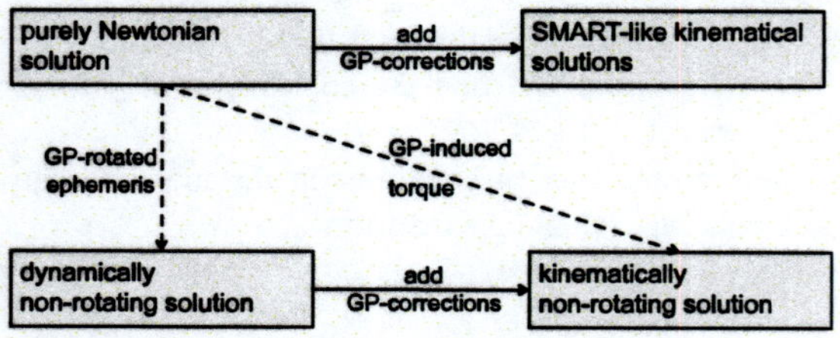

Fig. 46.3 Schematic representation of the differences in the standard and the correct way to treat geodetic precession. "GP" stands for geodetic precession/nutation. Each gray block represents a solution. A solid arrow means: add precomputed geodetic precession/nutation to a solution to get a new one. A dashed arrow means: recompute a solution with indicated change in the torque model

6 Difference to Existing GCRS Solutions

The difference between the Euler angles of the GCRS solution obtained in this work and the published kinematically non-rotating SMART solution (Bretagnon et al. 1998) is given in Fig. 46.4. Analysing the sources of these differences, one can identify three components:

– Influence of the rotation of the ephemeris shown in Fig. 46.2 due to the incorrect treatment of the geodetic precession.

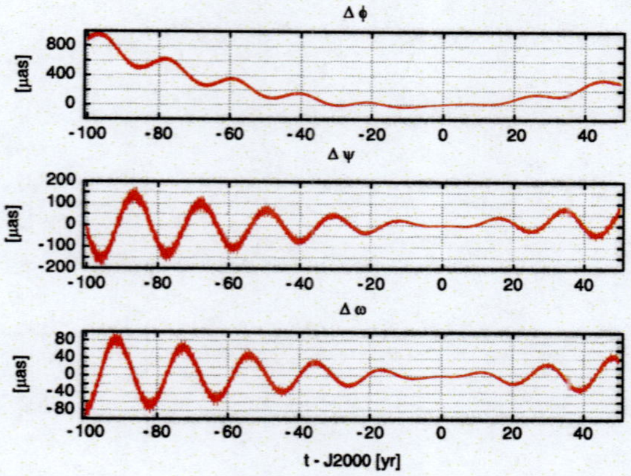

Fig. 46.4 Differences (in μas) between the published kinematically non-rotating SMART solution and the correct kinematically non-rotating solution derived in this study

- Sign error in the correction for geodetic precession in φ.
- Errors of the analytical SMART solution compared to the more accurate numerical integration, as already discussed in Bretagnon et al. (1998).

It should be remarked that the above-mentioned inconsistency is not only restricted to the SMART solution, which we used in this study for comparison, but is also contained in the IAU 2000A Precession-Nutation model as described in Sect. 5.5.1 of the IERS Conventions (McCarthy and Petit 2004). Therefore Fig. 46.4 allows us to conclude that the existing GCRS solutions for rigid Earth rotation are wrong by 1000 μas in φ, 200 μas in ψ and 100 μas in ω within 100 years from J2000. It can be shown that these differences cannot be eliminated by fitting the free parameters of our model, namely the moments of inertia, the initial Euler angles and their time derivatives.

Let us finally note that the effects of non-rigidity in the Earth rotation and the inaccuracies of the corresponding models, e. g. for the atmosphere and oceans, are significantly larger than the effects discussed in this work. Nevertheless to avoid a wrong geophysical interpretation of the observed Earth orientation parameters, the treatment of geodetic precession should be done along the lines presented in this paper.

References

Bretagnon P, Francou G, Rocher P, Simon JL (1998) SMART97: a new solution for the rotation of the rigid Earth. Astron. Astrophys. 329:329–338

Bretagnon P, Rocher P, Simon JL (1997) Theory of the rotation of the rigid Earth. Astron. Astrophys. 319:305–317

Brumberg VA, Bretagnon P, Francou G (1991) Analytical algorithms of relativistic reduction of astronomical observations. In: Journées 1991: Systèmes de Référence Spatio-temporels, pp 141–148

de Sitter W (1916) Einstein's theory of gravitation and its astronomical consequences. Mon. Not. R. Astron. Soc. 76:699–728

Klioner SA, Soffel MH, Xu C, Wu X (2001) Earth's rotation in the framework of general relativity: rigid multipole moments. In: The Celestial Reference Frame for the Future (Proc. of Journées 2007), N. Capitaine (ed.), Paris Observatory, Paris, pp 232–238

Klioner SA, Soffel MH, Le Poncin-Lafitte C (2008) Towards the relativistic theory of precession and nutation. In: Journées 2007: Systèmes de Référence Spatio-temporels, pp 139–142

Klioner SA, Gerlach E, Soffel MH (2010) Relativistic aspects of rotational motion of celestial bodies. In: Klioner SA, Seidelmann PK, Soffel MH (eds.) IAU Symposium 261, pp 112–123

McCarthy DD, Petit G (2004) IERS Conventions (2003). IERS Technical Note No.32, BKG, Frankfurt

Seidelmann PK (1982) 1980 IAU Nutation: The Final Report of the IAU Working Group on Nutation. Celest. Mech. Dyn. Astron. 27:79–106

Soffel MH, Klioner SA, Petit G, Wolf P, Kopeikin SM, Bretagnon P, Brumberg VA, Capitaine N, Damour T, Fukushima T, Guinot B, Huang T, Lindegren L, Ma C, Nordtvedt K, Ries JC, Seidelmann PK, Vokrouhlický D, Will CM, Xu C (2003) The IAU 2000 Resolutions for Astrometry, Celestial Mechanics, and Metrology in the Relativistic Framework: Explanatory Supplement. Astron. J. 126:2687–2706

47 Possibilities and Limits for Estimating a Dynamic and a Geometric Reference Frame Origin by the Integrated Approach Applied to the CHAMP–GRACE–GPS Constellation

Daniel König and Rolf König

Abstract

The integrated approach, i.e. the simultaneous processing of GPS satellites and Low Earth Orbiters (LEOs), for satellite orbit determination and estimation of Earth system parameters, is applied to the CHAMP–GRACE–GPS constellation for the purpose of deriving the dynamic and geometric origin of the Earth. The Earth system parameters comprise the low-degree gravity field coefficients and GPS ground station positions representing the center of mass of the Earth as well as the center of figure of the Earth. Based on simulated observations and dedicated changes of the reference models the possibilities and limits of solving and separating the parameters searched for are analyzed and the inherent datum defect can be qualified. The effects of different constraints imposed on the parameters and of various constellations of GPS satellites and LEOs are studied.

Keywords

Integrated approach • Geocenter • Datum defect • CHAMP • GRACE • GPS

1 Introduction

A geodetic Earth reference frame consists of a geometric and a dynamic part. Whereas the geometric frame is determined by the geometry of the Earth's surface, the dynamic frame is determined by the Earth's gravity field. Both reference frames require scale, origin, and orientation to be defined. The following is devoted to the determination of the origin of the geometric frame and that of the dynamic frame. Accordingly, both origins are denoted as geometric origin and dynamic origin or geocenter.

D. König (✉) · R. König
GFZ German Research Centre for Geosciences, c/o DLR, Muenchner Straße 20, 82234 Wessling, Germany
e-mail: dkoenig@gfz-potsdam.de

Theoretically, the geometric origin is defined as the integral center of figure of the entire Earth surface according to Dong et al. (1997), Dong et al. (2003), and Blewitt (2003). In practice, it is approximated by a set of coordinates of a ground station network distinctly distributed on the Earth's surface.

Given a spherical harmonic expansion of the Earth's gravity field the dynamic frame is established by the coefficients up to degree and order two. As revealed by their geophysical meaning, see Heiskanen and Moritz (1967), they represent the scale, the origin, and the orientation of the dynamic frame. The origin is the Earth's center of mass and its x-, y-, and z-component are given by the coefficients C_{11}, S_{11}, and C_{10}, respectively. As follows from the definitions the difference between both centers becomes clear: The dynamic origin is determined by the distribution of the Earth's

N. Sneeuw et al. (eds.), *VII Hotine-Marussi Symposium on Mathematical Geodesy*, International Association of Geodesy Symposia 137, DOI 10.1007/978-3-642-22078-4_47,

masses, and variations of its position are caused by mass motions. In contrast, variations of the geometric origin are coming from deformations of the Earth's surface.

Determining both a geometric frame and a dynamic frame is done highly efficient by precise orbit determination (POD) for GPS satellites and some GPS-tracked Low Earth Orbiters (LEOs) as the orbits establish a link between the geometric and the dynamic frame through GPS observations made at ground stations and in space. The method applied here is the integrated approach as proposed by Zhu et al. (2004). This method ensures both consistency of modeling and consideration of all correlations in space and time. This is not the case in the commonly applied approach where the parameter estimation for the GPS satellites and the LEOs is split into two or more steps. Furthermore, the high-orbiting GPS satellites provide stability to the geometric frame while the low-orbiting LEOs detect the dynamic frame with high sensitivity.

This study tries to quantify solvability and separability of the geometric and dynamic origin with the integrated approach by an empirical strategy. For this, ground and space-borne GPS data are simulated based on up-to-date a priori models acting as reference. By changing these a priori models systematically and introducing them into a recovery process for POD and parameter estimation the deviations of the estimated parameters of the geometric frame and the dynamic frame from the reference can be analyzed. This procedure leads also to the qualification of the intrinsic datum defect of the chosen solution.

2 The Testing Scenario

In the following the main elements of the testing scenario are described. The satellite constellation used consists of an outer layer of GPS satellites and an inner layer of one to three LEOs. In case only one LEO is used it is of CHAMP-type (Reigber et al. 1999), two LEOs are modeled as a GRACE-type (Tapley et al. 2004) satellite pair, and three LEOs are equivalent to a CHAMP and GRACE constellation. The ground station network is synthetic consisting of 60 regularly distributed stations on a 30° × 30° mesh. Using such an ideal ground station network is necessary in order to exclude disturbing effects coming from a nonuniform distribution of the stations. Investigating such disturbing effects would be worthwile but is out of the scope of this paper. As gravity field GFZ's EIGEN-GL04S1 (Foerste et al. 2007) is used, the initial elements for the GPS orbits are taken from IGS ultra-rapid orbits (IGS 2009) and for CHAMP and GRACE from routine predictions made by GFZ (Schmidt et al. 2003). Earth tides are modeled according to the IERS Conventions (2003) (McCarthy and Petit 2004), ocean tides according to FES2004 (Lyard et al. 2006). The non-conservative orbit perturbations are modeled according to accelerations resulting from CHAMP and GRACE POD based on real data. Fixing all a priori models, GPS observations are simulated as L3-linear combinations with a spacing of 30 s, a measurement sigma of 0.5 m (ground stations) and 0.3 m (LEOs) for code observations as well as 5.0 mm (ground stations) and 3.0 mm (LEOs) for phase observations. For the GRACE-type satellite pair also K-band range rate (KBRR) observations are simulated as an inter-satellite link with a spacing of 5 s and a measurement sigma of $0.3\,\mu m/s$. The observational data is simulated for any distinct satellite constellation. Each test covers the same period of 28 days consisting of daily solutions of 24 h length. For both simulations and the subsequent POD and parameter estimation the software used is GFZ's EPOS-OC. Its major functions are described in Zhu et al. (2004). The recoveries are carried out as combined POD and parameter estimation processes following the integrated approach using the simulated data and modified a priori models. The Earth system parameters solved for comprise the gravity field coefficients up to degree and order two and the ground station coordinates. Moreover, the initial elements of the satellite orbits are estimated as well as a set of parameters as usual in POD with real data modeling orbit perturbations and measurement errors. The a priori model changes are applied to C_{10}, C_{11}, S_{11} and to all station coordinates by adding 6 cm to each parameter of these parameter groups. I.e., the gravity field as a whole and the complete ground station network are translated in x, y, and z by this amount. 6 cm in all three spatial directions correspond to roughly 10 cm in 3D. Such a model change does not harm the linearity of the least-squares estimation but causes a clear effect.

Assessment of the results is done by comparing the estimated parameters to the unbiased model values. In case of the gravity field coefficients this is simply done by calculating the difference between the corresponding coefficients. The estimated ground station coordinates are compared to the reference coordinates by 7-parameter Helmert transformations per daily solution delivering global translations in x, y, and z (TX, TY, TZ in the following), a global scale, and global rotations around the x-, y-, and z-axis (RX, RY, RZ in the following). Analogously also the orbits are compared to the reference orbits by 7-parameter Helmert transformations. Those reference orbits are the orbits from the simulation process with unchanged a priori models. For the purpose of intuitive comparison, all non-metric parameters are rescaled to metric distances on the Earth surface.

3 Results and Discussion

The test cases are arranged in four groups, the results are presented in Tables 47.1–47.4. All parameters not considered here are resolved very well in all test cases. Concerning the Helmert transformations between the orbits, results of the transformations between the GPS orbits only are discussed as the results of the transformations between the LEOs' orbits do not differ significantly from the GPS results. All values given in the tables represent mean values over the 28-day period together with their standard deviations. As a reference for all test cases serves an initial test that is called 'standard case' in the following, i.e. an integrated

Table 47.1 Results: Standard case and ideal solution

Constraints sigmas	Deviations from reference (mean and standard deviation (mm))				
Initial Elements (mm)	Geocenter C10	Station Coord.		GPS orbits	
		TZ	RZ	TZ	RZ
–	-2.4 ± 68.5	-2.4 ± 68.6	$+28.1 \pm 205.7$	-2.5 ± 68.7	-28.1 ± 205.7
1.0	0.0 ± 0.2	0.0 ± 0.2	0.0 ± 0.0	0.0 ± 0.2	0.0 ± 0.1

Table 47.2 Results: Different constraints

Constraints sigmas	Deviations from reference (mean and standard deviation (mm))				
Station Coord. (cm)	Geocenter C10	Station Coord.		GPS orbits	
		TZ	RZ	TZ	RZ
–	-2.4 ± 68.5	-2.4 ± 68.6	$+28.1 \pm 205.7$	-2.5 ± 68.7	-28.1 ± 205.7
10	$+57.8 \pm 2.4$	$+57.9 \pm 2.4$	$+0.4 \pm 1.8$	$+57.8 \pm 2.5$	-0.4 ± 1.7

Table 47.3 Results: Additional LEOs

LEOs used	KBRR used	Deviations from reference (mean and standard deviation (mm))				
	y/n	Geocenter C10	Station Coord.		GPS orbits	
			TZ	RZ	TZ	RZ
C	n	-2.4 ± 68.5	-2.4 ± 68.6	$+28.1 \pm 205.7$	-2.5 ± 68.7	-28.1 ± 205.7
G	n	-3.2 ± 101.0	-3.2 ± 101.2	-77.8 ± 296.2	-3.3 ± 101.2	$+77.8 \pm 296.2$
C+G	n	-5.0 ± 47.8	-5.0 ± 47.9	$+24.3 \pm 122.9$	-5.0 ± 47.9	-24.3 ± 122.9
C+G	y	-8.5 ± 23.3	-8.5 ± 23.4	-9.5 ± 152.1	-7.7 ± 23.4	$+9.5 \pm 152.1$

C CHAMP, *G* GRACE, *C+G* CHAMP+GRACE KBRR: K-band range-rate

Table 47.4 Results: Additional GPS orbit plane

Additional GPS orbit plane	Deviations from reference (mean and standard deviation (mm))				
y/n	Geocenter C10	Station Coord.		GPS orbits	
		TZ	RZ	TZ	RZ
n	-2.4 ± 68.5	-2.4 ± 68.6	$+28.1 \pm 205.7$	-2.5 ± 68.7	-28.1 ± 205.7
y	-11.2 ± 55.7	-11.3 ± 55.8	$+21.8 \pm 169.9$	-11.3 ± 55.8	-21.7 ± 169.9

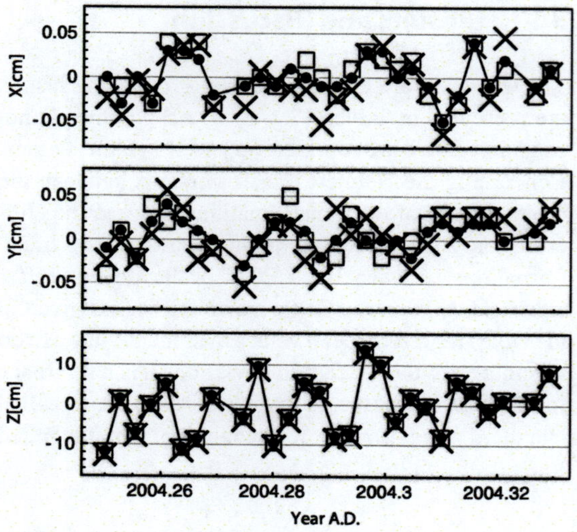

Fig. 47.1 Time series of dynamic origin (*circles*), geometric origin (*squares*), and translations of GPS orbits (*crosses*) for the standard case (no constraints)

solution of the CHAMP–GPS constellation without any a priori constraints on all parameters.

First of all it should be stated that for all test cases presented in the following the resolved values in the x- as well as in the y-component are very stable and accurate: their standard deviations do not exceed 0.3 mm, the mean values are between −0.1 and +0.1 mm. This means that both components are extremely well solvable simultaneously. Therefore the following discussion is focusing the z-component.

In order to give an idea of the character of the solutions, for the standard case the time series of the translational parameters are shown in Fig. 47.1. The maximum difference between the time series is 0.4 mm in the x-, 0.6 mm in the y-, and 0.7 mm in the z-component. From this and visual inspection it can be stated that the translational parameters tend to the same direction meaning that they represent the same physical effect.

The first group of tests, see Table 47.1, comprises the standard case as well as a solution where the initial elements of all satellite orbits are constrained with a sigma of 1.0 mm to their a priori values. Regarding the z-component of the translational parameters the standard case cannot resolve them as seen by the standard deviations between 68.5 and 68.7 mm corresponding to the modeled changes of 6 cm. The second case with perfect orbits delivers a stable day by day recovery of these parameters with standard deviations of 0.2 mm. On the other hand, for the standard case, the mean values of the z-translations deviate from the reference by −2.5 to −2.4 mm only. So by sampling over a certain time period an unbiased solution can be gained.

The whole constellation of ground stations and satellite orbits is rotationally unstable around the z-axis as can be seen by regarding the RZs in the standard case with standard deviations of 205.7 mm and biases of +28.1 and −28.1 mm. Again with perfect orbits the RZs could be resolved day by day with standard deviations of 0.1 mm maximum and no biases. Summarizing, the integrated CHAMP-GPS constellation solution delivers highly accurate daily x- and y-components of the geometric and dynamic origin simultaneously, however the z-translations need long-term sampling to become free of systematics. With perfect orbits (however out of reality) daily resolution of the z-translations would be possible and the rotational defect would disappear.

In the second group of tests, see Table 47.2, the standard case is compared to a case where 10-cm constraints are imposed on the station coordinates. Such constraints are in use for operational GPS processing where they are meant to prevent any impact of unexpected antenna changes on the overall solution. As revealed by the standard deviations of 2.4–2.5 mm the case of constrained station coordinates leads to much more stable time series of the z-translations than the standard case with standard deviations of 68.5 and 68.7 mm. On the other hand, in case of constrained station coordinates the biases of these time series, +57.8 to +57.9 mm, are on the level of the artificial error of 6 cm. This is certainly due to the constraints of 10 cm that are relatively strong compared to the changes of 6 cm. Regarding the RZ estimates, as opposed to the standard case, the case of constrained station coordinates causes the solutions to become rotationally stationary around the z-axis as revealed by standard deviations of 1.7 and 1.8 mm and biases of +0.4 and −0.4 mm. So in summary putting 10-cm a priori constraints on the station coordinates allows to efficiently care for the rotational stability of the solution. If the solution aims at resolving the geocenter motion with expected amplitudes in the millimeters, c.f. Dong et al. (1997), the 10-cm a priori constraints would allow the recovery of daily translations with a few millimeter accuracy.

The third group of tests, see Table 47.3, aims at clarifying the effect of making use of additional LEOs or of KBRR observations. Using a GRACE satellite pair without KBRR data leads to increased standard deviations of 101.0–101.2 mm in the z-translations in comparison to the standard CHAMP case. The reason being that correlations of environmental perturbations between both satellites are not taken into account here. In case of combining CHAMP and GRACE the standard deviations reduce to 47.8–47.9 mm being also smaller than the standard case. By using KBRR data the standard deviations improve further down to 23.3–23.4 mm. This can be expected as combining CHAMP and GRACE means to increase the number of observations, and using KBRR data means to introduce high-quality relative measurements. Compared to the standard case the biases increase up to approximately −8 mm, however staying well below the standard deviations.

Concerning the z-rotations the standard deviations behave as above going through the cases. The biases of the z-rotations stay again below significance. In summary combining CHAMP and GRACE as well as adding KBRR data help to improve the rotational stability around the z-axis.

In the fourth and last test group (see Table 47.4) the benefit of an additional near-polar GPS orbit plane of five equally distributed satellites is investigated whose keplerian elements are $a = 26{,}600$ km, $e = 0.005$, $i = 89°$, $\Omega = 85°$. Compared to the standard case the standard deviations of the translational parameters in Z reduce slightly. Also the standard deviations of the z-rotations reduced. The biases in the z-translations and -rotations are not significant. Overall, an additional near-polar GPS orbit plane would make the z-translations and -rotations more stable but would not allow an accurate solution in the end.

Conclusion

Based on simulated observations and an ideal ground station network empirical evidence is given that it is possible to determine simultaneously both the geometric and the dynamic reference frame origin applying the integrated approach to a CHAMP–GRACE–GPS constellation. Even in the absence of any a priori constraints it is possible to resolve simultaneously the geometric and the dynamic x- and y-components with daily resolution. The remaining errors in z-direction and around the z-axis indicate that there is a datum defect of two imminent to the chosen configuration. Averaging over a longer period leads to nearly unbiased z-coordinates of the origins. A perfect estimation of all the components of the origins would be possible in case the orbits would become a factor of 10 more accurate than current practice. Constraining the ground station coordinates removes the rotational freedom around the z-axis but leaves the z-translations unresolved. Extending the basic constellation by a GRACE-type satellite pair leads to improved z-translation accuracies and z-rotation stabilities. This improvement is largely pronounced if the K-band range-rate data are used. An additional near-polar GPS orbit plane would not lead to a thorough elimination of the deficiencies in the z-translations and -rotations.

Acknowledgment This study was funded within the Geotechnologien program under the grant 03G0728A of the German Federal Ministry of Education and Research (BMBF). IGS ultra-rapid orbits for the GPS satellites were provided by the International GNSS Service (IGS), IGS (2009).

References

Blewitt G (2003) Self-consistency in reference frames, geocenter definition, and surface loading of the solid Earth. J Geophys Res 108(B2):2203, doi:10.1029/2002JB002082

Dong D, Dickey JO, Chao Y, Cheng MK (1997) Geocenter variations caused by atmosphere, ocean and surface ground water. Geophys Res Lett 24:1867–1870

Dong D, Yunck T, Heflin M (2003) *Origin of the International Terrestrial Reference Frame.* J Geophys Res 108(B4):2200, doi:10.1029/2002JB002035.

Foerste C, Schmidt R, Stubenvoll R, Flechtner F, Meyer U, König R, Neumayer H, Biancale R, Lemoine JM, Bruinsma S, Loyer S, Barthelmes F, Esselborn S (2007) *The GeoForschungsZentrum Potsdam/Groupe de Recherche de Geodesie Spatiale satellite-only and combined gravity field models: EIGEN-GL04S1 and EIGEN-GL04C.* J Geodesy 82(6):331–346

Heiskanen, W.A. and H. Moritz (1967). *Physical Geodesy.* W.H. Freeman and Company, San Francisco and London.

IGS - International GNSS Service (2009). Internet site http://igscb.jpl.nasa.gov/.

König, D., R. König and N. Panafidina (2007). *Combination of Ground Observations and LEO Data.* Geotechnologien Science Report, No. 11, ISSN 1619-7399

Lyard, F., F. Lefevre, T. Letellier and O. Francis (2006). *Modelling the global ocean tides: modern insights from FES2004.* Ocean Dynamics, 56: 394–415, Springer

McCarthy, D.D. and G. Petit (2004). *IERS Conventions (2003).* IERS Technical Note, No. 32, ISSN 1019-4568.

Reigber, Ch., P. Schwintzer, and H. Lühr (1999). *The CHAMP geopotential mission.* Bolletino di Geofisica Teorica e Applicata, 40, 285–289

Schmidt, R., G. Baustert, R. König and C. Reigber (2003). *Orbit Predictions for CHAMP - Development and Status.* In: Reigber C., H. Lühr and P. Schwintzer (eds), First CHAMP Mission Results for Gravity, Magnetic and Atmospheric Studies. Springer, Berlin, Heidelberg, pp. 104-111.

Tapley, B.D., S. Bettadpur, M. Watkins and C. Reigber (2004). *The gravity recovery and climate experiment: Mission overview and early results.* Geophysical Research Letters, 31, L09607-1-4

Zhu, S., C. Reigber and R. König (2004). *Integrated Adjustment of CHAMP, GRACE, and GPS Data.* J Geodesy 78(1–2):103–108

Source Parameters of the September 10, 2008 Qeshm Earthquake in Iran Inferred from the Bayesian Inversion of Envisat and ALOS InSAR Observations

48

Pegah Faegh-Lashgary, Mahdi Motagh, Mohammad-Ali Sharifi, and Mohammad-Reza Saradjian

Abstract

This paper presents InSAR observations and source parameter results for September 10, 2008 Qeshm Island (Mw 6.1) earthquake in Iran. An ascending ALOS interferogram and two Envisat interferograms from ascending and descending orbits are used to derive coseismic displacement field of the earthquake. Interferometric observations show coseismic displacement of more than 10 cm in the northern part of the island. To obtain source parameters of the earthquake and their associated posteriori probability distribution we use the Bayesian approach. The inversion is done in a two-step process. First we derive, using the Simulated Annealing (SA) method, the posterior marginal distribution of parameters to determine their sensitivity to the observations. This step is important as it determines which parameters can be resolved reliably by the inversion of geodetic data and which ones need to be constrained by other sources. Following the above procedure, we use a Genetic Algorithm (GA) to find the maximum Bayesian probability for each parameter. We find that InSAR observations of 2008 Qeshm earthquake are consistent with an average thrust motion of 75 cm that occurred on an $\sim$13 $\times$ 3 km fault plane with a centroid depth of $\sim$5 km, striking 218° and dipping north at 33°.

1 Introduction

On September 10, 2008 an earthquake of magnitude 6.1 occurred in Qeshm island region in southern Iran (Fig. 48.1). The Harvard centroid-moment tensor (CMT) solution indicates predominantly thrust faulting with dip 33°, strike 234° and rake of 76° (http://www.globalcmt.org/; Event ID. 200809101100A). In this study we use radar interferometric measurements obtained from ALOS and Envisat satellites to assess precisely the geometry and location of the rupture process associated with this earthquake.

P. Faegh-Lashgary (✉) · M.-A. Sharifi · M.-R. Saradjian
Department of Surveying and Geomatics Engineering, University of Tehran, Tehran, Iran
e-mail: pflashgary@gmail.com

M. Motagh
Department of Surveying and Geomatics Engineering, University of Tehran, Tehran, Iran

Helmholtz Centre Potsdam, GFZ German Research Centre for Geosciences, Potsdam, Germany

2 INSAR Data

We use Space-borne Synthetic Aperture Radar interferometry (InSAR) to measure the coseismic displacement field associated with the 2008 Qeshm earthquake.

N. Sneeuw et al. (eds.), *VII Hotine-Marussi Symposium on Mathematical Geodesy*, International Association of Geodesy Symposia 137, DOI 10.1007/978-3-642-22078-4_48, © Springer-Verlag Berlin Heidelberg 2012

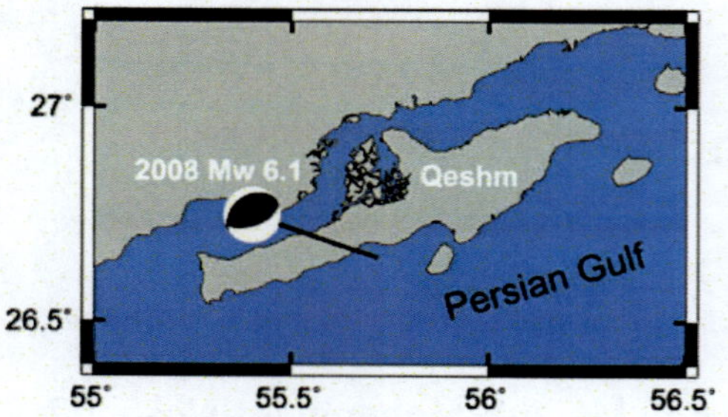

Fig. 48.1 Study area. The beach ball shows the focal mechanism of September 10, 2008 Qeshm earthquake

InSAR is a remote sensing technique that enables to calculate the interference pattern caused by the difference in phase between two images taken by a space-borne synthetic aperture radar over the same region from slightly different positions at two distinct times. The resulting interferogram is a contour map of the change in the distance between the ground and the radar instrument, which can be further processed to yield important parameters such as topography and the ground motion (Burgmann et al. 2000; Elachi and Jakob 2006).

Coseismic displacement from the September 10, 2008 Qeshm earthquake are measured using two C-band Envisat interferograms from ascending and descending orbits –hereafter called ifm1 and ifm2, and one L-band ascending interferogram from ALOS – hereafter called ifm3. Table 48.1 lists the data used in this study. The ascending Envisat images were acquired in I6 mode (incidence angle ~41), while the descending Envisat images were acquired in I2 mode (incidence angle ~23). The Envisat interferograms are processed using the DORIS software developed at Delft University of Technology (Kampes et al. 2003), while the ALOS interferogram was processed using the Sarscape software (www.sarmap.ch). The topographic contribution to interferometric phase in all interferograms was modeled using the 90-m digital elevation model (DEM), generated by the NASA Shuttle Radar Topography Mission (SRTM). The resulting differential interferograms are then filtered using a weighted power spectrum technique (Goldstein and

Table 48.1 Interferograms used in this study

Sensor	Envisat (ifm1)	Envisat (ifm2)	ALOS (ifm3)
Master	25.04.07	17.04.08	30.12.06
Slave	01.10.08	09.10.08	19.11.08
Heading angle (°)	350	190	344
Incidence angle (°)	40.7	22.4	38.73
Baseline	255.9	244	560.1

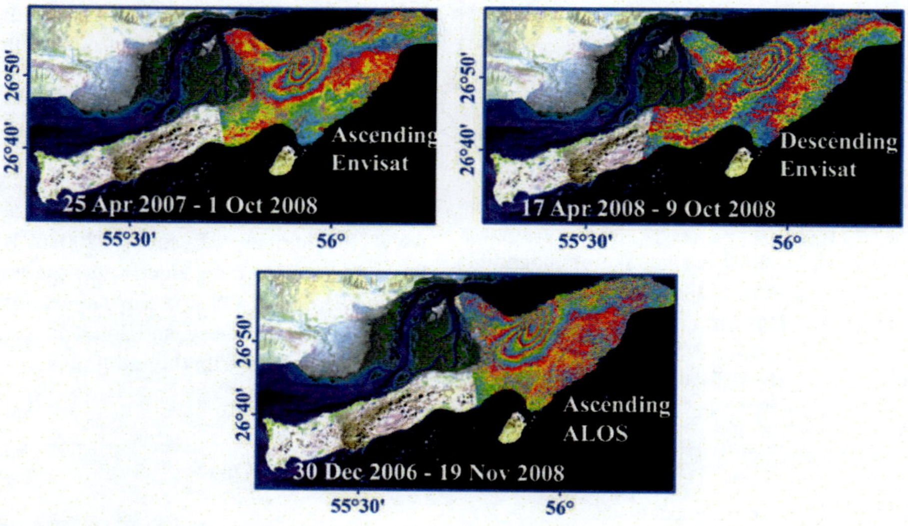

Fig. 48.2 InSAR observations of coseismic displacement caused by September 10, 2008 Qeshm earthquake in Iran. Data are unwrapped and rewrapped again with each color cycle (fringe) representing 4 cm of ground motion in the line-of-sight direction of the satellite

Werner 1998), unwrapped using the Snaphu software (Chen and Zebker 2001) and converted from phase cycles to centimeters of range changes in the satellite line-of-sight (LOS). The favorable environmental conditions for radar interferometry in southern Iran, which is dominated by dry climate and sparse vegetation, results in very high interferometric correlation and consequently, reliable unwrapping.

Figure 48.2 shows three coseismic interferograms associated with the September 10, 2008 Qeshm earthquake. The pattern of coseismic displacement field is remarkably similar in 3 interferograms, all showing an uplift signal of more than 10 cm at the northern part of the Qeshm Island.

The coseismic interferograms used in this study cover different time spans before (master image) and after (slave) the earthquake. Therefore, they may encompass not only coseismic but also interseismic and postseismic deformation. However, the analysis of independent interferograms before and after the earthquake shows no significant (cm-level) preseismic or postseismic displacement in the study area (not shown here). Therefore, the interferograms illustrated here reflect mainly coseismic signal associated with the 2008 Qeshm earthquake.

3 Bayesian Inversion for Fault Parameters

We use InSAR data to determine a simple fault model consisting of uniform slip on a rectangular fault embedded in uniform elastic half-space (Okada 1985). The relation between InSAR data and fault parameters can be expressed as $d = G(x) + e$ where d denotes displacement data obtained by InSAR, x is a vector of fault parameters describing its length, width, depth, dip, strike, horizontal coordinates of the midpoint of its upper edge, and the amount of slip in strike and dip directions, e is the observation error, and G is the nonlinear Green function in the LOS direction of the SAR satellite obtained by:

$$G(x) = d_v(x)\cos\theta_{inc} - \sin\theta_{inc} \times [d_e(x)\cos H - d_n(x)\sin H]$$

where θ is the look angle, H is the heading angle of the SAR satellite, and d_v, d_e and d_n are vertical, easting, and northing components of displacement field, respectively, obtained from (Okada 1985). To solve this nonlinear problem, we use the Bayesian probabilistic approach combined with a heuristic search approach of the Genetic Algorithm (Houck et al. 1996) as detailed below.

The Bayes' Rule is defined as (Tarantola 2005):

$$p(x \mid d) = \frac{p(x)p(d \mid x)}{\int p(x)p(d \mid x)dx} = \frac{1}{c}p(d \mid x)p(x)$$

where $p(x)$ denotes a priori probability distribution function (pdf) of unknown parameters (here fault parameters), $p(d \mid x)$ is the goodness of fit between the observed data and those computed from the forward model, called the likelihood expression, and $p(x \mid d)$ is the posterior probability of unknown parameters. The likelihood expression is computed as:

$$p(d \mid x) = 2\pi^{-\frac{n}{2}}|\Sigma_d|^{-\frac{1}{2}} \times \exp\left[-\frac{1}{2}(d - G(x))^T \Sigma_d^{-1}(d - G(x))\right]$$

where n is the number of data (observation points) and Σ_d is the covariance matrix of observations, describing the noise structure of the data. In this paper, we neglect the off-diagonal components of the covariance matrix and treat the InSAR data as independent observations with uniform weight.

For the inversion we follow a two-step process: First we obtain the sensitivity of individual parameters to the observations by computing their associated marginal distribution function, defined as (Moraes and Scales 2008):

$$p(x_i \mid d) = kp(x_i)\int p(x_n \mid x_i)p(d \mid x_i, x_n)dx_n$$

where x_i represents a single parameter, and x_n is the vector of all parameters excluding the x_i component. The marginal pdfs[1] can be determined by sampling the posterior space using an appropriate sampling tool. Here we use the simulated annealing algorithm with the acceptance probability criterion of Gibbs' distribution (Sen and Stoffa 1995) to sample posterior marginal pdfs (For details of the SA method please see e.g. Sen and Stoffa 1995). As the prior information, we assume a uniform distribution for parameters with lower and upper bounds as indicated in Table 48.2.

[1]Probability Density Function.

Table 48.2 Evaluating the sensitivity of fault parameters to InSAR datasets used in this study

	Bound	σ_f	σ_i	$P\%$
Length (km)	1–40	1.8	11.26	84
Width (km)	1–20	1.67	5.48	69
Depth (km)	0–20	0.77	5.77	87
Dip (°)	13–53	7.36	11.55	36
Strike (°)	214–254	3.72	11.55	68
Mid-Lat (°)	26.6–26.95	0.007	0.1	93
Mid-Lon (°)	55.6–56	0.006	0.12	95

The above procedure helps determine which parameters can be resolved reliably by the inversion of geodetic data and which ones need to be constrained by other sources (e.g. seismic data). Near uniform marginal distributions correspond to parameters with the least sensitivity to observations while the unimodal functions represent the highest sensitivity to observations.

Figure 48.3 shows the marginal distributions of the fault parameters associated with the 2008 Qeshm earthquake in Iran. As one can see in this figure, of the 7 fault geometry parameters the dip angle has the least sensitivity to the observations, evident from its near uniform marginal pdf. To better understand quantitatively to what extent our parameters are sensitive to observations, we list in Table 48.2 the percentage improvement in resolving the parameters by comparing the posterior standard deviations of them, computed from marginal pdfs, with their a priori ones, computed from the uniform distribution. The percentage improvement is defined as

$$P = \begin{cases} 0 & \sigma_f \geq \sigma_i \\ \frac{|\sigma_f - \sigma_i|}{\sigma_i} \times 100 & \sigma_f \prec \sigma_i \end{cases}$$

where σ_f and σ_i are the posterior and initial standard deviation of parameters, respectively.

As indicated in Table 48.2 most of the fault parameters are resolved very well in the inversion, which is indicated in their associated values of percentage improvement. Only the dip angle has a low percentage improvement of 36%, suggesting some constrains need to be defined for this parameter to narrow down the search space of the inverse problem.

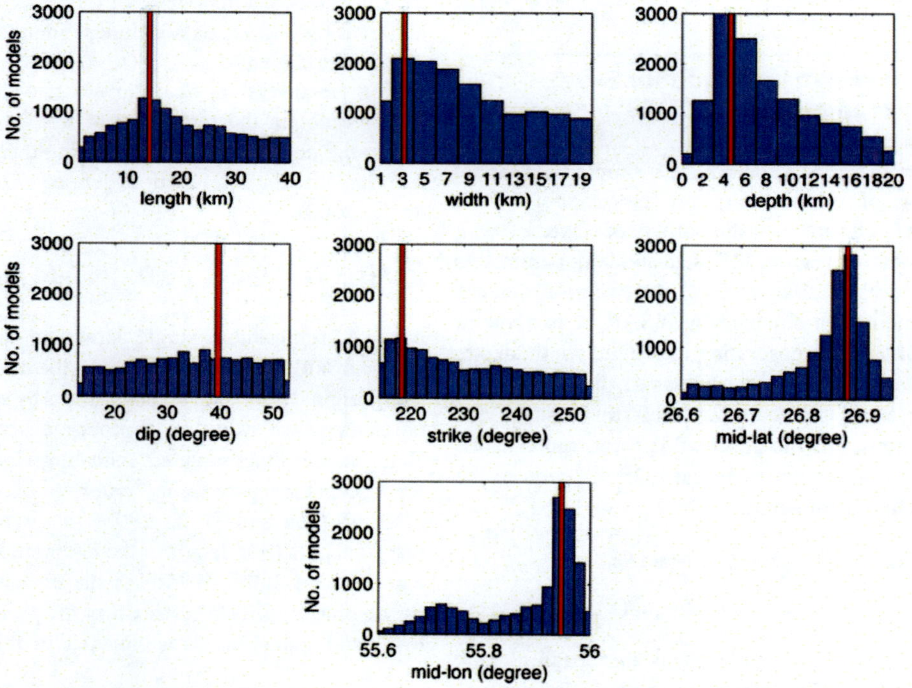

Fig. 48.3 Marginal frequency of fault parameters obtained by Gibbs' sampler. Red bars represent the mean of the distributions

Here we assume, as the a priori information, the dip angle has a normal distribution with the mean value of 33°- obtained from the CMT solution, and a standard deviation of 5°, while for other parameters a uniform distribution is assumed.

Following the above procedure to define the prior information on parameters, we use a Genetic Algorithm (GA) to find the maximum Bayesian probability for each parameter. We start the genetic algorithm by generating 1,000 random individuals (source models), which are selected in a way that each individual fulfills the condition of producing an earthquake with magnitude between 5.9 and 6.2. In this way we eliminate from the initial population of GA the unrealistic source parameters, allowing GA to better search for the most probable values of source parameters when computing posterior probability in each generation.

Table 48.3 Estimated parameters of the 2008 Qeshm earthquake

	ifm1	ifm2	ifm3	ifm1+ifm2+ifm3
Length (km)	12.22 ± 3.37	14.46 ± 2.51	14.07 ± 3.84	13.29 ± 1.6
Width (km)	7.09 ± 2.09	1.000 ± 0.68	3.36 ± 1.41	3.24 ± 0.71
Depth (km)*	3.73 ± 0.93	4.84 ± 0.72	4.17 ± 1.06	4.12 ± 0.45
Dip (°)	33.04 ± 3.97	33.12 ± 2.98	32.96 ± 3.99	33.21 ± 2.87
Strike (°)	224.25 ± 7.72	219.92 ± 4.96	216.04 ± 6.86	218.41 ± 3.54
Mid-Lat (°)	26.88 ± 0.005	26.88 ± 0.0004	26.88 ± 0.007	26.88 ± 0.0000
Mid-Lon (°)	55.95 ± 0.006	55.94 ± 0.0002	55.95 ± 0.006	55.94 ± 0.0002
Strike-slip (m)	−0.00 ± 0.07	−1.38 ± 0.64	−0.13 ± 0.15	−0.2 ± 0.08
Dip-slip (m)	0.38 ± 0.07	2.42 ± 0.24	0.58 ± 0.13	0.75 ± 0.05

*Depth of the upper edge of the fault plane

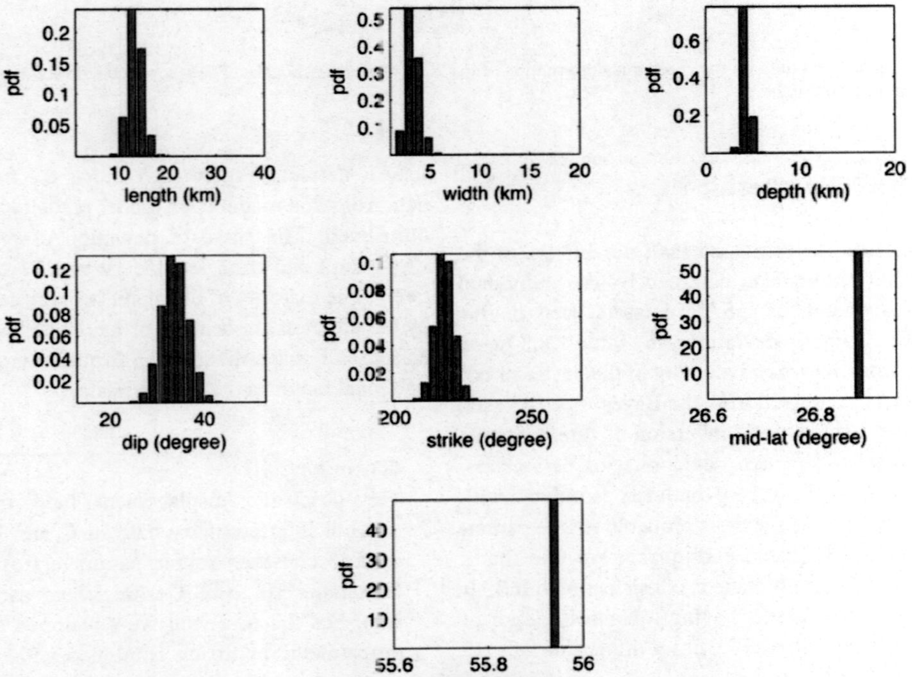

Fig. 48.4 Probability distribution of parameters associated with the 2008 Qeshm earthquake

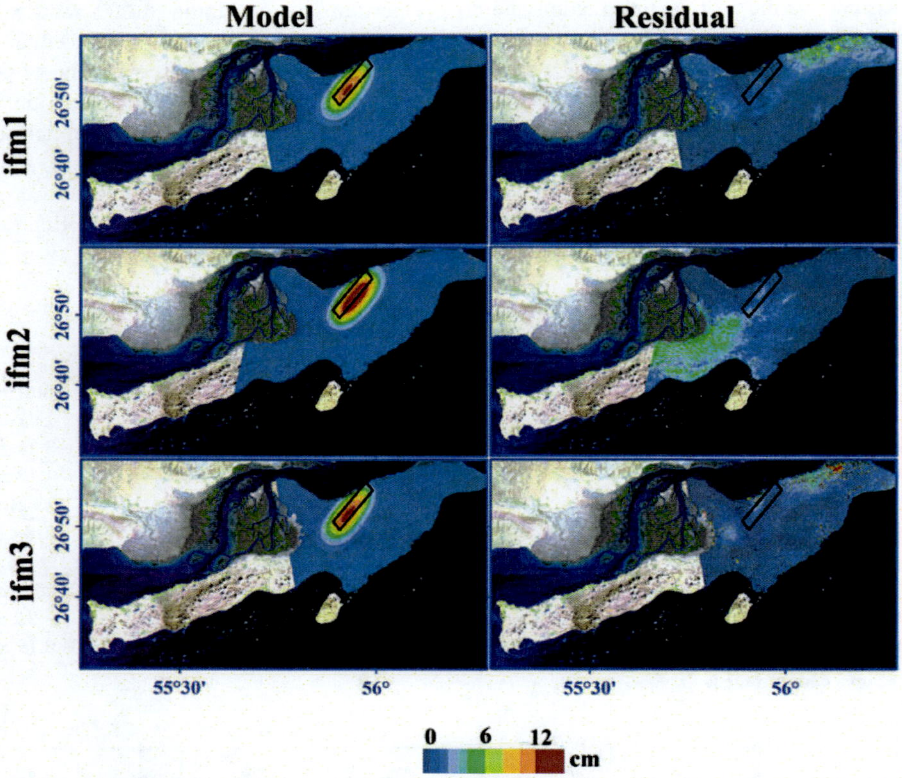

Fig. 48.5 Modeling results for the coseismic deformation of the 2008 Qeshm earthquake. The black rectangle shows the horizontal projection of the fault plane

4 Modeling Result

Table 48.3 lists the estimated fault parameters of the 2008 Qeshm earthquake, obtained by the individual and joint inversion of InSAR datasets used in this study. The standard deviations in Table 48.3 were estimated from the second moment of the posterior pdf of parameters as derived from the Bayes' rule. As seen in this table, using a joint inversion of three datasets result in smaller standard deviations for parameters. Therefore, we consider the parameters associated with this joint inversion as the most probable source parameters of the 2008 Qeshm earthquake. We find that a 33° north-dipping fault plane is consistent with InSAR observations. This plane, having a centroid depth of 5 km with a strike of ~218°, has a dimension of ~13 by 3 km (Fig. 48.5).

Figure 48.4 shows the posteriori distribution of parameters associated with the joint inversion of 3 InSAR datasets. Figure 48.5 shows the fit to InSAR data from this model. Our model predicts InSAR data quite well. The standard deviation of residuals for ifm1, ifm2 and ifm3 are 15, 19 and 22 mm, respectively. The majority of the misfit is likely caused either by the effect of atmosphere or by the deficiency in the elastic half-space model. The former seems to be the dominant factor in the far-filed data.

Conclusion

The coseismic displacement field mapped by satellite interferometry data in C and L bands is used to constrain source parameters of Mw = 6.1 September 10, 2008 Qeshm Island earthquake in Iran. For the inversion we developed a Bayesian probabilistic approach combined with a heuristic search algorithm. This allows us to overcome in a probabilistic way the non-uniqueness of the inverse problem and infer a meaningful posterior density for parameters from the a priori information used in the inversion. Our results indicate that InSAR observations are consistent with a shallow thrust faulting (centroid depth ~5 km) of about 75 cm on a plane dipping northward. The dip and strike are ~33° and 218°, respectively on a fault plane ~13 km long and 3 km wide.

Acknowledgements The authors acknowledge the comments of five anonymous reviewers, which led to significant improvements in the manuscript.

References

Burgmann R, Rosen PA, Fielding EJ (2000) Synthetic aperture radar interferometry to measure earth's surface topography and its deformation. Annu Rev Earth Planet Sci 128:169–209

Chen CW, Zebker HA (2001) Two-dimensional phase unwrapping with use of statistical models for cost functions in nonlinear optimization. J Opt Soc Am A 18(2):338–351

Elachi C, Jakob V (2006) Introduction to the physics and techniques of remote sensing. Wiley Series in Remote Sensing

Goldstein R, Werner C (1998) Radar interferogram filtering for geophysical applications. Geophys Res Lett 25:4035–4038

Houck CR, Joines JA, Kay MG (1996) A genetic algorithm for function optimization: a Matlab implementation. NCSU-IE TR 95–09

Kampes B, Hanssen R, Perski Z (2003) Radar interferometry with public domain tools. FRINGE, Frascati

Moraes FS, Scales JA (2008) Local bayesian inversion: theoretical developments. Geophys J Int 141:713–723

Okada Y (1985) Surface deformation due to shear and tensile faults in a half-space. Bull Seismol Soc Am:1135–1154

Sen M, Stoffa PL (1995) Global optimization methods in geophysical inversion. Elsevier, Amsterdam

Tarantola A (2005) Inverse problem theory and methods for model parameter. estimation. ISBN: 13:978–0–898715–72–9

Part VIII

Theory, Implementation and Quality Assessment of Geodetic Reference Frames

The Choice of Reference System in ITRF Formulation

49

Zuheir Altamimi and Athanasios Dermanis

Abstract

The problem of choosing an optimal reference system for the International Terrestrial Reference Frame (ITRF) is studied for both the rigorous solution which is a simultaneous stacking (removal of the reference system at each data epoch and implementation of a linear in time coordinate model) for all techniques, as well as for the usual numerically convenient separation into a set of individual stackings one for each technique and a final combination step for the derived initial coordinates and velocities. Two approaches are followed, an algebraic and a kinematic one. The algebraic approach implements the inner constraints, which minimize the sum of squares of the unknown parameters, as well as partial inner constraints, which minimize the sum of squares of a subset of the unknown parameters. In the kinematical approach the optimal minimal constraints are derived by requiring the minimization of the apparent coordinate variations: (a) with respect to the origin by imposing constant coordinates for the network barycenter, (b) with respect to orientation by imposing zero relative angular momentum for the network points conceived as mass points with equal mass and (c) with respect to the scale by imposing constant mean quadratic size (involving the distances of stations from their barycenter).

Keywords

Reference systems • ITRF • Minimal constraints • Inner constraints

1 Introduction

The implementation of an International Terrestrial Reference System (ITRS) by means of an International Terrestrial Reference Frame (ITRF) is based on the utilization of time series of station coordinates referring to different but overlapping subnetworks, one from each particular space technique (VLBI, SLR, GPS, DORIS). The object is to construct an optimal set of initial coordinates $\mathbf{x}_{0i}$ and velocities $\mathbf{v}_i$ for the stations P_i of the ITRF network which is the union of the subnetworks of all techniques. The adoption of the simple model of linear evolution in time

$$\mathbf{x}_i(t) = \mathbf{x}_{0i} + (t - t_0)\mathbf{v}_i \qquad (49.1)$$

Z. Altamimi
Institut Géographique NationalENSG/LAREG, Champs-sur-Marne, France

A. Dermanis (✉)
Department of Geodesy and Surveying, Aristotle University of Thessaloniki, University Box 503, 54124 Thessaloniki, Greece
e-mail: dermanis@topo.auth.gr

N. Sneeuw et al. (eds.), *VII Hotine-Marussi Symposium on Mathematical Geodesy*, International Association of Geodesy Symposia 137, DOI 10.1007/978-3-642-22078-4_49,

(Altamimi et al. 2007, 2008) imposes a smooth temporal variation in order to remove noise from the input data, although systematic effects of various geophysical origins remain in the final ITRF residual series. With respect to the input coordinate data it is assumed that not only each technique refers to its own reference system but even each coordinate epoch refers to a separate reference system. The aim of this last rather strong assumption is the removal of systematic coordinate variations due to the temporal instability in the reference system definition. Thus the model for the observed coordinates $\mathbf{x}_{T,i}(t_k)$ of station P_i from technique T at epoch t_k has the form

$$\mathbf{x}_{T,i}(t_k) = \mathbf{f}_{T,i}(\mathbf{p}_{T,k}, \mathbf{x}_{0i}, \mathbf{v}_i) \tag{49.2}$$

where in addition to the standard ITRF unknowns $\mathbf{x}_{0i}$ and $\mathbf{v}_i$, additional nuisance parameters appear, namely the transformation parameters $\mathbf{p}_{T,k}$ from the ITRF reference system to the one for each technique T and each coordinate input epoch t_k. In the most general case these involve 3 displacement components d_1, d_2, d_3, 3 rotation angles θ_1, θ_2, θ_3, and a scale factor $1+s$.

The determination of initial coordinates and velocities with simultaneous transformation of every epoch coordinates to a common reference system is usually referred to as "stacking" (Altamimi et al. 2007, 2008). Thus the ITRF formulation problem is in fact a simultaneous stacking for all the techniques, which involves a very large number parameters, most of which are the parameters $\mathbf{p}_{T,k}$, which are nuisance parameters in the ITRF formulation but they are needed for transforming earth orientation parameters from each epoch and technique to the reference system of the ITRF. In order to cope with the computational burden of a simultaneous stacking a two step approach is used instead (Altamimi et al. 2007, 2008). In the first step a stacking is performed for each technique T separately producing initial coordinates $\mathbf{x}_{T,0i}$ and velocities $\mathbf{v}_{T,i}$. In the second "combination" step the initial velocities from all techniques are combined to obtain the common ones of the ITRF. The model of the observations of the form

$$\mathbf{x}_{T,0i} = \mathbf{f}_{\mathbf{x}}(\mathbf{x}_{0i}, \mathbf{v}_i, \mathbf{q}_T), \quad \mathbf{v}_{T,i} = \mathbf{f}_{\mathbf{v}}(\mathbf{x}_{0i}, \mathbf{v}_i, \mathbf{q}_T) \tag{49.3}$$

involves the transformation parameters $\mathbf{q}_T$ from the ITRF reference system to that of the one of each technique T after its own stacking has been performed.

The observation models for the simultaneous stacking or the separate stackings and the combination have an inherent rank deficiency due to the lack of definition of the reference system. Indeed any change in the ITRF reference system by a particular transformation is counterbalanced by a change of the transformation parameters by the inverse transformation. If e.g. we write (49.2) in the form $\mathbf{x}_T = T(\mathbf{p})\mathbf{x}$ a coordinate transformation $\mathbf{x}' = T(\delta\mathbf{p})\mathbf{x}$ with inverse $\mathbf{x} = T^{-1}(\delta\mathbf{p})\mathbf{x}'$ leads to the model $\mathbf{x}_T = T(\mathbf{p}')\mathbf{x}'$ where the transformation parameters change from $\mathbf{p}$ to the ones $\mathbf{p}'$ implied by $T(\mathbf{p}') = T(\mathbf{p})T^{-1}(\delta\mathbf{p})$.

Therefore the rank deficiency must be overcome by choosing an optimal reference system among all possible ones. This is typically done in classical rigid geodetic networks by introducing additional constraints on the parameters which resolve the "choice of datum" problem. The fact that we are dealing with deformable networks requires the choice of an optimal reference system at each particular epoch among equivalent reference systems with coordinates connected by transformations $\mathbf{x}'(t) = T(\delta\mathbf{p}(t))\mathbf{x}(t)$ with parameters $\delta\mathbf{p}(t)$ which are smooth functions of time. A problem that arises in this respect is that general coordinate transformations $T(\delta\mathbf{p}(t))$ are not compatible with the linear time evolution model (49.1), since they transform coordinates $\mathbf{x}_i(t)$ into coordinates $\mathbf{x}'_i(t)$ which are not linear with respect to time t.

2 Observation Equations for Model-Preserving and Close to Identity Transformations

In order to overcome the above problems we follow the usual linearization procedure of replacing parameters with their corrections to known approximate values and neglecting of second and higher order terms. We also assume that the coordinate transformations involved are "close to the identity" so that only first order terms in the small coordinate transformation parameters $\mathbf{d} = [d_1\, d_2\, d_3]^T$, $\boldsymbol{\theta} = [\theta_1\, \theta_2\, \theta_3]^T$ and s are preserved. Since $\mathbf{R}(\boldsymbol{\theta}) \approx \mathbf{I} - [\boldsymbol{\theta}\times]$ a general coordinate transformation of the form $\mathbf{x}' = (1+s)\mathbf{R}(\boldsymbol{\theta})\mathbf{x} + \mathbf{d}$ becomes $\mathbf{x}' = \mathbf{x} + [\mathbf{x}\times]\boldsymbol{\theta} + s\mathbf{x} + \mathbf{d}$ or in terms of corrections $\delta\mathbf{x} = \mathbf{x} - \mathbf{x}^{ap}$ to approximate coordinates

$$\delta\mathbf{x}'(t) = \delta\mathbf{x}(t) + [\mathbf{x}^{ap}\times]\boldsymbol{\theta}(t) + s(t)\mathbf{x}^{ap} + \mathbf{d}(t). \tag{49.4}$$

The model (49.1) with $\mathbf{x}_i^{ap}(t) = \mathbf{x}_{0i}^{ap} + (t - t_0)\mathbf{v}_i^{ap}$ takes the form $\delta\mathbf{x}_i(t) = \delta\mathbf{x}_{0i} + (t - t_0)\delta\mathbf{v}_i$ and application of (49.4) to the general equation (49.2) yields the linearized observation equations for the stacking problem

$$\delta\mathbf{x}_i^k = \delta\mathbf{x}_{i0} + (t_k - t_0)\delta\mathbf{v}_i + s_k\mathbf{x}_{i0}^{ap} + [\mathbf{x}_{i0}^{ap}\times]\boldsymbol{\theta}_k + \mathbf{d}_k + \mathbf{e}_i^k \tag{49.5}$$

where the observational noise $\mathbf{e}_i^k$ has also been taken into account while $\delta\mathbf{x}_i^k \equiv \mathbf{x}_{T,i}(t_k) - \mathbf{x}_{T,i}^{ap}(t_k)$. Dependence on the particular technique T has been dropped, while the subscript k denotes evaluation at epoch t_k.

The observation equations for the combination step are somewhat more involved. It can be shown (Altamimi and Dermanis 2012) that only transformations preserving the linear model form (1) must be used which are the ones with parameters of the form

$$\mathbf{d}_T(t) = \mathbf{d}_{T0} + (t - t_0)\dot{\mathbf{d}}_T, \tag{49.6a}$$

$$\boldsymbol{\theta}_T(t) = \boldsymbol{\theta}_{T0} + (t - t_0)\dot{\boldsymbol{\theta}}_T, \tag{49.6b}$$

$$s_T(t) = s_{T0} + (t - t_0)\dot{s}_T. \tag{49.6c}$$

With such linear in time parameter functions the observation equations for the combination step of the general form (49.3) become (Altamimi and Dermanis 2012)

$$\delta\mathbf{x}_{T0i} = \delta\mathbf{x}_{0i} + \left[\mathbf{x}_{0i}^{ap}\times\right]\boldsymbol{\theta}_{T0} + s_{T0}\mathbf{x}_{0i}^{ap} + \mathbf{d}_{T0} + \mathbf{e}_{\mathbf{x}_{T0i}}, \tag{49.7a}$$

$$\delta\mathbf{v}_{Ti} = \delta\mathbf{v}_i + \left[\mathbf{x}_{0i}^{ap}\times\right]\dot{\boldsymbol{\theta}}_T + \dot{s}_T\mathbf{x}_{0i}^{ap} + \dot{\mathbf{d}}_T + \mathbf{e}_{\mathbf{v}_{Ti}}. \tag{49.7b}$$

where $\delta\mathbf{x}_{T0i} = \mathbf{x}_{T,0i} - \mathbf{x}_{0i}^{ap}$ and $\delta\mathbf{v}_{Ti} = \mathbf{v}_{T,i} - \mathbf{v}_i^{ap}$ are the reduced observations.

3 Constraints for the Introduction of the Optimal Reference System

For the realization of the ITRF solution by either a simultaneous stacking with observation equations (49.5) or a two-step approach with stacking per technique using (49.5) followed by combination using (49.7), it remains to determine the minimal constraints which define the optimal reference system without affecting the optimal network shape at any epoch, which is uniquely defined by the least squares adjustment principle. There are two possible approaches: The first is a kinematic one where the optimality criterion is introduced directly by requiring that the variation of the coordinates is minimized in a specific way. The second is an algebraic one based on the inner constraints which minimize the sum of squares of all unknown parameters, or the partial inner constraints where the sum of selected parameters is involved. The kinematic constraints follow by requiring that the network barycenter remains constant and zero without loss of generality (definition of origin)

$$\mathbf{x}_B(t) \equiv \frac{1}{N}\sum_i \mathbf{x}_i(t) = \mathbf{0}, \tag{49.8a}$$

that the relative kinetic energy of the network stations (visualized as mass points of equal mass) is minimized or equivalently that the relative angular momentum is vanishing (definition of orientation)

$$\mathbf{h}_R(t) = \sum_i [\mathbf{x}_i(t)\times]\frac{d\mathbf{x}_i}{dt}(t) = \mathbf{0} \tag{49.8b}$$

and that the mean quadratic size S of the network defined by

$$S^2(t) = \sum_i [\mathbf{x}_i(t) - \mathbf{x}_B(t)]^T[\mathbf{x}_i(t) - \mathbf{x}_B(t)] \tag{49.8c}$$

remains constant (definition of scale).

The above optimality criteria lead to the following minimal constraints (Altamimi and Dermanis 2012): For the definition of the system origin:

$$\frac{1}{N}\sum_i \delta\mathbf{x}_{0i} = -\bar{\mathbf{x}}_0^{ap} \equiv -\frac{1}{N}\sum_i \mathbf{x}_{0i}^{ap}, \tag{49.9a}$$

$$\frac{1}{N}\sum_i \delta\mathbf{v}_i = -\bar{\mathbf{v}}^{ap} \equiv -\frac{1}{N}\sum_i \mathbf{v}_i^{ap} \tag{49.9b}$$

For the definition of the system orientation:

$$\sum_i \left[\mathbf{x}_{0i}^{ap}\times\right]\delta\mathbf{v}_i = -\mathbf{h}_R^{ap} \equiv -\sum_i \left[\mathbf{x}_{0i}^{ap}\times\right]\mathbf{v}_i^{ap} \tag{49.10b}$$

For the definition of the system scale:

$$\sum_i \left(\mathbf{x}_{0i}^{ap} - \bar{\mathbf{x}}_0^{ap}\right)^T \delta\mathbf{x}_{0i} = 0 \tag{49.11a}$$

$$\frac{1}{N}\sum_i \left(\mathbf{x}_{0i}^{ap} - \bar{\mathbf{x}}_0^{ap}\right)^T \delta\mathbf{v}_i = \left(\bar{\mathbf{x}}_0^{ap}\right)^T \bar{\mathbf{v}}^{ap} - \frac{1}{N}\sum_i \left(\mathbf{x}_{0i}^{ap}\right)^T \mathbf{v}_i^{ap} \tag{49.11b}$$

Among the above constraints (49.9a) and (49.11a) define origin and scale, respectively, at the original epoch, while (49.9b), (49.10b) and (49.11b) define the rates of origin and scale, respectively. Note that the condition (49.8b) does not define the orientation at the initial epoch which must be also chosen by an additional constraint (to be borrowed from the next algebraic approach) in order to pick up a particular reference system orientation from an infinite number of dynamically equivalent ones satisfying (49.8b). Usually we choose $\mathbf{v}_i^{ap} = \mathbf{0}$ and if in addition approximate initial coordinates are chosen so that $\bar{\mathbf{x}}_0^{ap} = \mathbf{0}$ the constraints (49.9a), (49.9b), (49.10b), (49.11a), (49.11b) simplify, respectively, to

$$\sum_i \delta\mathbf{x}_{0i} = \mathbf{0}, \qquad (49.12a)$$

$$\sum_i \delta\mathbf{v}_i = \mathbf{0} \qquad (49.12b)$$

$$\sum_i [\mathbf{x}_{0i}^{ap}\times]\delta\mathbf{v}_i = \mathbf{0} \qquad (49.13b)$$

$$\sum_i (\mathbf{x}_{0i}^{ap})^T \delta\mathbf{x}_{0i} = 0 \qquad (49.14a)$$

$$\sum_i (\mathbf{x}_{0i}^{ap})^T \delta\mathbf{v}_i = \mathbf{0} \qquad (49.14b)$$

The algebraic approach follows the same general lines as in the case of rigid networks (Meissl 1965, 1969; Blaha 1971; Sillard and Boucher 2001; Dermanis 2003), with time independent coordinates. When observable quantities $\mathbf{y}$ are related to coordinate-related unknown parameters $\mathbf{x}$ by a linear(ized) model $\mathbf{y} = \mathbf{A}\mathbf{x}$, then the design matrix $\mathbf{A}$ has a rank deficiency equal to the number of coordinate transformation parameters which change $\mathbf{x}$ but leave $\mathbf{y}$ invariant (Grafarend and Schaffrin 1976). A coordinate transformation with parameters $\mathbf{p}$ transforms the unknown $\mathbf{x}$ into new ones $\mathbf{x}' = T(\mathbf{p})\mathbf{x}$, which depend on both $\mathbf{x}$ and $\mathbf{p}$ through a linear(ized) relation of the form $\mathbf{x}' = \mathbf{x} + \mathbf{E}\mathbf{p}$. The derived matrix $\mathbf{E}$ determines the additional inner constraints $\mathbf{E}^T\mathbf{x} = \mathbf{0}$, which yield the unknown values satisfying $\mathbf{x}^T\mathbf{x} = \min$. Splitting the unknowns and the inner constraints in two sets $\mathbf{E}^T\mathbf{x} = \mathbf{E}_1^T\mathbf{x}_1 + \mathbf{E}_2^T\mathbf{x}_2 = \mathbf{0}$, we obtain the partial inner constraints $\mathbf{E}_1^T\mathbf{x}_1 = \mathbf{0}$, which satisfy $\mathbf{x}_1^T\mathbf{x}_1 = \min$.

In the stacking problem the unknowns are the initial coordinates and velocities $\mathbf{a}_i = [\mathbf{x}_{0i}^T\ \mathbf{v}_i^T]^T$ for each station P_i and the transformation parameters $\mathbf{z}_k = [\mathbf{d}_k^T\ \boldsymbol{\theta}_k^T\ s_k]^T$. A coordinate change with parameters $\mathbf{p} = [\mathbf{g}_0^T\ \boldsymbol{\psi}_0^T\ \lambda\ \dot{\mathbf{g}}^T\ \dot{\boldsymbol{\psi}}^T\ \dot{\lambda}]^T$, transforms the unknowns into

$$\mathbf{a}'_i = \mathbf{a}_i + \mathbf{E}_{\mathbf{a}_i}\mathbf{p}$$

$$= \mathbf{a}_i + \begin{bmatrix} \mathbf{I} & [\mathbf{x}_{0i}^{ap}\times] & \mathbf{x}_{0i}^{ap} & \mathbf{0} & \mathbf{0} & \mathbf{0} \\ \mathbf{0} & \mathbf{0} & \mathbf{0} & \mathbf{I} & [\mathbf{x}_{0i}^{ap}\times] & \mathbf{x}_{0i}^{ap} \end{bmatrix}\mathbf{p} \qquad (49.15)$$

$$\mathbf{z}'_k = \mathbf{z}_k + \mathbf{E}_{\mathbf{z}_k}\mathbf{p}$$

$$= \mathbf{z}_k + \begin{bmatrix} -\mathbf{I} & -(t_k - t_0)\mathbf{I} \end{bmatrix}\mathbf{p} \qquad (49.16)$$

and the inner constraints $\sum_{i=1}^{N} \mathbf{E}_{\mathbf{a}_i}^T\mathbf{a}_i + \sum_{k=1}^{M} \mathbf{E}_{\mathbf{z}_k}^T\mathbf{z}_k = \mathbf{0}$ become

$$\sum_{i=1}^{N} \delta\mathbf{x}_{0i} - \sum_{k=1}^{M} \mathbf{d}_k = \mathbf{0} \qquad (49.17a)$$

$$\sum_{i=1}^{N} \delta\mathbf{v}_i - \sum_{k=1}^{M} (t_k - t_0)\mathbf{d}_k = \mathbf{0} \qquad (49.17b)$$

$$\sum_{i=1}^{N} [\mathbf{x}_{0i}^{ap}\times]\delta\mathbf{x}_{0i} + \sum_{k=1}^{M} \boldsymbol{\theta}_k = \mathbf{0} \qquad (49.18a)$$

$$\sum_{i=1}^{N} [\mathbf{x}_{0i}^{ap}\times]\delta\mathbf{v}_i + \sum_{k=1}^{M} (t_k - t_0)\boldsymbol{\theta}_k = \mathbf{0} \qquad (49.18b)$$

$$\sum_{i=1}^{N} (\mathbf{x}_{0i}^{ap})^T \delta\mathbf{x}_{0i} - \sum_{k=1}^{M} s_k = 0 \qquad (49.19a)$$

$$\sum_{i=1}^{N} (\mathbf{x}_{0i}^{ap})^T \delta\mathbf{v}_i - \sum_{k=1}^{M} (t_k - t_0)s_k = 0. \qquad (49.19b)$$

The partial inner constraints where only the parameters $\delta\mathbf{x}_{0i}$, $\delta\mathbf{v}_i = \mathbf{0}$ participate, are exactly the constraints (49.12a), (49.12b), (49.13b), (49.14a), (49.14b) plus the missing initial epoch orientation constraint (49.13a) which becomes

$$\sum_{i=1}^{N} [\mathbf{x}_{0i}^{ap}\times]\delta\mathbf{x}_{0i} = \mathbf{0} \qquad (49.13a)$$

In the combination problem the unknowns are again initial coordinates and velocities $\mathbf{a}_i = [\mathbf{x}_{0i}^T \ \mathbf{v}_i^T]^T$ as well as the transformation parameters $\mathbf{z}_T = [\mathbf{d}_{T0}^T \ \boldsymbol{\theta}_{T0}^T s_{T0} \ \dot{\mathbf{d}}_T^T \ \dot{\boldsymbol{\theta}}_T^T \ \dot{s}_T]^T$, which under a change of coordinate system transform according to

$$\mathbf{z}'_T = \mathbf{z}_T + \mathbf{E}_{\mathbf{z}_T}\mathbf{p} = \mathbf{z}_T - \mathbf{p} \tag{49.20}$$

and the inner constraints $\sum_i \mathbf{E}_{\mathbf{a}_i}^T \mathbf{a}_i + \sum_T \mathbf{E}_{\mathbf{z}_T}^T \mathbf{z}_T = \mathbf{0}$ with $\mathbf{E}_{\mathbf{a}_i}$ from (49.15) and $\mathbf{E}_{\mathbf{z}_T} = -\mathbf{I}$ become

$$\sum_{i=1}^{N} \delta\mathbf{x}_{0i} - \sum_{T=1}^{K} \mathbf{d}_{T0} = \mathbf{0} \tag{49.21a}$$

$$\sum_{i=1}^{N} \delta\mathbf{v}_i - \sum_{T=1}^{K} \dot{\mathbf{d}}_T = \mathbf{0} \tag{49.21b}$$

$$\sum_{i=1}^{N} [\mathbf{x}_{0i}^{ap}\times]\delta\mathbf{x}_{0i} + \sum_{T=1}^{K} \boldsymbol{\theta}_{T0} = \mathbf{0} \tag{49.22a}$$

$$\sum_{i=1}^{N} [\mathbf{x}_{0i}^{ap}\times]\delta\mathbf{v}_i + \sum_{T=1}^{K} \dot{\boldsymbol{\theta}}_T = \mathbf{0} \tag{49.22b}$$

$$\sum_{i=1}^{N} (\mathbf{x}_{0i}^{ap})^T \delta\mathbf{x}_{0i} - \sum_{T=1}^{K} s_{T0} = 0 \tag{49.23a}$$

$$\sum_{i=1}^{N} (\mathbf{x}_{0i}^{ap})^T \delta\mathbf{v}_i - \sum_{T=1}^{K} \dot{s}_T = 0. \tag{49.23b}$$

The partial inner constraints involving only initial coordinates and velocities are the same as in the stacking problem. The partial inner constraints involving only transformation parameters become

$$\sum_{T=1}^{K} \mathbf{d}_{T0} = \mathbf{0}, \quad \sum_{T=1}^{K} \dot{\mathbf{d}}_T = \mathbf{0}, \tag{49.24}$$

$$\sum_{T=1}^{K} \boldsymbol{\theta}_{T0} = \mathbf{0}, \quad \sum_{T=1}^{K} \dot{\boldsymbol{\theta}}_T = \mathbf{0} \tag{49.25}$$

$$\sum_{T=1}^{K} s_{T0} = 0, \quad \sum_{T=1}^{K} \dot{s}_T = 0. \tag{49.26}$$

Of the above constraints (kinematic, inner, partial inner) only the ones related to the actual deficiencies of the reference system must be implemented. For example the origin and origin rate constraints do not apply to the SLR case where the geocenter is the known system origin. Since all techniques have their own scale, scale or scale rate constraints appear to be redundant. However since each technique has a different scale due to the different time unit realized through a different set of atomic clocks, these constraints should be incorporated into the combination step.

Conclusion

In comparison to inner or partial inner constraints the kinematic minimal constraints have the advantage of being independent of the approximate values of the parameters used in the linearization of the observation equations. They involve only station related parameters (initial coordinates and velocities), while inner constrains involve both station and reference system transformation parameters. Partial inner constraints may be formulated for either station or transformation parameters. The ones for station parameters may coincide with the kinematical ones (and thus share their independence from approximate values) if care is taken so that the used approximate values of velocities are zero and the approximate values of the initial coordinates have zero mean.

References

Altamimi Z, Collilieux X, Legrand J, Garayt B, Boucher C (2007) ITRF2005: a new release of the International Terrestrial Reference Frame based on time series of station positions and Earth Orientation Parameters J Geophys Res 112:9401

Altamimi Z, Collilieux X, Boucher C (2008) Accuracy assessment of the ITRF datum definition. In: Peiliang X, Jingnan Liu, A. Dermanis (eds) VI Hotine-Marussi symposium on theoretical and computational geodesy, international association of geodesy symposia, vol 132, Springer, Berlin, pp 101–110

Altamimi Z, Dermanis A (2012) Theoretical foundations of ITRF determination. The algebraic and the kinematic approach. In: Volume in Honor of Prof. D. Vlachos, Publication of the School of Rural and Surveying Engineering, Aristotle University of Thessaloniki

Blaha G (1971) Inner adjustment constraints with emphasis on range observations Department of Geodetic Science Report 148, The Ohio State University, Columbus

Dermanis A (2003) The rank deficiency in estimation theory and the definition of reference systems In: Sansò F (ed) 2003 V Hotine-Marussi Symposium on Mathematical Geodesy, Matera, Italy June 17–21, 2003 International Association of Geodesy Symposia, vol 127, Springer Verlag, Heidelberg, pp 145–156

Meissl P (1965) Über die innere Genauigkeit dreidimensionaler Punkthaufen. Zeitschrift für Vemessungswesen 90(4): 109–118

Meissl P (1969) Zusammenfassung und Ausbau der inneren Fehlertheorie eines Punkthaufens Deutsche Geodätische Kommission, Reihe A Nr 61 8–21

Grafarend E, Schaffrin B (1976) Equivalence of estimable quantities and invariants in geodetic networks Zeitschrift für Vemessungswesen 101(11):485–491

Sillard P, Boucher C (2001) Review of algebraic constraints in terrestrial reference frame datum definition. J Geodesy 75:63–73

Some Pitfalls to be Avoided in Combining Simultaneous GNSS Networks

50

L. Biagi and F. Sansò

Abstract

Assume that a network, in this case a network of GNSS observables, has to be adjusted and one likes to split it into subnetworks, and adjust them separately; then the subnetworks have to be recomposed by using the cross observations connecting their points. Ideally, this should be done in a way such that the final solution is the same as that provided by a rigorous joint solution of all the observations. The problem is classical and has a rigorous solution, called Helmert blocking, for terrestrial networks, where the observations are typically uncorrelated. For the GNSS networks thought, where many times the phase double differences are treated as observations, the hypothesis of linear independence is by definition inconsistent: this gives rise to blunders in the solution that are illustrated in the paper by numerical examples. An alternative solution is envisaged to overcome these drawbacks.

Keywords

GPS networks adjustment; NEQ stacking

1 An Introduction to Network Splitting

In the general Least Squares (LS; Koch 1987) model, a vector of observations $\mathbf{y}_0$ and a vector of unknown parameters $\mathbf{x}$ are given; the deterministic and the stochastic models can be written as

$$\mathbf{y}_0 = \mathbf{y} + \boldsymbol{\nu}, \quad \mathbf{y} = \mathbf{A}\mathbf{x}, \quad E\{\mathbf{v}\} = 0, \quad \mathbf{C}_{vv} = \sigma_0^2 \mathbf{Q}$$

In the following, only the not rank deficient case will be considered: the parameters vector can be estimated by solving the normal system

$$\mathbf{N}\mathbf{x} = \mathbf{A}^T\mathbf{Q}^{-1}\mathbf{y}_0, \quad \mathbf{N} = \mathbf{A}^T\mathbf{Q}^{-1}\mathbf{A} \Rightarrow \hat{\mathbf{x}} = \mathbf{N}^{-1}\mathbf{A}^T\mathbf{Q}^{-1}\mathbf{y}_0 \tag{50.1}$$

Two particular cases can be built from the general model. In the pre-elimination model, the parameters vector is split into two subvectors

$$\mathbf{y} = \begin{bmatrix} \mathbf{A}_1 & \mathbf{A}_2 \end{bmatrix} \begin{bmatrix} \mathbf{x}_1 \\ \mathbf{x}_2 \end{bmatrix}$$

The relevant normal system is given by the

$$\begin{bmatrix} \mathbf{N}_{11} & \mathbf{N}_{12} \\ \mathbf{N}_{12}^T & \mathbf{N}_{22} \end{bmatrix} \begin{bmatrix} \mathbf{x}_1 \\ \mathbf{x}_2 \end{bmatrix} = \begin{bmatrix} \mathbf{b}_1 \\ \mathbf{b}_2 \end{bmatrix},$$

$$\mathbf{N}_{ij} = \mathbf{A}_i^T\mathbf{Q}^{-1}\mathbf{A}_j, \quad \mathbf{b}_i = \mathbf{A}_i^T\mathbf{Q}^{-1}\mathbf{y}_0$$

L. Biagi (✉) · F. Sansò
Politecnico di Milano, DIIAR, Laboratorio di Geomatica del Polo Territoriale di Como, via Valleggio 11, 2100 Como, Italy
e-mail: ludovico.biagi@polimi.it

N. Sneeuw et al. (eds.), *VII Hotine-Marussi Symposium on Mathematical Geodesy*, International Association of Geodesy Symposia 137, DOI 10.1007/978-3-642-22078-4_50, © Springer-Verlag Berlin Heidelberg 2012

Alternatively to a joint estimate, the two subvectors can be separately estimated, by the inversion of the two split, smaller, systems.

$$\mathbf{x}_i = \left(\mathbf{N}_{ii} - \mathbf{N}_{ij}\mathbf{N}_{jj}^{-1}\mathbf{N}_{ij}^T\right)^{-1}(\mathbf{b}_i - \mathbf{N}_{ij}\mathbf{N}_{jj}^{-1}\mathbf{b}_j)$$

Pre-elimination represents an obvious choice when only one subset of the unknowns is actually needed but in the model is mixed with a set of disturbance unknowns: indeed the numerical effort of its estimation can be significantly reduced with respect to the global estimation. In the sequential model, two vectors of independent observations and one common parameters vector are given

$$\mathbf{y}_1 = \mathbf{A}_1\mathbf{x}, \quad \mathbf{y}_2 = \mathbf{A}_2\mathbf{x},$$
$$\mathbf{C}_{y1y1} = \sigma_0^2\mathbf{Q}_{11}, \quad \mathbf{C}_{y2y2} = \sigma_0^2\mathbf{Q}_{22}, \quad \mathbf{C}_{y1y2} = \mathbf{0}$$

The two systems can also written as

$$\mathbf{y} = \mathbf{A}\mathbf{x}, \quad \mathbf{y} = \begin{bmatrix}\mathbf{y}_1\\ \mathbf{y}_2\end{bmatrix},$$
$$\mathbf{A} = \begin{bmatrix}\mathbf{A}_1\\ \mathbf{A}_2\end{bmatrix}, \quad \mathbf{C}_{yy} = \sigma_0^2\begin{bmatrix}\mathbf{Q}_{11} & \mathbf{0}\\ \mathbf{0} & \mathbf{Q}_{22}\end{bmatrix}$$

Two estimation approaches are equivalent: the obvious batch solution by (50.1), or two independent solutions computed by

$$\hat{\mathbf{x}}_i = \mathbf{N}_i^{-1}\mathbf{b}_i, \quad \mathbf{N}_i = \mathbf{A}_i^T\mathbf{Q}_{ii}^{-1}\mathbf{A}_i, \quad \mathbf{b}_i = \mathbf{A}_i^T\mathbf{Q}_{ii}^{-1}\mathbf{y}_{i_0}$$

and combined in

$$\hat{\mathbf{x}} = (\mathbf{N}_1 + \mathbf{N}_2)^{-1}(\mathbf{N}_1\hat{\mathbf{x}}_1 + \mathbf{N}_2\hat{\mathbf{x}}_2)$$

This is clearly equivalent to the batch solution. In the unified model two observations vectors depend both on individual and common parameters

$$\mathbf{y}_1 = [\mathbf{F}_1\ \mathbf{A}_1]\begin{bmatrix}\mathbf{x}_1\\ \boldsymbol{\xi}\end{bmatrix}, \quad \mathbf{y}_2 = [\mathbf{A}_2\ \mathbf{F}_2]\begin{bmatrix}\boldsymbol{\xi}\\ \mathbf{x}_2\end{bmatrix}$$
$$\mathbf{C}_{yy} = \sigma_0^2\begin{bmatrix}\mathbf{Q}_{11} & \mathbf{0}\\ \mathbf{0} & \mathbf{Q}_{22}\end{bmatrix}$$

A sequential estimation is straightforward: by pre-elimination the two individual parameters vectors are estimated

$$\mathbf{y}_1 \Rightarrow \hat{\mathbf{x}}_1,\ C_{\hat{x}_1\hat{x}_1}, \quad \mathbf{y}_2 \Rightarrow \hat{\mathbf{x}}_2,\ C_{\hat{x}_2\hat{x}_2} \qquad (50.2)$$

Moreover the two normal systems relevant to the common parameters vector are generated

$$\mathbf{y}_1 \Rightarrow \hat{\boldsymbol{\xi}}_1,\ \mathbf{N}_1,\ \mathbf{b}_1, \quad \mathbf{y}_2 \Rightarrow \hat{\boldsymbol{\xi}}_2,\ \mathbf{N}_2,\ \mathbf{b}_2 \qquad (50.3)$$

By sequential adjustment, the common parameters are estimated

$$\hat{\boldsymbol{\xi}} = (\mathbf{N}_1 + \mathbf{N}_2)^{-1}(\mathbf{N}_1\hat{\boldsymbol{\xi}}_1 + \mathbf{N}_2\hat{\boldsymbol{\xi}}_2)$$

Let suppose that two overlapping geodetic networks have been independently surveyed and adjusted; then (50.2), (50.3) is the obvious way to obtain a final solution for all their points. In the geodetic framework, this way to combine the adjustments of overlapping networks is called Normal Equation (NEQ) stacking.

2 The GNSS Permanent Network Standard Approach

The raw data of GNSS permanent networks are typically processed by daily sessions in order to estimate daily coordinates, ancillary unknowns and their covariances. Typically an open baseline graph is built and the relevant double differences are processed; in the case of local networks, the international guidelines (see for example, Kouba 2003) suggest to minimally constrain the reference frame by fixing the barycenter of a polyhedron of IGS (Beutler et al. 1999) fiducial stations included in the processing. The normal system to be inverted in the LS estimation has dimension equal to the square of the unknowns number, and this causes a limit to the number of stations that can be simultaneously processed. Hardware evolves very quickly, and the processing limits are continuously changing: at the present, a standard Windows server can easily estimate a batch solution for about 200 stations. Very few networks in the world exceed these dimensions and require a distributed adjustment, as IGS (igscb.jpl.nasa.gov) and some regional networks, for example EPN in Europe (www.epncb.oma.be). In any case, the distributed adjustment is a popular choice also for smaller networks. To implement a distributed adjustment (Davies and Blewitt 2000), at first the network is split into overlapping subnetworks, such that each station belongs to an assigned minimum number of subnetworks, n_{min}. The subnetworks are separately adjusted and the relevant NEQ's are then stacked to obtain a final network solution. For example, EPN is split into 16 subnetworks, each one adjusted by a Local Analysis Center. Subnetworks dimensions range from 30 stations to 70 stations and each station is included at least in three subnetworks. It is worth to note that in the GPS case, the subnetworks overlap is not strictly needed, because a consistent network solution can be obtained also by joining separate solutions: links and consistence are guaranteed by the raw observations to a common GNSS constellation and by using the same IGS products in the observations processing. In any case, the overlap is useful because it guarantees the cross check of the results. The standard implementation of the distributed adjustment is based on a daily process:

1. each subnetwork is daily processed by one processing facility;
2. the resulting daily NEQ file is saved and transmitted to a coordination center;
3. the daily NEQ's are stacked to produce a network solution, on a daily or weekly basis.

Note that a Processing Facility (PF) is simply a set of procedures installed on a server to adjust a daily session of a permanent network, while the usual name Analysis Center (AC) is referred to a set of PF's and obviously of personnel.

Let us consider the example of a network of N_S stations $(S_1, \ldots, S_{N_S})$ to be estimated and N_R reference frame stations $(R_1, \ldots, R_{N_R})$: typically $N_R \ll N_S$; suppose that each PF can process about $N_S/2$ stations and that $n_{\min} = 2$. Given the above requirements, 4 PF's are needed: in a typical implementation, supposing that all reference stations are shared between all the subnetworks, each day:

1.1 P_1 adjusts $\{S_1, \ldots, S_{|\frac{1}{2}N_S|}, R_1, \ldots, R_{N_R}\}$,
1.2 P_2 adjusts $\{S_{|\frac{1}{4}N_S|+1}, \ldots, S_{|\frac{3}{4}N_S|}, R_1, \ldots, R_{N_R}\}$,
1.3 P_3 adjusts $\{S_{|\frac{1}{2}N_S|+1}, \ldots, S_{N_S}, R_1, \ldots, R_{N_R}\}$;
1.4 P_4 adjusts $\{S_{|\frac{3}{4}N_S|+1}, \ldots, S_{N_S}, \ldots, S_{|\frac{1}{4}N_S|}, R_1, \ldots, R_{N_R}\}$;

Finally, the four daily NEQ's are then combined to produce a final daily solution; $|x|$ is an integer rounding on the stations indexes, for example the lower integer rounding.

As it is well known, this praxis builds artificial independent observations, either false repeated baselines or false closed polygons, because the same files are processed by more PF's but the correlations of the relevant NEQ's are ignored at the stacking level. In the following this distributed approach will be called COnstant in time Distribution (COD).

To understand COD limits, let us analyze an elementary example, a leveling triangle, composed by three points P_1, P_2, P_3 (h_1, h_2, h_3). The observed height differences and the relevant covariance matrix are

$$\begin{aligned} y_{1_0} &= dh_{12_0} = h_2 - h_1 + \nu_{12} \\ y_{2_0} &= dh_{13_0} = h_3 - h_1 + \nu_{13} \end{aligned}, \quad \mathbf{C} = \sigma_0^2 \begin{bmatrix} 1 & 1/2 \\ 1/2 & 1 \end{bmatrix}$$

Fix the reference frame, e.g. by putting $\bar{h}_1 = 0$:

$$\mathbf{y}_0 = \begin{bmatrix} y_{1_0} \\ y_{2_0} \end{bmatrix}, \quad \mathbf{Q}_{yy} = \begin{bmatrix} 1 & 1/2 \\ 1/2 & 1 \end{bmatrix}, \quad \mathbf{x} = \begin{bmatrix} h_2 \\ h_3 \end{bmatrix}$$

No redundancy exists and the estimates are given by the

$$\hat{h}_2 = \bar{h}_1 + dh_{12_0} = h_2 + \nu_{12}, \quad \hat{\sigma}_2 = \sigma_0$$
$$\hat{h}_3 = \bar{h}_1 + dh_{13_0} = h_3 + \nu_{13}, \quad \hat{\sigma}_3 = \sigma_0$$

Now, let us combine the original observations to close the triangle, $y_{3_0} = y_{2_0} - y_{1_0} = h_3 - h_2 + \nu_{13} - \nu_{12}$, and consider the new observations vector $\mathbf{y}_0 = [y_{1_0}\, y_{2_0}\, y_{3_0}]^T$; its covariance matrix is given by

$$\mathbf{C} = \sigma_0^2 \begin{bmatrix} 1 & 1/2 & -1/2 \\ 1/2 & 1 & 1/2 \\ -1/2 & 1/2 & 1 \end{bmatrix} = \sigma_0^2 \mathbf{Q}$$

The new observations are linearly dependent and $\mathbf{Q}$ is singular: although the closed vector can't be adjusted by ordinary LS, let us go on ignoring the correlations, and putting $\mathbf{Q} = \mathbf{I}$. In this case the observations vector can be adjusted to estimate heights and variances:

$$\hat{h}_2 = h_2 + \nu_{12}, \quad \hat{\sigma}_2 = (2/3)\sigma_0$$
$$\hat{h}_3 = h_3 + \nu_{13}, \quad \hat{\sigma}_3 = (2/3)\sigma_0$$
$$\hat{\sigma}_2 = 0$$

Heights are correct but standard deviations are underestimated; the estimated residuals and the resulting a posteriori variance are zero.

3 An Alternative Approach for GNSS Permanent Networks

An alternative approach is possible to distribute the adjustment of a network into overlapping subnetworks. Each day the network is split into subnetworks that share just one station (the connecting station, R_C): in this way, all the possible independent double differences are processed and the resulting daily baselines graph is a connected open graph, exactly as in a rigorous batch adjustment of the whole network. To guarantee overlaps, the daily subnetworks configuration varies in a cycle over more days: the cycle is such that overlaps exist between subnetworks of different days. At the end of the whole cycle, all the daily NEQ's are stacked and, in this way, true closures and repetitions and cross checks for all the stations are obtained. Consider the previous network and the previous requirements; they can be accomplished by a very simple cycle based on two PF's and the odd/even day scheme: in odd days:

D1.1 P_1 adjusts $\{S_1, \ldots, S_{|\frac{1}{2}N_S|}, R_1, \ldots, R_{|\frac{1}{2}N_R|}, R_C\}$,

D1.2 P_2 adjusts $\{S_{|\frac{1}{2}N_S|+1}, \ldots, S_{N_S}, R_{|\frac{1}{2}N_R|}, \ldots, R_{N_R}, R_C\}$,

In even days:

D2.1 P_1 adjusts $\{S_1, S_3, \ldots, S_{N_S-1}, R_1, R_3, \ldots, R_{N_R-1}, R_C\}$,

D2.2 P_2 adjusts $\{S_2, S_4, \ldots, S_{N_S}, R_2, R_4, \ldots, R_{N_R}, R_C\}$,

With this simple scheme, each two days four NEQ's are available and can be stacked in a two days final solution; clearly even more complex schemes can be implemented, if a greater $n_{\min}$ is required, or the split in more PF's is needed. In the following this approach will be called VAriable in time Distribution (VAD): with respect to COD, the approach based on a cycle over more days imposes a significant coordination work between the PF's, in order to define the cycle and the configuration of the network splitting for each day of the cycle. On the other hand, less computational effort is needed: each day only the connecting station is included in all the subnetworks, and $N_S + N_R + N_{PF} - 1$ daily files are processed; with COD, at least $n_{\min}(N_R + N_S)$ files have to be processed. In any case, despite the coordination difficulties, the approach based on a cycle over more days allows a rigorous combination of different overlapping subnetworks that, at the daily scale, is equivalent to the batch adjustment of the whole network; in the daily subnetworks stacking, just the correlation due to the connecting station is neglected.

4 A Numerical Test

In order to numerically compare the results provided by COD and VAD, a simple network of 30 stations has been analyzed (Fig. 50.1): it is composed of 6 IGS official Reference Frame stations and 24 stations belonging to the new official zero order network of the italian Istituto Geografico Militare (www.igmi.org); two weeks of data have been processed: GPSW's 1459 and 1462. At first a batch adjustment of the network has been computed and the relevant results represent the benchmark (BA); then COD and VAD approaches have been simulated and compared with BA. All processing have been performed by the Bernese Software 5.0 (BSW5.0, Dach et al. 2007), by following the

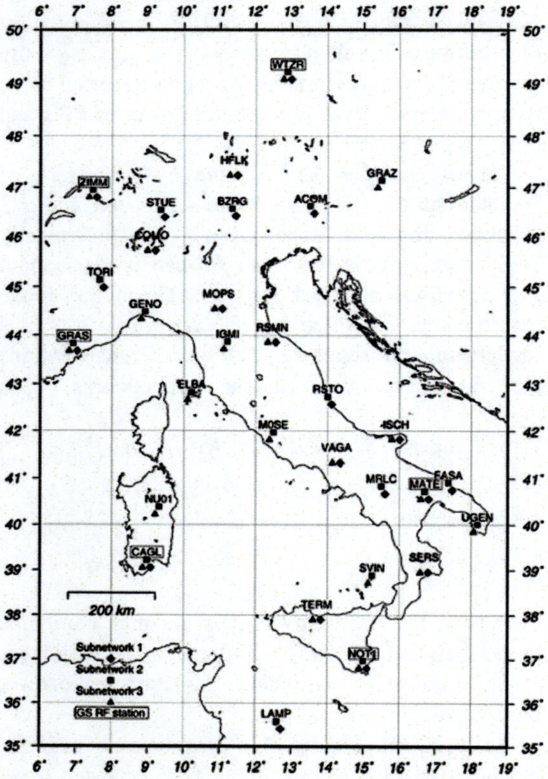

Fig. 50.1 COD approach: stations belonging to subnetworks 1, 2 and 3; IGS RF stations are included in all the three subnetworks

international guidelines and the strategies summarized in (Benciolini et al. 2008). Particularly, in BA a minimally constrained NEQ has been produced for each daily data processing. In COD and VAD, daily loosely constrained NEQ's are computed for each subnetwork; then, they are combined in a minimally constrained daily NEQ: for each adjustment approach, 14 daily NEQ's are available, that can be stacked to produce a final solution. In total 418 daily RINEX files have been acquired out of the 420 expected and only four files have been rejected due to the presence of significant data problems. For BA, the QIF ambiguities fixing percentages are of about 87–90%, while the RMS's of the daily solutions are of about 0.9–1.1 mm: on this regards, the data general quality is completely satisfying. In total, 29 × 14 baselines have been adjusted, and 3,764,630 observations have been used to estimate 90 coordinates, 11,975 tropospheric parameters and 2,470 not fixed ambiguities. To implement COD, three PF's have been simulated (Fig. 50.1): RF's stations are shared by all the three relevant subnetworks, all the other stations are shared by two subnetworks. Each PF processes 16 + 6 stations; in this case the apparent total number of observations is 8,187,915, the not fixed ambiguities are 5,294 while the coordinates and troposphere parameters numbers are the same.

To simulate VAD, a 2-day (odd/even) cycle has been applied, as in the above example (Fig. 50.2). In odd days, the connecting station is GRAS while in even days is CAGL: each other station is included in two subnetworks. In this case 3,784,070 observations have been used and the not fixed ambiguities are 2,248.

Note that to be consistent with the COD example of Sect. 4, four PF's should be simulated, each one adjusting a very small subnetwork of 18 stations, but we have preferred a more realistic splitting: in any case, the present numerical example is consistent with the usual implementation of the COD. The subnetworks, both in COD and VAD, have been designed on order to guarantee their quasi homogeneous distribution: for this reasons, connecting stations are different for VAD even and odd days.

Coordinates differences of the three approaches (Table 50.1) are negligible. Daily repeatabilities (Table 50.2) of BA and VAD are almost equal, while COD (over) overlapping improves slightly the Up repeatabilities (Tables 50.3 and 50.4).

Although not realistic, the final standard deviations of BA are rigorously estimated; COD false redundancies add further underestimations. VAD provides correct estimates: only the values of the two connecting stations are underestimated, as their correlations in the subnetworks are not taken into

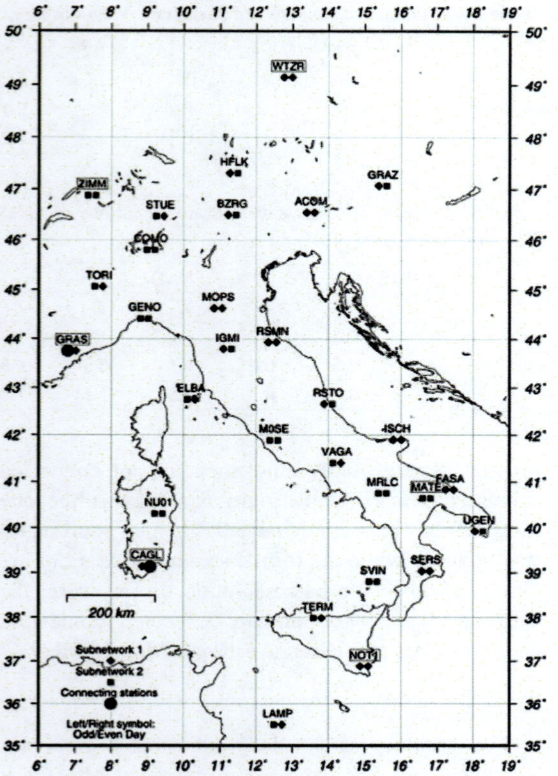

Fig. 50.2 VAD approach. Stations belonging to subnetworks 1 and 2 and connecting stations

Table 50.1 Differences of COD and VAD results with respect to BA. Values in mm

	COD			VAD		
	X	Y	Z	X	Y	Z
Mean	0.0	0.0	0.0	0.1	0.1	0.1
σ	0.5	0.2	0.5	0.6	0.2	0.6
Min	−1.1	−0.5	−0.8	−0.8	−0.5	−0.7
Max	1.3	0.4	1.4	1.6	0.6	1.6

Table 50.2 Repeatabilities of daily coordinates of BA, COD and VAD. Values in mm

	BA			COD			VAD		
	E	N	U	E	nN	U	E	N	U
Mean	1.1	1.1	4.1	1.1	1.1	3.8	1.1	1.1	4.1
Worst	2.2	3.1	7.5	2.1	3.1	6.3	2.3	3.4	7.2

Table 50.3 Estimated BA standard deviations. Values in mm

	BA		
	X	Y	Z
Mean	0.1	0.1	0.1
Max	0.2	0.1	0.2

Table 50.4 Ratios of estimated standard deviations of COD and VAD with respect to BA

	COD/BA			VAD/BA		
	X	Y	Z	X	Y	Z
Mean	0.7	0.7	0.7	1.0	1.0	1.0
Min	0.6	0.6	0.6	0.8	0.8	0.8
Max	0.7	0.7	0.7	1.0	1.1	1.0

account. So, the daily adjustment of not connected subnetworks is a possible solution: along this line, one more test has been implemented: the final coordinates don't change while the formal standard deviations are almost equal to the batch solution. In any case, the splitting of the network into not connected subnetwork causes the loss of a baseline estimated for each day.

Conclusion

The problem of the daily adjustment of a permanent network has been discussed. Normally, it should be implemented by a batch solutions, i.e. all the observations of all the stations should be used to estimate coordinates and other unknowns in a joint adjustment. In some cases, the network splitting into overlapping subnetworks, the separate processing of each subnetwork and the final stacking of the relevant NEQ's are needed or useful. The standard praxis foresees a constant configuration of overlapping subnetworks, that are separately processed on a daily basis and finally combined: in this way, the correlations due to the overlaps are ignored and false independent redundancies are introduced; this error doesn't affect coordinates estimates but involves a significant underestimation of the covariances.

An alternative approach has been discussed, based on a variable configuration of the subnetworks: these are connected each day just by one station, but their configuration changes according to a cyclic scheme over several days, in order to obtain the desired overlap at the end of each cycle. The approach requires a significant coordination activity between the processing facilities but decreases the numerical requirements of each daily processing. The results on a test network show that it provides accuracies estimates very similar to those obtained by the batch adjustment.

Acknowledgements The research has been supported by the "Galileo and the modernized satellite positioning" Italian PRIN 2006 project.

References

Benciolini B, Biagi L, Crespi M, Manzino A, Roggero M (2008) Reference frames for GNSS positioning services: some problems and proposals. J Appl Geodesy N2

Beutler G, Rothacher M, Schaer S, Springer TA, Kouba J, Neilan RE (1999) The International GPS Service (IGS): an interdisciplinary service in support of earth sciences. Adv Space Res 23(4)

Dach R, Hugentobler U, Fridez P, Meindl M (2007) Bernese GPS Software Version 5.0. Astronomical Institute, University of Berne, Berne

Davies P, Blewitt G. (2000) Methodology for global geodetic time series estimation: a new tool for geodynamics. J Geophys Res 105(B5)

Koch KR, (1987) Parameter estimation and hypothesis testing in linear models. Springer Verlag, New York

Kouba J (2003) A guide to using International GPS Service (IGS) products, IGS website

Toward a Dense Italian GPS Velocity Field: Data Analysis Strategies and Quality Assessment

51

R. Devoti, E. Flammini, G. Pietrantonio, F. Riguzzi, and E. Serpelloni

Abstract

A test combination procedure of loose constrained SINEX solutions has been set up with the aim to integrate different network solutions (clusters of GPS site coordinates) processed by different analysis groups. Two solution types have been combined on a daily basis, one cluster of about 60 sites processed with the Bernese software and a second cluster of about 30 sites processed with the Gamit software. The chosen networks consists in 15 overlapping EUREF sites on which the combination test is figured out. In origin each solution is obtained in a loose constrained reference frame and archived in SINEX format, the full time series span a period of 2.5 years. If the reference frame constraints, if present, are conveniently relaxed in the original solutions, the combination of daily solutions can be carried out without the removal of relative rotations between reference frames and the different contributing SINEX solutions may be merged (in a least squares sense) into a combined consensus solution. The corresponding velocity fields, computed for the common sites, are consistent with respect to each other and the combined time series show a repeatability of a few millimeters in the vertical and 1–2 millimeters in the horizontal components respectively.

Keywords

Combination of geodetic solutions • Loosely constrained solutions • GPS coordinates

1 Introduction

In 2004 the Istituto Nazionale di Geofisica e Vulcanologia (INGV) started the construction of a continuous GPS network in the Italian region (RING) to investigate the details of the geodynamics of the area. At present the network consists of about 130 stations whose data are transmitted continuously to the data acquisition center (http://ring.gm.ingv.it) that performs the quality check and the data storage. Currently the daily RINEX files of 36 sites are freely provided on the web site. The same data center archives also all the available GPS data in the Italian area in cooperation with local governmental and private agencies thus providing over 400 RINEX files per day. The processing of GPS observations is carried out by different analysis

R. Devoti (✉) · E. Flammini · G. Pietrantonio · F. Riguzzi · E. Serpelloni
Istituto Nazionale di Geofisica e Vulcanologia, sez. Centro Nazionale Terremoti, via di Vigna Murata, 605–00143 Roma, Italy
e-mail: roberto.devoti@ingv.it

N. Sneeuw et al. (eds.), *VII Hotine-Marussi Symposium on Mathematical Geodesy*, International Association of Geodesy Symposia 137, DOI 10.1007/978-3-642-22078-4_51, © Springer-Verlag Berlin Heidelberg 2012

groups at INGV using different softwares and procedures. For this reason and given the large amount of processed sites a comparison and validation procedure based on partially overlapping networks is required.

We implement a convenient combination strategy of geodetic solutions provided by the different analysis centers, dealing with loose constrained solutions (Davies and Blewitt 2000).

This work represents a first step toward an integrated strategy of data analysis and combination, starting from loosely constrained SINEX files processed with two different softwares, Bernese (http://www.bernese.unibe.ch/) and Gamit (http://www-gpsg.mit.edu/~simon/gtgk/).

2 GPS Data Analysis

2.1 Bernese Processing

The GPS data processing is performed by the *Bernese Processing Engine* (BPE) ver. 5.0 (Beutler et al. 2007) forming double difference observables. The GPS orbits and the Earth's orientation parameters are fixed to the combined IGS products and an a priori error of 10 m is assigned to all site coordinates. The pre-processing phase, used to clean up the raw observations, is carried out in a baseline by baseline mode. Independent baselines are defined by the criterion of maximum common observations. The elimination of gross errors, cycle slips and the determination of new ambiguities are computed automatically using the triple-difference combination. The a posteriori normalized residuals of the observations are checked for outliers, too. These observations are marked for the final parameter adjustment. The elevation-dependent phase centre corrections are applied including in the processing the IGS phase centre calibrations (absolute calibrations). The troposphere modeling consists in an a priori dry-Niell model fulfilled by the estimation of zenith delay corrections at 1-h intervals at each site using the wet-Niell mapping function. The ionosphere is not modeled a priori, it is removed by applying the ionosphere-free linear combination of L1 and L2. The ambiguity resolution is based on the QIF baseline-wise analysis. The final network solution is solved with back-substituted ambiguities, if integer; otherwise ambiguities are considered as real valued measurement biases. Each solution is realized in an intrinsic reference frame defined by the observations itself, differing from day to day only for rigid network translations but keeping the site inter-distances always well determined.

2.2 Gamit Processing

To analyze code and phase data with the GAMIT software (Version 10.33; Herring et al. 2006), we adopt standard procedures for the analysis of regional networks (e.g., McClusky et al. 2000; Serpelloni et al. 2006) applying loose constraints to the geodetic parameters.

The GAMIT software uses double-differenced, ionosphere-free linear combinations of the $L1$ and $L2$ phase observations, to generate weighted least square solutions for each daily session (Schaffrin and Bock 1988; Dong and Bock 1989). An automatic cleaning algorithm (Herring et al. 2006) is applied to post-fit residuals, in order to repair cycle slips and to remove outliers. The observation weights vary with elevation angle and are derived individually for each site from the scatter of post-fit residuals obtained in a preliminary GAMIT solution. The effect of solid-earth tides, polar motion and oceanic loading are taken into account according to the IERS/IGS standard 2003 model (McCarthy and Petit 2004). We apply the ocean-loading model FES2004 and use the IGS (International GNSS Service) absolute antenna phase center table for modeling the effective phase center of the receiver and satellites antennas. We use orbits provided by the Scrips Orbit Permanent Array Center (SOPAC).

Estimated parameters for each daily solution include the 3D Cartesian coordinates for each site, the six orbital elements for each satellite (semi-major axis, eccentricity, inclination, longitude of ascending node, argument of perigee, and mean anomaly), Earth Orientation Parameters (pole position and rate and UT1 rate), and integer phase ambiguities. We also estimate hourly piecewise-linear atmospheric zenith delays at each station to correct the poorly modeled troposphere, and 3 east-west and north-south atmospheric gradients per day, to account for azimuth asymmetry; the associated error covariance matrix is also computed and saved in SINEX format.

3 Combination Procedure

The strategy adopted in this work foresees the combination of geodetic solutions without the estimation and removal of the relative rotations between their reference frames and is based on the assumption of solutions with large uncertainties on the reference frame, i.e. loosely constrained solutions (Heflin et al. 1992; Davies and Blewitt 2000). Loosely constrained solutions can be combined regardless of the datum definition of each contributing solution. The solution reference frame is defined stochastically by the input data and changes from day to day, therefore since the datum is treated as a stochastic noise, it is not necessary to estimate or to apply relative rigid transformations (rotations, translations and scale) between reference frames and this naturally leads to a combined solution not distorted by any a priori constraint or transformation.

Denoting with $X_c(t)$ and $X_i(t)$ the array of the combined daily coordinates and the contributing solution coordinates at time t, respectively, the merging of daily site coordinates is achieved by solving in a least squares sense the following design equation. The design matrix, $\mathbf{P_i}$, being simply the reorder matrix of the i-th contributing solution.

$$X_C(t) = \mathbf{P}_i X_i(t)$$

Usually each contributing solution X_i originates from different processing schemes and processing software and each processing centre is responsible for assuring the best up-to-date solution, the only forced requirement is that the a priori covariance does not represent any specific inner constraint. The solutions and their associated covariance matrices have to be computed in the data-defined reference frame and have to be loosely constrained in translation, rotation and scale. Furthermore since each daily covariance matrix is known apart from a solution-dependent variance factor, a scale factor has been also estimated together with the combined solution. This assures that each solution contribution to the total χ^2 is equally balanced, and the estimated relative scaling factors f_i fulfil the following conditions:

$$R_i^T (f_i C_i)^{-1} R_i = R_j^T (f_j C_j)^{-1} R_j$$

$$\frac{1}{N} \sum f_i = 1$$

where $R_i = X_i - X_C$ are the *i-th* solution residuals, and N is the number of contributing solutions.

To assess the constraint treatment of the GPS solutions we first compute the *looseness* of the daily covariance matrix, defined as the Helmert uncertainties of the given solution:

$$C_\vartheta = (A^T C^{-1} A)^{-1}$$

where ϑ are the 7 parameters of the Helmert transformation (translations, rotations and scale) and A is the Helmert design matrix. Hence by definition the *looseness* represents the uncertainty of the 7 Helmert parameters when the solution is transformed into an error-free reference frame and indicates how much the reference frame in which the solution is represented is relaxed. It turns out that the Bernese solutions have an average looseness of 7–8 mm in translation, about 0.07 milli-arc-second (mas) in rotation and 10^{-10} in scale. On the other hand, the Gamit solutions show 0.5 m in translation, 200 mas in rotations and 10^{-9} in scale. The observed differences in the looseness reflect the peculiar approaches adopted in the two processing schemes. The Bernese approach foresees a fixed a priori celestial reference frame (precise GPS orbits and Earth Orientation Parameters, EOP) and relaxed a priori coordinates, whereas the Gamit processing scheme relax both the celestial frame and the site coordinates at the metre level. To deal with different constraints, and hence with different covariance structures, we analyze the rank deficiency of the solution covariance matrices. Figure 51.1 shows the first lowest eigenvalues (averaged over all the solutions) of the normal matrix (C^{-1}) for the Bernese and Gamit solutions respectively. Low eigenvalues reflect the original rank deficiency of the network solution, which has been stochastically

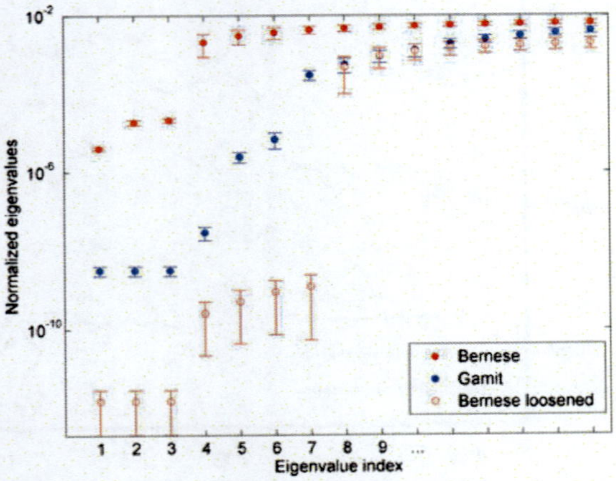

Fig. 51.1 Looseness of the Bernese and Gamit solutions

constrained imposing loose a priori constraints. In general geodetic problems, the kernel of the normal equation matrix (eigenvectors corresponding to zero eigenvalues) is the consequence of invariance in the observation equations, in particular each zero eigenvalue corresponds to an invariance of the observation equation, e.g. in small regional networks three translational degeneracies arise from the threefold rank deficiency of the GPS observable (e.g. Biagi and Sansò 2003). Therefore low, but non zero eigenvalues represent the remnant rank deficiency of the network solution that permits to define a loosely constrained reference frame (i.e. highly uncertain rigid transformation parameters). The Bernese solutions show a clear threefold null space, a fourth highly scattered eigenvalue that could represent a remnant scale uncertainty whereas the Gamit solutions display a broader null space (three very low eigenvalues and three intermediate values) that reflect a well relaxed translational and rotational degeneracy. Thus we conclude that the two solution types are not equally constrained and the Bernese solutions would prevail in the combination because of its smaller null-space dimension. Therefore we apply a loosening transformation (Davies and Blewitt 2000) to the Bernese covariance matrices in order to remove the Helmert parameter constraints and forcing the seven parameters to be uncertain at the meter level. Figure 51.1 shows the effect on the average eigenvalues of the loosening transformation, the first seven eigenvalues are now quite reduced.

4 Discussion

A first test combination has been carried out on two solution clusters obtained respectively by the Bernese and Gamit processors containing 15 overlapping sites (see Fig. 51.2) and spanning a time period of 2.5 years (2007–2009.5). The three solutions (Bernese, Gamit and Combined) have been then transformed into the

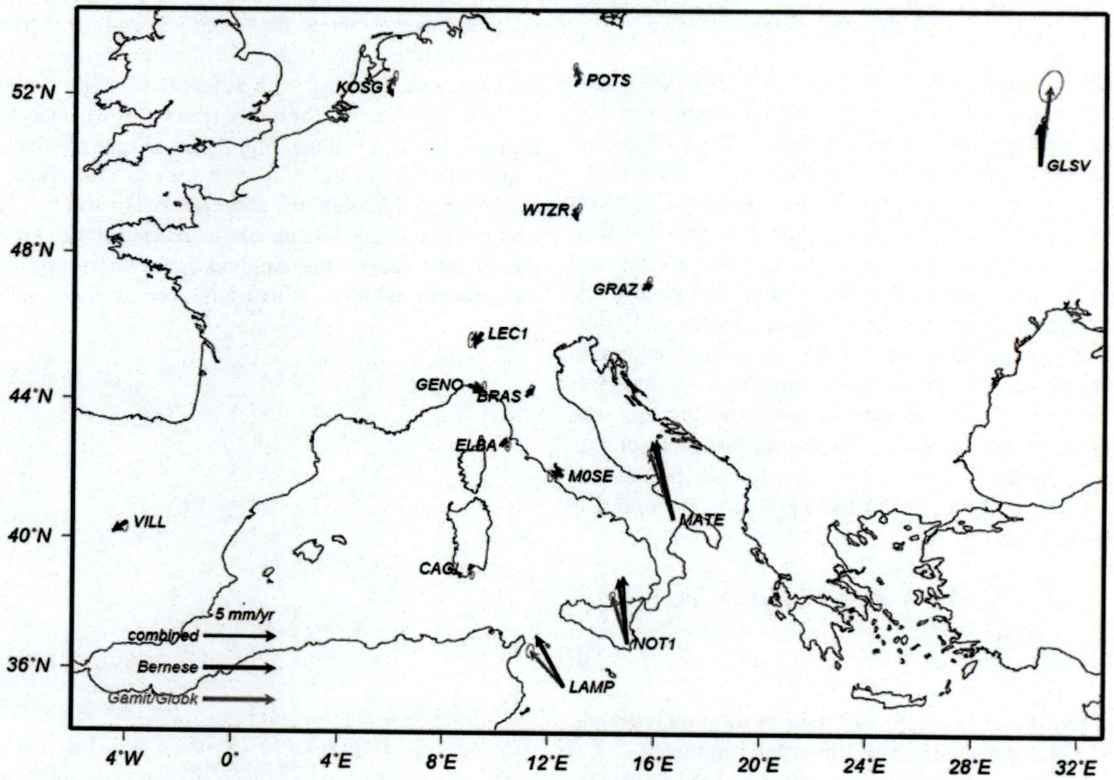

Fig. 51.2 Estimated velocity fields (Bernese, Gamit and Combined) with respect to the Eurasian plate, after a rigid rotation of the Bernese and Gamit velocities on the Combined velocity field

ITRF2005 reference system (Altamimi et al. 2007) and a velocity field has been estimated consistently for all three time series (Bernese ITRF core sites: 9–11; Gamit core sites 5–6; Combined core sites: 10–11). Secular trends and sporadic offsets have been estimated in a network least squares run, in which all the coordinates are modelled by linear trends:

$$X(t) = (\mathbf{I}(t - t_0)\mathbf{I}) \begin{pmatrix} X(t_0) \\ \dot{X} \end{pmatrix}$$

Figure 51.2 shows the three estimated velocity fields with respect to the Eurasian plate, after a rigid rotation of the Bernese and Gamit velocities on the combined velocity field in order to accommodate systematic differences due to the ITRF frame definition. Although the time span is too short to give a reliable velocity field, the weighted mean velocity residuals with respect to the combined solution are below 0.03 mm/year in each dimension, thus indicating that the three solutions lie in the same reference frame.

Whereas the velocity differences in magnitude are on the average 0.2 mm/year for Bernese-Combined and 0.4 mm/year for Gamit-Combined, indicating that the Bernese solution is closer to the combined solution than the Gamit one. In principle this could imply that the Bernese solution is slightly overcoming the combination process but also other effects could explain this behaviour e.g. seasonal signals in the time series that distort the secular drifts as well as the uneven distribution of the ITRF core sites that may cause instabilities in the reference frame realization in the two different clusters. Since these residuals are still within the 1-sigma level, we should extend the combination in time and spatial extension in order to get a better understanding of the ongoing systematic errors.

The time series of residuals reveal a general broader dispersion for the Gamit solution, especially in the horizontal components (see e.g. Matera station in Fig. 51.3) that, again, may be an indication of the reference frame instability for the Gamit solution.

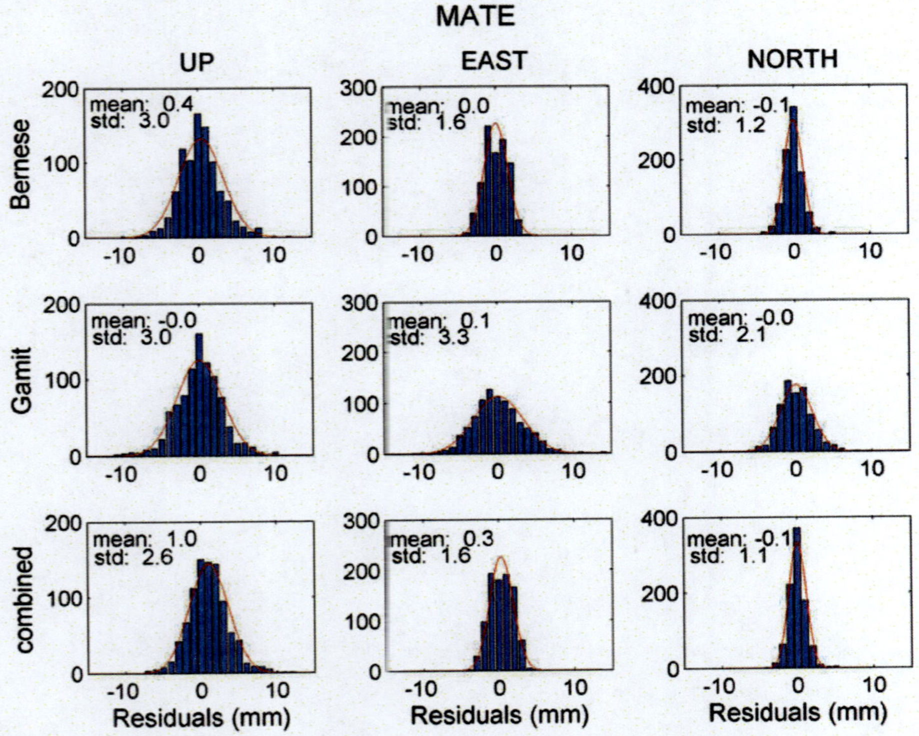

Fig. 51.3 Residual distributions of each solution (Bernese, Gamit and Combined) for each component with respect to the linear model

References

Altamimi Z, Collilieux X, Legrand J, Garayt B, Boucher C (2007) ITRF2005: A new release of the International Terrestrial Reference Frame based on time series station positions and Earth Orientation Parameters. J Geophys Res 112:B09401. doi:10.1029/2007JB004949

Beutler G et al. (2007) Bernese GPS software. In: Dach R, Hugentobler U, Fridez P, Meindl M (eds) Astronomical Institute University of Bern, Bern

Biagi L, Sansò F (2003) Sistemi di riferimento in Geodesia: algebra e geometria dei minimi quadrati per un modello con deficienza di rango. Bollettino di Geodesia e Scienze Affini 4(1–2) 2004

Davies P, Blewitt G (2000) Methodology for global geodetic timeseries estimation: a new tool for geodynamics. J Geophys Res 105(B5):11083–11100

Dong D, Bock RW (1989) GPS network analysis with phase ambiguity resolution applied to crustal deformation studies in California. J Geophys Res 94:3949–3966

Heflin MB et al (1992) Global geodesy using GPS without fiducial sites. Geophys Res Lett 19(2):131–134

Herring T, King RW, McClusky S (2006) GAMIT reference manual, Release 10.3. Department of Earth, Atmospheric, and Planetary Sciences, Massachussetts Institute of Technology, Cambridge

McCarthy DD, Petit G (2004) IERS Conventions (2003), IERS Technical Note 32, Verlag des Budesamts fur Kartographie und Geodasie, Frankfurt

McClusky S et al (2000) Global Positioning System constraints on plate kinematics and dynamics in the eastern Mediterranean and Caucasus. J Geophys Res 105(B3): 5695–5719

Schaffrin B, Bock Y (1988) A unified scheme for processing GPS phase observations. Bull Geodesique 62: 142–160

Serpelloni E, Casula G, Galvani A, Anzidei M, Baldi P (2006) Data analysis of permanent GPS networks in Italy and surrounding regions: application of a distributed processing approach. Ann Geophys 49:897–928

A Review on Non Iterative Closed Form Configuration Matching and Rotations Estimation

52

Mattia De Agostino, Chiara Porporato, and Marco Roggero

Abstract
Orthonormal matrices, Procrustes and quaternion analysis are closed form solutions of the configuration matching problem, common in geodesy as in the datum transformation problem. Literature reports more Procrustes based geodetic applications than Quaternions, which are more used in other application fields, such as aerospace navigation, robotics and computer vision. The large popularity of Procrustes in geodesy is mainly due to its capability to take into account a priori observation weighting in a simple way.

Keywords
Rotations • Quaternions • Orthonormal matrices • Procrustes

1 Introduction

A rotation is a transformation of the Euclidean space that rigidly moves objects leaving fixed at least one point (the origin of Euclidean space). In geodesy rotations are involved in many problems, especially for solving the transformation between reference frames. More in detail, a rotation is an isometry of an Euclidean space that preserves the orientation, and it is described by an orthogonal matrix. In a Euclidean space of two or three dimensions each orthogonal matrix expresses a rotation around a point or an axis, a reflection, or a combination of these two transformations.

As mentioned above, the representation of a rotation expresses the orientation of an object with respect to a reference system, or the relative orientation of two or more reference systems. Euler's theorem shows that any space rotation can be decomposed into the product of the three rotations $R^i_\psi R^j_\theta R^k_\phi$, where $i \neq j \neq k$ and $i, j, k \in \mathbb{R}^3$, and where R^i_α indicates a rotation of α radians counterclockwise around the i axis. According to Euler's theorem, the attitude of a rigid body can be described by a rotation around only one axis. Furthermore, this rotation can be defined uniquely by a minimum of three parameters, such as the directors cosines matrices, that represent the most widespread method for estimating rotations into geodetic networks.

However, for various reasons, there are several ways to represent rotations, making use of a number of parameters even higher than three, although also those redundant representations have always only three degrees of freedom.

M. De Agostino
Department of Land, Environment and Geoengineering, Politecnico di Torino, Corso Duca degli Abruzzi 24, 10129 Torino, Italy
e-mail: mattia.deagostino@polito.it

C. Porporato (✉) · M. Roggero
Department of Human Settlements Science and Technology, Politecnico di Torino, Viale Mattioli 39, 10125 Torino, Italy
e-mail: chiara.porporato@polito.it; marco.roggero@polito.it

N. Sneeuw et al. (eds.), *VII Hotine-Marussi Symposium on Mathematical Geodesy*, International Association of Geodesy Symposia 137, DOI 10.1007/978-3-642-22078-4_52, © Springer-Verlag Berlin Heidelberg 2012

In particular, for three-dimensional datum transformation, quaternions, orthonormal matrices and Procrustes Algorithm, presented over the years, are hereinafter described and compared.

Further details about the methods (in particular, about the Procrustes Algorithm) and a detailed bibliography about this argument can be found in Awange and Grafarend (2004) and in Grafarend (2006).

2 The Quaternion-Based Approach

What Sir William Rowan Hamilton wrote on the 16th of October 1843 on a stone of Brougham Bridge in Dublin, is simply (Fig. 52.1):

$$\mathrm{i}^2 = \mathrm{j}^2 = \mathrm{k}^2 = \mathrm{ijk} = -1 \tag{52.1}$$

The combination:

$$\dot{q} = q_0 + iq_1 + jq_2 + kq_3 \tag{52.2}$$

where q_0, q_1, q_2 and q_3 are real numbers, defines the generic quaternion. Quaternions satisfy all the laws of algebra, except the multiplication commutative law. In fact:

$$\mathrm{ij} = -\mathrm{ij},\ \mathrm{jk} = -\mathrm{kj},\ \mathrm{ki} = -\mathrm{ik} \tag{52.3}$$

which represents a serious violation of the commutative law ab = ba. It is also apparent:

$$\begin{aligned} &\mathrm{ij} = \mathrm{k} \quad \mathrm{jk} = \mathrm{i} \quad \mathrm{ki} = \mathrm{j} \\ &\mathrm{ji} = -\mathrm{k} \quad \mathrm{kj} = -\mathrm{i} \quad \mathrm{ik} = -\mathrm{j} \end{aligned} \tag{52.4}$$

The fundamental values of quaternions, i, j and k, can be handled as three mutually perpendicular clockwise axes in a common three-dimensional Euclidean space. Each unit quaternion, in particular, defines a rotation in $\mathbb{R}^3$ space. These rotations are given by the conjuge:

$$\dot{r}' = \dot{q}\dot{r}\dot{q}^{-1} = \dot{q}\dot{r}\dot{q}^* \tag{52.5}$$

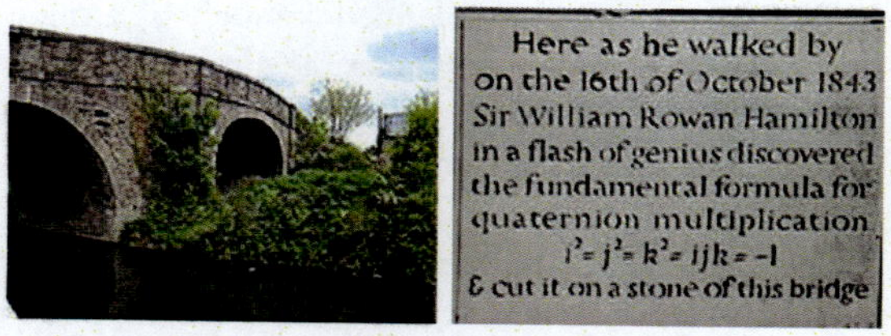

Fig. 52.1 The Brougham Bridge in Dublin and the Hamilton commemorative plaque

It can be verified that if $\dot{r}$ is purely imaginary (the real part is equal to zero), also $\dot{r}'$ is purely imaginary; therefore it can be defined an action of the group of unit quaternions on $\mathbb{R}^3$. Each action defined in this way is indeed a rotation, since it preserves the norm:

$$|\dot{r}'| = |\dot{q}\dot{r}\dot{q}^*| = |\dot{q}||\dot{r}||\dot{q}|^* = |\dot{r}| \tag{52.6}$$

It is possible to show the equivalence between the conjuge and the product of the 3x3 Rodriguez rotation matrix and a real vector. In fact, the operation:

$$\dot{q}\dot{r}\dot{q}^* = (Q\dot{r})\dot{q}^* = \bar{Q}^T(Q\dot{r}) = (\bar{Q}^T Q)\dot{r} \tag{52.7}$$

produces a 4×4 rotation matrix $\bar{Q}^T Q$, whose lower right 3×3 sub-matrix is orthonormal, and it is the rotation matrix that takes $\dot{r}$ to $\dot{r}'$.

Given the three-dimensional rotation between two frames $\dot{r}$ and $\dot{s}$ with a scale-change ρ:

$$\dot{s} = \rho\, \dot{q}\dot{r}\dot{q}^* \tag{52.8}$$

where $\dot{q}\dot{q}^* = 1$, the measurement equation is:

$$\dot{v}_i = \dot{s}_i - \rho\dot{q}\dot{r}_i\dot{q}^* \tag{52.9}$$

where v_i are still purely imaginary quaternions. In the following, we will present two methods in the literature to solve the problem (52.9), respectively by minimizing the residual vector $\dot{v}$ (Sansò 1973) or alternatively maximizing the scalar product $\dot{s} \cdot \dot{r}$ (Horn 1987).

2.1 Residual Vector Minimization

In accordance with the least squares approach, we must compute the minimum of the function:

$$\Phi(\dot{q}, \rho) = \sum_i \dot{v}_i^* \dot{v}_i \tag{52.10}$$

where $|\dot{q}|^2 = \dot{q}\dot{q}^* = 1$. Differentiating Φ with respect to ρ and $\dot{q}$ and introducing a real Lagrange multiplier α, we obtain, after some mathematical steps:

$$\left[\sum_i |\dot{s}_i|^2 + \rho^2 \sum_i |\dot{r}_i|^2 + \alpha\right]\dot{q} + 2\rho \sum_i \dot{s}_i \dot{q}\dot{r}_i = 0 \tag{52.11}$$

It is possible to show that:

$$\sum_i |\dot{s}_i|^2 = \rho^2 \sum_i |\dot{r}_i|^2 - \alpha \tag{52.12}$$

Using (52.12) in (52.11), through some mathematical steps we reach the final equation:

$$\left(\sum_i \dot{r}_i \dot{s}_i\right) \dot{q} = \left(-\rho \sum_i |\dot{r}_i|^2\right) \dot{q}$$

$$A\dot{q} = \lambda \dot{q} \tag{52.13}$$

As it is possible to see, the unknowns λ and $\dot{q}$ are respectively an eigenvalue and eigenvector of a symmetric matrix A, that can be built directly from the data r_i and s_i. In particular, using the products expansion rules of quaternions (and in particular the cyclic equality $\dot{r}_i\dot{s}_i\dot{q} = \dot{r}_i\dot{q}^*\dot{s}_i = \dot{q}\dot{r}_i\dot{s}_i = \dot{q}\dot{s}_i^*\dot{r}_i = \dot{s}_i\dot{q}\dot{r}_i$), we find:

$$\sum_i \dot{r}_i\dot{s}_i\dot{q} = \sum_i \dot{s}_i\dot{q}\dot{r}_i = \sum_i (S_i\dot{q})\dot{r}_i = \sum_i R_i^T(S_i\dot{q})$$

$$= \left(\sum_i R_i^T S_i\right)\dot{q} = \left(\sum_i A_i\right)\dot{q} \tag{52.14}$$

from which it is possible to derive the expressions of A_i and A.

2.2 Scalar Product Maximization

In this second approach, we seek the quaternion $\dot{q}$ to maximize the scalar product:

$$\sum_{i=1}^{n} (\dot{q}\dot{r}_i\dot{q}^*) \cdot \dot{s}_i = \sum_{i=1}^{n} \dot{r}'_i \cdot \dot{s}_i \tag{52.15}$$

Reminding the geometric meaning of the scalar product of vectors, we have:

$$\dot{r}'_i \cdot \dot{s}_i = |\dot{r}'_i||\dot{s}_i| \cos\theta \tag{52.16}$$

where θ is the angle subtended. Since $|\dot{r}'_i| = |\dot{r}_i|$ and $|\dot{s}_i|$ are constants, the maximization of the scalar product is equal to minimize the parameter θ (or maximize $\cos\theta$). Using the above results, we can rewrite the scalar product as:

$$\sum_{i=1}^{n} (\dot{q}\dot{r}_i) \cdot (\dot{s}_i\dot{q}) \tag{52.17}$$

The products $(\dot{q}\dot{r}_i)$ and $(\dot{s}_i\dot{q})$ can be expressed by means of the matrices R and S, therefore:

$$\sum_{i=1}^{n} (\dot{q}\dot{r}_i) \cdot (\dot{s}_i\dot{q}) = \sum_{i=1}^{n} \dot{q}^T \bar{R}_i^T S_i \dot{q}$$

$$= \dot{q}^T \left(\sum_{i=1}^{n} \bar{R}_i^T S_i\right) \dot{q}$$

$$= \dot{q}^T \left(\sum_{i=1}^{n} \mathring{A}_i\right) \dot{q} = \dot{q}^T \mathring{A} \dot{q} \tag{52.18}$$

It is now simple to derive the values of the sub-matrices $\mathring{A}_i$, and consequently of the matrix $\mathring{A}$, where $\mathring{A}_i^T = \mathring{A}_i$ and $\mathring{A}^T = \mathring{A}$. Recalling that we were seeking:

$$\underset{\dot{q}}{\mathrm{Max}} \left(\sum_{i=1}^{n} \dot{r}'_i \cdot \dot{s}_i\right) = \underset{\dot{q}}{\mathrm{Max}}(\mathring{A}) \tag{52.19}$$

It is possible to note that $\mathring{A} = -A$.

3 The Orthonormal Matrices Approach

Among the existing ways to represent rotation we present one that is most often used in photogrammetry: the orthonormal matrices. Again r is the position vector in the original RS and s the position vector in the final RS. The aim is to find the rotation that minimizes the residual errors. Therefore, we have to find the orthonormal matrix B, 3 × 3 matrix, that maximize

$$\sum_{i=1}^{n} s_i(Br_i) = \sum_{i=1}^{n} s_i^T B r_i \tag{52.20}$$

Being:

$$a^T B b = TR(B^T a b^T) \tag{52.21}$$

it is possible to write the (52.20) as:

$$\mathrm{Tr}\left(B^T \sum_{i=1}^{n} s_i r_i^T\right) = \mathrm{Tr}\left(B^T M\right) \tag{52.22}$$

where $M = \sum_{i=1}^{n} s_i r_i^T$ and:

$$M = \begin{bmatrix} \sum_i s_i^1 r_i^1 & \sum_i s_i^1 r_i^2 & \sum_i s_i^1 r_i^3 \\ \sum_i s_i^2 r_i^1 & \sum_i s_i^2 r_i^2 & \sum_i s_i^2 r_i^3 \\ \sum_i s_i^3 r_i^1 & \sum_i s_i^3 r_i^2 & \sum_i s_i^3 r_i^3 \end{bmatrix} \quad (52.23)$$

It follows that the rotation that minimize the residual errors corresponds to the orthonormal matrix B that maximizes $Tr(B^T M)$.

A square matrix M could always be decomposed into the product of an orthonormal matrix U and a positive semi-definite matrix S. When M is non singular, the matrices U and S are univocally determined and it allows to write

$$M = \frac{M}{(M^T M)^{1/2}} (M^T M)^{1/2} = US \quad (52.24)$$

In this expression, $U = M(M^T M)^{1/2}$ is an orthonormal matrix and $S = (M^T M)^{1/2}$ is the square root positive semi-definite of the symmetric matrix $M^T M$. It is possible to write this matrix $M^T M$ using its eigenvalues $\{\lambda_i\}$ and eigenvectors $\{\hat{u}_i\}$ as following:

$$M^T M = \lambda_1 \hat{u}_1 \hat{u}_1^T + \lambda_2 \hat{u}_2 \hat{u}_2^T + \lambda_3 \hat{u}_3 \hat{u}_3^T \quad (52.25)$$

Since $M^T M$ is positive semi-definite, its eigenvalues are positive and their square root is real and it is possible to write the symmetrical matrix S

$$S = \sqrt{\lambda_1} \hat{u}_1 \hat{u}_1^T + \sqrt{\lambda_2} \hat{u}_2 \hat{u}_2^T + \sqrt{\lambda_3} \hat{u}_3 \hat{u}_3^T \quad (52.26)$$

As the eigenvectors are orthogonal, it follows that $S^2 = M^T M$. This expression of the S matrix is allowed also when some eigenvectors are null. For this reason, the result is positive semi-definite instead of positive definite. If all the eigenvectors are positive, S becomes

$$S^{-1} = \frac{1}{\sqrt{\lambda_1}} \hat{u}_1 \hat{u}_1^T + \frac{1}{\sqrt{\lambda_2}} \hat{u}_2 \hat{u}_2^T + \frac{1}{\sqrt{\lambda_3}} \hat{u}_3 \hat{u}_3^T \quad (52.27)$$

It is useful to calculate the U matrix $U = MS^{-1} = M(M^T M)^{-1/2}$. It is possible to note that the sign of the determinant of U is the same of the determinant of M matrix. In fact

$$\det(U) = \det(MS^{-1}) = \det(M)\det(S^{-1}) \quad (52.28)$$

And the $\det(S^{-1})$ is positive because its eigenvalues are positive. The U matrix is a rotation when $\det(M) > 0$ and it represents a reflection if $\det(M) < 0$. It is necessary to minimize this expression $Tr(B^T M) = \det(B^T US)$ that, substituting the expression (52.26) becomes

$$Tr(B^T US) = \frac{1}{\sqrt{\lambda_1}} Tr(B^T U \hat{u}_1 \hat{u}_1^T) + \frac{1}{\sqrt{\lambda_2}} Tr(B^T U \hat{u}_2 \hat{u}_2^T) + \frac{1}{\sqrt{\lambda_3}} Tr\left(B^T U \hat{u}_3 \hat{u}_3^T\right) \quad (52.29)$$

For each X and Y matrices such that the XY and YX products are square, it follows that $Tr(XY) = Tr(YX)$ and:

$$Tr(B^T U \hat{u}_i \hat{u}_i^T) = Tr(\hat{u}_i^T B^T U \hat{u}_i) = Tr(B\hat{u}_i U \hat{u}_i) = B\hat{u}_i U \hat{u}_i \quad (52.30)$$

Since $\{\hat{u}_i\}$ is a unit vector and both U and B are orthogonal transformations, it is verified that $B\hat{u}_1 U \hat{u}_1 \leq 1$. It follows that

$$Tr(B^T US) \leq \sqrt{\lambda_1} + \sqrt{\lambda_2} + \sqrt{\lambda_3} = Tr(S) \quad (52.31)$$

And there is the maximum value of $Tr(B^T US)$ when $B^T U = I$ or $B = U$. The sought orthonormal matrix is the one that arises from the decomposition of M decomposed into the product of an orthonormal matrix and symmetric one. When M is non-singular matrix, then

$$B = M(M^T M)^{-1/2} \quad (52.32)$$

The presented method is a closed-form solution using orthonormal matrices and their eigenvalues-eigenvector decomposition.

4 Procrustes Approach

The minimization problem known as "Procrustes" is the technique of matching one configuration R into another configuration S. This method has been used in photogrammetric applications (Crosilla 2003) and is successfully used also in 3D datum transformation problems (Grafarend 2006).

In order to produce a measure of match an orthogonal transformation matrix T is used, minimizing the sum of squares of the residual matrix $E = RT - S$:

$$\Phi = \mathrm{Tr}(E^T E) = \min \quad (52.33)$$

Expanding the product $E^T E$, the Φ can be expressed as function of T:

$$\Phi = 2\mathrm{Tr}(T^T R^T S) + \mathrm{Tr}(T^T R^T RT + S^T S) \quad (52.34)$$

which partial derivative with respect to T is:

$$\frac{\partial \Phi}{\partial T} = 2\mathrm{Tr}(M^T) + 2\mathrm{Tr}(T^T R^T R) = \mathrm{Tr}(MT^T + R^T R) \quad (52.35)$$

The condition $||RT - S||^2 = \min$ it is equivalent to $\mathrm{Tr}(S^T RT) = \mathrm{Tr}(M^T T) = \max$. Be UDV^T the singular value decomposition of M where $D = \mathrm{diag}(\sigma_1, \sigma_2, \sigma_3)$, then

$$\mathrm{Tr}(T^T UDV^T) = \mathrm{Tr}(V^T T^T UD) = \mathrm{Tr}(ZD) = \sum_i z_{ii}\sigma_i \le \sum_i \sigma_i \quad (52.36)$$

where $Z = V^T T^T U$, that has a maximum in $Z = I$. Finally $T = UV^T$ is the optimal rotation matrix. Further details about the procedure and the possible applications can be found in the papers cited in bibliography.

5 Different Approaches Comparisons

The methods that are presented in this paper are widely used in geodesy, photogrammetry, robotics and computer graphic. Here we can underline some common or different aspects for each approach.

The presented approaches differ in term of rotation representation and optimization method, while the optimization criteria is always least squares.

The two quaternion approach investigated are formally equivalent. In fact, it was shown that the two matrix A and $\mathring{A}$ defined respectively for the residual vector minimization approach and for the scalar product maximization one, are related by $\mathring{A} = -A$. Moreover, it is find that these matrices depend only from the data sets: $\dot{r}$ and $\dot{s}$.

The orthonormal matrix approach has been studied in a closed-form solution performed using orthonormal matrices. This method requires the computation of the square root of a symmetric matrix to solve the rotation problem.

Finally the Procrustes approach is often used in photogrammetry, in order to solve the orientation problem. When the weight (both in the data set and for each 3D component) in the transformation is introduce we are dealing with the Generalized Procrustes Analysis. The Procrustes technique can also solve the 3D datum transformation problem. It is possible to shown that the used orthonormal matrix T is equivalent to the Rodriguez matrix.

The described algorithm have been implemented in a FORTRAN90 software and numerically verified on real data examples. The source code is available contacting the authors.

References

Akca D (2003a) Full automatic registration of laser scanner point clouds. ETH Swiss Federal Institute of Technology Zurich, Institute of Geodesy and Photogrammetry

Akca D (2003b) Generalized procrustes analysis and its applications in photogrammetry. ETH, Swiss Federal Institute of Technology Zurich, Institute of Geodesy and Photogrammetry

Awange JL, Grafarend EW (2004) Solving algebraic computational problems in geodesy and geoinformatics. Springer, New York. ISBN 3-540-23425-X

Beinat A, Crosilla F (2002) A generalized factored stochastic model for the optimal global registration of LIDAR range images. Int Arch Photogrammetry Remote Sensing Spatial Information Sci 34:36–39

Bertone C (2006) Quaternioni e rotazioni. Lecture handouts (in Italian)

Commandeur JJF (1998) A new approach to the subspace matching of configurations, including object weights. RR-98-03, Department of Education – University of Leiden

Crosilla F (2003) Procrustes analysis and geodetic sciences. Invited Lecturers at International Centre for Mechanical Sciences (CISM), October 13–14

Gower JC (1992) Coffee images: a study in the simultaneous display of multivariate quantitative and categorical variables for several assessors. RR-92-06, Department of Data Theory – University of Leiden

Grafarend E (2006) The nonlinear problem of the 3D transformation and the procrustes algorithm, linear and nonlinear models: fixed effects, random effects, and mixed models. de Gruyter GmbH, pp 431–444

Gruen A, Akca D (2005) Least squares 3D surface and curve matching. ISPRS J Photogrammetry Remote Sensing 59(3):151–174

Horn BKP (1987) Closed-form solution of absolute orientation using unit quaternions. J Optic Soc Am A-4:629–642

Horn BKP, Hugh MH, Shaxhriar N (1988) Closed-form solution of absolute orientation using orthonormal matrices. J Optic Soc Am A-5:1127–1135

Horn BKP (2001). Some notes on unit quaternions and rotation. Lecture handouts.

Lorusso A, Eggert DW, Fisher RB (1997) A comparison of four algorithms for estimating 3D rigid transformations. Special Issue Perform Characteristics Vision Algorithms 9(5): 272–290

Ottaviani G Introduzione alla geometria dei quaternioni. Lecture handouts (in Italian)

Penrose R (2005) Numeri ipercomplessi. La strada che porta alla realtà. Rizzoli, pp 198–216. ISBN 88–17–00785–4 (in Italian)

Rivest LP, Chang T (2006) Regression and correlation for 3 × 3 rotation matrices. Can J Stat 34(2):187–202

Sansò F (1973) An exact solution of the roto-translation problem. Photogrammetria 29:203–216

Sansò F (2006) Navigazione geodetica e rilevamento cinematico. Polipress, Devon. ISBN 8873980198 (in Italian)

Titterton DH, Weston JL (2004) Strapdown inertial navigation technology, 2nd edn. The American Institute of Aeronautics and Astronautics, Reston. ISBN 1 56347 693 2

Part IX

Temporal Variations of Deformation and Gravity

Co-seismic Gravity Change of M_W 7.9 Wenchuan Earthquake and Pre-Seismic Gravity Anomaly Detection by Superconducting Gravimeter at Hsinchu, Taiwan, from April to June 2008

53

S.C. Lan, T.T. Yu, C. Hwang, and R. Kao

Abstract

Earthquakes are caused by crustal movement that releases a huge amount of energy and leads to deformation. The gravity near the epicenter is changed by the deformation of rock materials. Precursors may be detected during the period of seismogenic activity. On 12 May, 2008, a M_w 7.9 earthquake struck near Wenchuan in China. A co-seismic gravity offset of 10.5 μGal and a pre-seismic gravity anomaly about 2 days before the main shock was found in gravity records of the superconducting gravimeter (SG) at Hsinchu, Taiwan. SG records from April to June 2008 were used in this study. During this period of time, 8 gravity anomaly events caused by earthquakes were found. Sonograms of time-frequency analysis show that the perturbations occurred in 1–2 major bands (0.1–0.15 Hz and 0.2–0.3 Hz). The results of canonical correlation analysis show a significant positive correlation between the duration of gravity perturbation before the main shock and the magnitude of the earthquake. This study suggest that SG is a promising instrument for the analysis of earthquake events.

Keywords

Superconducting gravimeter • Earthquake • Co-seismic gravity change • Earthquake precursor

1 Introduction

The first superconducting gravimeter (SG) in Taiwan was installed in a tunnel in 18 Peaks Hill, east of downtown Hsinchu. The SG (GWR Serial No.048) was supported by the Ministry of Interior, Taiwan, the Center of Measurements, for the purpose of launching projects involving various scientific and governmental missions. It has been recording data since April 2006 at a sampling rate of 1 s.

Co-seismic gravity changes were observed for the 1964 M_w 9.2 Alaska earthquake (Barnes 1966). In recent years, high frequency sampling and highly sensitive SGs have provided new opportunities for detecting small co-seismic gravity changes. For the 2004 Kii Peninsula earthquake (M7.1 foreshock and M7.4 main shock), a 1 μGal co-seismic gravity change was

S.C. Lan · T.T. Yu (✉)
Department of Resources Engineering, National Cheng Kung University, No.1, Ta-Hsueh Road, 701 Tainan, Taiwan
e-mail: lanshyhchin@gmail.com; yutt@mail.ncku.edu.tw

C. Hwang · R. Kao
Department of Civil Engineering, National Chiao Tung University, No.1001 University Road, 300 Hsinchu, Taiwan
e-mail: cheinway@mail.nctu.edu.tw; ricky.kao@gmail.com

N. Sneeuw et al. (eds.), *VII Hotine-Marussi Symposium on Mathematical Geodesy*, International Association of Geodesy Symposia 137, DOI 10.1007/978-3-642-22078-4_53, © Springer-Verlag Berlin Heidelberg 2012

detected with the SG at the Inuyama Station of Nagoya University, central Japan (Nawa et al. 2009). On 12 May, 2008, a strong earthquake with a moment magnitude (M_w) of 7.9 struck Wenchuan (31.0°N, 103.4°E), east of the Tibetan plateau and northwestern Sichuan province in China. The epicenter was shallow, with a depth of 19 km.

Hao et al. (2008) observed a gravity disturbance about 2 days before the Wenchuan earthquake at the Wuhan National Field Scientific Station for Geodesy with a LacosteET-20 gravimeter. Our review of the SG records from April to June 2008 also found this correlation. Earthquake precursors of gravity anomalies during this period of time are analyzed in the present study.

2 Co-seismic Gravity Changes of Wenchuan Earthquake

The observed residual gravity values were obtained from the raw SG gravity records which were corrected for the effects of the body and ocean tides as suggested by Hwang et al. (2009). Before the tidal analysis, the SG data were despiked, filtered and decimated to hourly records using the program TSoft (Van Camp and Vauterin 2005). The program ETERNA (Wenzel 1996) was used for the tidal analysis. The NAO.99b tide model was suggested as the best fit for the SG observations of all tidal components. The atmospheric pressure effect (δg_b, μGal) is estimated from (53.1).

$$\delta g_b = f_a(P_a - 1013) \quad (53.1)$$

where P_a is the pressure in hPa and f_a is the gravity-atmosphere admittance. According to Hwang et al. (2009), f_a was set to $-0.350\,\mu\text{Gal}\,\text{h}\,\text{Pa}^{-1}$. Because of the large amplitudes during the first few hours of the earthquake, the data were separated into two parts. Part 1 starts at 1,200 UT on 8 May, 2008, and ends at 0628 UT on 12 May, 2008, about 40 s before the earthquake. Part 2 starts at 1,000 UT on 12 May, 2008, and ends at 1,200 UT on 17 May, 2008. The two data parts were respectively fitted by the quadratic function shown in (53.2) (Kim et al. 2009).

$$y(t) = a + bt + ct^2 \quad (53.2)$$

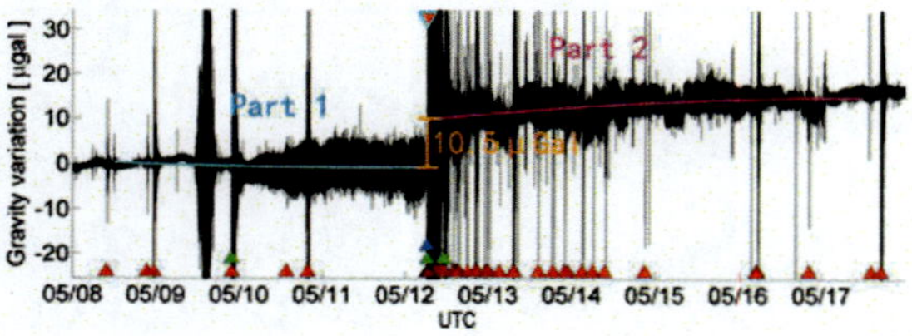

Fig. 53.1 Wenchuan earthquake recorded at Hsinchu SG station and the fitted functions before and after the earthquake

Then, the two parts were respectively used to extrapolate the expected gravity values at 06:28:40 UT on 12 May, 2008. The co-seismic gravity change was determined using the difference between the two values. The calculated value of co-seismic gravity was an increase of 10.5 μGal after the Wenchuan earthquake. Figure 53.1 shows the recorded Wenchuan earthquake and fitted curves. The gravity offset was caused by the deformation and movements near the fault plane in the crust. The distance from the epicenter of the Wenchuan earthquake to the SG station in Hsinchu, Taiwan, is more than 1,800 km. At this distance, the co-seismic gravity offset of 10.5 μGal is unrealistic. The maximum co-seismic gravity change of the 2004 Sumatra-Andaman earthquake detected by GRACE is only about 10–15 μGal (de Linage et al. 2009). An unavoidable estimation error was caused by the vibrations of the pre-seismic and post-seismic gravity anomalies. Therefore, further refinements are required.

3 Pre-seismic Gravity Anomaly Detection

Besides the co-seismic gravity change shown in Fig. 53.1, a gravity perturbation event started about 2 days before the Wenchuan earthquake. The amplitude of the gravity perturbation increased before the earthquake and gradually decreased after the earthquake. This is consistent with the report of Hao et al. (2008). Thus, the SG records from April to June 2008 were inspected in this study. A total of 8 gravity perturbation events caused by earthquakes were found. The parameters of correlated earthquakes are summarized in Table 53.1. The fault-plane solutions and distribution of the earthquakes are shown in Fig. 53.2.

Table 53.1 Summary of gravity perturbations caused by earthquake events in superconducting gravimeter records from April to June 2008

Earthquake event	Date (year/M/D)	Time (UTC)	Lat. (°N)	Lon. (°E)	M_w	Depth (km)	Linear distance to SG (km)	Duration of gravity perturbation (offset days)			A^a_{max} (μGal)
								Before (b)	After (a)	Ratio (a/b)	
I	2008/04/09	12:46:12	−20.07	168.89	7.3	33.0	6,786	1.09	2.74	2.52	2.3
II	2008/04/18	05:58:36	26.00	128.41	5.1	28.0	757	1.00	2.47	2.48	3.5
III	2008/04/23	18:28:41	22.88	121.62	6.0	10.0	222	1.07	1.56	1.46	5.2
IV[b]	2008/05/05	09:41:56	22.85	121.69	3.4	19.4	227	0.28	0.65	2.34	4.4
V[c]	2008/05/12	06:28:01	31.00	103.32	7.9	19.0	1,858	2.22	4.11	1.85	9.4
VI	2008/06/01	01:57:23	20.12	121.35	6.3	31.0	518	1.19	1.66	1.39	7.0
VII	2008/06/17	05:14:34	31.64	104.16	4.7	10.0	1,807	0.82	1.37	1.67	3.9
VIII	2008/06/24	17:02:35	29.00	142.45	5.0	42.0	2,162	1.22	2.77	2.26	8.3

Data source: U.S. Geological Survey Earthquake Data Base

[a] A_{max}: Max Amplitude of pre-seismic gravity anomaly. The duration from the start of pre-seismic anomaly to the earthquake time were separated into eight sections. After excluding the sections with spike values, the A_{max} was estimated from the maximum half difference value between 98 and 2 percentile gravity signals of each section

[b] No published data from Taiwan Central Weather Bureau

[c] Wenchuan Earthquake

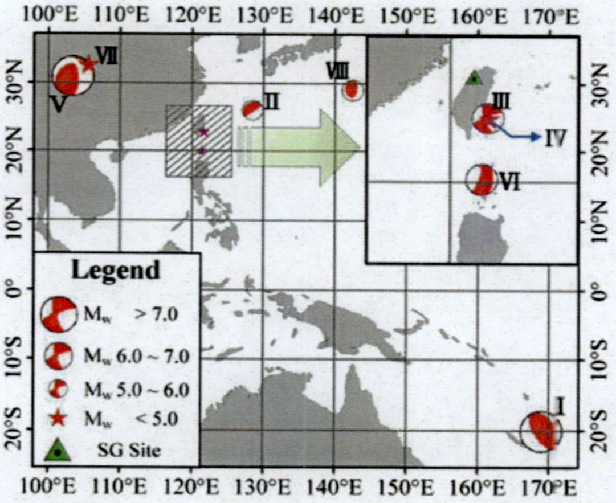

Fig. 53.2 Fault plane solutions and distribution of earthquakes listed in Table 53.1. Fault plane solutions are available from the Global CMT catalog (http://www.globalcmt.org,2009)

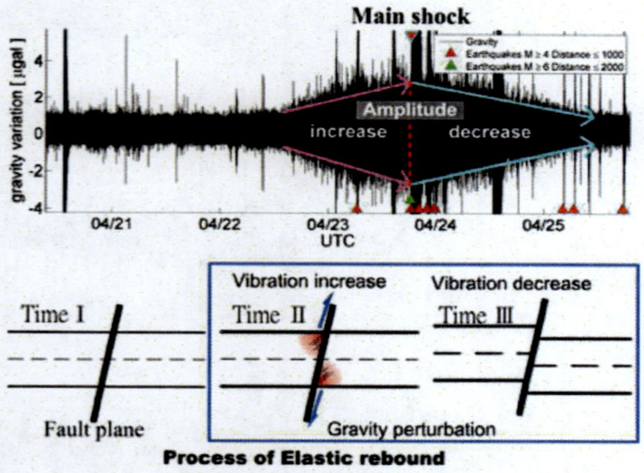

Fig. 53.3 Comparison of gravity perturbations and the process of elastic rebound

3.1 Data Processing and Observation

SG records obtained at a 1 s sampling rate were used in this study. The tidal signal was removed from the raw data using Tsoft and ETERNA software packages. Then, frequencies below 0.005 Hz were filtered to make sure that the tide did not affect the residual gravity value. Fig. 53.4 shows the records of SG gravity after data processing and the results of time-frequency analysis. An increasing amplitude before the main shock and a gradually decreasing amplitude after the main shock were found in the gravity records as shown in Fig. 53.3. The increasing amplitude suggests a concentration of energy before the main shock. This is called the seismogenic process. After the main shock, the energy gradually stabilized. Therefore, the amplitude of gravity perturbation decreased. This can be well explained by the process of elastic rebound, as shown in Fig. 53.3 (Koseluk and Bischke 1981). It is worth noting that the durations of gravity perturbation after the main shock are always approximately

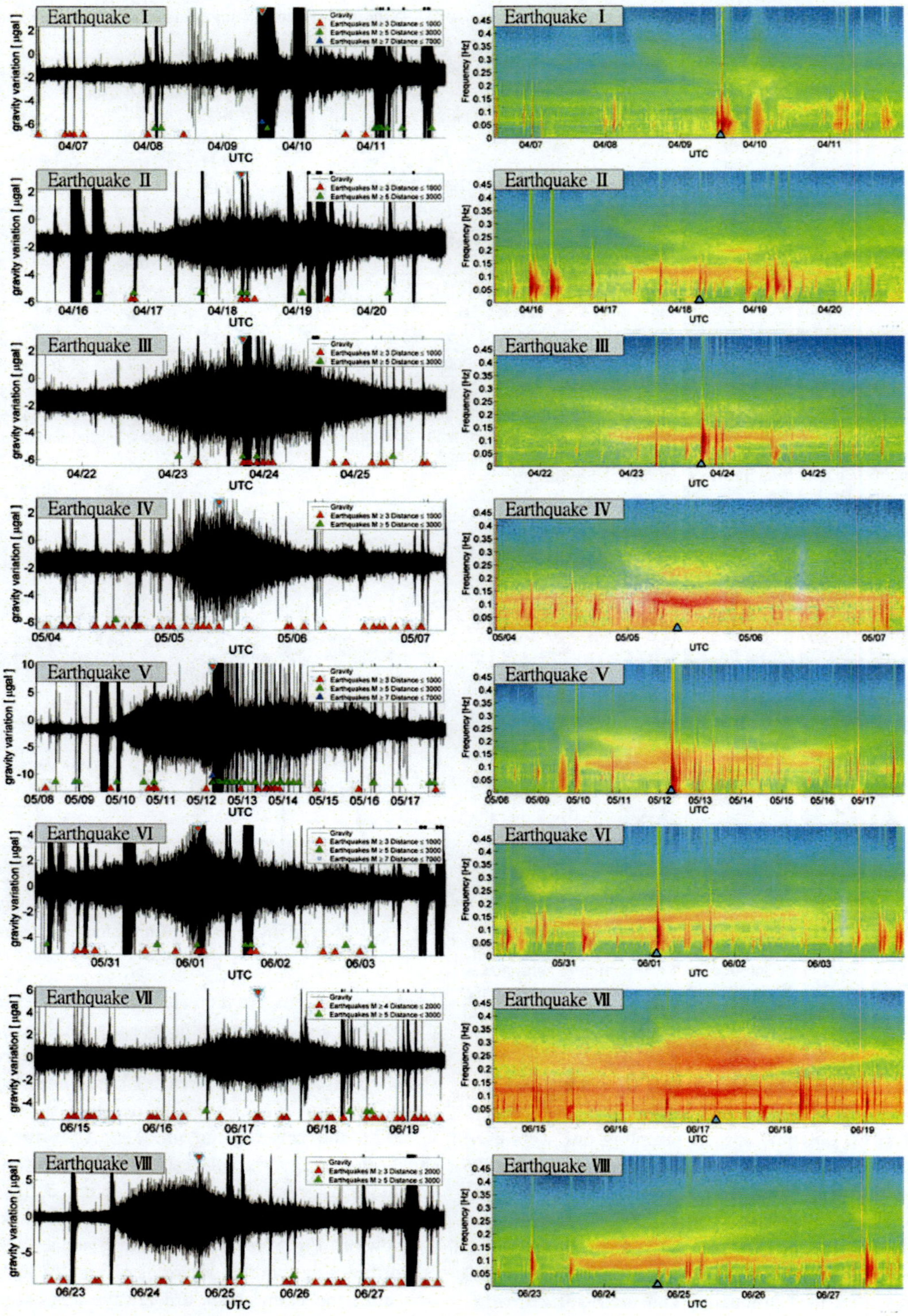

Fig. 53.4 Waveform and time-frequency analysis of gravity perturbation events

Table 53.2 Canonical correlation analysis summary

	Canonical variables	
	1	2
Canonical correlations	0.956	0.747
Tests of significance (p values)	0.042	0.196
Canonical loadings for input variables		
Perturbation before earthquake (days)	−0.920	0.077
Perturbation after earthquake (days)	−0.715	0.452
Amplitude of gravity perturbation (μGal)	−0.507	−0.527
Canonical loadings for output variables		
Moment magnitude (M_w)	−0.928	0.373
Linear distance (km)	−0.129	0.992

p values of <0.05 were considered statistically significant

1.39–2.52 times longer then those before the main shock. This finding indicates that the period of tectonic recovery is much longer than the period of the stress build-up on opposite sides of the fault plane before the main shock.

3.2 Time-Frequency Analysis

Sonograms of the time-frequency analysis are shown in Fig. 53.4. The results of time-frequency analysis of earthquakes II-VI reveal that the major spectral peaks are in the range of 0.1–0.15 Hz, and that a significant amount of energy is present in the range of 0.2–0.3 Hz band during the period of gravity perturbations. However, the two major bands for earthquake VIII are 0.05–0.1 Hz and 0.15–0.25 Hz, respectively. This difference is possibly due to the depth and location of the epicenter. Further research is required.

3.3 Canonical Correlation Analysis

The concepts of canonical correlation analysis were first introduced by Hotelling (1936). Canonical correlation analysis is used for finding the linear combination of each of two sets of variables which have the maximum correlation (Glahn 1968). Table 53.2 summarizes the results of canonical correlation analysis. A p value of less than 0.05 was considered statistically significant in this study. Thus, the results of the significance test show that only canonical variable set 1 is statistically significant at 0.042 with a correlation coefficient of 0.956. Of the three canonical loadings, the input variables of the duration of gravity perturbation before the earthquake and the amplitude were highly positively correlated with the moment magnitude of the earthquake. The duration of gravity perturbation before the earthquake is the most effective factor. In other words, the larger an earthquake's magnitude, the longer the required period of energy concentration. This relationship, while significant, is moderate in strength.

Conclusion

In this paper, a co-seismic gravity change of 10.5 μGal was found for the Wenchuan earthquake in the gravity records of SG in Hsinchu, Taiwan. However, because of the vibrations of pre-seismic and post-seismic gravity anomalies, this result is probably biased and further refinements are requied. Eight gravity perturbation events from April to June 2008 were analyzed in this study. The three major findings are: (1) the developmental sequence of gravity perturbations fits the process of elastic rebound, and the ratio of duration after/before the earthquake events indicates that the period of tectonic recovery is longer than the period of the stress build-up on opposite sides of the fault plane before the mainshock; (2) the gravity perturbation exhibited 1–2 major bands of spectral peaks (0.1–0.15 Hz and 0.2–0.3 Hz) in the sonograms of time-frequency analysis; (3) the duration of gravity perturbation before the earthquake shows a maximum positive correlation with the moment magnitude of the earthquake. The evolution of earthquake processes was as expected. This study suggests that SG can be a useful resource for earthquake precursor analysis.

Acknowledgements This work was supported in part by the National Science Council of Taiwan under grants NSC-93-2119-M-006-006, NSC-94-2116-M-006-016, NSC-95-2116-M-006-004, and NSC-97-2116-M-006-013.

References

Barnes DF (1966) Gravity changes during Alaska earthquake. J Geophys Res 71(2):451–456

de Linage C, Rivera L et al. (2009) Separation of coseismic and postseismic gravity changes for the 2004 Sumatra–Andaman earthquake from 4.6 yr of GRACE observations and modelling of the coseismic change by normal-modes summation. Geophys J Int 176(3):695–714

Glahn HR (1968) Canonical correlation and its relationship to discriminant analysis and multiple regression. J Atmos Sci 25(1), 23–31

Hao XG, Hu XG, Xu HZ, Zhong M, Fang J, Hao XH, Liu M, Liu GY, Xue HP (2008) Gravity disturbance before Wenchuan Ms8.0 earthquake. J Geodes Geodyn 28(3): 129–131 (in Chinese)

Hwang C, Kao R et al. (2009) Results from parallel observations of superconducting and absolute gravimeters and GPS at the Hsinchu station of Global Geodynamics Project, Taiwan. J Geophys Res 114

Kim JW, Neumeyer J, Kim TH, Woo I, Park H-J, Jeon J-S, Kim K-D (2009) Analysis of superconducting gravimeter measurements at MunGyung station, Korea. J Geodyn 47(4):180–190

Koseluk RA, Bischke RE (1981) An elastic rebound model for normal-fault earthquakes. J Geophys Res 86(NB2): 1081–1090

Nawa K, Suda N, Yamada I, Miyajima R, Okubo S (2009) Coseismic change and precipitation effect in temporal gravity variation at Inuyama, Japan: A case of the 2004 off the Kii peninsula earthquakes observed with a superconducting gravimeter. J Geodyn 48(1):1–5

Van Camp M, Vauterin P (2005) Tsoft: graphical and interactive software for the analysis of time series and Earth tides. Comput Geosci 31(5):631–640

Wenzel HG (1996) The nanoGal software: Earth tide processing package ETERNA 3.30 Bulletin d'Informations Mareés Terrestres 124:9425–9439

Optimal Cross-Validation of Different Surveying Techniques

54

Maddalena Gilardoni, Fernando Sansò, and Giovanna Venuti

Abstract

When monitoring deformations by means of different sensors, one has to be sure that the various observations do see the same variations in time of the earth surface. As an example one can think of a deformation as seen by the SAR technique and the deformation of the same surface as seen by GPS. To this aim a hypothesis testing procedure has to be set up (Koch 1999). The first question is how to compare the different data sets, which usually do not refer neither to the same positions in space nor to the same time. The standard prediction of one set of variables from the other, for instance, is not always the best solution. It is better to use both observation sets to predict one and the same functional of the "random field" describing the deformation pattern and to evaluate the difference between the two predictions. This difference has to be small on condition that the signal we try to estimate has a fixed amplitude in mean quadratic sense. The problem is formally solved and a few examples are illustrated.

Keywords

Cross-validation • Optimal statistics

1 Definition of the Problem

The problem we want to discuss, described in abstract form, is as follows: let $\{u(t)\}$, $t \in T$, be a random field, defined on some set T ($T \subseteq R^n$), with zero mean

$$E\{u(t)\} = 0, \tag{54.1}$$

and covariance function

$$E\{u(t)u(s)\} = C(t,s) \tag{54.2}$$

M. Gilardoni (✉) · F. Sansò · G. Venuti
DIIAR, Politecnico di Milano, P.zza Leonardo da Vinci 32, Milan, Italy
e-mail: maddalena.gilardoni@mail.polimi.it

and assume that there is a vector of "observations" performed on $u(t)$, which are linear functionals of $u(t)$ with some additive noise

$$\underline{Y} = \underline{L}(u) + \underline{\nu}. \tag{54.3}$$

For (54.3) to be meaningful, we need the components L_i ($i = 1, \ldots, N$) of $\underline{L}$ to be bounded linear stochastic functionals; namely, we shall assume that, adopting Krarup's notation (Krarup 1969),

$$E\{L_i(u)^2\} = L_{it}\{L_{is}[C(t,s)]\} = C(L_i, L_i) < +\infty; \tag{54.4}$$

in addition we assume that

$$E\{\underline{\nu}\} = 0, \quad E\{\underline{\nu}\underline{\nu}^+\} = C_\nu \tag{54.5}$$

N. Sneeuw et al. (eds.), *VII Hotine-Marussi Symposium on Mathematical Geodesy*, International Association of Geodesy Symposia 137, DOI 10.1007/978-3-642-22078-4_54,

where C_ν is a known covariance matrix, and finally that u and ν are linearly independent, implying

$$E\{\underline{L}(u)\underline{\nu}^+\} = 0. \tag{54.6}$$

Note that, under the above conditions, the covariance of $\underline{Y}$ is known and given by

$$C_Y = C(\underline{L}, \underline{L}^+) + C_\nu. \tag{54.7}$$

Suppose that a similar situation is independently duplicated; namely, there is another random field $\{w(t)\}$, $t \in T$, generally not centered, and another vector of observations

$$\underline{Z} = \underline{K}(w) + \underline{\eta} \tag{54.8}$$

with $\underline{K} = \{K_i;\ i = 1, \ldots, M)$ and

$$C(K_i, K_i) < +\infty, \tag{54.9}$$

$$E\{\underline{\eta}\} = 0, \quad E\{\underline{\eta}, \underline{\eta}^+\} = C_\eta, \quad E\{\underline{K}(w)\underline{\eta}^+\} = 0. \tag{54.10}$$

We further assume that $\underline{\nu}$ and $\underline{\eta}$ are linearly independent of one another and, both, of $u(t)$ and $w(t)$. We would like to test the hypothesis

$$H_0 : u = w \tag{54.11}$$

i.e., that the two sets of measurements $\underline{Y}$ and $\underline{Z}$ refer to one and the same random field. Among other things, this implies that there is no bias between u and w. So, the problem becomes that of finding a suitable statistic to verify (54.11). To be simple, we will try to find a linear function of the observables that becomes zero in the average when (54.11) is satisfied and that is as small as possible in variance, in a suitable sense, such as to be very sensitive to all departures from (54.11). To understand the possible applications of a scheme like that, we make two examples:

Example 54.1.1. Assume you want to build a digital elevation model (DEM) for a portion of the earth surface and you have data collected for instance by SAR surveying (SRTM) and photogrammetry. The two techniques will produce elevations on grids with different knots so that the hypothesis that there is no bias between them is not immediately verifiable. Even more, SAR will give a grid of mean heights, while the photogrammetric model refers to almost point-wise observations.

Example 54.1.2. Assume a landslide to be monitored by SAR and GPS. The first technique observes the on going deformation along the line of sight (LOS) at some highly coherent points; the second observes the deformation vector at some other points in the same area. In this case, by projecting the GPS 3D deformation vector along the SAR LOS, $u(\underline{t})$ will be the LOS component of the deformation pattern. The hypothesis is now that the two techniques see the same pattern, so that one can be used to validate the other.

In the next section those problems will be formalized through a suitable target function that will be minimized. In Sect. 3, we shall develop example 54.1.2. A short discussion will conclude the paper.

2 The Optimization Problem

As we said, under the assumptions of Sect. 1, the sensible thing to do is to try to estimate a linear functional $M(\cdot)$ from both the observation sets and then to take the difference of the estimates. As we know that, by a Wiener-Kolmogorov optimal predictor, no information can be drawn on any subspace of functionals orthogonal to both $\underline{Y}$ and $\underline{Z}$, instead of defining our problem for a general $M(\cdot)$, we rather define directly the statistic

$$S = \underline{\lambda}^+\underline{Y} - \underline{\mu}^+\underline{Z} \tag{54.12}$$

that we shall try to make as small as possible in mean quadratic sense. Under the assumption (54.1) and hypothesis (54.11), S has zero mean and its variance, representing the departure from zero, is

$$F(\underline{\lambda}, \underline{\mu}) = \sigma^2(S) = E\{S^2\}$$
$$= \underline{\lambda}^+ C_Y \underline{\lambda} + \underline{\mu}^+ C_Z \underline{\mu} - 2\underline{\lambda}^+ C_{YZ} \underline{\mu} \tag{54.13}$$

where (cf. (54.7))

$$C_Y = C(\underline{L}, \underline{L}^+) + C_\nu,$$
$$C_Z = C(\underline{K}, \underline{K}^+) + C_\eta, \quad (54.14)$$
$$C_{YZ} = C(\underline{L}, \underline{K}^+).$$

Obviously, we cannot look for an unconstrained minimum of the variance (54.13), because this is indeed obtained at $\underline{\lambda} = 0$, $\underline{\mu} = 0$. Therefore, we must keep the couple $(\underline{\lambda}, \underline{\mu})$ on a surface far away from zero by imposing a suitable condition. The theory gets its simplest and most elegant form, if we choose the following normalization condition:

$$G(\underline{\lambda}, \underline{\mu}) = E\{(\underline{\lambda}^+ \underline{Y} + \underline{\mu}^+ \underline{Z})^2\}$$
$$= \underline{\lambda}^+ C_Y \underline{\lambda} + \underline{\mu}^+ C_Z \underline{\mu} + 2\underline{\lambda}^+ C_{YZ} \underline{\mu} = 1. \quad (54.15)$$

It is interesting to remark that exactly the same equations would be obtained by imposing separately the two variables $\underline{\lambda}^+ \underline{Y}$ and $\underline{\mu}^+ \underline{Z}$ to have a unit variance. The minimization of the Lagrange target function

$$\mathcal{L}(\underline{\lambda}, \underline{\mu}) = F(\underline{\lambda}, \underline{\mu}) - \gamma G(\underline{\lambda}, \underline{\mu}) \quad (54.16)$$

leads to the following normal system

$$\begin{pmatrix} C_Y & -C_{YZ} \\ -C_{ZY} & C_Z \end{pmatrix} \begin{pmatrix} \underline{\lambda} \\ \underline{\mu} \end{pmatrix} = \gamma \begin{pmatrix} C_Y & C_{YZ} \\ C_{ZY} & C_Z \end{pmatrix} \begin{pmatrix} \underline{\lambda} \\ \underline{\mu} \end{pmatrix}. \quad (54.17)$$

This is a generalized eigenvalue problem: for each eigenvalue γ_n, we have the corresponding eigenvector $[\underline{\lambda}_n^+, \underline{\mu}_n^+]^+$, which can be normalized with condition (54.15). By multiplying (54.17) by $[\underline{\lambda}_n^+, \underline{\mu}_n^+]$, and taking into account (54.15), one gets

$$\gamma_n = F(\underline{\lambda}_n, \underline{\mu}_n), \quad (54.18)$$

and the original problem becomes that of finding the minimum eigenvalue of (54.17) and the corresponding eigenvector. This is similar to canonical analysis discussed in literature (Wackernagel 1995). We observe, for future use, that our problem can even be dimensionally reduced. Getting, for instance, $\underline{\lambda}$ from the first row of (54.17)

$$\underline{\lambda} = \alpha C_Y^{-1} C_{YZ}\, \underline{\mu} \quad (54.19)$$

and substituting in the second one, it results

$$(C_Z - \alpha^2 C_{ZY} C_Y^{-1} C_{YZ}) \underline{\mu} = 0 \quad (54.20)$$

with

$$\alpha = \frac{1+\gamma}{1-\gamma}. \quad (54.21)$$

Note that (54.21) implies

$$\gamma(\alpha) = \frac{\alpha - 1}{\alpha + 1} \quad (54.22)$$

and that

$$\gamma(-\alpha) = \frac{1}{\gamma(\alpha)}. \quad (54.23)$$

Since $F(\underline{\lambda}, \underline{\mu})$ is positive by definition, (cf. (54.13)), due to (54.18), γ is positive too. Then, from (54.22) we see that either $\alpha > 1$ or $\alpha < -1$, which is complying with the fact that the values of α from (54.20) come in couples with the same modulus. Since (cf. (54.22)) $\alpha > 1$ implies $\gamma < 1$, then from (54.23) we have that $\alpha < -1$ implies $\gamma > 1$. It follows that we have to find the minimum positive α, such that α^2 is an eigenvalue of (54.20), and the corresponding eigenvector $\underline{\mu}$. There on, we compute $\underline{\lambda}$ from (54.19). It is not difficult to see that, if $(\alpha, \underline{\lambda}, \underline{\mu})$ is the triple corresponding to γ_{min}, then $(-\alpha, \underline{\lambda}, -\underline{\mu})$ is the triple corresponding to γ_{max}.

Remark 54.2.1. Let us consider a particular case, that we shall develop in the next paragraph: this is when Z is one dimensional. In this case, we have

$$C_Z = \sigma_Z^2 = C(K, K) + \sigma_\eta^2 \quad (54.24)$$

$$C_{YZ} = \begin{vmatrix} \vdots \\ C(L_i, K) \\ \vdots \end{vmatrix} \equiv \underline{c}_Z \quad (54.25)$$

and the equation for the eigenvalues is simply

$$\sigma_Z^2 - \alpha^2 \underline{c}_Z^+ C_Y^{-1} \underline{c}_Z = 0, \quad (54.26)$$

with the two solutions:

$$\alpha = \pm \frac{\sigma_Z}{[\underline{c}_Z^+ C_Y^{-1} \underline{c}_Z]^{1/2}}. \quad (54.27)$$

Getting the positive root, which is the one that provides the value $\gamma_{\min}$, the corresponding eigenvector is

$$[\underline{\lambda} = \alpha\mu C_Y^{-1}\underline{c}_Z, \mu], \qquad (54.28)$$

where μ has to be fixed so as to satisfy the normalization condition (54.15). Indeed, this last condition is quite irrelevant, while the interesting thing is that by using (54.28) one finds the sought statistic

$$S = \mu(\alpha\underline{c}_Z^{+}C_Y^{-1}Y - Z). \qquad (54.29)$$

In turn, by using (54.27), this can be written as

$$S = \mu\sigma_Z\left[\frac{\underline{c}_Z^{+}C_Y^{-1}Y}{\sqrt{\underline{c}_Z^{+}C_Y^{-1}\underline{c}_Z}} - \frac{Z}{\sigma_Z}\right]. \qquad (54.30)$$

The multiplicative constant in S has not a particular meaning, in fact it is one and the same scale for both S and its standard deviation. What is in parenthesis, though, is quite suggestive: S is basically the best linear predictor of Z from Y, normalized to have variance 1, minus Z, also normalized to variance 1. That seems a good solution to our validation problem. It is interesting also to apply (54.30) when Y too is one dimensional. To make it simple, we assume even that $\sigma_Y = \sigma_Z = 1$ and $\rho > 0$. By applying (54.30) we find

$$S = \mu(Y - Z).$$

3 A Case Study

We report here a real case of deformation monitoring in central Italy: the Assisi landslide. The area extends for about 1.2 km × 1.4 km. It is monitored by means of a GPS network of 25 points (cf. Fig. 54.1) whose position is determined every year by the University of Perugia (Cilli et al. 2002). The difference in time of the local cartesian coordinates give a sample of the 3d displacement field under analysis. On the same area, DInSAR time series of LOS deformations are available on 55 highly coherent points (cf. Fig. 54.1); those series come from images of the ENVISAT SAR sensor and were processed by the CNR IREA center

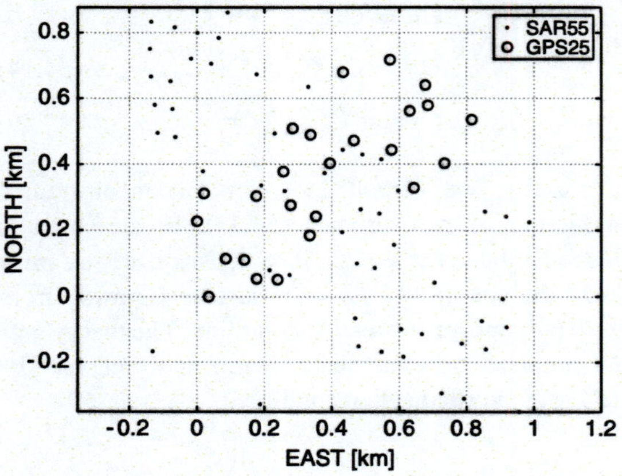

Fig. 54.1 GPS and SAR points on the monitored landslide

of Naples (Berardino et al. 2002). Note that the LOS is considered constant over the area. Within the Italian Space Agency project MORFEO, the Politecnico di Milano has to perform the DInSAR data validation by means of GPS; the optimal statistic (54.30) was applied, by comparing the multi-dimensional SAR data set, denser than the GPS one, with each GPS point, that is, performing 25 different tests. The GPS data come from the 2006 and 2007 campaigns, so that, from the DInSAR deformations series we retrieved those related to the same time span. The field u, defined on $t \in T \equiv R^2$, observed by both GPS and SAR, apart from the mean, is the LOS deformation field, that we assumed to have a homogeneous and isotropic variogram

$$\gamma(|s-t|) = \frac{1}{2}E\{[u(t) - u(s)]^2\}, \qquad (54.31)$$

which is related to the covariance function by the following relation

$$\gamma(|s-t|) = C(t,t) + C(s,s) - 2C(s,t). \qquad (54.32)$$

From SAR data we estimated the covariance function of u, by computing first their empirical variogram (cf. Fig. 54.2) and then exploiting relation (54.32) (Sansó et al. 2008). Even more, the estimate of the nugget effect gave us an estimate of the noise variance σ_ν^2 of the SAR data, which can not be directly computed by the SAR data analysis. We found

$$C(|s-t|) = 0.13\exp[-15.31|s-t|], \qquad (54.33)$$

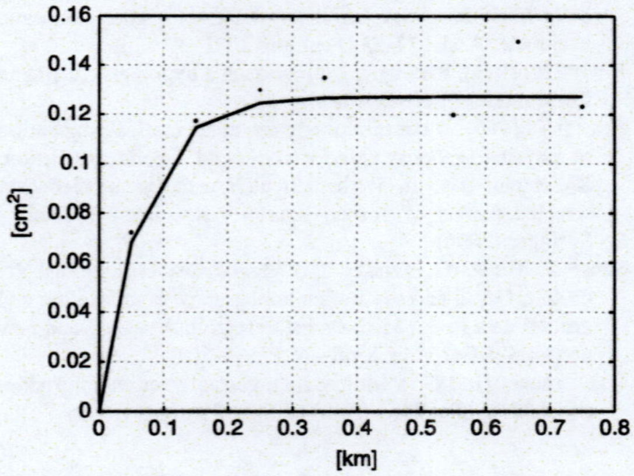

Fig. 54.2 Empirical variogram and interpolated conditionally definite negative model

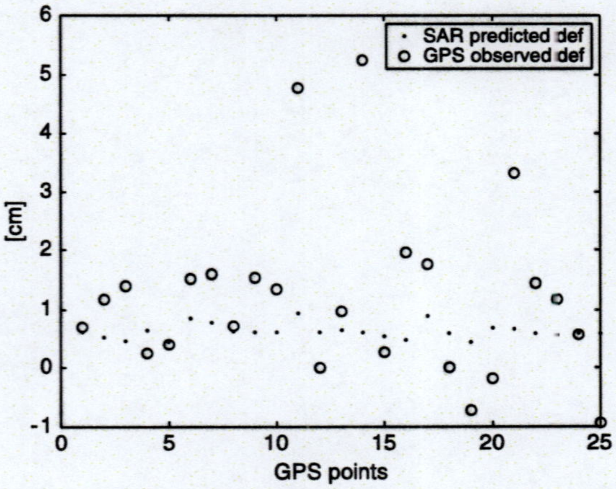

Fig. 54.3 SAR predicted LOS deformations and GPS LOS projected observations

while the nugget effect resulted to be negligible. The values predicted from SAR and the GPS LOS projected observations are shown in Fig. 54.3.

The $\underline{Y}$ vector is now that of LOS SAR deformations after their estimated mean removal, while Z is the GPS LOS projected deformation again after the same mean removal. The GPS LOS deformations reduced by their average with the accordingly propagated standard deviation and the predicted values (from the SAR data, after the mean removal) with the prediction error standard deviations are reported in Table 54.1. In the last column of the same table we report the corresponding statistic S values.

Table 54.1 Residual LOS deformations predicted from SAR ($\underline{Y}$) – Standard deviation of the estimated prediction error (std) – Residual GPS LOS deformations (Z) – Standard deviation of the residual GPS LOS deformation (σ_Z)

No	$\underline{Y}$ cm	std cm	Z cm	σ_Z cm	S
1	0.00	0.36	0.09	0.66	0.01
2	−0.08	0.31	0.57	0.54	−1.13
3	−0.15	0.27	0.79	0.33	**−2.22**
4	0.04	0.34	−0.36	1.31	0.43
5	−0.16	0.29	−0.21	0.52	−0.39
6	0.25	0.31	0.91	0.45	−0.16
7	0.17	0.27	0.99	0.87	−0.25
8	0.02	0.36	0.11	0.73	0.89
9	0.00	0.36	0.93	0.88	−0.63
10	0.00	0.36	0.74	1.08	0.05
11	0.34	0.29	4.17	1.22	−1.29
12	0.01	0.36	−0.61	0.88	1.41
13	0.04	0.36	0.37	0.67	0.75
14	−0.00	0.36	4.64	0.95	**−3.27**
15	−0.06	0.36	−0.34	0.82	−0.45
16	−0.13	0.31	1.35	0.79	−1.82
17	0.29	0.30	1.15	1.58	0.62
18	−0.01	0.34	−0.60	1.17	0.29
19	−0.17	0.32	−1.33	0.55	0.79
20	0.08	0.32	−0.79	1.01	0.92
21	0.06	0.34	2.72	1.41	−0.97
22	−0.02	0.34	0.84	1.15	−0.65
23	−0.05	0.35	0.57	0.84	−0.81
24	−0.01	0.36	−0.04	0.41	−0.35
25	−0.02	0.36	−1.54	0.71	1.06

By assuming S to be normally distributed, and by using a significance level $\alpha = 5\%$, as we can see in Table 54.1, for all the points but two, the hypothesis is accepted.

Conclusion

The conclusion about "equivalence" of SAR and GPS is somehow obscured by the relatively large variance of noise in GPS data. Anyway, by performing a proper noise propagation, the result is that the two data sets must be considered compatible but for two points, which have to be monitored separately. As a general remark, the assumption of zero average for $u(t)$ is critical. The full use of a kriging approach, which is under implementation, would show that basically the disagreement within the two data sets is in the average.

Acknowledgements This research has been conducted with the support of the Italian Space Agency in the framework of the MORFEO project.

References

Berardino P, Fornaro G, Lanari R, Sansosti E (2002), A new algorithm for surface deformation monitoring based on small baseline differential SAR interferograms, IEEE Trans GeoSc and Remote Sens, 40 (11): 2375–2383

Cilli ME, Radicioni F, Stoppini A (2002), Analisi dei risultati di recenti campagne di misura nella rete di monitoraggio GPS della frana di Assisi, La difesa della montagna. Convegno nazionale, Assisi 11-12 Dicembre 2002

Koch KR (1999), Parameter estimation and hypothesis testing in linear models. Springer, Berlin

Krarup T (1969), A contribution to the mathematical foundation of physical geodesy. Meddelelse N.44 GeodaetiskInstitut, Kobehavn. Also in Mathematical Foundation of Geodesy: selected papers of Torben Krarup - Kai Borre (Editor) - Springer (2006)

Sansó F, Venuti G, Tziavos IN, Vergos GS, Grigoriadis VN (2008), Geoid and sea surface topography from satellite and ground data in the Mediterranean region. A review and new proposal, BGG Vol LXVII - N.3: 155–201

Wackernagel H (1995), Multivariate geostatistics. An introduction with applications. Third edition. Springer, Berlin

Ocean Loading in Brittany, Northwest France: Impact of the GPS Analysis Strategy

55

Joëlle Nicolas, Stéphane Durand, Mathilde Vergnolle, Laurent Morel, Stavros Melachroinos, Marie-Noëlle Bouin, and François Fund

Abstract

In this contribution, we analyze the impact of different GPS processing strategies on ocean tide loading estimation. We use continuous GPS data acquired during a 4-month campaign performed in 2004 in Brittany, Northwest France. Since the expected geodynamical signal in the estimated positions is exceeding the typical GPS data analysis noise, this data set can be used to compare the results obtained with different analysis software packages. Moreover, in this specific case we need short sub-daily solutions to study short-period signals instead of classical 24 h-solutions. The GPS capability for measuring 3D ocean tide loading deformation has already been assessed, but since we are looking for the finest signal as the one induced by the shallow water constituents, it is essential to be sure that the position time series represent a geodynamical signal and are not biased by the data processing strategy used. To analyze the possible effect of the methodology used on the geodynamical results, we compare different solutions computed with different strategies (Double Differencing and Precise Point Positioning) with various GPS analysis software packages (Bernese, GAMIT, GINS, and GIPSY/OASIS). We show that the different solution consistency is at the level of 1–3 mm. We also show that the data processing strategy has a mean effect of about 10–20% of the ocean tide loading signal amplitude.

Keywords

GPS • Ocean tide loading • Software impact

1 Introduction

Ocean Tide Loading (OTL) induces surface displacements which can reach several centimeters in the vertical component and one-tenth to one-third of the vertical displacements in the horizontal components

J. Nicolas (✉) · S. Durand · L. Morel · F. Fund
Laboratoire de Géodésie et Géomatique, 1 boulevard Pythagore, 72000 Le Mans, France

M. Vergnolle
Laboratoire de Géophysique Interne et Tectonophysique Maison des Géosciences, BP 53, 38041 Grenoble Cedex 9, France

S. Melachroinos
GRGS/DTP/CNES, 18 avenue Edouard Belin, 31400 Toulouse, France

M.-N. Bouin
ENSG/LAREG/IGN 6 et 8, av. Blaise Pascal, 77455 Marne-la-Vallée, France

N. Sneeuw et al. (eds.), *VII Hotine-Marussi Symposium on Mathematical Geodesy*, International Association of Geodesy Symposia 137, DOI 10.1007/978-3-642-22078-4_55, © Springer-Verlag Berlin Heidelberg 2012

in coastal regions. These periodic crustal deformations are detectable by all modern space geodetic techniques and can be well resolved from GPS data analysis (see, e.g., Khan and Scherneck 2003; Yun et al. 2007). GPS data can also be used to validate ocean tide models (see, e.g., Urschl et al. 2005; Penna et al. 2008).

The purpose of this study is to assess the impact of the GPS data analysis strategy on the OTL signal observation. For this, we seek direct observations of OTL Displacements (OTLD) from GPS position time series obtained from the same data set with different analysis software packages: Bernese, GAMIT, GINS, and GIPSY. These comparisons are essential in order not to misinterpret the geodetic analysis error as geophysical signal.

Several studies show that any mis-modelled or un-modelled sub-daily periodic ground displacements can propagate into spurious longer wavelength signals in GPS position time series (see, e.g., Penna et al. 2007; King et al. 2008). These induced aliased signals could then result in incorrect geophysical interpretation from GPS position time series. Moreover, recent studies (see, e.g., Teferle et al. 2008) demonstrated from the analysis of position time series obtained by different analysis centers that the strategy used has a non negligible impact on velocity estimates.

Our area of study is the continental shelf of Brittany and Cotentin, in north-western France, a region where the ocean tide amplitudes are known to be amongst the highest in the world and can reach 14 m. So, the OTLD can be as large as 15 cm in vertical and about one-third of this deformation in the horizontal components. This strong OTLD signal is suitable to quantify the differences and the errors from one strategy to another.

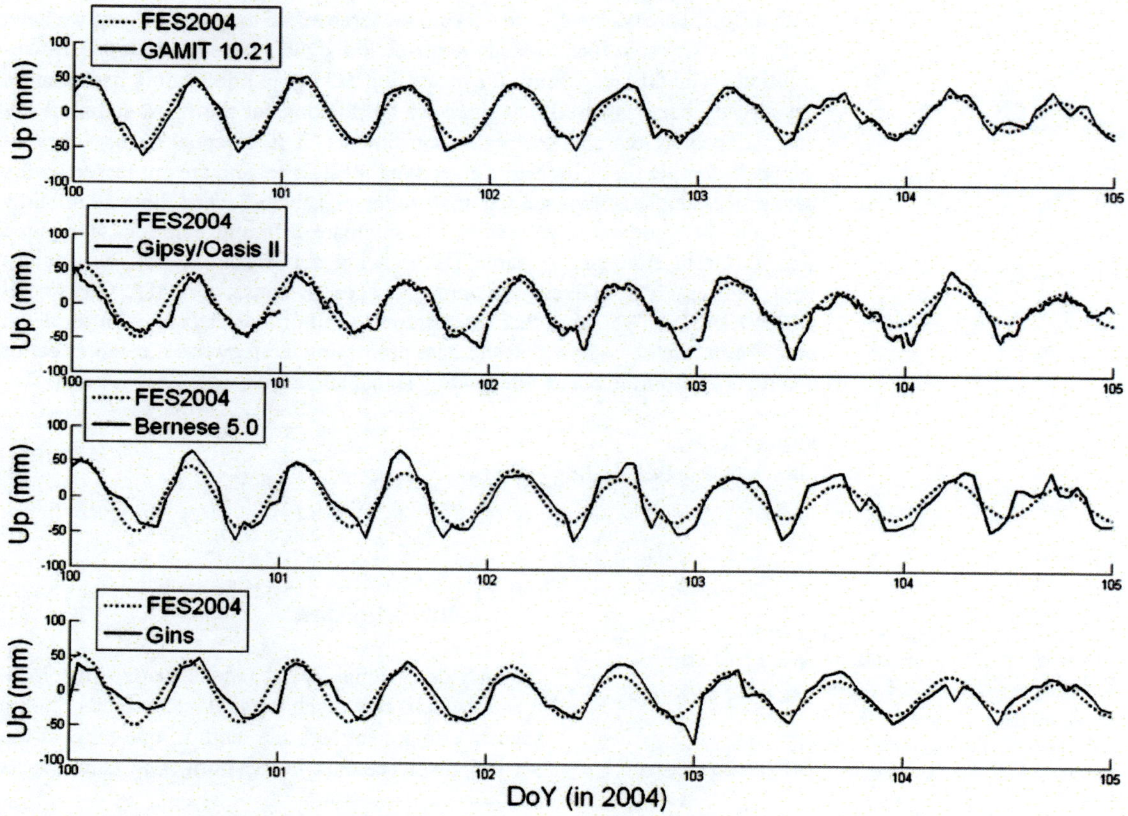

Fig. 55.1 Location of the studied area. Triangles represent the campaign GPS stations and squares represent permanent GPS stations belonging to the French GPS network (RGP network)

Notice that our purpose is not to evaluate OTL models which has been done in (Vergnolle et al. 2008) and in (Melachroinos et al. 2008).

2 Data Set

In 2004, an ambitious multi-technique campaign was realized (Llubes et al. 2008) to study and to characterize ocean loading effects in Brittany, Northwest France, and to validate new ocean models such as FES2004 (Lyard et al. 2006). For this, several geodetic techniques were used simultaneously with a monitored area covering the whole northwest part of France: tide gauge, GPS, Satellite Laser Ranging (SLR), tiltmeter, absolute and relative gravimeters, and barometers. In this study, we focus on the GPS part of this campaign. In addition to the local permanent GPS stations of the global French Permanent Network (RGP, http://rgp.ign.fr), a set of 12 GPS stations were installed in the studied area from March to June. They were mainly located on the north coast where ocean tides are huge. Few stations were installed on the south coast and inland to observe the earth's crust response far away of the main load (see Fig. 55.1).

3 Data Analysis Strategies

To process the GPS data, we used different software packages using different processing strategies. We processed the 105 days of data obtained during the campaign previously described. We used four GPS data processing software packages: Bernese 5.0 (Dach et al. 2007), GAMIT 10.21 (King and Bock 2005), GINS/DYNAMO (Lemoine et al. 2007), and GIPSY/OASIS II (Zumberge et al. 1997). For the first three software packages, double-differenced GPS phase measurements are used in order to resolve carrier phase ambiguities, whereas GIPSY/OASIS PPP (Precise Point Positioning) strategy is used allowing to determine absolute positioning.

All the software packages are used with common processing parameters in order to do valuable comparison: 10° cut-off angle, Niell (1996) hydrostatic and wet tropospheric mapping functions and the Saastamoinen (1972) zenithal tropospheric delay a priori model, and IERS conventions 2003 for Earth Orientation Parameters and for the Earth and polar tides (McCarthy and Petit 2004). International GNSS Service (IGS) final orbits and satellites clocks, Earth rotation parameters, and absolute phase center variations are used such as recommended by IGS (Dow et al. 2009).

For the first three software packages, the estimation strategy is very similar and makes use of the least squares approach. Station coordinates are estimated using 2-hour sessions. We estimated that a 2-hour session is sufficient to sample correctly the movements and achieve a sufficient precision on the 3D coordinates over the 105 days to obtain a good estimation of amplitude and phase of the main OTL components. The typical precision obtained from this type of positioning is 3–7 mm and 10–15 mm for the horizontal and vertical components, respectively. To express the solution in the ITRF2000 reference frame (Altamimi et al. 2002), the coordinates of 16 IGS stations in Europe have been heavily constrained to their ITRF2000 values. These permanent sites are strongly constrained in the different software strategies (Addneq2 for Bernese, GlobK for GAMIT, and Dynamo for GINS) and corrected for OTL displacements by using the FES2004 model. One Zenithal Tropospheric Delay (ZTD) per site is processed every 30 min.

The parameter estimation strategy in GIPSY is different since a Kalman filter is used allowing the parameters to have a stochastic behaviour: station positions and ZTD are estimated every 5 min as a random walk process. After several tests, the rates of change of the process noise covariance is fixed to 30 mm/h for the station position and to 10 mm/h for ZTD which is a standard value in the GIPSY software package (see, e.g., Larson et al. 2001). For this type of positioning the typical precision is 5–10 mm and 10 mm for horizontal and vertical components, respectively.

4 Analysis Results

Using these processing strategies, GPS position time series are obtained. For each position time series, the larger outliers were removed and the mean and the linear trend were subtracted from each solution. A representative example is given in Fig. 55.2 for the different software solutions of the vertical component at PAIM station superimposed to the FES2004

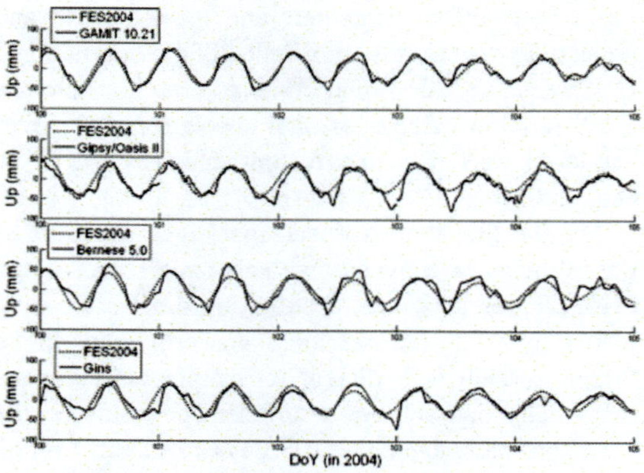

Fig. 55.2 Example of GPS position time series obtained from the four different software packages for the vertical component of station PAIM (Paimpol). In addition the FES2004 OTL displacements are plotted for comparisons

Table 55.1 Means differences (top) and their RMS (bottom) over all the stations between FES2004 and the estimated M2 amplitude (A in mm) and Greenwich phase (ϕ in degrees) for the north, east, and vertical components

	Mean differences to FES2004					
	North		East		Vertical	
	A (mm)	ϕ (deg)	A (mm)	ϕ (deg)	A (mm)	ϕ (deg)
Bernese	−1.1	−14.5	−0.3	−5.9	1.1	5.5
Gamit	0.6	−6.5	−0.4	−3.8	0.5	4.7
Gins	−1.4	−11.5	−0.9	−6.9	−7.4	4.9
Gipsy	−1.3	3.3	0.0	−17.4	−3.5	2.6
	RMS to the mean differences to FES2004					
	North		East		Vertical	
	A (mm)	ϕ (deg)	A (mm)	ϕ (deg)	A (mm)	ϕ (deg)
Bernese	0.7	6.5	0.3	3.2	1.0	2.7
Gamit	0.3	1.7	0.2	2.8	0.8	1.9
Gins	0.6	6.9	0.3	3.3	2.9	1.6
Gipsy	0.5	10.6	0.8	7.7	2.5	2.6

predicted displacements. The four solutions seem to be globally consistent and all software solutions seem to well agree with the model.

For each component, the correlation coefficients between each solution and the predicted FES2004 time series were computed. All the stations show nearly the same values of correlation, except MANS for the three coordinates for the four solutions. The mean correlation values between each solution and FES2004, without MANS, are of 82% for Bernese, 89% for GAMIT, 77% for GINS, and 61% for GIPSY for the vertical component. GINS and GIPSY solutions are less correlated compared to Bernese and GAMIT solutions which are in better agreement with the predicted signal. For the vertical component, the cross-correlation values between each solution and the GAMIT one amount 87% for Bernese, 80% for GINS, and 64% for GIPSY. The general lower correlation of the GIPSY solution can be explained by the quite different processing strategy used in this case (PPP instead of double differences).

The spectral analysis of each GPS position time series was performed using the T-TIDE toolbox from (Pawlowicz et al. 2002). This toolbox is able to estimate, using least squares, the amplitude and Greenwich phase of tidal constituents from position time series. From this spectral analysis, we extracted the signal of each tidal component from each position time series. Table 55.1 shows for the M2 tide the differences between each solution and FES2004 in terms of mean and RMS to the mean value over all the stations. For this tide, the amplitude estimation error is lower than 1 mm for each solution and the Greenwich phase estimation error is between 1 and 15°. The amplitude results show a fair good agreement between each solution on the north and east components at the milimeter level, with a RMS lower than 1 mm.

For the vertical component, Bernese and GAMIT are in good agreement and exhibit the same behavior. Nevertheless, GIPSY and GINS solutions exhibit large discrepancies to the model (several millimeters), do not agree with the other solutions and have larger RMS values (about 3 mm).

For the Greenwich phase estimation, GAMIT, Bernese and GINS solutions are in good agreement for the east and vertical components (at a 2° level), with RMS of about 2–3°. On the north component, these three software packages agree at a 10° level, but Bernese and GINS exhibit larger RMS values. Results from the GIPSY solution show discrepancies to the FES2004 model and to other solutions, with larger RMS (up to 11° on the north component). Nevertheless, these discrepancies are at the same level as the phase estimation error for this software package.

To perform a reliable comparison between the different solutions, we also computed the vector length between each software package solution and the FES2004 model. It allows to take into account

Table 55.2 Vector length values (in mm) for the M2 tide on the horizontal and vertical components between each software package solution and FES2004 model at different stations

	Horizontal				Vertical			
	Bernese	GAMIT	GINS	GIPSY	Bernese	GAMIT	GINS	GIPSY
BRST	2.3	1.4	2.2	4.7	3.4	2.4	11.6	8.1
CHER	2.0	0.9	2.4	2.6	2.8	2.6	4.3	2.6
COUT	2.1	0.8	2.5	2.5	2.5	2.5	4.7	2.4
DIBE	2.8	1.7	2.6	3.5	3.7	2.7	9.9	7.2
MANS	2.4	1.2	2.4	3.3	2.3	1.4	4.5	1.6
PAIM	2.7	1.6	2.7	3.2	2.9	2.4	8.1	2.7
TREV	2.3	1.4	2.4	2.6	1.9	2.0	11.1	4.2
YGEA	2.5	1.5	2.9	3.3	2.2	2.0	7.1	2.1
Mean	2.4	1.3	2.5	3.2	2.7	2.3	7.7	3.9
RMS	0.3	0.3	0.2	0.7	0.6	0.4	3.0	2.3

simultaneously the differences in terms of amplitude and phase. We present for the most relevant subset of stations the results concerning the vertical component for M2 (amplitude between 8 and 40 mm), S2 (2–14 mm), N2 (2–8 mm), and for the horizontal OTLD induced by M2 (4–8 mm in North and 6–11 mm in East), which are the main OTLD constituents. For the M2 tide the vector lengths are displayed in Table 55.2 for different stations for both horizontal and vertical components.

Focussing on M2, in planimetry, mean vector lengths (between each solution and FES2004 model) are 2 mm for Bernese solution, 1 mm for GAMIT, 2 mm for GINS, and 3 mm for GIPSY. These vector lenghts (lower than 2 mm) are smaller than the difference between all the GPS solutions and the FES2004 model values.

For the vertical component, no general pattern can be extracted from this, except that GINS shows a larger mean vector length due to an amplitude underestimation at each site relatively to all the other solutions. Mean values are 2 mm for GAMIT, 3 mm for Bernese, 4 mm for GISPY, and 8 mm for GINS. All the four solutions exhibit a general agreement with the model predictions at the level of 2–3 mm. The different solutions agree at the level of 1–2 mm except for GINS. Bernese and GAMIT solutions are in very fair agreement and GINS and GIPSY solutions present larger vector lengths for the BRST and DIBE stations.

Looking to the S2 vertical component, all the solutions agree at the level of 2–4 mm. The GAMIT solution is the closest to the model, nevertheless relative to the tide amplitude it is twice less (3 mm) in agreement than for M2. The three other solutions present larger discrepancies relatively to the model: 6–7 mm for GINS and Bernese solutions and 10 mm for GIPSY solution. Finally, concerning the N2 vertical component, whose amplitude is twice smaller than the M2 one, the agreement of the four solutions is as good as the one obtained for M2. All the solutions agree with each other at the level of 1 mm or even better, depending of the considered site, and agree at the sub-mm level with the predicted FES2004 values.

Conclusion

This study demonstrated the impact of the GPS analysis strategy on the OTLD signal observed in the position time series. Our results indicate a general agreement between the different solutions obtained with different processing software packages at a few millimeter level. Nevertheless, our analysis evidences some significant differences between the software packages solutions (up to 10 mm) showing that the data processing strategy can have a mean effect up to 10–20% of the ocean tide loading signal amplitude. Our results show that discrepancies between software solutions are below or equal to the discrepancies relative to the model. So, in this area, we suggest that the differences between GPS solutions and the model correspond to a geodynamical signal not included in FES2004, particularly for M2 vertical component.

Further investigations are to be performed to well understand these discrepancies. Thus, we demonstrated that GPS data analysis strategy artifacts are not negligible. It is also important to well understand and control these aspects for further studies combining different GNSS systems such as GPS, GLONASS, and in the near future GALILEO.

Acknowledgements We are grateful to all the people who contributed to the GPS campaign used in this study. Thanks to the GDR G2 (Geodesy and Geophysics) who supported the campaign, to the INSU (Institut National des Sciences de l'Univers) and to the LDL (Laboratoire de Dynamique de la Lithosphère) for providing the GPS receivers.

References

Altamimi Z, Sillard P, Boucher C (2002) ITRF2000: a new release of the International Terrestrial Reference Frame for Earth science applications, J Geophys Res 107, B10, 2214, doi:10.1029/2001JB000561

Dach R, Hugentobler U, Fridez P, Meindl M (2007) Bernese GPS Software, Version 5.0, Astronomical Institute, University of Bern

Dow JM, Neilan RE, Rizos C (2009) The International GNSS Service in a changing landscape of global navigation satellite systems. J Geodes 83:191–198, DOI: 10.1007/s00190-008-0300-3

Khan SA, Scherneck H-G (2003) The M2 ocean tide loading wave in Alaska: vertical and horizontal displacements, modelled and observed. J Geodes 77(3-4):117–127

King RW, Bock Y (2005) Documentation for GAMIT GPS analysis software, Release 10.2, Mass. Inst. of Technol. Cambridge, MA

King M, Watson C, Penna N, Clarke C (2008) Subdaily signals in GPS observations and their effect on semiannual and annual periods. Geophys Res Lett 35, L03302

Larson KM, Cervelli P, Lisowski M, Miklius A, Segall P, Owen S (2001) Volcano monitoring using the global positioning system: filtering strategies. J Geophys Res 106(B9), 19, 453–19, 464

Lemoine J-M, Bruisma S, Loyer S, Biancale R, Marty J-C, Perosanz F, Balmino G (2007) Temporal gravity field models inferred from GRACE data. J Adv Space Res 39(10): 1620–1629

Llubes M, Florsch N, Boy J-P, Amalvict M, Bonnefond P, Bouin M-N, Durand S, Esnoult M-F, Exertier P, Hinderer J, Lalancette M-F, Masson F, Morel L, Nicolas J, Vergnolle M, Wöppelmann G (2008) Multi-technique monitoring of ocean tide loading in northern France. Compt Rendus Geosci 340:379–389

Lyard F, Lefèvre F, Letellier T, Francis O (2006) Modelling the global ocean tides: a modern insight from FES2004. Ocean Dyn 56:394–415

McCarthy DD, Petit G (2004) IERSConventions 2003, (IERSTechnical Note; 32) Frankfurt am Main: Verlag des Bundesamts für Kartographie und Geodäsie

Melachroinos S, Biancale R, Llubes M, Perosanz F, Lyard F, Vergnolle M, Bouin M-N, Masson F, Nicolas J, Morel L, Durand S (2008) Ocean tide loading (OTL) displacements from global and local grids: comparisons to GPS estimates over the shelf of Brittany, France. J Geodes 82(6): 357–371

Niell AE (1996) Global mapping functions for the atmosphere delay at radio wavelengths. J Geophys Res 101: 3227–3246

Pawlowicz R, Beardsley B, Lentz S (2002) Classical tidal harmonic analysis including error estimates in MATLAB using T-TIDE. Computers and Geosciences 28(8)

Penna N, Bos MS, Baker T, Scherneck H-G (2008) Assessing the accuracy of predicted ocean tide loading displacement values. J Geodes 82(12):893–907

Penna N, King M, Stewart MP (2007) GPS height time series: Short-period origins of spurious long-period signals. J Geophys Res 112, B02402

Saastamoinen J (1972) Atmospheric correction for the troposphere and stratosphere in radio ranging of satellite, Geophysical Monograph 15, The use of artificial satellites for geodesy, American Geophysical Union, Washington, D.C.

Teferle FN, Williams S, Kierulf H, Bingley R, Plag H-P (2008) A continous GPS position time series analysis strategy for high-accuracy vertical land movements. Phys Chem Earth 33:205–216

Urschl C, Dach R, Hugentobler U, Schaer S, Beutler G (2005) Validating ocean tide loading models using GPS. J Geodes 78(10):616–625

Vergnolle M, Bouin M-N, Morel L, Masson F, Durand S, Nicolas J, Melachroinos AS (2008) GPS estimates of ocean tide loading in NW-France: Determination of ocean tide loading constituents and comparison with a recent ocean tide model. Geophys J Int 173(2):444–458

Yun H-S, Lee D-H, Song D-S (2007) Determination of vertical displacements over the coastal area of Korea due to ocean tide loading using GPS observations. J Geodyn 43: 528–541

Zumberge JF, Heflin MB, Jefferson DC, Watkins MM, Webb FH (1997) Precise Point Positioning for the efficient and robust analysis of GPS data from large networks. J Geophys Res 102(B3):5005–5017

Repeated Geodetic Measurements in the Tatra Mountain

56

M. Mojzes, J. Papco, and M. Valko

Abstract

For detection of horizontal and vertical movements of the Earth's surface the repeated geodetic measurements were organized in the Tatra Mountain. Non-permanent GPS measurements with 4–5 days observation time were provided from 1998 to 2008 yearly. Near the 3 non-permanent stations the GPS permanent stations and absolute gravity stations were established for better understanding and detecting of vertical movements. The analyze of GPS permanent measurements detected subsidence effect approximately −0.72 mm/year at the sites GANO and −0.98 mm/year at the site SKPL and uplift effect +2 mm/year at the site LIES. The station LIES is located 6 km from a water dam. The repeated absolute gravity measurements detected subsidence +0.12 μGal/year at the station SKPL, +0.27 μGal/year at the station GANO and +0.85 μGal/year at the station LIES. The paper presents the analysis of GPS and absolute gravity measurements.

Keywords

Non permanent and permanent GPS measurements • Absolute gravity measurements • Time series analysis

1 Introduction

The project CERGOP-2, A Multipurpose and Interdisciplinary Sensor Array for Environmental Research in Central Europe (acronym CERGOP-2/Environment) was primarily addressed to monitoring of crust movements in the central part of Europe, covering 15% of the continent, with the aim to determine the velocities of selected points in seismic active areas over decades. Investigation of the change of velocities leads to the description of the underlying driving forces and of energy transfers leading to earthquakes. The methodology basically relies on the use of GPS data for monitoring environmental parameters in the area of 14 European countries. The reference GPS network consists of about 60 stations (50% of them are operated on permanent basis). In the frame of the CERGOP-2/Environment project as Work Package 10.5, the GPS and absolute gravity measurements in the Tatra Mountain area for local deformation studies of the Earth's surface have been organized. The area of the Tatra Mountain takes approximately $2,400\,km^2$ (60 km WE direction and 40 km SN direction). The physical height

M. Mojzes (✉) · J. Papco · M. Valko
Department of Theoretical Geodesy, Slovak University of Technology in Bratislava, Radlinskeho 11, 813 68 Bratislava, Slovak Republic
e-mail: marcel.mojzes@stuba.sk; juraj.papco@stuba.sk; milos.valko@stuba.sk

N. Sneeuw et al. (eds.), *VII Hotine-Marussi Symposium on Mathematical Geodesy*, International Association of Geodesy Symposia 137, DOI 10.1007/978-3-642-22078-4_56,

of the earth surface varies between 700 m and 2 660 m above the sea level. The GPS non permanent network was established in 1998 and consists of 11 special monument sites (brass module is mounted in concrete pillars or directly to the rocks). The GPS observation Campaigns were provided every year by Slovak University of Technology in Bratislava and Warsaw University of Technology started from 1998 to 2009 with Trimble dual fervency receivers. The absolute gravity measurements started in 2003 and have been realized by the Research Institute of Geodesy Topography and Cartography of Czech Republic (Palinkas and Kostelecky jr.) by FG5, No. 215 instrument in different months (September, October or November). The comparison of non-permanent and permanent GPS measurements and absolute gravity measurements is presented in the paper.

2 Motivations

The repeated classical geometric leveling was used for detection of vertical uplift or subsidence in Nordic countries by Ekman and Mäkkinen (1996), see Fig. 56.1 and in Central Europe by Arabadzijski et al. (1986), see Fig. 56.2. The Fennoscandian land uplift was also detected by the absolute gravity measurements and presented by Ekman and Mäkkinen (1996), Wilmes (2006) and Timmen et al. (2008).

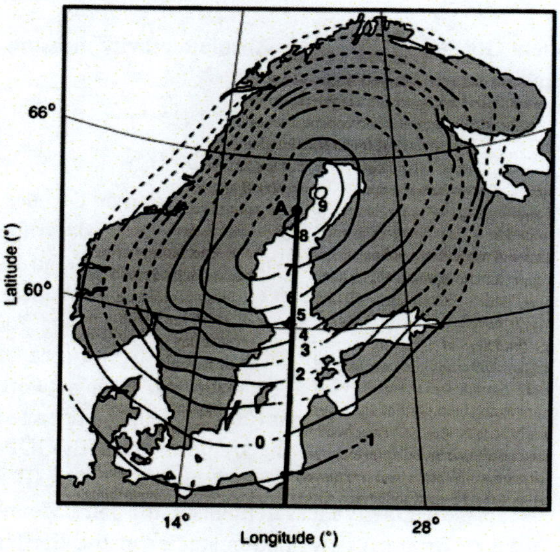

Fig. 56.1 Fennoscandian uplift observed by classical geometric leveling methods after Ekman and Mäkkinen (1996). The isoclines represent rates in mm/year. *Strong line* represents profile

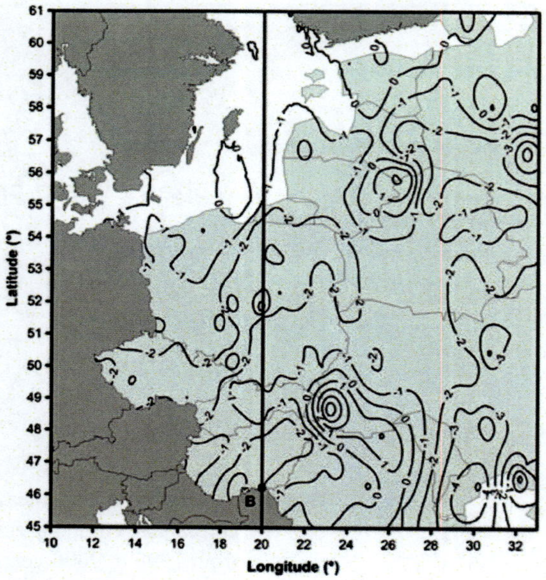

Fig. 56.2 Central Europe uplift/subsidence rates determined by classical geometric leveling method compiled by Arabadzijski et al. (1986). The isoclines represent rates in mm/year. *Strong line* represents profile, see Fig. 56.3

The Fig. 56.3 presents the Fennoscandian uplift rates and Central Europe uplift/subsidence rates along the profile AB compiled from repeated geometric leveling. The land subsidence rates in the Tatra Mountain area are from −0.1 mm/year to −1 mm/year (see Fig. 56.3).

For verification of land subsidence rates the geodetic network for GPS and absolute gravity measurements have been established in the Tatra Mountain.

The gravity changes coupled with earthquakes are of particular importance as precursor phenomena. The epicentres of earthquake activity in the area of the Tatra Mountain are presented on Fig. 56.4 (compiled from the data of Geophysical Institute, Slovak Academy of Science).

3 Physical Model

In present time the vertical temporal change of the Earth surface is determined by repeated GPS and absolute gravity observations. Temporal gravity

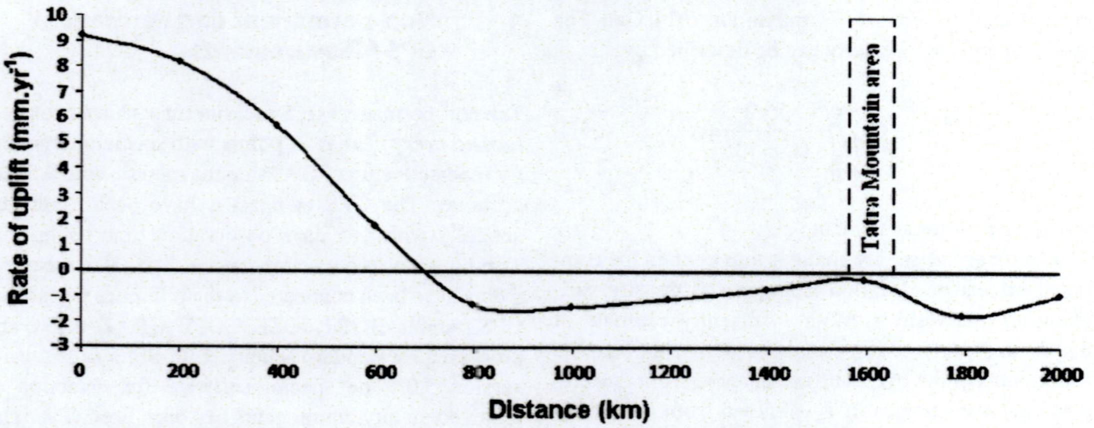

Fig. 56.3 Uplift and subsidence profile AB in central Europe compiled from classical geometric leveling ($\varphi = 64.0°$, $\lambda = 20.0°$ start point, $\varphi = 46.0°$, $\lambda = 20.0°$ end point)

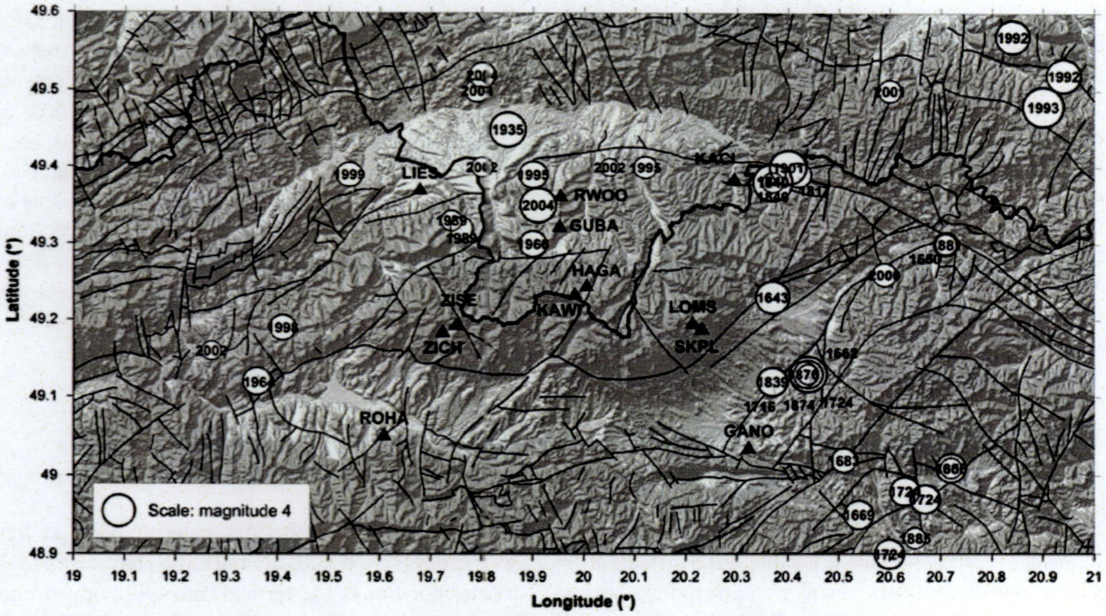

Fig. 56.4 Earthquake activity in the area of the Tatra Mountain. The *number in circle* represents the year of the earthquake and the *diameter of circle* represents the magnitude of the earthquake

changes discovered by repeated observations represent the important information of terrestrial mass displacements of global, regional and local type, Torge (1989). The gravity vector is a function of the position vector **r** and the density function $\rho(\mathbf{r}')$

$$\mathbf{g}(P) = \mathbf{g}(G, \omega, \mathbf{r}, \rho(\mathbf{r}')), \tag{56.1}$$

where G is the Newtonian gravity constant, ω is the angular velocity of Earth's rotation, in our model it is treated as a constant because the effect of temporal change of ω on gravity is maximum 0.1 μGal. The temporal change of gravity can be described by

$$\dot{g} = \frac{\partial g}{\partial r}\dot{r} + G \iiint\limits_{earth} \frac{\dot{\rho}(\mathbf{r}')}{(\mathbf{r}' - \mathbf{r})^2} d\tau, \quad (56.2)$$

where $d\tau$ is volume element.

The observed gravity changes thus contain the combined effects of elevation changes and density variations or mass shifts. With sufficiently estimate of $\partial g/\partial r$ and $\dot{\rho}(\mathbf{r}')$, it is possible to derive the vertical movements of the Earth surface in space from the temporal gravity changes. If $\dot{r}$ is known from GPS measurements, the gravity changes represent constraints for the determination of density changes.

The observed height and gravity changes support modelling of the generating thermo-mechanic processes. In postglacial uplift and recent subsidence areas, observed variations provide indications of the rheology (viscosity) of the upper mantle.

A linear relationship between the gravity change and vertical movement of the Earth surface can be assumed for limited area. The differential gravity-height ratio change is approximated by observed gravity changes δg and observed physical height changes δH or observed ellipsoidal height changes δh or observed geocentric distance changes δr

$$\frac{\dot{g}}{\dot{H}} \approx \frac{\delta g}{\delta H} \approx \frac{\delta g}{\delta h} \approx \frac{\delta g}{\delta r}. \quad (56.3)$$

The ellipsoidal height change δh is practically the same as geocentric distance change δr. The value of $\delta g/\delta h$ can exhibit large variations but is mostly between −0.15 μGal/mm and −0.35 μGal/mm, Jachens (1978). The free-air value −0.31 μGal/mm, which mostly occurs locally, corresponds to a vertical surface movement without a mass change. For large area variations as Fennoscandian area, the Bouguer relation −0.20 μGal/mm is frequently found, Mäkkinen et al. (2004), which, in addition, indicates internal mass displacements. The values of $\delta g/\delta H$ can by very uncertain due to the small values of gravity and heights changes and due to the present measurement uncertainties, Torge (1989).

4 Non-permanent and Permanent GPS Measurements

The non-permanent GPS measurements have been performed every year at 11 points with special monument for repeated setup of GPS antenna module with 0.5 mm accuracy. The GPS campaigns have been organized annually with 4–5 days observation time using only Trimble dual frequency receivers. The GPS measurements have been connected to the reference permanent GPS stations: BOR1, JOZE, GOPE, GRAZ, PENC and processed by standard scheme with Bernese software, version 5.0. The special software for detection of relative velocity components has been used. The relative horizontal and vertical velocities determined from observation time interval 1998–2003 are presented on Figs. 56.5 and 56.6.

For better determination of horizontal and vertical movements near the three non-permanent GPS stations (GANO, LOMS and LIES) the permanent GPS stations have been established. The measurements of permanent stations started at GANO in 2003, at LONS in 2004 and at LIES in 2007 and have been processed in the local analytical centre Slovak University of Technology in Bratislava (SUT) in frame of the European Permanent Network (EPN). The stations BOR1, BUCU, GOPE, GRAZ, MATE, PENC, SOFI and WTZR have been used as reference permanent GPS station in processing of EPN.

5 Repeated Absolute Gravity Measurements

The repeated absolute gravity measurements at tree gravity sites (GANO, SKPL and LIES) have been used for determination of the vertical land uplift/subsidence. The one day measurements (from 12 to 24 hours) have been performed at GANO in 2003, 2005, 2006, 2007 and 2008, at SKPL in 2005, 2006, 2007 and 2008 and at LIES in 2004, 2006, 2007 and 2008. The absolute gravity data have been processed by Micro-g software assuming polar motion, atmospheric and tide corrections. The results of absolute gravity measurements are presented in Table 56.1. All reference heights of

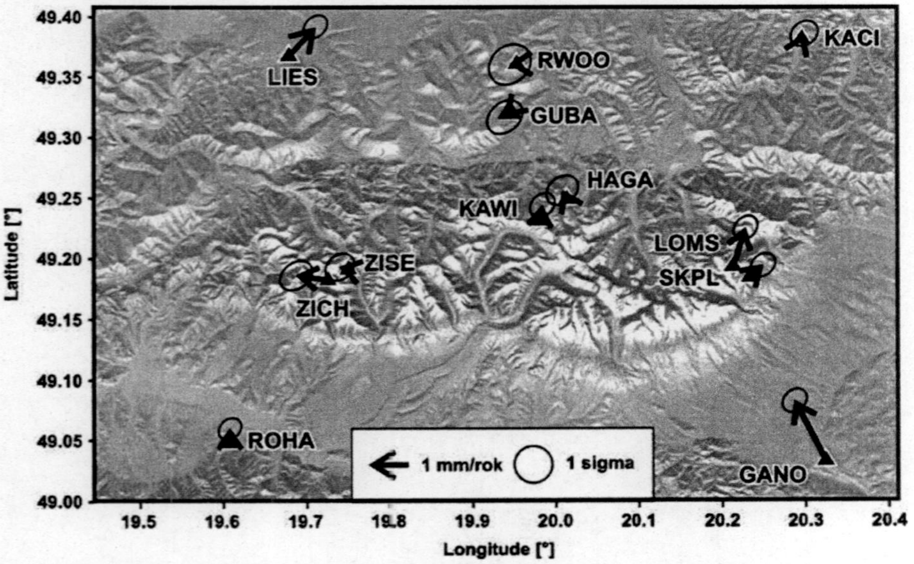

Fig. 56.5 Relative horizontal velocities of non-permanent GPS sites, an *arrow* represents the horizontal velocity and an *ellipse* represents the 1-sigma accuracy of the velocity

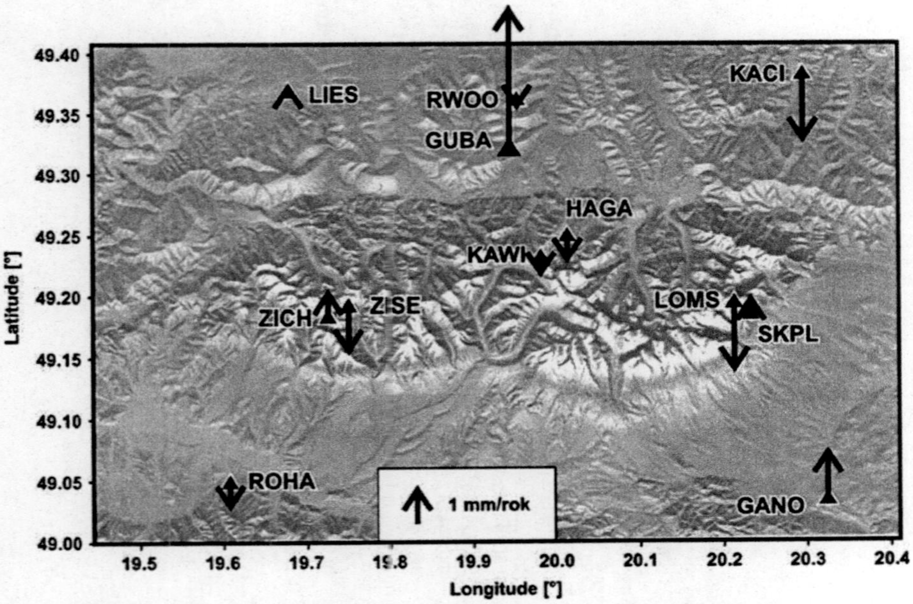

Fig. 56.6 Vertical velocities of non-permanent GPS sites, an *arrow* represents the vertical velocity of particular points

Table 56.1 Measured absolute gravity at the reference heights

GANO				SKPL				LIES			
Date	g(μGal)	Sigma (μGal)	Ref. height (m)	Date	g(μGal)	Sigma (μGal)	Ref. height (m)	Date	g(μGal)	Sigma (μGal)	Ref. height (m)
6–7.10.2003	980794745.5	0.18	1.2085	2–3.10.2005	980576110.1	0.23	1.2208	15–16.9.2004	980804318.3	0.16	1.2111
3–4.10.2005	980794748.2	0.20	1.2186	22–23.9.2006	980576109.7	0.17	1.2222	12–13.9.2007	980804321.2	0.18	1.2128
23–24.9.2006	980794747.6	0.15	1.2172	13–14.9.2007	980576118.0	0.25	1.2208	28–29.5.2008*	980804303.4	0.38	1.2727
11–12.9.2007	980794748.7	0.16	1.2183	4–5.11.2008	980576108.4	0.25	1.2206	5–6.11.2008	980804322.0	0.25	1.2120
3–4.11.2008	980794749.3	0.27	1.2169								
Gradient	−310 (μGal m^{-1})				−309 (μGal m^{-1})				−278.8 (μGal m^{-1})		

* Measured by FG5 230

absolute gravity values were transformed to the first reference height for minimizing the vertical gradient error.

6 Comparison of the Results

The time series of ellipsoidal heights determined by GPS measurements and absolute gravity values on the points GANO, SKPL and LIES are presented on Fig. 56.7. The trends of ellipsoidal changes and the absolute gravity changes, presented in Table 56.2 and Fig. 56.7, are very similar at the points GANO and SKPL but they differ slightly at the point LIES.

Table 56.2 Trends of velocities

Point	$\dot{g}(\mu\text{Gal} \cdot \text{year}^{-1})$	$\dot{h}(\text{mm} \cdot \text{year}^{-1})$	$\frac{\dot{g}}{\dot{h}}(\mu\text{Gal} \cdot \text{mm}^{-1})$
GANO	+0.272	−0.72	−0.38
SKPL	+0.127	−0.98	−0.13
LIES	+0.850	+2.30	0.36

Conclusion

The permanent GPS measurements in stable coordinate system using the same GPS antennas per point gives a reliable results of vertical changes of ellipsoidal heights in millimetres level. The non-permanent GPS measurements are influenced by

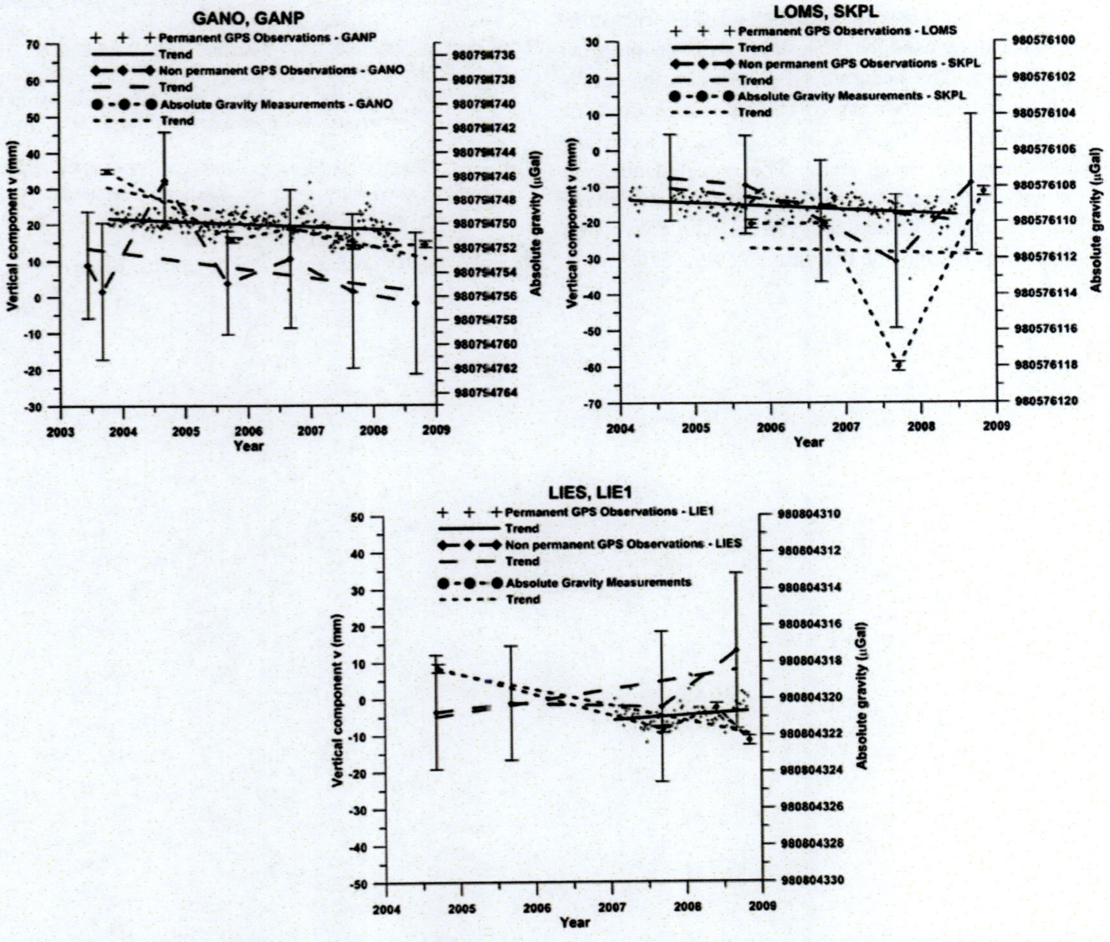

Fig. 56.7 Comparison of GPS non-permanent, permanent and absolute gravity time series at the points GANO, SKPL and LIES

repeated GPS antenna setup and the changes of antennas.

The repeated absolute gravity observations can give the information on the vertical changes of the earth surface, but the gravity signal is influenced by atmospheric and water mass changes effects. The determination of water mass variations especially in the mountain areas is very complicated and modelling of this phenomenon requires further study. Our future investigation activities will be oriented to modelling of water mass variation in local, regional and global scales.

The subsidence effect determined by classical geometric leveling is the same as determined by our experiments performed using GPS method in the Tatra Mountain.

The values of $\partial g/\partial r$ derived empirically from the simultaneous absolute gravity and GPS measurements can by used for the testing of the geophysical modelling. The accurate GPS and absolute gravity measurements seem to be the best method for a detailed investigation of recent Earth's surface movements in large areas. The repeated absolute gravity measurements can also be used effectively for the control of vertical systems. The conversion factor $\partial g/\partial r$ must be determined empirically.

Acknowledgements The authors thank the European Commission and Slovak Grant Agency of Science for financial support of the project and all reviewers for detail reading and comments to the paper.

References

Arabadzijski D et al (1986) Karta sovremennych vertikalnych dvizenij zemnoj kory na teritorii Bulgariji, Vengriji, NDR, Polsi, Rumuniji, ZSSR, Cechoslovakiji. Glovnoe upravlenie geodezii kartografiji pri Sovete Ministrov ZSSR. Moskva 1986

Ekman M, Mäkkinen J (1996) Recent postglacial rebound, gravity change and mantle flow in Fennoscandian. Geophys J Int 126:229–234

Jachens RC (1978) Temporal gravity changes as applied to studies of crustal deformations. In: Proceedings of conference VII stress and strain measure related to earthquake prediction, 7–9.9.1978, U.S. Geol. Survey Open-File Report 79–370

Mäkkinen J, Engfeldt A, Harsson BG, Ruotsalainen H, Strykowski G, Oja T, Wolf D (2004) The fennoscandian land uplift gravity lines 1966–20034. In: Symposium 129: geoid and space missions. Springer, New York, pp 328–332

Timmen L, Gitlein O, Müller J, Strykowski G, Forsberg R (2008) Absolute gravimetry with the hannover meters JILAG-3 and FG5–220 and their deployment in a Danisch-German Cooperation. ZfV 3/2008

Torge W (1989) Gravimetry. Walter de Gruyter, Berlin

GRACE-Derived Linear and Non-linear Secular Mass Variations Over Greenland

57

Oliver Baur, Michael Kuhn, and Will E. Featherstone

Abstract

In the past, GRACE (Gravity Recovery And Climate Experiment) monthly gravity field solutions have mainly been exploited to derive secular and seasonal mass changes on the Earth's surface. After seven years in operational mode, the satellite mission makes accelerated and decelerated mass variations detectable. Here we investigate the temporal characteristics of secular trends by fitting both linear and higher-order polynomials to the mass-change time-series. Our findings have been derived from GRACE gravity field time-series provided by CSR, GFZ and JPL. As a case study, we look at recent ice-mass variations over Greenland. Based on various model selection criteria (Akaike and Bayesian information criterion, cross-validation, hypotheses testing), our investigations show that linear regression is unable to describe recent deglaciation. Instead, the secular trend is best represented by a second-order polynomial, confirming accelerated deglaciation of the Greenland ice sheets, which increased by 250% between April 2002 and March 2009.

Keywords

Ice-mass balance • Information criteria • Satellite gravimetry • Time-series • Trend estimation

1 Introduction

In recent years, many GRACE studies focused on secular ice-mass variations in the Earth's polar regions (e.g., Chen et al. 2006; Luthcke et al. 2006; Velicogna and Wahr 2006; Wouters et al. 2008; Baur et al. 2009). In this context, Greenland plays a dominant role; it exhibits the strongest mass-variation signals observed by GRACE (Baur et al. 2009). Typically, mass variations derived within a certain sample period are expressed as annual change rates, assuming a linear progress over time. However, is this assumption of linearity justified? To answer the question of adequate description, this contribution compares and contrasts linear with higher-order polynomial mass-change models for estimation of the secular ice-mass loss over Greenland.

O. Baur (✉)
Institute of Geodesy, University of Stuttgart, 70174 Stuttgart, Geschwister-Scholl-Str. 24D, Germany
e-mail: O.Baur@gis.uni-stuttgart.de

M. Kuhn · W.E. Featherstone
Western Australian Centre for Geodesy & The Institute for Geoscience Research, Curtin University of Technology, Perth, WA 6845, Australia

N. Sneeuw et al. (eds.), *VII Hotine-Marussi Symposium on Mathematical Geodesy*, International Association of Geodesy Symposia 137, DOI 10.1007/978-3-642-22078-4_57, © Springer-Verlag Berlin Heidelberg 2012

The increased lifespan (now over seven years) of the GRACE satellite mission makes accelerated and decelerated mass variations detectable. Velicogna and Wahr (2006) demonstrated accelerated deglaciation over Greenland between April 2002 to April 2004 and May 2004 to April 2006. They derived their results by simply fitting linear trends to the two analysis periods. Comparing the slopes of the regression lines indicated a 250% increase (acceleration) in the ice-loss rate.

In this contribution, apart from the comparison of linear trends taken over different analysis periods, we reveal accelerated melting over Greenland by means of non-linear regression. In particular, we approximate the mass variations (here, residuals relative to the temporal mean) by higher-order polynomials. In order to find the best amongst the candidate regression models, we consider various selection criteria. The criteria are subject to least-squares residuals evaluation, parameter space complexity and significance of estimated model parameters.

2 Data

Monthly gravity field solutions are typically released to the public in terms of fully normalized spherical harmonic coefficients of the Earth's external gravitational potential. We used release four (RL04) GRACE-only gravity field estimates provided by CSR, GFZ and JPL. In order to prevent aliasing effects of strong seasonal signals from falsifying our time-series analysis, we chose the total time span of the gravity field series to cover an integer number of years. Table 57.1 presents the sequences used. The sample periods were selected according to available monthly GRACE solutions at the time of the study.

We have followed the methodology of Wahr et al. (1998) to derive surface mass densities from (residual) gravity field coefficients. In this context, we applied spatial averaging in terms of Gaussian smoothing with a radius $R = 500$ km. The isotropic filter damps errors in the high-degree GRACE coefficients. Scaling surface mass densities by water density yields equivalent water height (EWH) values. Finally, we transformed EWH to mass changes.

Table 57.1 Monthly GRACE gravity field time-series used

Centre	Period	# Years	# Sol'ns	# Gaps
CSR	04/2002 – 03/2009	7	80	4
CSR	08/2002 – 07/2008	6	71	1
GFZ	08/2002 – 07/2008	6	67	5
JPL	08/2002 – 07/2008	6	70	2

We did not consider global isostatic adjustment (GIA) effects on our results. The GIA impact is of minor concern for our investigations because we are primarily interested in the temporal progress of mass changes rather than absolute deglaciation amounts.

3 Methods

The residual time-series (reduced by the temporal mean) of GRACE-derived mass variations were approximated by polynomials of different order according to

$$y = \beta_0 + \beta_1 x + \beta_2 x^2 + \ldots + \beta_p x^p, \qquad (57.1)$$

where x indicates the individual month of data availability and y is the corresponding residual (Greenland-averaged) ice-mass value. The polynomial coefficients are denoted by β_i with $i_{\max}$ equal to the polynomial order p. For each candidate model, the coefficients were estimated in terms of an unweighted least-squares (LS) fit to the data.

When incorporating different data fits, the main issue is to find the optimum amongst the candidate models. Actually, the term "optimum" has to be attributed to some model selection criteria. Here we consider (1) the Akaike information criterion (AIC), (2) the Bayesian information criterion (BIC), (3) cross validation (CV), and (4) hypotheses testing. The main purpose of introducing several criteria is to investigate whether they provide comparable results. Importantly, such criteria allow judging models relative to one another; they are unable to detect the superior model in an absolute sense.

3.1 Akaike Information Criterion

The philosophy behind the AIC is to perform model selection according to maximum log-likelihood (Akaike 1974). The theoretical foundation bases on Kullback-Liebler distance minimization (Burnham and Anderson 1998). Nonetheless, sometimes the criterion is referred to as a purely heuristic approach, seeking a model characterized by both a good data fit and few parameters. In the case of LS estimation, the AIC is defined as

$$\text{AIC} = n \ln(\hat{\sigma}^2) + 2k, \qquad (57.2)$$

where $\hat{\sigma}^2 = n^{-1} \sum_{i=1}^{n} \hat{e}_i^2$ indicates the maximum likelihood estimate of σ^2, $\hat{e}_i$ are the estimated LS residuals for a particular candidate model, and n is the number of samples (or observations). The integer k denotes the total number of estimated parameters, including the intercept and σ^2, i.e., $k = p + 1 + 1$ holds true.

The first term in (57.2) accounts for the criterion of a good statistical fit, whereas the second term incorporates the doctrine of parsimony [In this context, parsimony is the adoption of the simplest assumption in the interpretation of data, especially in accordance with the rule of Ockham's razor]. Both terms are balanced between one another.

Absolute AIC values have no relevance; therefore, AIC differences $\Delta\text{AIC} = \text{AIC} - \min(\text{AIC})$ over all candidate models are commonly used for interpretation. The model with the lowest AIC, ΔAIC-value, is considered to be the best approximation to the data, relative to the other models considered (in the sense of parsimony).

An important refinement to the AIC is its adaption to small sample sizes (e.g., Hurvich and Tsai 1989). In this case, the corrected AIC should be used instead,

$$\text{AIC}_c = \text{AIC} + \frac{2k(k+1)}{n-k-1}. \qquad (57.3)$$

3.2 Bayesian Information Criterion

For LS applications, the BIC (occasionally also referred to as the Schwarz criterion, SC) is defined as

$$\text{BIC} = n \ln(\hat{\sigma}^2) + k \ln(n) \qquad (57.4)$$

(Schwarz 1978), where all terms are as defined above.

The BIC/SC's structure is very similar to the AIC, even though the deviations of the criteria underlie differing motivations. The BIC penalises complex models more than the AIC does (for fixed n and increasing k, the penalty term $k \log(n)$ increases faster opposed to the term $2k$). BIC differences are computed by $\Delta\text{BIC} = \text{BIC} - \min(\text{BIC})$. Again, the model with the lowest BIC, ΔBIC-value, is the best model; it should be selected to describe the data.

3.3 Cross Validation

We implemented the special case of leave-one-out CV (cf. Featherstone and Sproule 2006). The model parameter estimates are based on all-but-one data points. The leave-one-out CV estimator is used to predict the observation that has been omitted within the estimation process. The squared differences between the real observations and their predictions are summed up to define the CV functional

$$\text{CV} = n^{-1} \sum_{i=1}^{n} \left(y_i - \hat{y}_i^{\text{CV}}\right)^2, \qquad (57.5)$$

where $\hat{y}_i^{\text{CV}}$ indicates the prediction of the i-th observation (or data point) y_i. The model with lowest CV value is considered to be the best model, so it should be selected to describe the data.

3.4 Hypotheses Testing of Polynomial Coefficients

For each regression coefficient β_i ($i = 0, \ldots, p$) separately, hypotheses testing balances the null hypothesis $H_0 : \beta_i = 0$ against the alternative hypothesis $H_1 : \beta_i \neq 0$. The test provides evidence on the statistical significance of model parameters. As such, the null hypothesis is rejected if the Student-distributed test value

$$T = \frac{\hat{\beta}_i}{\sigma_{\beta_i}} \sim t_{1-\alpha/2,\, n-(p+1)} \qquad (57.6)$$

lies outside the confidence interval. Otherwise, H_0 is accepted. This is akin to the development of multiple regression equations (MRE) used in geodetic coordinate transformations (e.g., NGA 2004).

Here, we chose the level of significance to be $\alpha = 5\%$ (95% confidence). Moreover, in (57.6), $\hat{\beta}_i$ denotes the LS estimate of β_i and σ_{β_i} its standard deviation.

4 Results

Recent deglaciation of the Greenland ice sheets has been reported by a variety of previous studies. The major objective of this investigation is to shed some light on the temporal dynamics of the ablation (ice-mass loss) process. In this framework, we now address linear versus non-linear ice-mass balance models.

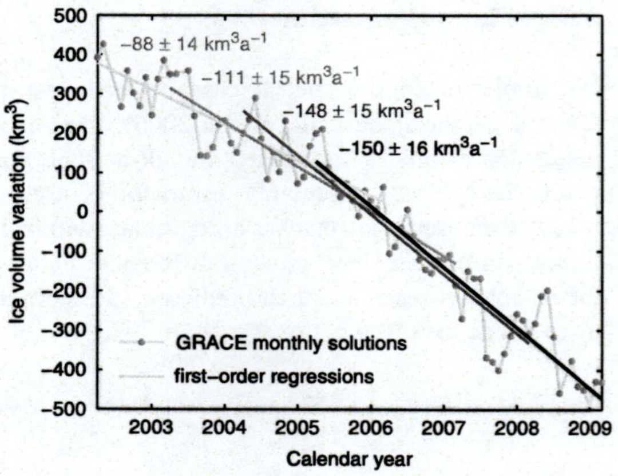

Fig. 57.1 Greenland ice-volume variations from April 2002 to March 2009 (CSR solutions). The monthly residual values are reduced by the temporal mean. The linear regression lines refer to four 4-year periods, each starting one year after the previous one. The linear ice-volume trend decreases continuously (from $-88 \pm 14\,\mathrm{km^3\,a^{-1}}$ to $-150 \pm 16\,\mathrm{km^3\,a^{-1}}$), suggesting accelerated ice-mass decline in recent years (no GIA correction applied)

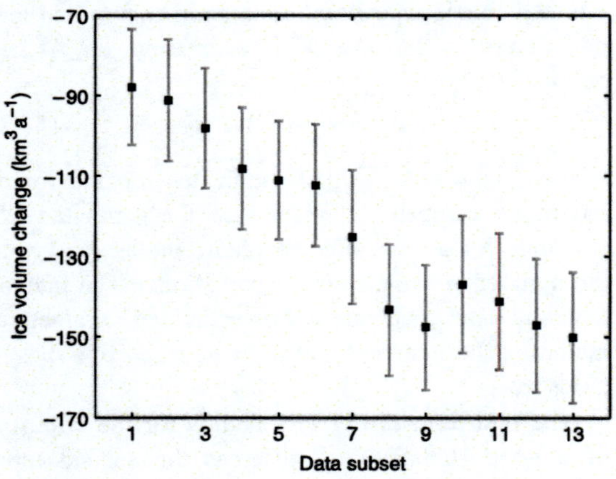

Fig. 57.2 Greenland ice-volume change rates from April 2002 to March 2009 (CSR solutions). Each data subset refers to four 4-year periods. Every period has an offset of 3 months to the previous one (i.e., the change rates provided in Fig. 57.1 correspond to the data subsets 1, 5, 9 and 13, respectively). The error bars indicate the standard deviations of the estimates. They were derived from the LS residuals of the polynomial fit

In a first experiment, we analyzed CSR monthly gravity field solutions from April 2002 to March 2009. In particular, we LS-estimated the linear trends of 13 four-year periods. Each period has a 3-month offset from the previous one. Figure 57.1 shows a representative sample of four of the epochs. The slopes of the linear regression lines increase almost continuously (Fig. 57.2). The overall rise is about 70%. Hypotheses testing shows the slopes of the last few epochs to differ statistically significantly from the earlier ones, indicating accelerated ablation. The analysis of five-year periods yields comparable results (not displayed here).

In a second experiment, we analysed the whole seven-year gravity field time-series at once, rather than splitting it into subsets as above. For this purpose, we fitted polynomials up to fifth order to the data. In order to detect the best amongst the models used, for each of the polynomial fit, Table 57.2 summarizes the four model selection criteria introduced in Sect. 3.

From Table 57.2, the linear approximation shows the poorest results, followed by the third-order and fifth-order polynomials. The second-order polynomial performs best. Furthermore, hypothesis testing reveals statistical significance for all model parameters only up to the second-order. For any higher-order polynomial, only the intercept is significant. Hence, the Greenland GRACE data trend is best represented by a second-order polynomial (Fig. 57.3). Note that the total ice-volume variations are largely independent of the model (cf. the last column in Table 57.2).

Table 57.2 Model selection criteria and total ice-volume changes over the period April 2002 to March 2009 (CSR solutions), dependent on the order of polynomial fit

Order	ΔAIC	ΔAIC$_c$	ΔBIC	CV	Total change (km³)
1	10.7	10.5	8.3	5100.7	-858 ± 70
2	0	0	0	4430.0	-853 ± 65
3	10.4	1.3	3.4	4488.7	-815 ± 64
4	0	0.6	4.7	4407.3	-822 ± 63
5	1.11	2.2	8.3	4465.3	-858 ± 63

Next, we tackle the question "Does the superiority of the second-order approximation opposed to linear regression depend on the gravity field time-series investigated?" This is what we tried to determine in our third experiment. Here, we considered monthly GRACE solutions from three of the "independently" operating data centres: CSR, GFZ and JPL. To make the results comparable, we chose a consistent six-year analysis period (cf. Table 57.1).

The model selection criteria in Table 57.3 reveal the predominance of the quadratic fit over the linear model for the Greenland case. Only the BIC for the

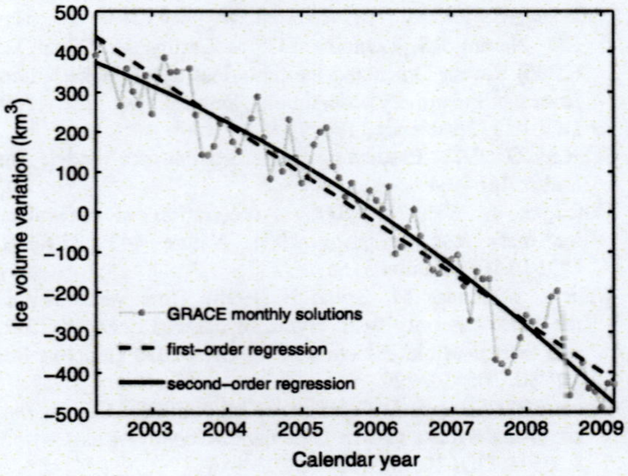

Fig. 57.3 Greenland ice-volume variations from April 2002 to March 2009 (CSR solutions). The residual monthly values are reduced by the temporal mean. Both the linear and quadratic fit provide comparable variations over the whole period ($-858 \pm 70\,km^3$, $-853 \pm 65\,km^3$ respectively). Model selection criteria (Table 57.2) show the quadratic model to be superior to the linear fit over Greenland

Table 57.3 Model selection criteria and total ice-volume changes over the period August 2002 to July 2008, dependent on the order of polynomial fit and monthly gravity field series

Order	Center	ΔAIC	ΔAIC$_c$	ΔBIC	CV	Total change (km^3)
1	CSR	6.1	5.8	3.8	5252.7	-706 ± 71
2	CSR	0	0	0	4864.8	-705 ± 67
1	GFZ	8.3	8.0	6.1	5334.8	-587 ± 71
2	GFZ	0	0	0	4676.1	-574 ± 66
1	JPL	2.0	1.7	−0.3	2451.1	-255 ± 48
2	JPL	0	0	0	2387.1	-255 ± 47

JPL solutions shows a slight preference for the linear compared to the second-order fit. All other criteria are in favour of the quadratic model.

5 Discussion and Conclusion

Most previous studies addressing ice-mass balance over Greenland have focussed on the estimation of total mass-loss numbers within certain sample periods. In contrast, here we address the temporal progress of deglaciation. Therefore, the correction of change-rates by disturbing effects such as signal leakage and GIA is ignored (disturbing effects mainly scale change-rates).

In order to specify temporal characteristics adequately, we suggest the adoption of suitable criteria. These allow more objective judgement among various candidate models relative to each other. The Akaike information criterion, its modified version for small samples, and the Bayesian information criterion provide comparable results. Concerning a seven-year sample period, all these criteria clearly discard the linear approximation ($\Delta > 8$); however, they attribute substantial support to higher-order polynomial models ($\Delta < 2$). In short, the second-order polynomial performs best over Greenland. This result is also supported by hypotheses testing of the polynomial coefficients.

Similar conclusions hold for the investigation of a six-year period based on GRACE gravity field solutions provided by CSR, GFZ and JPL. Again, the quadratic model is mostly superior to the linear fit. As the time-series now consists of fewer samples compared to the seven-year investigation, the criterion differences are smaller. Actually, the largest value is $\Delta = 8.3$. Nonetheless, we want to highlight that the model selection results are almost independent of the GRACE gravity field time-series used.

In conclusion, we proved that the linear model is insufficient to describe long-term Greenland secular mass changes. Indeed, the trend is better represented by a second-order polynomial. Within a seven-year sample period, the ablation rate increased by 250%. This result confirms accelerated contemporary melting of the Greenland ice sheets, as suggested originally by Velicogna and Wahr (2006), but now supported by various metric criteria and using longer analysis periods.

Extended time-series will pinpoint the validity of present-day results for future predictions. Furthermore, in order to substantiate these GRACE-derived change-rate accelerations, comparison with alternative independent data or in-situ measurements is desirable.

Acknowledgements WEF and MK would like to thank the Australian Research Council for funding through project grants DP0663020 and DP0877381. This is TIGeR publication no. 213.

References

Akaike H (1974) A new look at the statistical model identification. IEEE Trans Autom Control 19(6): 716–723, DOI 10.1109/TAC.1974.1100705

Baur O, Kuhn M, Featherstone WE (2009) GRACE-derived ice-mass variations over Greenland by accounting for leakage effects. J Geophys Res 114, B06407, DOI 10.1029/2008JB006239

Burnham KP, Anderson DR (1998) Model selection and inference – a practical information-theoretic approach. Springer, New York

Chen JL, Wilson CR, Tapley BD (2006) Satellite gravity measurements confirm accelerated melting of Greenland ice sheet. Science 313: 1958–1960, DOI 10.1126/science.1129007

Featherstone WE, Sproule DM (2006) Fitting AUSGeoid98 to the Australian height datum using GPS data and least squares collocation: application of a cross-validation technique. Survey Review 38(301): 573–582

Hurvich CM, Tsai CL (1989) Regression and time series model selection in small samples. Biometrika 76: 297–307

Luthcke SB, Zwally HJ, Abdalati W, Rowlands DD, Ray RD, Nerem RS, Lemoine FG, McCarthy JJ, Chinn DS (2006) Recent Greenland ice mass loss by drainage system from satellite gravity observations. Science 314: 1286–1289, DOI 10.1126/science.1130776

Schwarz G (1978) Estimating the dimension of a model. Ann Stat 6: 461–464

Velicogna I, Wahr J (2006) Acceleration of Greenland ice mass loss in spring 2004. Nature 443: 329–331, DOI 10.1038/nature05168

Wahr J, Molenaar M, Bryan F (1998) Time variability of the Earth's gravity field: Hydrological and oceanic effects and their possible detection using GRACE. J Geophys Res 103:30, 205–30, 229

Wouters B, Chambers D, Schrama EJO (2008) GRACE observes small-scale mass loss in Greenland. Geophys Res Lett 35, L20501, DOI 10.1029/2008GL034816

A New Continuous GPS Network to Monitor Deformations in the Iberian Peninsula (Topo-Iberia Project). First Study of the Situation of the Betic System Area

58

M.C. de Lacy, A.J. Gil, J.A. García Armenteros, A.M. Ruiz, M. Crespi, A. Mazzoni, and Topo-Iberia GPS Team

Abstract

Topo-Iberia is a project funded by the Spanish Research Council that shares interests with Topo-Europe project. Its objective is to understand the interactions in the Iberian Peninsula (SW Europe) between deep, shallow and atmospheric processes, through a multidisciplinary approach linking Geology, Geophysics and Geodesy. In order to achieve the observational goals of the project, three main working groups have been set up: seismic, magneto-telluric and geodetic. The first task of the GPS geodetic working group was to design the new Continuous GPS network to complement other GPS networks which are already operating under the supervision of different institutions. The Betic System is the most active tectonically area in the Iberian Peninsula. As this area in under a NNW-SSE to NW-SE compression regime, two station alignments (from NE to SW and NW to SE) were built, including continuous GPS (CGPS) stations in Northern Africa. Nowadays the Topo-Iberia GPS network includes twenty-six operational stations: 22 installed in different regions of Spain and 4 located in strategically chosen places in Morocco. In this work the state of the art of the CGPS network in the Betic System Area is presented.

Keywords

Topo-Iberia · GPS · time series · deformation

1 Introduction

Topo-Iberia is a research project that involves researchers from 10 different groups (www.igme.es/Internet/Tpoiberia). It reflects the willingness and interest of the Spanish scientific community to establish an integrated framework to develop multidisciplinary geoscientific studies in Spain. The "micro-continent" formed by the Iberian Peninsula and its margins constitutes a most suitable natural laboratory, well identified by the international scientific community, to develop innovative, frontier research on its topography and 4-D evolution. The objective of Topo-Iberia is to understand the interaction between deep, superficial and atmospheric processes, by integrating research on geology, geophysics, geodesy and geotechnology. The knowledge on the relief

M.C. de Lacy (✉) · A.J. Gil · J.A. García Armenteros · A.M. Ruiz · M. Crespi · A. Mazzoni · Topo-Iberia GPS Team
Departamento de Ingeniería Cartográfica, Geodésica y Fotogrametría, Universidad de Jaén, Spain

DITS – Area di Geodesia e Geomatica, Università di Roma La Sapienza, Italy

N. Sneeuw et al. (eds.), *VII Hotine-Marussi Symposium on Mathematical Geodesy*, International Association of Geodesy Symposia 137, DOI 10.1007/978-3-642-22078-4_58, © Springer-Verlag Berlin Heidelberg 2012

changes and their causes is of great social impact regarding climate change and evaluation of natural resources and hazards. Three major domains of research have been identified: the southern and northern borders of the Iberian plate (the Betic-Rif system and the Pyrenean-Cantabrian system) and its central core (Meseta and Central-Iberian systems).

In this paper we are focusing on the status of the geodetic project in the Betic Area. In particular, in Sect. 2 a general description of Topo-Iberia is shown by including a short review of the CGPS network. In Sect. 3 first tests of processing CGPS data are explained.

2 Description of Topo-Iberia Project

Topo-Europe (Cloetingh et al. 2007) has identified up to 6 highly relevant natural laboratories on which the future research in the European region should be focused, one of them being the "Iberian microcontinent". The scientific research of Topo-Europe is supported by a fundamental technological component, a multidisciplinary observation platform EuroArray, that is a multi-sensor pool of instrumentation "Terrascope" of seismics, GPS, magnetotellurics, etc. EuroArray has decided to promote a first thematic initiative, named PICASSO (Program to Investigate Convective Alboran Sea System Overturn, http://www.geophysics.dias.ie/projects/PICASSO/) in the interaction area between the Iberian and African plates.

Topo-Iberia is a multidimensional programme, with interrelated and interactive components and integration of results to establish a topography and evolution model of the natural environment. Actions to be developed include experimental studies making use of seismic techniques, deformation measurements by GPS measurements, magnetotellurics, gravimetric, magnetism and paleomagnetism and satelite imaging, etc. These actions will be carried out in three areas: South, Centre and North. In the South: the Betic-Rif and Atlas orogens, and the marine domains in-between: Alboran Sea and Gulf of Cadiz (initiative PICASSO). In the Centre: Meseta, Central system and Iberian system, including Catalan ranges and eastern basins. In the North: Pyrenean and Cantabrian ranges, including Duero and Ebro basins and Cantabrian and Galician margins. In particular, the University of Jaen (UJA) along with the University of Barcelona (UB) and the Real Instituto y Observatorio de la Armada (ROA) are in charge of estimating deformations from GPS measurements.

2.1 Continuous GPS Network

GPS can be used as a tool for the measurement of the active deformation providing a way of monitoring active tectonics and the detection of relative movements when the recorded deformation is only of cm/year, as is the case of the Iberian Peninsula. A fundamental objective of this project involves the determination of the present day deformation vectors of the crust.

As the study area in under a NNW-SSE to NW-SE compression regime, two station alignments (from NE to SW and NW to SE) were built, including continuous CGPS stations in Northern Africa, (Fig. 58.1). In particular, an array of new permanent stations has been installed complementing other GPS networks already deployed by different Institutions, EPN (EUREF Permanent Network) (Bruyninx 2004), IGS (International GNSS Service) (Dow et al. 2009) and regional governmental agencies. After an initial period of designing, purchasing new equipments, monumentation and deploying stations, nowadays the Topo-Iberia GPS network includes twenty-six operational stations, six of them (AREZ, PALM, PILA, NEVA, TGIL, VILA) under the control of the University of Jaen.

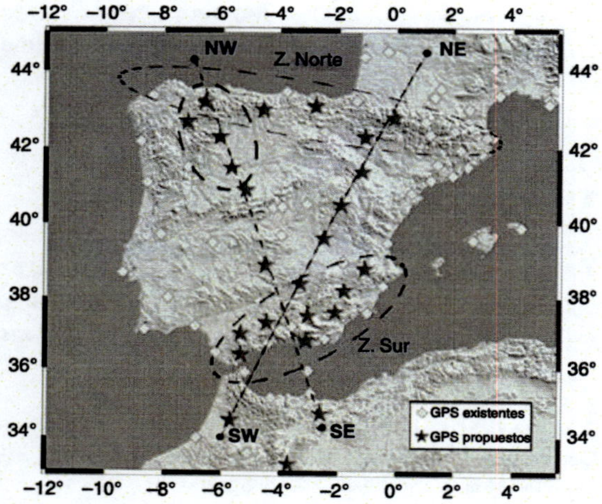

Fig. 58.1 Continuous GPS stations

The data Analysis of the whole set of new and existing CGPS stations is performed at three different analysis centers: ROA, UB and UJA. Different approaches to processing GPS data by using different programs are being carried out: GIPSY-OASIS (Zumberge et al. 1997) is used by ROA, GAMIT (Herring et al. 2007) by UB and Bernese (Beutler et al. 2007) by UJA. A further analysis of the whole set of coordinate time series should give a nice view of relative displacements, deformations and stresses all over the frame of interest, (Kierulf et al. 2008).

In this paper, only first results provided by Bernese and computed at the University of Jaen are shown.

3 Status of Topo-Iberia in the Betic Area

One of the main objectives of the project is the establishment of a common work platform, called Iber-Array, in the form of an integrated network of research groups and centres. This implies the development of a multitask platform to allow to manage new multidisciplinary data-sets. A database, called SITOPO, is being built with files from existing permanent stations with good quality data. Therefore every analysis centre must implement a GPS data management allowing to check their quality before submitting them to SITOPO. After that every analysis centre will compute and analyze the corresponding time series of the CGPS stations, close to 70 stations in total. At this moment, not all analysis centres check GPS data downloaded from the stations. Therefore, the UJA has checked the quality of all GPS data used in this study.

The UJA has developed an own software written in Visual Basic for Applications implemented in Microsoft Excel to carry out the data manegement automatically. RINEX files are downloaded from six stations under the control of the University of Jaen via Internet/GPRS; from the EPN and IGS data centres and SITOPO via ftp. TEQC program (Estey and Meertens 1999) (http://facility.unavco.org/software) is used on all downloaded observations to check the data quality. Thresholds of quality parameters are fixed following the recommendations of IGS stations (http://igscb.jpl.nasa.gov/network/dataplots.html). These parameters are related to the number of observations per day, the RMS of the MP1 and MP2 linear combinations and the observations per slip.

After that, all necessary files are prepared to compute a daily solution with Bernese Processing Engine. The time series analysis is carried out by the software KINADGPS (Barzaghi et al. 2004). KINADGPS is a program written in C language and takes into account temporal covariances of solutions. This fact is basic to a correct estimation of GPS-derived coordinate and velocity precisions when GPS permanent stations are used to infer geodynamical interpretations. After linear detrending and outlier removal the time series of station coordinates behave as a second order stationary process and their autocovariance functions depend on the time only. When the signal variance is significant with respect to the noise variance, the fitting with a proper model function is carried out. The program is able to find the most correct fitting function among eight different autocovariance models (Barzaghi et al. 2003):

1. $y = a\ exp(-b|x|)$ exponential
2. $y = a\ exp(-bx^2)$ normal
3. $y = a\ exp(1 - cx^2)\ exp(-b|x|)$ exp-parable
4. $y = a\ exp(1 - cx^2)\ exp(-bx^2)$ normal-parable
5. $y = a\ exp(-b|x|)\ cos(cx)$ exp-cos
6. $y = a\ exp(-bx^2)\ cos(cx)$ normal-cos
7. $y = a\ exp(-b|x|)\ sin(c|x|)/(c|x|)$ exp-sin x/x
8. $y = a\ exp(-bx^2)\ sin(c|x|)/(c|x|)$ normal-sin x/x

3.1 First Tests

First tests of the stations controlled by UJA are carried out in last months. In particular, GPS data belonging to stations of AREZ, PALM, PILA, NEVA, TGIL, VILA are processed with Bernese software and first time series have been computed with available GPS data, one-year since may 2008.

The basic observable are dual-frequency GPS carrier phase observations. They have been preprocessed in a baseline by baseline mode using triple-differences. During the final estimation, based on ionosphere-free double differences, a 10 degrees elevation angle cut off was used and elevation-depending weighting scheme was applied. Receiver and satellite antennae phase center calibrations were applied using absolute antenna phase center corrections from the IGS05 model. The a priori tropospheric refraction was modeled using the Dry-Niell model (Niell 1996) and the remaining wet part was estimated hourly for each station, using the wet-Niell mapping function (Niell 1996)

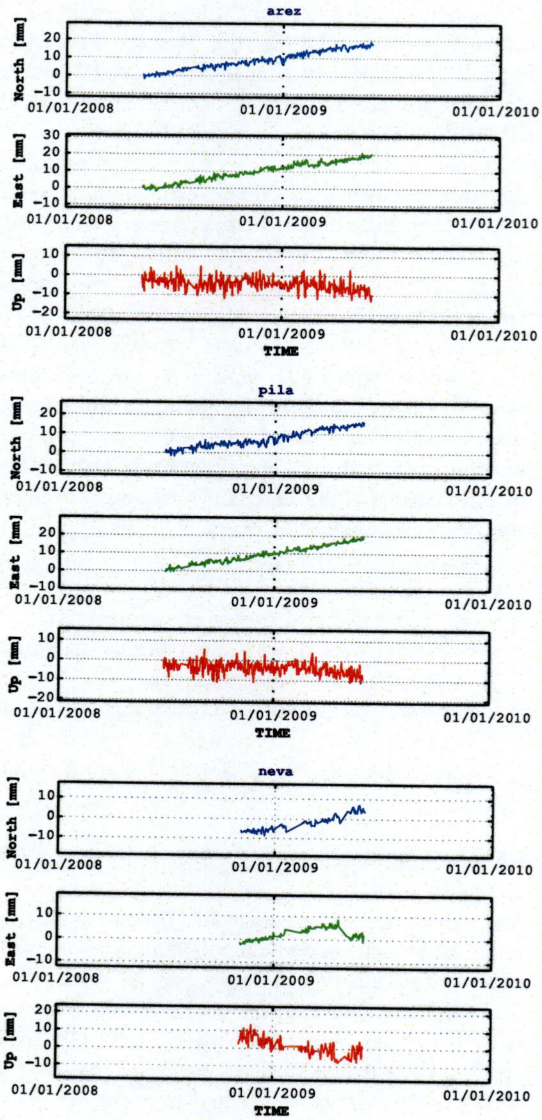

Fig. 58.2 Time series of the North, East and Up components of AREZ, PILA and NEVA in the IGS05 reference frame obtained by the University of Jaen

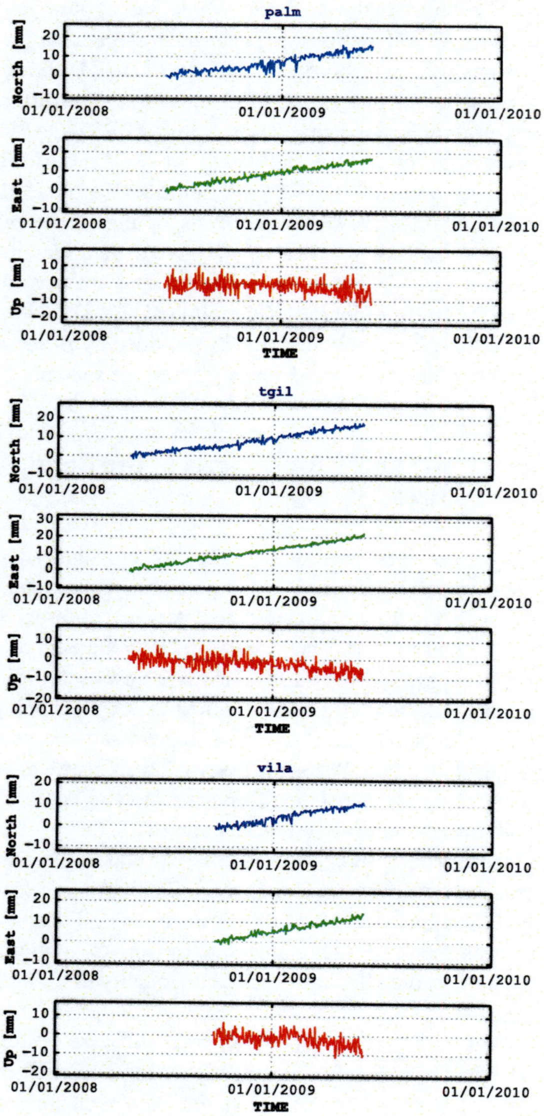

Fig. 58.3 Time series of the North, East and Up components of PALM, TGIL and VILA in the IGS05 reference frame obtained by the University of Jaen

without a priori sigmas. A horizontal gradient parameter is estimated for each day and station (TILTING) without introducing a priori constraints. Phase ambiguities which have been fixed to their integer values in a previous step are now introduced as known parameters.

Datum definition was carried out by five stations ALAC, CASC, SFER, RABT, VILL constrained to 2 mm in North and East components and 4 mm in Up component to the IGS05 (Ferland 2006) reference frame. Their IGS05 coordinates and velocities are computed by linear interpolation from the last 52

Table 58.1 Preliminary velocities estimated by KINADGPS

Station	V_{East} (mm/year)	V_{North} (mm/year)
ALAC	18.98 ± 0.13	16.20 ± 0.15
AREZ	19.01 ± 0.14	17.12 ± 0.16
CASC	17.32 ± 0.14	16.61 ± 0.16
NEVA	18.45 ± 0.58	16.05 ± 0.67
PALM	17.06 ± 0.15	15.82 ± 0.17
PILA	19.28 ± 0.14	17.43 ± 0.17
RABT	16.25 ± 0.47	17.46 ± 0.53
SFER	15.93 ± 0.11	16.84 ± 0.12
TGIL	19.58 ± 0.13	16.57 ± 0.15
VILA	19.79 ± 0.48	15.69 ± 0.57
VILL	17.33 ± 0.11	16.96 ± 0.13

weekly IGS solutions to separate any annual periodic effect and take account of eventual datum change, (Benciolini et al. 2008).

First time series of daily solutions are shown in Figs. 58.2 and 58.3. The velocities estimated by KINADGPS are reported in Table 58.1. It is important to underline that these values are preliminary due to one year is a short period to estimate them in a reliable way.

A year is a very short period to come to conclusions but the NEVA time series clearly indicate the presence of problems at that station. Its GPS antenna has been changed recently to solve them. North and East components of all time series show a trend in NE direction in agreement with Eurasian plate movement.

Different strategies are being tested in order to establish a final procedure. In particular a loose solution (deconstrained network in terms of coordinates without any reference frame) was computed. At the moment the differences between this loose solution and the solution with a defined reference frame are being studied (Kawar et al. 1998).

4 Summary

In this paper, the status of the CGPS network to monitor deformations inside the research project "Topo-Iberia" is presented. In particular, a first study of the status of the situation in the Betic Area was carried out by the University of Jaen. In this study a year of GPS data coming from the stations controlled by the UJA are processed and the corresponding coordinate time series computed by the program KINADGPS. Different strategies are being tested in order to establish a final procedure. The other groups are computing their first results as well. In the near future, the solutions coming from ROA, UJA and UC will be compared to analyze the differences and to study a methodology to combine all of them over the Iberian Peninsula.

Acknowledgements This work has been funded by the Spanish Government through the CONSOLIDER Topo-Iberia Research Project. Ref: CDS2006-00041

References

Barzaghi R, Borghi A, Crespi M, Pietrantonio G, Riguzzi F (2003) The GPS Permanent Network Solutions: the Impact of Temporal Correlations. V Hotine-Marussi Symposium on Mathematical Geodesy. IAG Symposium no. 127, 179–183

Barzaghi R, Borghi A, Crespi M, Giannone F, Pietrantonio G, Riguzzi F (2004) The Impact of Auto and Cross-correlations in Daily Solution Time Series of GPS Permanent Stations. Celebrating a decade of the International GPS Service workshop & symposium 2004. AIUB, Switzerland M. Meindl, Editor IGS Proceedings Berne, Switzerland March 1–5

Benciolini B, Biagi L, Crespi M, Manzino A M, Roggero M (2008) Reference Frames for GNSS positioning services: Some problems and proposed solutions. J of Appl Geodes (2):53–62

Beutler G, Bock H, Brockmann E, Dach R, Fridez P, Gurtner W, Habrich H, Hugentobler U, Ineichen D, Jaeggi A, Meindl M, Mervart L, Rothacher M, Schaer S, Schmid R, Springer T, Steigenberger P, Svehla D, Thaller D, Urschl C, Weber R (2007) BERNESE GPS software version 5.0. ed. Urs Hugentobler R, Dach P, Fridez M, Meindl, University of Bern, Switzerland

Bruyninx C (2004) The EUREF Permanent Network: a multidisciplinary network serving surveyors as well as scientists. GeoInformatics, Vol. 7, pp. 32–35

Cloetingh S, Ziegler P, Bogaard P, Andriessen P, Artemieva I, Bada G, van Balen R, Beekman F, Ben-Avraham Z, Brun J, Bunge H, Burov E, Carbonell R, Facenna C, Friedrich A, Gallart G, Green A, Heidbach O, Jones A, Matenco L, Mosar J, Oncken O, Pascal C, Peters G, Sliaupa S, Soesoo A, Spakman W, Stephenson R, Thybo H, Torsvik T, de Vicente G, Wenzel F, Wortel M, TOPO-EUROPE Working Group. TOPO-EUROPE: The geoscience of coupled deep Earth-surface processes. Global and Planetary Change 58, pp. 1–118

Dow JM, Neilan RE, Rizos C (2009) The International GNSS Service in a changing landscape of Global Navigation Satellite Systems. J Geodes 83:191–198, DOI: 10.1007/s00190-008-0300-3

Estey LH, Meertens CM (1999) TEQC: the multipurpose toolkit for GPS/GLONASS data. GPS Solutions 3(1):42–49

Ferland R (2006) IGSMAIL-5447: Proposed IGS05 Realization

Herring TA, King RW, McClusky SC (2007) Introduction to GAMIT/GLOBK, Release 10.3. Massachusetts Institute of Technology, Cambridge, MA

Kawar R, Blewitt G, Davies P (1998) The Newcastle Global Network Associate Analysis Center Annual Report. International GPS Service for Geodynamics 1998 Annual Report, pp. 139–147.

Kierulf HP, Plag HP, Bingley RM, Teferle N, Demir C, Cingoz A, Yildiz H, Gárate J, Davila JM, Silva CG, Zdunek R, Jaworski L, Martinez Benjamin JJ, Orus R, Aragon A (2008) Comparison of GPS analysis strategies for high-accuracy vertical land motion. Physics and Chemistry of the Earth. Parts A/B/C., 33(3-4):194–204

Niell A (1996) Global Mapping Functions for the Atmosphere Delay at Radio Wavelengths, J Geophys Res 101(B2): 3227–3246

Zumberge J, Heflin M, Jefferson D, Watkins M, Webb F (1997) Precise Point Positioning for the efficient and robust analysis of GPS data from large networks. JGR 102:5005–5010.

Implementation of the Complete Sea Level Equation in a 3D Finite Elements Scheme: A Validation Study

59

G. Dal Forno, P. Gasperini, and G. Spada

Abstract

We describe the implementation of the complete Sea Level Equation (SLE) in a Finite Element (FE) self-gravitating 3D model. The procedure, originally proposed by Wu (2004), consists of iterating the solution of the SLE starting from a non self-gravitating model. At each iteration, the perturbation to the gravitational potential due to the deformation at the density interfaces is determined, and the boundary conditions for the following iteration are modified accordingly. We implemented the computation of the additional loads corresponding to the perturbations induced by glacial and oceanic forcings at the same iteration at which such forcings are applied. This implies an acceleration of the convergence of the iterative process that occurs actually in three to four iterations so that the complete procedure, for a 6,800 elements FE grid, can be run in about two hours of computing time, on a four-core 2.2 GHz Linux workstation. This spherical and self-gravitating FE model can be employed to simulate the deformation of the Earth induced by any kind of load (non necessarily of glacial origin) acting on the surface and/or internally.

1 Introduction

The modeling of the response of the Earth to surface ice loads was usually performed in the past by the normal-mode method (Peltier 1974) that is suitable to deal with linear rheologies and simple spherically symmetric geometries. Such method is not suitable when the problem involves non-linear rheologies or lateral variations of material properties. In these cases the Finite Element (FE) method represents one of the possible approaches. Several papers employed FE to study the relaxation of a linear mantle with lateral viscosity variations (Gasperini and Sabadini 1989; Kaufmann et al. 1997, 2000; Paulson et al. 2005; Spada et al. 2006) or a purely non-linear (Wu 1992, 1999; Giunchi and Spada 2000; Wu and Wang 2008) or composite (linear plus non-linear) mantle (Gasperini et al. 1992, 2004; Dal Forno et al. 2005; Dal Forno and Gasperini 2007). In these studies, the gravitational potential perturbations due to the variations of the loads and to the deformation of the solid Earth has been neglected, which prohibits a reliable modeling of relative sea level observations. More recently, Wu (2004) developed a mixed approach that allows to deal with self-gravitation and the complete Sea

G. Dal Forno (✉) · P. Gasperini
Dipartimento di Fisica Università di Bologna, Viale Berti Pichat 8, I-40127 Bologna, Italy
e-mail: giulio.dalforno@envia.it; paolo.gasperini@unibo.it

G. Spada
Dipartimento di Scienze di Base e Fondamenti Urbino University "Carlo Bo", Via Santa Chiara 27, I-61029 Urbino, Italy
e-mail: giorgio.spada@gmail.com

N. Sneeuw et al. (eds.), *VII Hotine-Marussi Symposium on Mathematical Geodesy*, International Association of Geodesy Symposia 137, DOI 10.1007/978-3-642-22078-4_59, © Springer-Verlag Berlin Heidelberg 2012

Level Equation (SLE, see Farrell and Clark, 1976) by FE codes through spherical harmonics expansion of involved quantities.

In this work we implemented such procedure that schedules the iteration of the solution starting from a non-self-gravitating model with eustatic sea level, using the commercial code ABAQUS (Simulia 2009). The effects of gravitational potential perturbations due to ice and water mass variations and to deformation of density interfaces is simulated by the application of equivalent pressure loads. The sea level variations at each iteration are computed from potential perturbations as well. The procedure is coded in a load routine linked with ABAQUS and in a post-processing module that extracts displacements computed by previous FE iteration and computes potential perturbations and relevant differential loads for the successive iteration. We tested our codes with the results of program SELEN by Spada and Stocchi (2007) that implements the complete sea level equation by solving analytically the self-gravitating relaxation of a linear Earth. This procedure would allow to treat the problem consistently even in the case of non-linear or composite rheologies and of lateral variations of material properties.

2 Outline of the Iterative Procedure

The spherical harmonics (SH) coefficients of the load associated with ice thickness and sea level variations are computed by the pixelization technique developed by Tegmark (1996)

$$\iota_\ell^m = \frac{\rho_{ice}\gamma}{N_p}\sum_{i=1}^{N_p} I(\theta_i,\lambda_i)\, Y_\ell^m(\theta_i,\lambda_i) \tag{59.1}$$

$$s_\ell^m = \frac{\rho_w\gamma}{N_p}\sum_{i=1}^{N_p} S(\theta_i,\lambda_i)\, Y_\ell^m(\theta_i,\lambda_i), \tag{59.2}$$

where θ is colatitude and λ is longitude, N_p is the number of pixels, ρ_{ice} and ρ_w are the ice and water densities, γ is the reference gravity field at the surface, $I(\theta_i,\lambda_i)$ and $S(\theta_i,\lambda_i)$ are the ice height variation (i. e., the difference between the height at 18 ka B.P. and at the time increment) and the sea level change at the i-th pixel respectively, and $Y_\ell^m(\theta_i,\lambda_i)$ are the real SHs at the i-th pixel, according to the "Geodesy" convention of SHTOOLS package (Wieczorek 2005).

For ease of notation we do not indicate explicitly the time dependence. All the computations are repeated at various time increments at which load amplitudes are updated by the FE code and the ensuing displacements are computed.

At the first iteration, sea level change coincides with the melted ice equivalent sea level

$$S(\theta_i,\lambda_i) = -\frac{M_{ice}}{\rho_w A_o} O(\theta_i,\lambda_i), \tag{59.3}$$

where M_{ice} is the melted ice mass, A_o the area of oceans and $O(\theta_i,\lambda_i)$ the ocean function at i-th pixel

$$O(\theta_i,\lambda_i) = \begin{cases} 1 \text{ if wet} \\ 0 \text{ if dry.} \end{cases} \tag{59.4}$$

According to Wu (2004), the SH coefficients of incremental potential due to the load $\rho_w S + \rho_{ice} I$ at each of the $N+1$ spherical density interfaces with radius r_p (with p ranging from 0 to N), are

$$\Theta_\ell^m(r_p) = \frac{4\pi G}{2\ell+1}\frac{\sigma_\ell^m}{\gamma} r_p \left(\frac{r_p}{a}\right)^{\ell-1}, \tag{59.5}$$

where $\sigma_\ell^m = \iota_\ell^m + s_\ell^m$ and G is the Newton's constant. Here r_0 is the Earth's core radius and $r_N = a$ is the Earth's radius. Equation (59.5) is equivalent to the terms of potential perturbation due to the load of (16a, b, c) in Wu (2004). The SH coefficients of differential load due to such potential perturbation is

$$\Delta\sigma_\ell^m(r_p) = (\rho_{p+1} - \rho_p)\,\Theta_\ell^m(r_p), \tag{59.6}$$

where ρ_p is density of the p-th layer (ρ_0 is the density of the core and $\rho_N \approx 0$ is air density). Then the total load SH coefficients

$$t_\ell^m(r_p) = \begin{cases} \Delta\sigma_\ell^m(r_p) & \text{if } p < N \\ \sigma_\ell^m + \Delta\sigma_\ell^m(r_p) & \text{if } p = N. \end{cases} \tag{59.7}$$

are stored to be used by ABAQUS. The load routine reads total load coefficients and computes the actual load at each n-th FE integration point as

$$T(r_p,\theta_n,\lambda_n) = \sum_{\ell=2}^{L_{max}}\sum_{m=-\ell}^{\ell} t_\ell^m(r_p)\, Y_\ell^m(\theta_n,\lambda_n), \tag{59.8}$$

where L_{max} is the truncation degree of the SH expansion. The total surface load does not include harmonic degree 0, owing to mass conservation. Harmonic degree 1, which accounts for displacement of the solid Earth with respect to the center of mass of the system, is neglected here.

After first FE iteration, the post-processor code reads, from the output of ABAQUS, the displacements at the nodes of all the $N+1$ density interfaces of the FE grid and computes their values $U(r_p, \theta_i, \lambda_i)$ at all of the pixels, by linear interpolation. The SH coefficients of displacements at each interface are computed as

$$U_\ell^m(r_p) = \frac{1}{N_p} \sum_{i=1}^{N_p} U(r_p, \theta_i, \lambda_i)\, Y_\ell^m(\theta_i, \lambda_i). \tag{59.9}$$

The potential perturbation due to the displacement at each interface is

$$\Phi_\ell^m(r_p) = \frac{4\pi G}{2\ell+1}\left[\sum_{i=0}^{p} U_\ell^m(r_i)(\rho_i - \rho_{i+1})\, r_i \left(\frac{r_i}{r_p}\right)^{\ell+1} + \sum_{i=p+1}^{N} U_\ell^m(r_i)(\rho_i - \rho_{i+1})\, r_p \left(\frac{r_p}{r_i}\right)^{\ell-1}\right]. \tag{59.10}$$

It can be shown, after some algebra, that (59.10) coincides with the terms of potential perturbation due to the displacement of (16a, b, c) in Wu (2004). The perturbed sea level is computed at the ith pixel of the Earth's surface as

$$S(\theta_i, \lambda_i) = \left\{\left[\frac{\Psi(a, \theta_i, \lambda_i)}{\gamma} - U(a, \theta_i, \lambda_i)\right] - \frac{M_{ice}}{\rho_w A_o} - \frac{C}{A_o}\right\} O(\theta_i, \lambda_i), \tag{59.11}$$

where the total incremental potential at the pixel is

$$\Psi(a, \theta_i, \lambda_i) = \sum_{\ell=0}^{L_{max}} \sum_{m=-\ell}^{\ell} [\Phi_\ell^m(a) + \Theta_\ell^m(a)] Y_\ell^m(\theta_i, \lambda_i) \tag{59.12}$$

and

$$C = 4\pi a^2 \sum_{\ell=0}^{L_{max}} \sum_{m=-\ell}^{\ell} \left[\frac{\Phi_\ell^m(a) + \Theta_\ell^m(a)}{\gamma} - U_\ell^m(a)\right] o_\ell^m \tag{59.13}$$

is the integral of sea level variation over the oceans, where o_ℓ^m are the SH coefficients of the ocean function.

The SH coefficients of updated ocean load and incremental potential due to total load are recomputed according to (59.2) and (59.5) respectively, while those of the updated differential load due to potential perturbations become

$$\Delta\sigma_\ell^m(r_p) = (\rho_{p+1} - \rho_p)\left[\Theta_\ell^m(r_p) + \Phi_\ell^m(r_p)\right]. \tag{59.14}$$

Finally, total load SH coefficients are recomputed according to (59.7) and stored for the next ABAQUS iteration. Iterations are repeated until displacements do not vary significantly with respect to the previous iteration. To speed-up computations, SH load coefficients are computed only every ten time-increments. The coefficients at other increments are interpolated linearly with time. We have verified that this approximation does not affect appreciably the results but reduces of a factor of ten the post-processing time.

3 Spherical Mesh Test

As a case study, we have implemented the axisymmetric spherical grid of Fig. 59.1. The integration volume is discretized by 10 layers across the mantle, 34 elements along colatitude and 20 repetitions along longitude.

The axisymmetric grid, composed of 6,400 eight-nodes and 400 six-nodes (around the polar axis) linear flat elements, is finer within a radius of 3,000 km from the North pole where the ice load is applied. The Earth model, described in Table 59.1, is incompressible

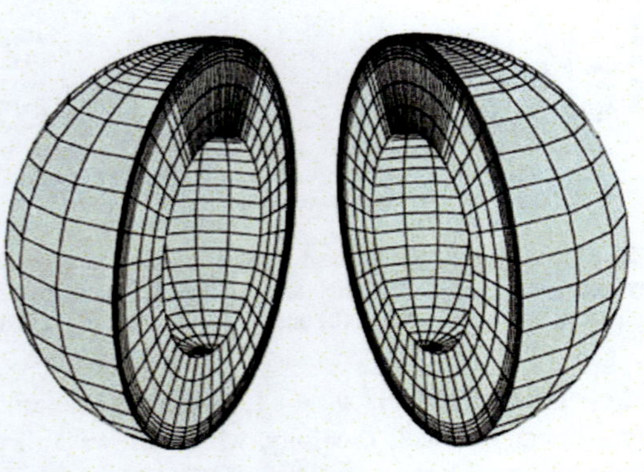

Fig. 59.1 Exploded view of the spherical mesh used for reproducing the SELEN outputs

Table 59.1 Parameters of Earth model employed for the numerical test

Layer	Thickness (km)	Density (kg/m^3)	Young modulus (Pa)	Viscosity (Pa s)	Gravity at the top (m/s^2)
Lithosphere	90	4,120	2.19×10^{11}	1.0×10^{30}	9.7075
Upper mantle	330	4,120	2.85×10^{11}	5.0×10^{20}	9.6725
Transition zone	250	4,220	3.30×10^{11}	1.0×10^{21}	9.5710
Lower mantle	2,201	4,508	6.00×10^{11}	2.0×10^{21}	9.5052
Core	3,480	10,925	–	–	10.6222

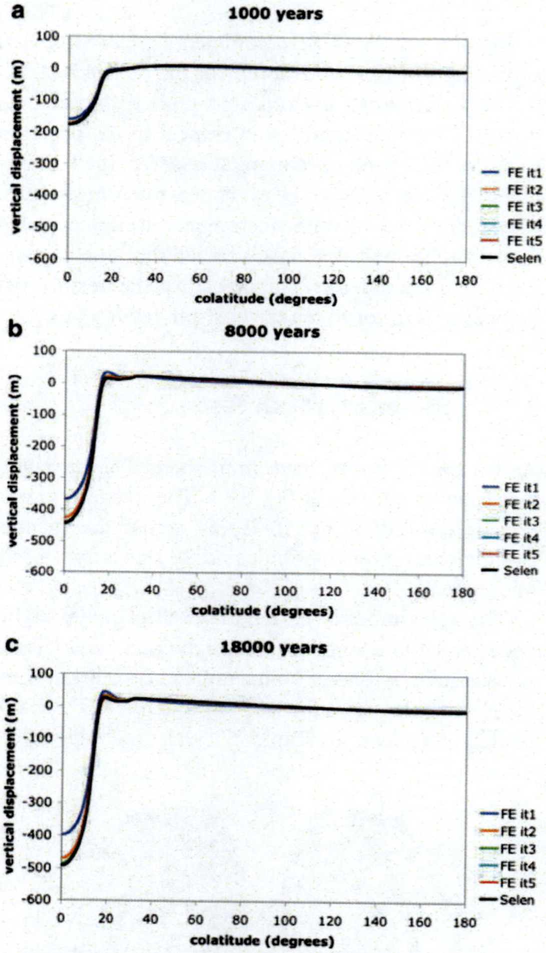

Fig. 59.2 Vertical displacement along a meridian as computed by the spectral code SELEN and by the FE code for five iterations computed 1 (**a**), 8 (**b**) and 18 (**c**) kyr after deglaciation

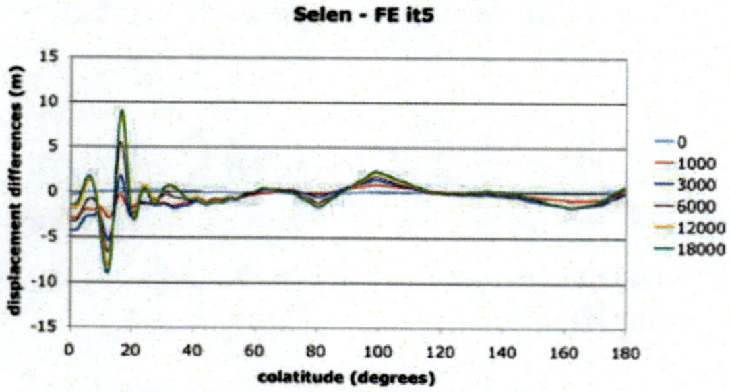

Fig. 59.3 Displacement differences along a meridian, at different times after deglaciation, between SELEN and FE solutions (after five iterations)

(i. e., Poisson ratio is $\nu = 1/2$) and characterized by a linear Maxwell rheology, which allows a direct comparison with the outcomes of SELEN (Spada and Stocchi 2007). The lithosphere is 90-km thick. The inviscid core is simulated by appropriate boundary conditions according to Wu (2004). The load is circular, with a radius of 15° and a parabolic profile. Load thickness at the center is 2,480 m. The water load is applied across a complementary ocean defined by $\theta \geq 30°$, assuming fixed shorelines. The FE analysis spans from 18 ka B.P. to present with a time grid composed of 360 increments of 50 years. The loads are imposed instantaneously at time $t = 18$ ka B.P. Rotational feedbacks are ignored. The SH is truncated at degree $L_{max} = 72$, while the resolution of pixelization is 40 that corresponds to about 64,000 pixels over every interface (Tegmark 1996). Winkler fundations (Williams and Richardson 1991) are applied at every density interface, modeled by a bed of springs with strength $\Delta\rho\gamma$ where $\Delta\rho$ is density jump and γ is gravity (see Table 59.1).

Fig. 59.2 shows that the convergence at different times after deglaciation is quite fast and is reached after the fourth iteration. From Fig. 59.3 we observe that the maximum deviations do not exceed a few meters except close to the edge of the load, where the discrepancies are smaller than 10 m. We can argue that, far from the load, they are related to inaccuracy of the interpolation of displacements on large integration elements (with size of $\sim$1,000 km in colatitude), while close to the load edge they probably reflect a sort of Gibbs effect due to load discretization over the FE grid.

Conclusion

We developed a procedure to implement the solution of the Sea Level Equation inside a finite-element model. Following previous schemes (Wu 2004), we introduced the pixelization of the spherical surface of the FE model and the interpolation of the harmonic coefficients. The implementation of the entire procedure on a Linux platform took advantage of the parallelization feature of ABAQUS that improves the computing speed almost proportionally to the number of available cpus. We validated our solution against the classical pseudo-spectral method (Spada and Stocchi 2007) by applying a circular parabolic load. The discrepancies are less than 10 m and can be attributed to the coarseness of our prototype FE mesh. Differently than previous authors, we have outlined more in detail some key-issues, such as the implementation of self-gravitation and the spatial discretization, which are important in order to facilitate the reproduction of the results obtained from this approach to GIA modeling. In this framework, the results presented are to be seen in the perspective of providing, in a follow up study, a full validation of the FE approach to GIA by benchmark comparisons with available pseudo-spectral codes and the release of the source codes so far developed to the GIA community.

Acknowledgements G.S. acknowledges COST Action ES0701 "Improved Constraints on Models of Glacial Isostatic Adjustment".

References

Dal Forno G, Gasperini P, Boschi E (2005) Linear or non-linear rheology in the mantle: a 3D finite-element approach to postglacial rebound modeling. J Geodyn 39:183–195

Dal Forno G, Gasperini P (2007) Modelling of mantle post-glacial relaxation in axisymmetric geometry with a composite rheology and a glacial load interpolated by adjusted spherical harmonics analysis. Geophys J Int 169:1301–1314

Farrell WE, Clark JA (1976) On postglacial sea level. Geophys J R Astron Soc 46:647–667

Gasperini P, Sabadini R (1989) Lateral heterogeneities in mantle viscosity and postglacial rebound. Geophys J 98: 413–428

Gasperini P, Yuen D, Sabadini R (1992) Postglacial rebound with a non-Newtonian upper mantle and a Newtonian lower mantle rheology. Geophys Res Lett 19:1711–1714

Gasperini P, Dal Forno G, Boschi E (2004) Linear or non-linear rheology in the Earth's mantle: the prevalence of power-law creep in the postglacial isostatic readjustment of Laurentia. Geophys J Int 157:1297–1302

Giunchi C, Spada G (2000) Postglacial rebound in a non-newtonian spherical Earth. Geophys Res Lett 27: 2065–2068

Kaufmann G, Wu P, Wolf D (1997) Some effectes of lateral heterogeneities in the upper mantle on postglacial land uplift close to continental margins. Geophys J Int 128:175–187

Kaufmann G, Wu P, Li G (2000) Glacial isostatic adjustment in Fennoscandia for a laterally heterogeneous earth. Geophys J Int 143:262–273

Paulson A, Zhong S, Wahr J (2005) Modelling post-glacial rebound with lateral viscosity variations. Geophys J Int 163:357–371

Peltier WR (1974) The impulse response of a Maxwell earth. Rev Geophys Space Phys 12:649–669

Simulia Inc (2009) ABAQUS Version 6.7 User's Manual, Providence, RI

Spada G, Antonioli A, Cianetti S, Giunchi C (2006) Glacial isostatic adjustment and relative sea-level changes: the role of lithospheric and upper mantle heterogeneities in a 3-D spherical Earth. Geophys J Int 165:692–702

Spada G, Stocchi P (2007) SELEN: a Fortran 90 program for solving the "sea-level equation". Comp Geosc 33: 538–562

Tegmark, M. (1996) An icosahedron-based method for pixelizing the celestial sphere. ApJ Lett 470:L81

Wieczorek M (2005) SHTOOLS conventions. accessed on 13 October 2011, http://www.ipgp.fr/~wieczor/SHTOOLS/www/conventions

Williams CA, Richardson RM (1991) A rheologically layered three-dimensional model of the San Andreas fault in central and southern California. J Geophys Res 96:16, 597–16, 623

Wu P (1992) Deformation of an incompressible viscoelastic flat earth with power law creep: a finite element approach. Geophys J Int 108:136–142

Wu P (1999) Modeling postglacial sea-levels with power law rheology and realistic ice model in the absence of ambient tectonic stress. Geophys J Int 139:691–702

Wu P (2004) Using commercial finite element packages for the study of earth deformations, sea levels and the state of stress. Geophys J Int 158:401–408

Wu P, Wang H (2008) Postglacial isostatic adjustment in a self-gravitating spherical earth with power-law rheology. J Geodyn 46:118–130

Author Index

N. Sneeuw et al. (eds.), *VII Hotine-Marussi Symposium on Mathematical Geodesy*, International Association of Geodesy Symposia 137, DOI 10.1007/978-3-642-22078-4, © Springer-Verlag Berlin Heidelberg 2012